Progress in the Chemistry of Organic Natural Products

The volumes of this classic series, now referred to simply as "Zechmeister" after its founder, Laszlo Zechmeister, have appeared under the Springer Imprint ever since the series' inauguration in 1938. It is therefore not really surprising to find out that the list of contributing authors, who were awarded a Nobel Prize, is quite long: Kurt Alder, Derek H.R. Barton, George Wells Beadle, Dorothy Crowfoot-Hodgkin, Otto Diels, Hans von Euler-Chelpin, Paul Karrer, Luis Federico Leloir, Linus Pauling, Vladimir Prelog, with Walter Norman Haworth and Adolf F.J. Butenandt serving as members of the editorial board.

The volumes contain contributions on various topics related to the origin, distribution, chemistry, synthesis, biochemistry, function or use of various classes of naturally-occurring substances ranging from small molecules to biopolymers.

Each contribution is written by a recognized authority in the field and provides a comprehensive and up-to-date review of the topic in question. Addressed to biologists, technologists, and chemists alike, the series can be used by the expert as a source of information and literature citations and by the non-expert as a means of orientation in a rapidly developing discipline.

All contributions are listed in PubMed.

Progress in the Chemistry of Organic Natural Products

Volume 119

A. Douglas Kinghorn · Heinz Falk ·
Simon Gibbons · Yoshinori Asakawa · Ji-Kai Liu ·
Verena M. Dirsch
Editors

Ancistrocladus Naphthylisoquinoline Alkaloids

By Doris Feineis and Gerhard Bringmann

Springer

Editors
A. Douglas Kinghorn
College of Pharmacy
Ohio State University
Columbus, OH, USA

Heinz Falk
Institute of Organic Chemistry
Johannes Kepler University of Linz
Linz, Austria

Simon Gibbons
Centre for Natural Products Discovery
Liverpool John Moores University
Liverpool, UK

Yoshinori Asakawa
Faculty of Pharmaceutical Sciences
Tokushima Bunri University
Tokushima, Japan

Ji-Kai Liu
School of Pharmaceutical Sciences
South Central University for Nationalities
Wuhan, China

Verena M. Dirsch
Department of Pharmaceutical Sciences
University of Vienna
Vienna, Austria

ISSN 2191-7043 ISSN 2192-4309 (electronic)
Progress in the Chemistry of Organic Natural Products
ISBN 978-3-031-10456-5 ISBN 978-3-031-10457-2 (eBook)
https://doi.org/10.1007/978-3-031-10457-2

This Springer imprint is published by the registered company Springer Nature Switzerland AG
The registered company address is: Gewerbestrasse 11, 6330 Cham, Switzerland

Dedicated to Prof. Lutz Friedjan Tietze, Institut für Organische und Bioorganische Chemie der Georg-August-Universität Göttingen, on the occasion of his 80th birthday

About This Book

This volume describes a unique class of secondary metabolites, the mono- and dimeric-naphthylisoquinoline alkaloids. They exclusively occur in lianas of the paleotropical Ancistrocladaceae and Dioncophyllaceae plant families. Their unprecedented structures include stereogenic centers and rotationally hindered, and therefore stereogenic, axes. Extended recent investigations on six *Ancistrocladus* species from Asia, as reported in this contribution, shed light on their fascinating phytochemical productivity, with over 100 intriguing natural products. This high chemodiversity arises from a similarly unique biosynthesis from acetate–malonate units, following a novel polyketidic pathway to plant-derived isoquinoline alkaloids. Some of the compounds show most promising antiparasitic activities. Additionally, strategies for the regio- and stereoselective total synthesis of the alkaloids, including the directed construction of the chiral axis, are also presented.

Contents

Contributors

Bringmann Gerhard Institute of Organic Chemistry, University of Würzburg, Würzburg, Germany

Feineis Doris Institute of Organic Chemistry, University of Würzburg, Würzburg, Germany

Asian *Ancistrocladus* Lianas as Creative Producers of Naphthylisoquinoline Alkaloids

Doris Feineis and Gerhard Bringmann

Contents

D. Feineis · G. Bringmann (✉)
Institute of Organic Chemistry, University of Würzburg, Am Hubland, 97074 Würzburg, Germany
e-mail: bringman@chemie.uni-wuerzburg.de

D. Feineis
e-mail: feineis@chemie.uni-wuerzburg.de

A. D. Kinghorn, H. Falk, S. Gibbons, Y. Asakawa, J.-K. Liu, V. M. Dirsch (eds.), *Ancistrocladus Naphthylisoquinoline Alkaloids* Progress in the Chemistry of Organic Natural Products 119, https://doi.org/10.1007/978-3-031-10457-2_1

1 Introduction

In the early 1970s, Govindachari's brilliant pioneering work on the secondary metabolites of the Indian liana *Ancistrocladus heyneanus* led to the discovery and complete structural elucidation of ancistrocladine (**1a**)[1] [1–5] as the first naphthylisoquinoline. With this discovery, the large class of isoquinoline alkaloids [6–12], with over 2000 structurally characteristic and pharmaceutically useful natural products, was extended by a new subclass of representatives with many unique additional facets. Naphthylisoquinoline alkaloids [13–18] have so far been isolated exclusively from plants belonging to the monogeneric family Ancistrocladaceae [19–21], which are indigenous to the rainforests of paleotropical Asia and Africa, or from the related Dioncophyllaceae [22] from West Africa. As already indicated by their name, the naphthylisoquinoline alkaloids consist of a naphthalene and an isoquinoline subunit, linked by a biaryl axis, which is—in most cases—rotationally hindered and, thus, displays the fascinating phenomenon of atropisomerism [13–15, 17, 18]. Three main structural subtypes are known (Fig. 1): By far the highest number of naphthylisoquinoline alkaloids (ca. 220 compounds) are monomeric C,C-coupled compounds [13–18] like ancistrocladine (**1a**) and its atropo-diastereomer hamatine (**1b**) [1–5, 23–25], ancistrocladisine (**2a**) [5, 26–28], and ancistrocladidine (**3a**) [5, 27–29]. Most of them are usually equipped with up to two

[1] For naphthylisoquinoline alkaloids appearing as two rotational isomers, a numbering system is applied in which the two atropo-diastereomers are denoted as **a** and **b**, like, e.g., for ancistrocladine (**1a**) and hamatine (**1b**), see Fig. 1.

stereogenic centers in the isoquinoline portion, at C-1 and/or C-3. Some naphthylisoquinoline alkaloids have been identified with a non-hydrogenated isoquinoline portion, so that the rotationally hindered biaryl axis is the only element of chirality and one prominent example is ancistrobenomine A (**4**) [30]. A second, smaller subfamily (50 compounds) comprises naphthylisoquinoline dimers [17] (all C,C-coupled), which may undergo further cyclization reactions, leading to fascinating molecular architectures with up to eight stereogenic elements, among them compounds with three consecutive chiral biaryl axis like shuangancistrotectorine C (**5**) [17, 31]. The smallest subclass (twelve representatives), showing an even higher degree of structural originality, are naphthylisoquinolines coupled via an unprecedented stereogenic *N*-iminium-aryl axis such as ancistrocladinium A (**6a**) [18, 32], also comprising further cyclized N,C-coupled compounds like ancistrocyclinone A (**7**) [18, 33] (Fig. 1).

The first review on naphthylisoquinoline alkaloids in 1977 [5] covered six representatives of this class of natural biaryl compounds, all isolated from the Indian liana *A. heyneanus* and the Sri Lankan species *Ancistrocladus hamatus*. Since then, the number of such alkaloids has tremendously increased to a total of about 280 examples known from Nature [13–18]. After an early comprehensive review by the authors' group in 1995 [13], three further reviews on

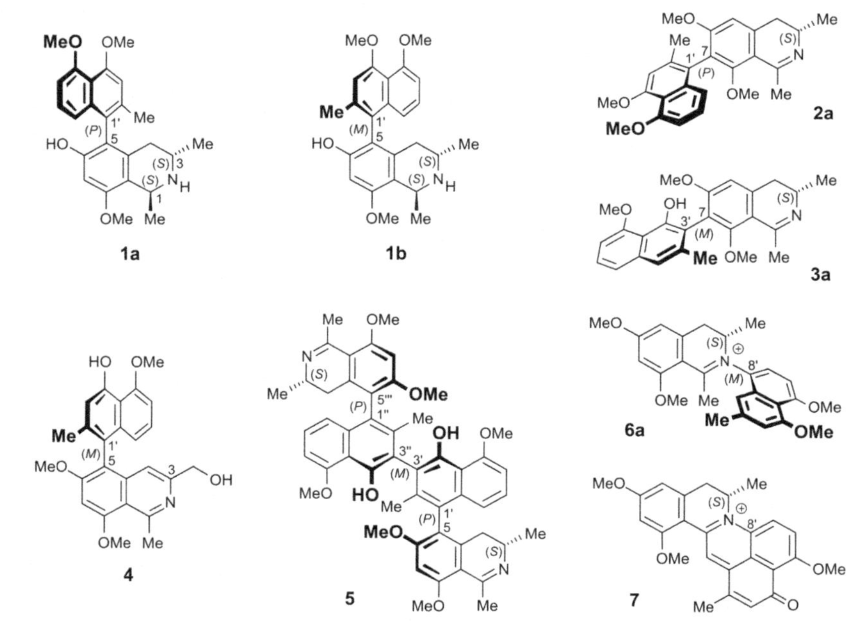

Fig. 1 The C,C-coupled monomers ancistrocladine (**1a**), its atropo-diastereomer hamatine (**1b**), ancistrocladisine (**2a**), ancistrocladidine (**3a**), and ancistrobenomine A (**4**), the dimer shuangancistrotectorine C (**5**), and the N,C-coupled alkaloids ancistrocladinium A (**6a**) and ancistrocyclinone A (**7**), representing the three main structural subtypes of naphthylisoquinoline alkaloids

naphthylisoquinoline alkaloids appeared in the literature in the course of the past decade. In 2015, Ibrahim and Mohamed [16] provided an overview on mono- and dimeric naphthylisoquinoline alkaloids, in particular focusing on their bioactivities. A comprehensive review [17] by the authors' group published in 2019, summarized key results on naphthylisoquinoline dimers, as predominantly discovered in African *Ancistrocladus* species. More recently, in 2021, a third review [18] was presented by the authors' group, which dealt with the occurrence, stereostructures, and total synthesis of N,C-coupled naphthylisoquinoline alkaloids.

The present review will supplement the above-mentioned overviews by drawing the readers' attention to phytochemical, phylogenetic, and taxonomic investigations on *Ancistrocladus* lianas from Asia with special emphasis on their botanical peculiarities (Sect. 3) [19, 21, 34], the diversity of their metabolite profiles (Sects. 4–9) [13, 16–18], and their difficult and (in part) still uncertain taxonomic treatment (Sect. 8) [19, 21, 34–36]. Most Asian *Ancistrocladus* plants are found in dry to wet, lowland to submontane evergreen forests in Southeast Asia, widely occurring from Malaysia, Singapore, Sumatra, and (in part) Borneo, through China (Hainan Island), Vietnam, Laos, Cambodia, Thailand, Myanmar, and the Andaman and Nicobar Islands of India, with two outliers in India and Sri Lanka [19, 21, 34–36]. More than 100 naphthylisoquinoline alkaloids have been discovered to date in six Asian species, comprising *A. heyneanus* and *A. hamatus* (Sect. 4) [1–5, 23, 24, 26–29, 37–41], *Ancistrocladus griffithii* from Thailand (Sect. 5) [42, 43], *Ancistrocladus benomensis* from Malaysia (Sect. 6) [30, 44], *Ancistrocladus cochinchinensis* from Vietnam (Sect. 7) [45–48], and *Ancistrocladus tectorius* collected from habitats in Thailand, Laos, Vietnam, Malaysia, and the Chinese Island Hainan (Sect. 8) [31, 33, 49–64]. Among these species, *A. tectorius* attracted special attention, since—in view of highly similar botanical features of *Ancistrocladus* plants collected all over Southeast Asia, and consequently considerable difficulties in distinguishing vegetative material. *A. tectorius* was, for a long time, regarded as the only valid taxon from this region [21, 34–36]. The species shows an extremely large chemodiversity with respect to naphthylisoquinoline alkaloids and related compounds identified in plants from different regions of Southeast Asia [31, 33, 34, 36, 49–64]. This metabolic diversity gave rise to the assumption that the taxon might consist of several distinct species, although morphologically similar, thereby indicating that the current taxonomic concept for the genus *Ancistrocladus*, with only four recognized species in Southeast Asia, namely *A. griffithii*, *Ancistrocladus attenuatus*, *A. benomensis*, and *A. tectorius* [21, 34], should possibly be revised. The results from a genetic study [36] revealing a large variability of sequence data obtained from the analysis of the cpDNA *trn*K intron, the nuclear internal transcribed spacers (ITS), and intersimple sequence repeat (ISSR) fingerprints, additionally support the existence of further distinct taxa yet not recognized.

The Asian *Ancistrocladus* lianas have in general proven to be a rewarding source of the structurally most diverse mono- and dimeric naphthylisoquinoline alkaloids and related compounds. Some of them—depending on their individual structures—exhibit strong and selective biological activities (Sects. 4–8) [12, 16, 31–33, 47,

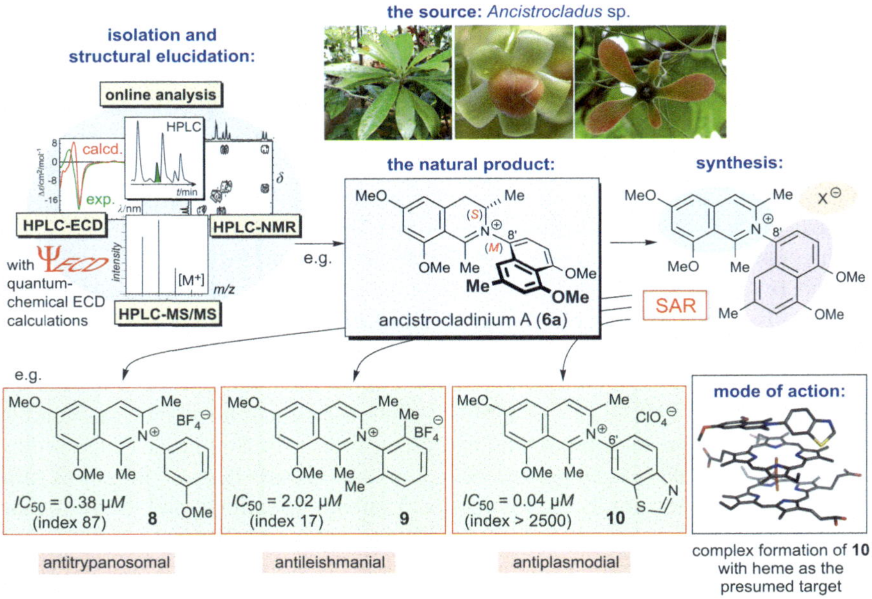

Fig. 2 Identification of the N,C-coupled alkaloid ancistrocladinium A (**6a**) online, in Asian *Ancistrocladus* species, by HPLC-MS/MS-NMR-ECD, assisted by quantum-chemical ECD calculations, along with structure–activity relationship (SAR) studies to design simplified analogues, such as **8–10**, with pronounced and specific antitrypanosomal, antileishmanial, or antiplasmodial activities, and with reduced toxicities. Further improvements can be made from a search for the mode of action and the identification of the molecular target. Shown in the photographs are *Ancistrocladus* lianas from Asia: young plant with a rosette of leaves (left), fully developed flower (center) and fruit (right) (Photographs: A. Irmer)

60–62, 64–72]. Intense research efforts were directed toward the identification of the pharmacophore, including the mode of action of N,C-coupled naphthylisoquinoline alkaloids (Fig. 2), since ancistrocladinium A (**6a**) showed noteworthy anti-infective properties (in particular against *Leishmania* parasites) with a comparably low cytotoxicity against mammalian cells [32, 66–72]. In an attempt to design promising candidates for the development of new agents against severe and widespread tropical diseases such as leishmaniasis, African sleeping sickness, Chagas' disease, and malaria, the selectivity of structurally simplified analogues against a given parasite was specifically improved by the directed modification of selected structural parameters [66, 67, 70–72] (Sect. 4.9).

This contribution deals mainly with the assignment of the complete stereostructures of naphthylisoquinoline alkaloids isolated from Asian *Ancistrocladus* lianas (Sects. 4–8). For that purpose, a repertoire of complementary—in part even newly developed—spectroscopic, chemical, and computational methods was applied, including one- and two-dimensional nuclear magnetic resonance spectroscopy (NMR) [73–75], electronic circular dichroism (ECD) spectroscopy in combination with quantum-chemical ECD calculations [76–78], and an efficient oxidative degradation procedure [79, 80] to give easy-to-analyze amino acids of known absolute configuration (Sects. 4–8). The hyphenation of high-performance liquid chromatography (HPLC) with NMR spectroscopy, electrospray ionization tandem mass spectrometry, and ECD spectroscopy, also known as the 'analytical triad' LC-NMR-MS/MS-ECD (Sect. 5) [74, 75, 78, 81–83], has become a most valuable tool for the rapid identification of known naphthylisoquinoline alkaloids, but also for the full stereochemical assignment of new metabolites such as ancistrocladinium A (**6a**), directly from crude plant extracts (Fig. 2) [32, 47].

The bulky molecular framework of naphthylisoquinoline alkaloids with up to four *ortho*-substituents next to the C,C- or N,C-axis, thus exhibiting a high degree of steric hindrance, has been a strong motivation—and thrilling challenge—for the authors' group [69, 84–101] and other scientists [102–121] to develop efficient methods for their regio- and stereoselective total synthesis. Some instructive examples regarding different synthesis pathways to prepare naphthylisoquinoline alkaloids are described in Sects. 4–8. Among the target molecules (Fig. 1) are C,C-coupled compounds e.g., ancistrocladine (**1a**) [84, 90–92], ancistrocladisine (**2a**) [84, 85, 90, 93, 94], and ancistrocladidine (**3a**) [99, 115, 116], but also N,C-linked alkaloids such as ancistrocladinium A (**6a**) [69, 101, 118, 119].

This contribution starts with a general introduction (Sect. 2) into the classification and structural variability of naphthylisoquinoline alkaloids discovered in Asian and African Ancistrocladaceae plants [12–18]. Furthermore, the botanical and phytochemical peculiarities of the only other plant family that also produces naphthylisoquinoline alkaloids, the Dioncophyllaceae [22] from Coastal West Africa, will be presented in Sect. 2, including an overview on the occurrence of dioncophyllaceous naphthylisoquinoline dimers in *Ancistrocladus* lianas [13, 14, 18, 122–124] and on the synthesis of structurally related merely artificial representatives [86, 97, 125–130]. Section 2 focuses on the unprecedented biosynthetic origin of naphthylisoquinoline alkaloids from acetate-malonate units following a novel pathway to isoquinolines in plants [131–134]. The two molecular portions of these biaryl compounds are formed from identical polyketide precursors, different from all the other known di- and tetrahydroisoquinolines previously investigated, which in turn, arise from aromatic amino acids [6, 10, 135–140]. The two portions of naphthylisoquinoline alkaloids are then linked to each other by regio- and stereoselective biaryl coupling via a phenol-oxidative mechanism [141]. The occurrence and special morphological features of *Ancistrocladus* species from Asia are described in Sect. 3. Sections 4–8 report on the structure elucidation, biotesting, and total synthesis of selected naphthylisoquinoline alkaloids and related compounds identified in Asian *Ancistrocladus* lianas. Section 9 summarizes the most important results and presents a complete list of all naphthylisoquinoline alkaloids discovered in *Ancistrocladus* lianas from Asia, categorized according to their coupling sites and their classification into the various subgroups of these natural products.

2 Naphthylisoquinoline Alkaloids, a Fascinating Class of Axially Chiral Biaryl Natural Products

2.1 *Structural Diversity and Classification of Naphthylisoquinoline Alkaloids—A Short Overview*

Mono- and dimeric naphthylisoquinoline alkaloids currently comprise approximately 280 representatives, all exclusively isolated from the two paleotropical plant families Ancistrocladaceae and Dioncophyllaceae [13–18]. These alkaloids have received increasing interest because of the complex diversity of their molecular architectures [17, 18], their exceptional biosynthesis from acetate-malonate units [131–134], and their promising pharmacological properties, in particular against the pathogens causing malaria tropica, leishmaniasis, Chagas' disease, and African sleeping sickness [12, 16, 66–72, 142–151]. The broad range of structural features, among them the different coupling sites between the two molecular halves, the substitution patterns, the often rotationally hindered—and thus chiral—biaryl axes and the stereogenic centers in the tetra- or dihydroisoquinoline portion, constitute thrilling challenges for structure elucidation and total synthesis.

Virtually all naphthylisoquinoline alkaloids that have so far been isolated consist of a 4,5-dioxy-2-methylnaphthalene[2] building block and a 1,3-dimethyl-8-oxy- or 1,3-dimethyl-6,8-dioxyisoquinoline moiety (in most cases in the form of a di- or a tetrahydroisoquinoline). This heterocyclic portion may additionally be equipped with a methyl group at N-2 [13–18] and with an oxygen function at the C-3 methyl group [30, 64], sometimes also at C-4 [13, 14, 152, 153], and can then even have three stereogenic centers. The oxygen substituents are either free hydroxy functions or methoxy groups (or, in rare cases, they bear sugar entities [154]). In most of these compounds, three stereoelements are present; two centers and the axis, being stereogenic when rotationally hindered due to the presence of three or four substituents in the *ortho*-positions.

It is mainly the different possible positions of the biaryl linkage that strongly contribute to the broad structural variability of naphthylisoquinoline alkaloids, by significantly accounting for their molecular shapes. According to their biosynthetic formation through regio- and stereoselective phenol-oxidative coupling, the two molecular portions, which are separately formed beforehand (see Sect. 2.3), can be linked in either the *ortho*- or *para*-positions relative to the phenolic oxygen functions, i.e., via C-1′, C-3′, C-6′, or C-8′ in the naphthalene moiety and via C-5 or C-7 in the isoquinoline part. Of the eight resulting possible C,C-coupling types, seven have so far been identified in Asian and African Ancistrocladaceae species. Typical examples are ancistrobenomine B (**11**) (5,1′-coupling) [64], ancistrotanzanine A (**12a**) (5,3′-coupling) [60, 155], korupensamine A (**13a**) (5,8′-coupling) [156],

---

[2] For an easier comparison of naphthylisoquinoline alkaloids with different coupling positions, a 2-methyl-4,5-dioxy substitution pattern is throughout applied in the numbering of the naphthalene moiety, regardless of the coupling site.

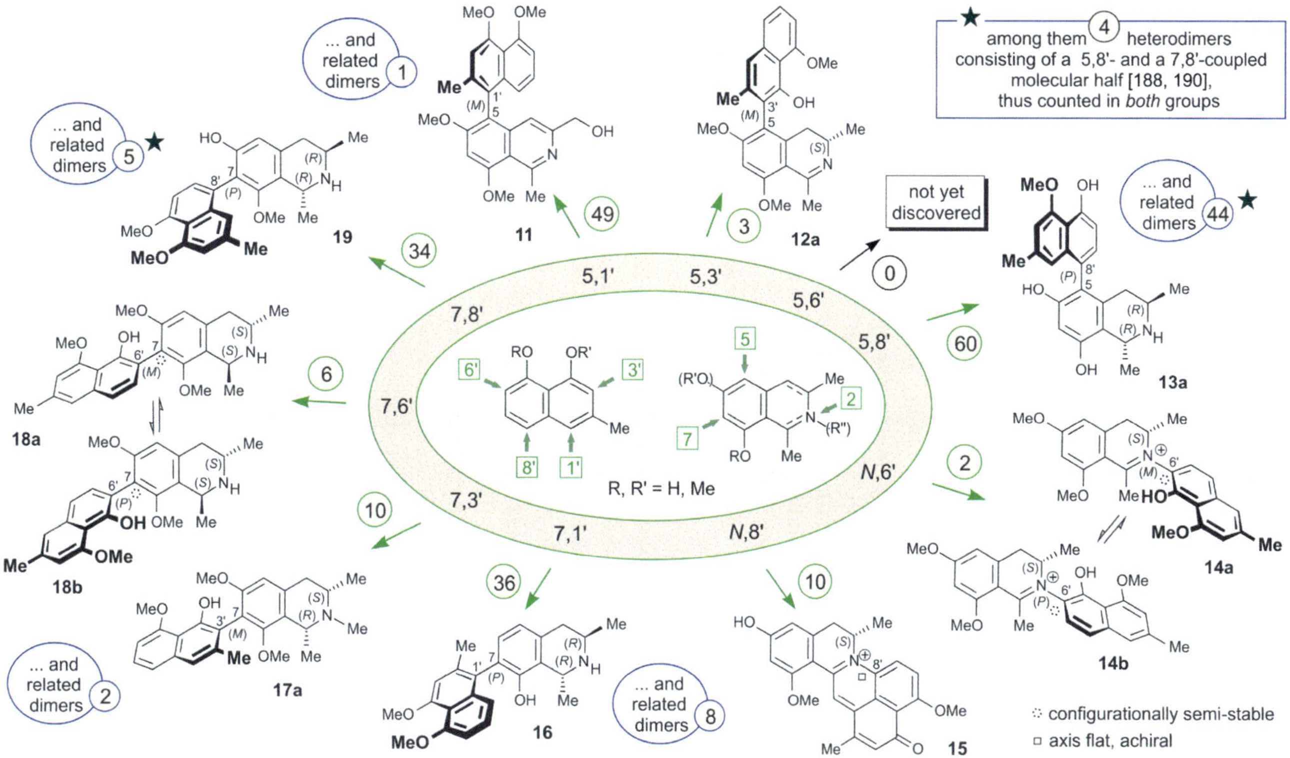

Fig. 3 Exemplified for selected representatives is a structural variety of C,C- and N,C-coupled naphthylisoquinoline alkaloids isolated from Asian and African Ancistrocladaceae plants. In most cases, the biaryl axis is configurationally stable and constitutes an additional element of chirality, with exceptions being, e.g., ancistrotectorine B (**18a/18b**) and ancistrocladinium B (**14a/14b**), which occur as pairs of configurationally semi-stable and thus slowly interconverting atropo-diastereomers

dioncophylline A (**16**) (7,1′-coupling) [13, 14, 157, 158], ancistrotectorine (**17a**) (7,3′-coupling) [51, 159], ancistrotectorine B (**18a/18b**) (7,6′-coupling) [60], and ealamine A (**19**) (7,8′-coupling) [160] (Fig. 3). Naphthylisoquinoline alkaloids displaying a 5,6′-coupling site, however, have not so far been found in Nature. The spectrum of possible coupling types has been further enlarged by the discovery of N,C-linked naphthylisoquinolinium salts [18] such as ancistrocladinium B (**14a/14b**) (N,6′-coupling) [32, 161] and ancistrocyclinone B (**15**) (N,8′-coupling), which have undergone an additional cyclization reaction [33].

About 18 *Ancistrocladus* species are currently accepted as valid taxa of the monogeneric Ancistrocladaceae plant family, with their natural habitats mainly being located in the rain and swamp forests of West and Central Africa, and in Southeast Asia, with outliers in East Africa, India, and Sri Lanka [21, 34, 162]. The naphthylisoquinoline patterns of African and Asian *Ancistrocladus* plants comprise representatives of all C,C- and N,C-coupling types discovered so far (Fig. 3). Remarkably, 5,8′-linked naphthylisoquinoline alkaloids are by far the largest group of constituents isolated from Central African *Ancistrocladus* species [13, 17, 156, 161, 163–182]. Among them are many monomeric compounds (60 alkaloids), e.g., korupensamine A (**13a**) [156] and most of the naphthylisoquinoline dimers (44 examples) [17] known from Nature. In Asian taxa, by contrast, this coupling site occurs much less frequently and only five examples have as yet been found [56, 60, 62].

The strongly dominating subgroups of naphthylisoquinoline alkaloids isolated from Asian species are 5,1′-coupled naphthyltetra- and naphthyldihydroisoquinolines that are structurally related to the widely occurring alkaloids ancistrocladine (**1a**) and its atropo-diastereomer hamatine (**1b**) (Fig. 1) [1–5, 23, 24, 30, 31, 37, 38, 43, 45–50, 53–61, 64]. Non-hydrogenated alkaloids have also been identified, such as ancistrobenomine B (**11**) (Fig. 3), which has an otherwise rarely found hydroxymethylene group at C-3 [64]. In total, 49 alkaloids with a 5,1′-biaryl linkage are known to be produced by African and Asian Ancistrocladaceae plants.

The alkaloid patterns of Asian and West African *Ancistrocladus* species, among them *A. tectorius* from Southeast Asia and *Ancistrocladus abbreviatus* from Coastal West Africa, are characterized by the presence of several 7,1′-coupled naphthylisoquinolines, in some cases even representing the main metabolites of the respective plants [13, 17, 26–28, 31, 43–46, 56, 60–62, 124, 183–187]. A considerable number of 7,1′-linked monomeric compounds (36 representatives) have as yet been discovered. These include one of the most prominent examples of naphthylisoquinoline alkaloids, dioncophylline A (**16**) (Fig. 3) [13, 122–124, 157, 158, 183], along with eight related dimers [31, 122–124]. For many years, only a few 7,8′-linked naphthylisoquinolines have been known to occur in *Ancistrocladus* plants, until most recently a unique series of such alkaloids (13 compounds, among them three dimers) have been discovered in the Congolese species *A. ealaensis* [160, 188]. This naphthylisoquinoline subgroup now comprises an impressive number of 34 representatives from *Ancistrocladus* lianas. Prior to this finding, in all of the other taxa that produce 7,8′-coupled alkaloids [13, 39, 42–44, 61, 73, 154, 182, 186, 187], up to three such compounds had been detected at best.

Naphthylisoquinolines displaying a 7,3′-coupling site, for example, ancistrotectorine (**17a**) (Fig. 3) [51, 159], on the contrary, have only rarely been identified in Asian and African *Ancistrocladus* species. So far, ten 7,3′-linked alkaloids have been discovered in Ancistrocladaceae lianas [5, 13, 17, 27–29, 31, 41, 51, 52, 63, 159, 189], with the majority of them isolated from *A. tectorius* from the Chinese Hainan Island. In this plant, no less than eight 7,3′-linked alkaloids [31, 51, 52, 63] were found and among them two are dimers [17, 31].

Another striking feature of the metabolite pattern of *A. tectorius* is the occurrence of 7,6′-coupled naphthylisoquinoline alkaloids like ancistrotectorine B (**18**) (Fig. 3) [60]. Three such 7,6′-linked compounds have so far been isolated from Ancistrocladaceae plants [59, 60] and all were detected in *A. tectorius*, exclusively. Due to the presence of only three—and not too bulky—*ortho*-substituents next to the biaryl axis (OH, OMe, OMe), **18** occurs as a mixture of configurationally semi-stable atropo-diastereomers, **18a** and **18b**, which slowly interconvert at room temperature [60]. Prior to the isolation of two further representatives from the Chinese liana *A. tectorius* [60], ancistrotanzanine A (**12a**) (Fig. 3) from the East African species *Ancistrocladus tanzaniensis* [155] was the only naphthylisoquinoline alkaloid showing the unusual and extremely rare 5,3′-coupling. Furthermore, twelve naphthylisoquinoline alkaloids with an unprecedented iminium-aryl N,C-axis have been identified in Ancistrocladaceae lianas [18]. Such N,C-coupled representatives, although small in number, are widespread and part of the alkaloid patterns of most Southeast Asian [32, 33, 40, 47] and Central African [32, 161, 176] *Ancistrocladus* lianas. Ancistrocladinium B (**14a**) and its rotational isomer **14b** (Fig. 3) are of the N,6′-coupling type, showing slow rotation about the central N,C-axis. Consequently, as in the case of the 7,6′-linked ancistrotectorine B (**18a/b**) [60], they occur as a mixture of two configurationally semi-stable atropo-diastereomers [18, 32, 161]. Ancistrocyclinone B (**15**) (Fig. 3), a further cyclized and pentacyclic naphthylisoquinoline alkaloid, represents a new subtype of naphthylisoquinoline alkaloids, with a molecular framework reminiscent of that of berberine alkaloids. However, **15** arises from a different biosynthesis pathway, apparently by N,8′-coupling and further oxidative condensation, demonstrating the 'chemical creativity' of Nature [18, 33].

As outlined in Fig. 3, naphthylisoquinoline alkaloids show broad chemodiversity, yet the plants follow strict biosynthesis principles, which are of useful relevance for a reliable botanical and phylogenetic classification of the respective Ancistrocladaceae species from Asia and Africa [13–18]. From Asian and East African *Ancistrocladus* species (3*S*)-configured and 6-oxygenated, the so-called 'Ancistrocladaceae-type' alkaloids (Fig. 4) [13, 15, 17] have been isolated nearly exclusively, such as ancistrocladine (**1a**). This finding clearly reveals distinct chemo- and geotaxonomic relationships between the only two known taxa from East Africa, *A. tanzaniensis* [155, 189] (from Tanzania) and *A. robertsoniorum* [25, 190] (from Kenya), and the Asian species such as *A. heyneanus* [1–5, 26–29, 37–41] (from India), *A. hamatus* [5, 23, 24] (from Sri Lanka), *A. griffithii* [42, 43, 82] (from Thailand), or *A. tectorius* [31, 33, 49–64] (distributed all over Southeast Asia).

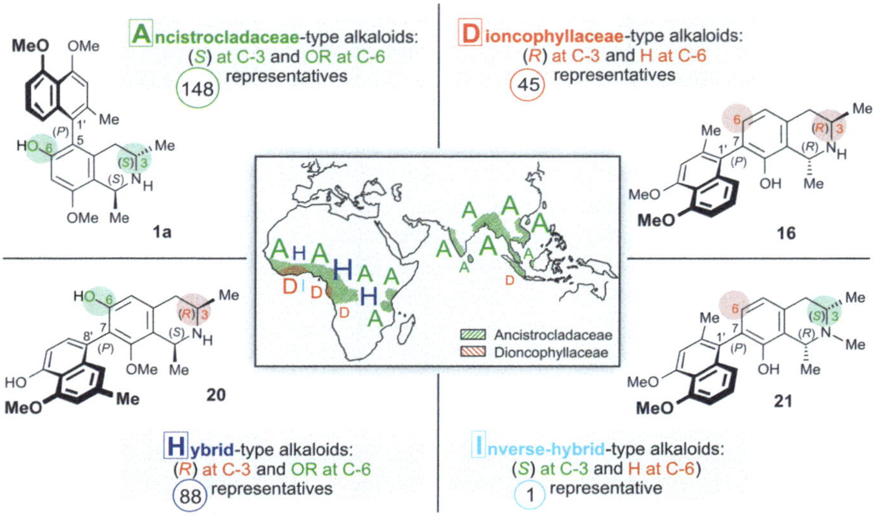

Fig. 4 Occurrence of Ancistrocladaceae-, Dioncophyllaceae-, and hybrid-type naphthylisoquinoline alkaloids, and of one single known inverse-hybrid-type compound from Ancistrocladaceae and Dioncophyllaceae species, serving as chemo- and geotaxonomic markers, as exemplified by the four selected alkaloids ancistrocladine (**1a**), dioncophylline A (**16**), ealamine G (**20**), and dioncoline A (**21**)

The only other plant family that likewise produces naphthylisoquinoline alkaloids, the West African Dioncophyllaceae [22] (see Sect. 2.3), however, is characterized by the exclusive occurrence of 'Dioncophyllaceae-type' alkaloids (Fig. 4) [13, 14] such as dioncophylline A (**16**) [157, 158], with (*R*)-configuration at C-3 and lacking an oxygen function at C-6. West and Central African *Ancistrocladus* species like *A. abbreviatus* [13, 124, 184–187, 191–193] and *A. barteri* [13, 194] (from Ivory Coast), *A. ileboensis* [123, 124, 182] (from Southern DR Congo), and a so far botanically undescribed species from the Congo Basin [122] seem to be closely related to the Dioncophyllaceae, since they contain both, typical Dioncophyllaceae- and Ancistrocladaceae-type alkaloids, and, in addition, hybrid forms thereof.

Most of the Central African lianas by contrast, for example *Ancistrocladus korupensis* [13, 154, 156, 163–165, 195] from Cameroon, and *Ancistrocladus likoko* [166–169] and *A. ealaensis* [160, 174, 176, 179, 188] from the Congo Basin, predominantly produce mixed, Ancistrocladaceae/Dioncophyllaceae hybrid-type naphthylisoquinoline alkaloids such as ealamine G (**20**) [160] (Fig. 4), with (*R*)-configuration at C-3 and an oxygen function at C-6, along with some Ancistrocladaceae-type compounds. Inverse-hybrid-type alkaloids with opposite characteristics (6-H, (3*S*)), for example, dioncoline A (**21**) (Fig. 4), are also possible though very rare, and as yet only discovered in *A. abbreviatus* [124].

Thus, phytochemical studies on the Ancistrocladaceae and Dioncophyllaceae have revealed a clear geographical dependence of the occurrence of naphthylisoquinoline alkaloids with certain structural features in certain plants. The classification—'Ancistrocladaceae'- versus 'Dioncophyllaceae'-type alkaloids—clearly marks a chemotaxonomic borderline between these two plant families, and strongly implies infrageneric relationships within the genus *Ancistrocladus* in Southeast Asia and East Africa on the one hand, and in Central and West Africa on the other.

The main topic of this review deals with the structural diversity of the nearly 90 Ancistrocladaceae-type mono- and dimeric naphthylisoquinoline alkaloids produced by *Ancistrocladus* species endemic to the rainforests of India, Sri Lanka, and Southeast Asia (see Sects. 4–8). Much smaller is the subclass of dioncophyllaceous naphthylisoquinoline alkaloids, with 45 representatives isolated so far from Dioncophyllaceae, but also from Ancistrocladaceae lianas. Section 2.2 will focus more closely on the structural peculiarities and bioactivities of mono- and dimeric representatives of Dioncophyllaceae-type alkaloids, and will report on the synthesis and biological evaluation of structurally related, but merely artificial quateraryls.

2.2 Mono- and Dimeric Dioncophyllaceae-Type Naphthylisoquinolines and Related Compounds

2.2.1 Occurrence of Dioncophyllaceous Alkaloids in Nature

The very small paleotropical plant family Dioncophyllaceae comprises only three monotypic genera and, thus, three species: *Triphyophyllum peltatum*, *Dioncophyllum thollonii*, and *Habropetalum dawei* [22, 196, 197]. These most peculiar lianas, which are endemic to the coastal regions of West Africa from Sierra Leone to the Gulf of Guinea, are fascinating to both botanists and natural product chemists, since they possess unusual botanical features (Fig. 5) and produce a series of structurally unique naphthylisoquinoline alkaloids (Fig. 6) with most promising bioactivities [13, 14, 16, 145, 146, 148, 150, 182, 198–207].

T. peltatum [22], mainly occurring in the evergreen rainforests of Sierra Leone, Liberia, and the Ivory Coast, is the most widespread and by far the phytochemically best investigated species of the Dioncophyllaceae. The adult plant forms short leaves with a pair of recurved hooks (Fig. 5b) helping the climbing liana to attach to branches and twigs in the rain forest. The large thin seed is unique with a circular membranous wing (Fig. 5e), which during maturation exceeds the size of the fruit from which it is borne. *T. peltatum* has also attracted attention because it shows the phenomenon of 'part-time carnivory' [208–214]. At the end of the juvenile growth period, just before it starts climbing, the liana forms leaves with stalked glands and reduced lamina (Fig. 5c, d), which capture and digest invertebrates. From viable seeds and adult sterile plants, the authors' group managed to establish in vitro organ cultures such as shoot [214, 215] and root cultures [216] as well as alkaloid-

Fig. 5 West African liana *Triphyophyllum peltatum* Airy Shaw (Dioncophyllaceae), cultivated in the greenhouse of the Botanical Garden of the University of Würzburg (Photographs **a**, **c–e**), and the Institute of Horticultural Production Systems, University of Hannover (Photograph **b**). The name *Triphyophyllum* means "three kinds of leaves", which characterize this particular species. **a** Juvenile plant with a rosette of small recurved lanceolate leaves; **b** hooked adult leaf specifically formed on elongated shoots, helping the plant to climb into the canopy of high trees; **c**, **d** young plant showing "part-time carnivory" based on glandular carnivorous leaves with a length of up to 25 cm, solely formed in the juvenile phase; **e** young fruits showing a rapid formation of large papery disks (with a diameter of 5–12 cm), which enable them to be dispersed by the wind (Photographs: **a**, **e**—H. Rischer; **b**—T. Winkelmann; **c**, **d**—A. Irmer)

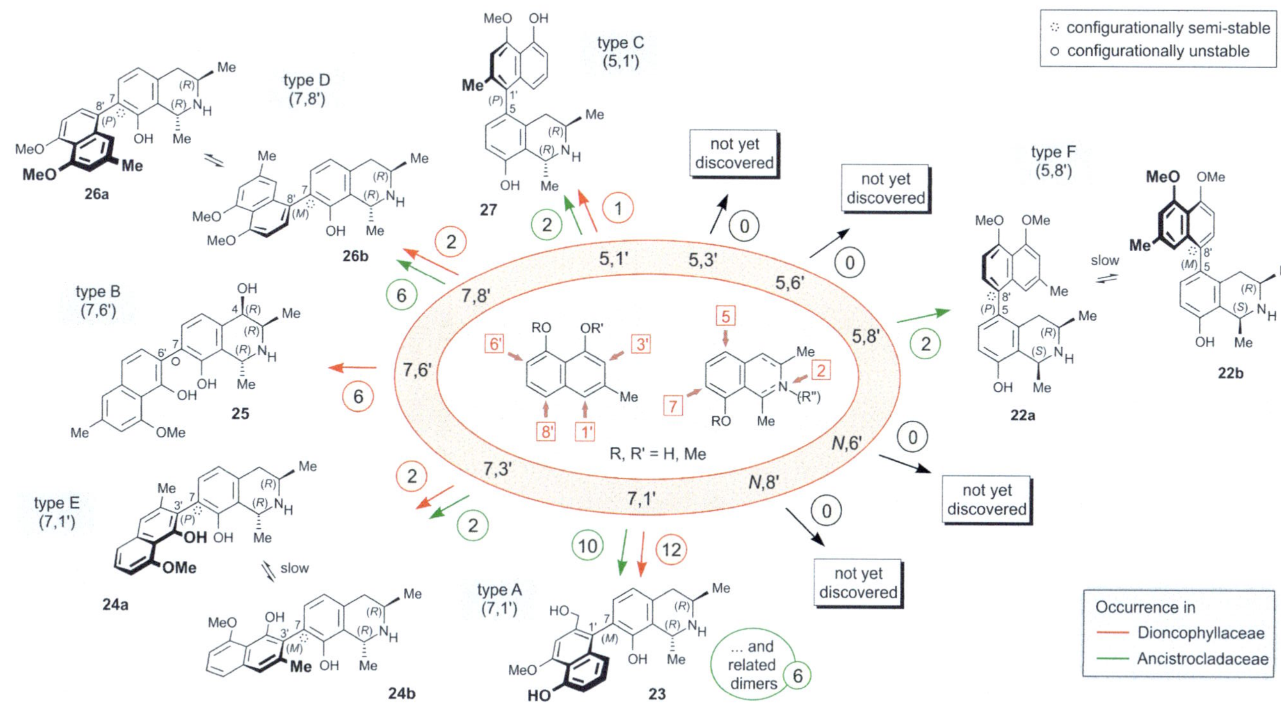

Fig. 6 Selected dioncophyllaceous naphthylisoquinoline alkaloids (6-H, (3*R*)), representing the six coupling types A (7,1′), B (7,6′), C (5,1′), D (7,8′), E (7,3′), and F (5,8′), and their occurrence in Dioncophyllaceae (red arrows) and Ancistrocladaceae (green arrows) species. In most cases, the biaryl axis is stable configurationally and constitutes an additional element of chirality, with exceptions being dioncophyllinol B (**25**), for which the axis can freely rotate, and 5′-*O*-methyldioncophylline D (**26a/26b**), dioncophylline E (**24a/24b**), and dioncophylline F (**22a/22b**) that occur as pairs of configurationally semi-stable and thus slowly interconverting atropo-diastereomers

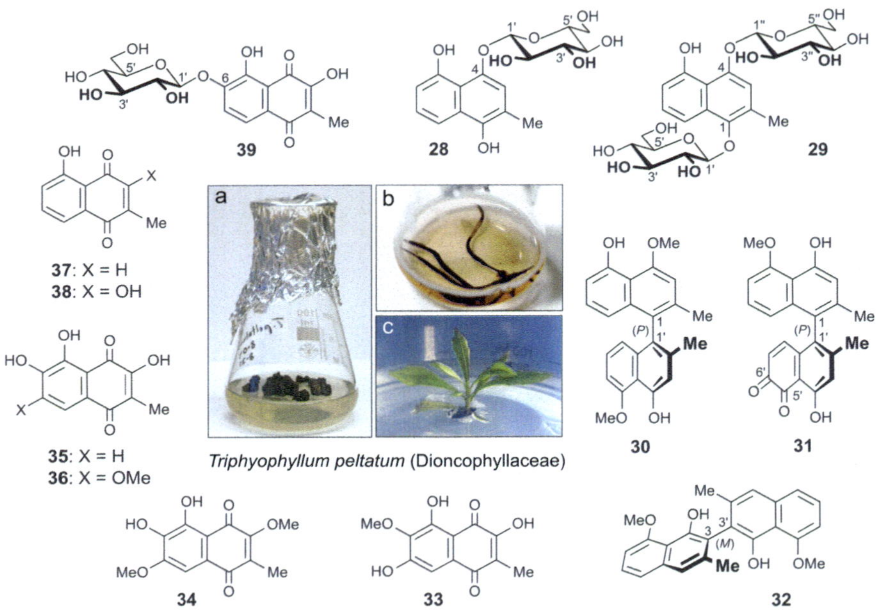

Fig. 7 New secondary metabolites related to the naphthalene moiety of naphthylisoquinoline alkaloids, produced by the callus and root cultures of *T. peltatum*, but not detected in the plant itself: dioncoquinones A (**39**), B (**35**), C (**36**), D (**34**), and E (**33**), dioncoquinoside A (**29**), triphyoquinols A_1 (**30**) and B (**32**), and triphyoquinone A (**31**), isolated along with the known metabolites plumbagin (**37**), droserone (**38**), and plumbaside A (**28**). The sterile shoots as presented in photograph (**c**) were used as the parent material for inducing **a** axenic cell and **b** root cultures (Photographs: **a** and **c**—A. Irmer; **b**—K. Fester)

producing cell cultures [131, 132, 214, 217]. This was a real milestone permitting for the first time the in-depth investigation of the exceptional biosynthesis of naphthylisoquinoline alkaloids from acetate-malonate units [131–134] (see Sect. 2.3). It also allowed the induction of novel natural products in callus [217, 218] and root cultures [216] (Fig. 7) that had not been observed in the intact plants before.

Extensive isolation work by the authors' group on *T. peltatum* revealed this peculiar liana to be a rich source of approximately twenty structurally diverse naphthylisoquinoline alkaloids [13, 14, 133, 152, 153, 157, 158, 216–224]. *T. peltatum* produces alkaloids of four different coupling types (Fig. 6), and most belong to the type-A coupling, as realized, e.g., in dioncopeltine A (**23**) [221], with its stereogenic axis being located between the 7-position and C-1′ of its two molecular portions. Dioncophyllinol B (**25**) [153], by contrast, possesses a configurationally unstable biaryl axis, thus it does not form stable atropisomers, but undergoes a rapid rotation about its 7,6′-located biaryl axis (type-B coupling), since it is neighbored by only two small hydroxy groups.

With the 5,1′-position of its biaryl axis (type-C coupling), dioncophylline C (**27**) [220] is configurationally stable; it is the only representative of this coupling type detected so far in a Dioncophyllaceae plant. Dioncophylline C (**27**) is highly active against different strains of the malaria parasite *Plasmodium falciparum* in vitro [198–201, 205–207], displaying a strong activity, and even a significant curative potential in vivo [204]. In view of its excellent therapeutic concentration range, and its good oral availability, dioncophylline C (**27**) has been regarded as a promising antimalarial lead [204, 205, 207].

5′-*O*-Methyldioncophylline D (**26a**) and its epimer **26b** [133, 216] were isolated from callus and root cultures of *T. peltatum*. They are the only representatives of the 7,8′-coupling type (type-D coupling) identified so far in a Dioncophyllaceae plant. A further new dioncophyllaceous alkaloid, dioncophylline E (**24a/b**) [225], possessing a 7,3′-linked biaryl axis (type-E coupling), has so far only been found to occur in the leaves and twigs of *D. thollonii*. Dioncophylline F (**22a/b**) (Fig. 6) [182], the first—and so far only—representative of a Dioncophyllaceae-type alkaloid showing the 5,8′-coupling type (type-F coupling), has as yet not been discovered in any of the three Dioncophyllaceae species *T. peltatum*, *D. thollonii*, or *H. dawei*, but was exclusively isolated from the Ancistrocladaceae liana *A. ileboensis* from the southern Congo Basin. As a consequence of the lack of an *ortho*-methyl group next to the biaryl axes, **22** [182], **24** [225], and **26** [133, 216] occur as pairs of configurationally semi-stable and, thus, slowly interconverting atropo-diastereomers. These metabolites, like all the other naphthylisoquinoline alkaloids isolated from Dioncophyllaceae plants, have in common the characteristic combination of an (*R*)-configuration at C-3 and the absence of an oxygen function at C-6—without even a single exception in any of the three Dioncophyllaceae species. As mentioned already in Sect. 2.1, these structural features classify them as 'Dioncophyllaceae-type' alkaloids, in contrast to the 'Ancistrocladaceae-type' naphthylisoquinolines, which are (*S*)-configured at C-3 and oxygenated at C-6 (Fig. 4).

Besides the plant itself, callus cultures of *T. peltatum* are also promising sources of novel natural products (Fig. 7). They comprise entirely undifferentiated cells, but also root cultures, which represent differentiated tissues of a complete organ. Cultivation of *T. peltatum* as sterile callus cultures resulted in a largely increased formation of plumbagin (**37**) and droserone (**38**) [134, 217, 226]. These two naphthoquinones are phytoalexins produced by intact plants as a reaction to chemical, physical, or biotic stress [13]. In particular plumbagin (**37**) is a potent bioactive compound displaying strong cytostatic, antioxidant, anti-inflammatory, and immunosuppressive properties, but also antiprotozoal, antibacterial, antifungal, and antiviral activities have been reported [227–233]. Solidification of the medium, which prevents the callus cultures from sinking into the agar, hence exposing them to aerobic and oxidative conditions, favored the additional formation of structurally related, but highly oxygenated naphthoquinones such as dioncoquinones A (**39**), B (**35**), C (**36**), D (**34**), and E (**33**) (Fig. 7) [217, 218].

The dioncoquinones A (**39**) and B (**35**) show high and specific activities against *Leishmania major*, while no or only weak activities are observed against protozoan parasites belonging to the genera *Plasmodium* and *Trypanosoma* [217].

Furthermore, the dioncoquinones A (**39**), B (**35**), and C (**36**) strongly act against human tumor cells derived from two different B cell malignancies, namely, B cell lymphoma and multiple myeloma (MM), where they induce apoptosis [217, 218]. Their effective concentrations, in particular against MM cell lines, are comparable to those of melphalan [234], a well-known DNA-alkylating agent used in the standard chemotherapy against B cell malignancies, but **39** and **35** do not display any significant toxicity toward normal peripheral mononuclear blood cells (PBMCs) [217, 218]. Specifically, dioncoquinones B (**35**) and C (**36**) could become promising agents and potential leads for the development of novel anti-MM candidates [217, 218].

Sterile root cultures of *T. peltatum*, by contrast, furnished a series of structurally new, axially chiral binaphthalenes like triphyoquinols A_1 (**30**) and B (**32**), and naphthylnaphthoquinones such as triphyoquinone A (**31**), but also related naphthalene glucosides such as dioncoquinoside A (**29**) and the known [235–237] plumbaside A (**28**) [216] (Fig. 7). All of these compounds possess structural patterns that resemble those of the isocyclic parts of naphthylisoquinoline alkaloids. They were identified only in the root cultures, not in the plants themselves and—except for triphyoquinol B (**32**)—not in callus cultures of *T. peltatum*. The highly oxygenated dioncoquinones A (**39**), B (**35**), C (**36**), D (**34**), and E (**33**) were detected exclusively in callus cultures on solidified medium [217, 218], but not in the root cultures [216].

LC-MS/MS investigations on a series of extracts from root, bark, twig, and leaf material of the three West African Dioncophyllaceae plants, *T. peltatum*, *D. thollonii*, and *H. dawei* gave no hints at the occurrence of dimeric naphthylisoquinoline alkaloids in these lianas [17]. Metabolic screening studies on the root bark extract of a botanically as yet undescribed *Ancistrocladus* plant from the Central Congo Basin, however, led to the discovery of the first Dioncophyllaceae-type dimer in Nature, jozimine A_2 (**40a**) (for *jozi* = pair in *Swahili*) (Fig. 8) [122]. This alkaloid consists of two identical molecular halves, which are coupled through the sterically more constrained C-3′ and C-3″ positions in the two naphthalene subunits, giving rise to an additional stable stereogenic element. Since the two outer axes are chiral, **40a** possesses three consecutive chiral biaryl axes. From the West African liana *A. abbreviatus*, three further Dioncophyllaceae-type dimeric naphthylisoquinoline alkaloids, jozibrevines A–C (**40b**–**40d**), were isolated, along with their parent compound jozimine A_2 (**40a**) [124]. The molecular halves of these four structurally intriguing quateraryls are the 7,1′-coupled naphthylisoquinoline alkaloids 4′-*O*-demethyldioncophylline A (**42a**) and/or its atropisomer, 7-*epi*-4′-*O*-demethyldioncophylline A (**42b**). They are known to occur in *A. abbreviatus* as well [124]. The dimers **40a**–**40d** have identical constitutions and the same absolute configurations at the four stereogenic centers, but differ in their axial chirality [124]. Such a broad, nearly complete series of atropo-diastereomeric dimers (four out of six possible rotational isomers) had never been found in any *Ancistrocladus* species before.

Jozimine A_2 (**40a**) displays an excellent antiplasmodial activity in the low nanomolar range (IC_{50} = 1.4 n*M*) against the NF54 strain of *P. falciparum* (sensitive to all known drugs), significantly higher than that of nearly all other naturally

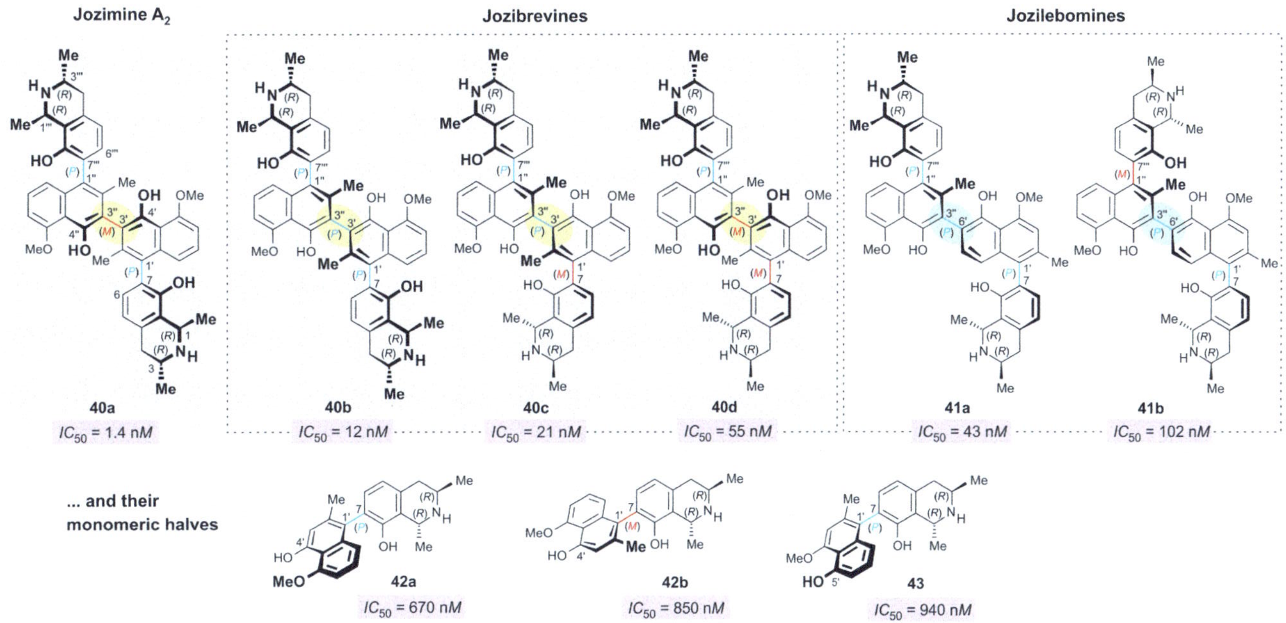

Fig. 8 Antiplasmodial Dioncophyllaceae-type naphthylisoquinoline dimers with three consecutive chiral biaryl axes isolated from the West African liana *A. abbreviatus* [124], from the Congolese species *A. ileboensis* [123], and from an as yet botanically undescribed *Ancistrocladus* species from the South-Central Congo Basin [122]: the parent compound jozimine A_2 (**40a**) (underlaid in gray) and the jozibrevines A–C (**40b**–**40d**) (also highlighted in gray), with all of them displaying a central 3′,3″-biaryl linkage, and the two 6′,3″-coupled dimers jozilebomines A (**41a**) and B (**41b**): also presented are the corresponding monomers, 4′-*O*-demethyldioncophylline A (**42a**), its atropo-diastereomer 4′-*O*-demethyl-7-*epi*-dioncophylline A (**42b**), and the regioisomeric 5′-*O*-demethyldioncophylline A (**43**). Half-maximum inhibitory concentration (IC_{50}) values of the compounds against the NF54 strain of *P. falciparum* are underlaid in light pink

occurring monomeric and dimeric naphthylisoquinoline alkaloids tested so far [122]. Moreover, **40a** possesses a very high selectivity index of >11,400 [122]. According to the TDR/WHO guidelines [238], it can therefore be considered as an antiplasmodial lead compound. Moreover, the inhibitory activities of jozimine A_2 (**40a**) against *Plasmodium* parasites seem to be highly pathogen-specific: biological tests against the protozoan parasites causing visceral leishmaniasis (*Leishmania donovani*), Chagas' disease (*Trypanosoma cruzi*), and African sleeping sickness (*Trypanosoma brucei rhodesiense*) revealed **40a** to display only weak activities [122].

The jozibrevines A–C (**40b–d**) likewise exhibit pronounced activities against *P. falciparum*, with only moderate cytotoxicities against rat skeletal myoblast (L6) cells, again similarly to their parent compound jozimine A_2 (**40a**), showing a high stereostructure-dependent specificity of their antiplasmodial activity [124]. The strongest effects against the NF54 strain of *P. falciparum* are displayed by the jozibrevines A (**40b**) and B (**40c**), with half-maximum inhibitory concentrations (IC_{50}) of 0.012 μM (for **40b**) and 0.021 μM (for **40c**). Their inhibitory effects are however, less pronounced than that of the (*P*,*M*,*P*)-atropisomer jozimine A_2 (**40a**). These findings clearly demonstrate the strong impact of axial chirality on the bioactivity of jozimine-A_2-type dimers [124].

Isolation work on *A. ileboensis* from the South-Central Congo Basin provided two further Dioncophyllaceae-type dimers, the jozilebomines A (**41a**) and B (**41b**) (Fig. 8), again along with jozimine A_2 (**40a**) [123]. The new dimers **41a** and **41b** are apparently formed through phenol-oxidative cross-coupling of two constitutionally different 7,1′-linked naphthylisoquinoline monomers, both from 5′-*O*-demethyldioncophylline A (**43**) as one of the molecular units and from 4′-*O*-demethyldioncophylline A (**42a**) as the second part in the case of **41a** and 4′-*O*-demethyl-7-*epi*-dioncophylline A (**42b**) in the case of **41b** (Fig. 8). The jozilebomines A (**41a**) and B (**41b**) are connected through a 6′,3″-coupled axis in their binaphthalene core, and not, as in jozimine A_2 (**40a**), via the C-3 positions in both naphthalene portions. Thus, different from the C_2-symmetric **40a**, they are constitutionally unsymmetric. The jozilebomines A (**41a**) and B (**41b**) exert strong activities against *P. falciparum*, but only low or no inhibitory effects against *Leishmania* or *Trypanosoma* pathogens [123].

Previous studies had already hinted at the importance of the number of free hydroxy groups for the potency of naphthylisoquinoline alkaloids toward *P. falciparum* [86, 125, 129]. This was again corroborated by the fact that the three bisphenolic monomeric precursors of jozimine-A_2-type dimers, 4′-*O*-demethyldioncophylline A (**42a**) (IC_{50}, 0.67 μM), its 7-epimer **42b** (IC_{50}, 0.85 μM) [124], and the regioisomeric 5′-*O*-demethyldioncophylline A (**43**) (IC_{50}, 0.94 μM) [223], each displayed distinctly higher antiplasmodial activities than dioncophylline A (**16**) itself (IC_{50}, 3.81 μM) [198], which possesses only one free hydroxy function. Oxidative dimerization and doubling of the number of free hydroxy groups increased the antiplasmodial activities. The resulting jozimine-A_2-type dimers **40a–40d**, now equipped with even four free hydroxy groups, showed by far the highest effects against *P. falciparum* within this series of Dioncophyllaceae-type mono- and dimeric naphthylisoquinoline alkaloids [124].

2.2.2 Artificial Dimers of Dioncophyllaceae-Type Monomers as Promising Antiplasmodial Agents: Design and Total Synthesis

Synthesis efforts to extend the scope of antiplasmodial naphthylisoquinolines by dimerization of naturally occurring Dioncophyllaceae-type monomeric compounds that are easily available by isolation from the plants such as dioncophyllines A (**16**) and C (**27**), provided new, artificial dimers (not yet found in Nature), with some displaying good or even strong inhibitory activities against *P. falciparum* in vitro.

Based on a biomimetic oxidative dimerization strategy [84, 86, 96, 125] (see also the photometric test [42] for the search of naphthylisoquinoline dimers in Sect. 5.2), the respective naphthylisoquinolines were (where required) specifically protected, usually by selective O,N-dibenzylation in the isoquinoline moiety, and then submitted to phenol-oxidative coupling using silver(I) oxide as the oxidant. Hydrogenolytic reduction of the resulting (deeply violet-colored) overoxidized diphenoquinone intermediate in the presence of Pd/C, with simultaneous cleavage of the *O*- and *N*-benzyl protective groups, finally provided the novel-type naphthylisoquinoline dimers in good overall yields [86, 97, 125, 127–130].

As a first example, jozimine C (**46**) [97], the homo-coupled unnatural dimer of the antimalarial natural alkaloid dioncophylline C (**27**) [204, 220] (Fig. 9), was easily accessible from the monophenolic *O,N*-dibenzyl-protected derivative **44** with a free hydroxy group specifically at C-5′, i.e., *ortho* to the desired coupling position C-6′. Coupling with silver(I) oxide and in situ reduction of the diphenoquinone intermediate **45**, with simultaneous deprotection, yielded the 6′,6″-coupled jozimine C (**46**) in an overall yield of 66% [97].

The new dimer **46** shows a strong structural resemblance to the likewise C_2-symmetric quateraryl michellamine A (**47a**), a widely occurring naphthylisoquinoline dimer in *Ancistrocladus* species from Central Africa [17, 163, 164, 167, 173]. Due to the 5,1′-coupled "outer axes" in the naphthylisoquinoline moieties of **46** however, the overall structure of this dimer is more linear than that of **47a**, showing the possibility to rationally design the molecular shapes of such dimers. Similar to **47a**, the newly generated central axis, which connects the respective 6′-positions of the two molecular halves of the merely artificial dimer **46**, can freely rotate and does not constitute an additional element of chirality [97]. Jozimine C (**46**) displays a good antiplasmodial activity (IC_{50}, 0.60 μ*M*; NF54 strain)—yet far weaker than that of its monomeric precursor, dioncophylline C (**27**) (IC_{50}, 38 n*M*; NF54 strain) [97].

Biomimetic phenol-oxidative coupling of 5′-*O*-demethyldioncophylline A (**43**), a main metabolite of the two Dioncophyllaceae lianas *T. peltatum* [223] and *D. thollonii* [74, 225], afforded the C_2-symmetric dimer jozimine A_3 (**49**) (Fig. 10) [130]. Treatment of the *O,N*-dibenzyl-protected derivative **48** with silver(I) oxide provided a deeply violet-colored reaction mixture, thereby hinting at a successful

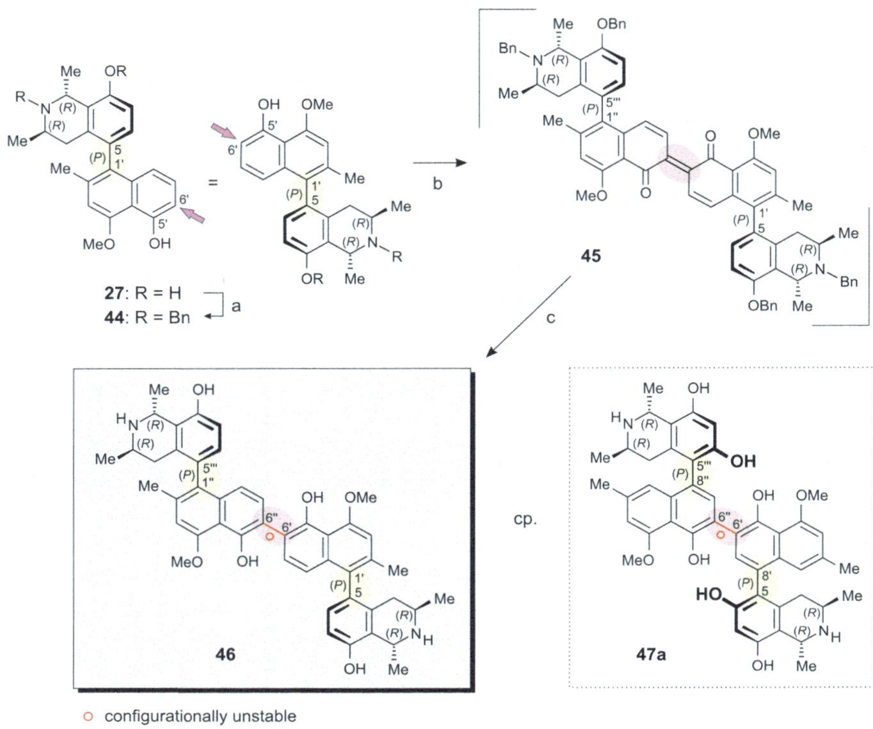

Fig. 9 Synthesis of the artificial dimer jozimine C (**46**), structurally closely related to the naturally occurring, yet more "compact", likewise 6′,6″-coupled michellamine A (**47a**), by phenol-oxidative dimerization of the antiplasmodial dioncophylline C (**27**). Reaction conditions: (a) BnBr, acetone, K_2CO_3, reflux, 95%; (b) Ag_2O, 0.2% Et_3N in $CHCl_3$; (c) Pd/C (10%), H_2, EtOH, 66%

dimerization reaction with the formation of the corresponding diphenoquinone [130]. Reduction of this intermediate diketone by hydrogenation in the presence of Pd/C, followed by addition of HCl to complete the removal of the benzyl protecting groups finally gave rise to jozimine A_3 (**49**) (Fig. 10), affording a new 6′,6″-coupled dimer with a configurationally unstable central biaryl axis between the two constitutionally identical molecular halves [130].

As a by-product, a C(sp^3)-C(sp^2)-linked, non-symmetric 6′,8″-coupled artificial dimer was likewise produced from 5′-*O*-demethyldioncophylline A (**43**). This compound, named jozimine A_4 (**50**), had an additional stereogenic center at C-8″ next to the axis, due to a newly formed tetralone moiety (Fig. 10). Most probably, an unsymmetric phenol-oxidative coupling led to C,C-bond formation between C-6′ and C-8″, i.e., in *ortho*- and *para*-positions relative to the 5′-hydroxy groups in the naphthalene rings of benzyl-protected 5′-*O*-demethyldioncophylline A (**43**), respectively. In view of the large steric hindrance of the 8″-coupled axis, further electronically facilitated by the adjacent second naphthalene ring, the dimerization was followed by an overhydrogenation under the reductive work-up conditions,

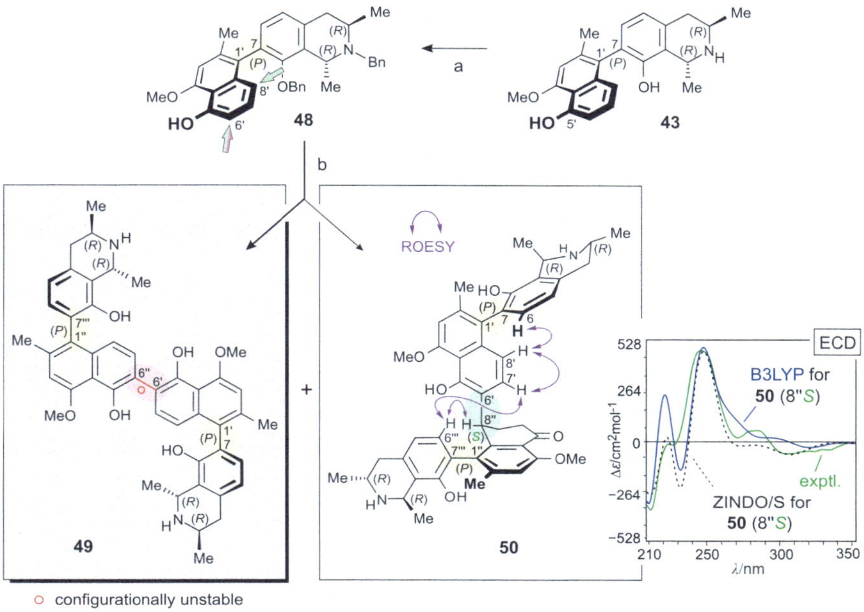

Fig. 10 Synthesis of jozimines A_3 (**49**) and A_4 (**50**) from 5′-*O*-demethyldioncophylline A (**43**), by phenol-oxidative coupling, assignment of the location of the central linkage between the two molecular subunits by ROESY interactions, and attribution of the absolute configuration of jozimine A_4 as (8″*S*)-**50** by ECD investigations. Reaction conditions: (a) BnBr, Cs_2CO_3, acetone, reflux, 85%; (b) Ag_2O, NEt_3, $CHCl_3$, r.t., → H_2, Pd/C (10%), EtOH, 2 N HCl, r.t., 27% (for **49**) and 13% (for **50**)

thus leading to a decrease of the steric hindrance by converting the sp^2-configured C-8′ into sp^3. The structure of this new unsymmetric dimer **50** with the central linkage located at C-6′ for its naphthalene moiety and at C-8″ for the tetralone unit, was clearly evidenced from the ROESY sequence 6-H ↔ 8′-H ↔ 7′-H ↔ 6‴-H ↔ 8″-H (Fig. 10) [130].

For the assignment of the absolute configuration of the newly induced stereogenic center at C-8″ of jozimine A_4, conformational investigations for the two possible epimers, **50** (8″*S*) and 8″-*epi*-**50** (8″*R*) (DFT-optimized global minima), followed by UV and ECD calculations, were performed. The semi-empirical ZINDO/S-CIS excited-state calculations showed an excellent agreement of the ECD spectrum predicted for **50** (8″*S*) with the experimental one (Fig. 10). The ECD curve calculated for 8″-*epi*-**50** (8″*R*), on the contrary, displayed an almost mirror-image like behavior (data not shown), therefore corroborating the strong influence of this central stereogenic center at C-8″ on the whole chromophoric system of the otherwise identical stereogenic elements in the monomeric halves of the dimers **50** and 8″-*epi*-**50**. TDDFT calculations using the B3LYP functional

[239, 240] in combination with Pople's 6-13G* basis set [241] gave additional proof of the (8″*S*)-configuration of jozimine A_4 (**50**) [130].

Jozimine A_3 (**49**) exhibits an excellent inhibitory activity (IC_{50}, 4 n*M*) against the chloroquine-resistant K1 strain of *P. falciparum*, much better than its monomeric half 5′-*O*-demethyldioncophylline A (**43**) (IC_{50}, 0.94 μ*M*), while exerting only weak cytotoxicity against rat skeletal myoblast (L6) cells, with an IC_{50} value of 0.78 μ*M*, which leads to a selectivity index of 195 [130]. According to the TDR/WHO guidelines [238], this dimer can be considered as an antiplasmodial lead. Jozimine A_4 (**50**) by contrast is far less active against the K1 strain (IC_{50}, 0.94 μ*M*) [130], evidencing the relevance of the intact central binaphthalene system for the bioactivity.

An example of an artificial dimer of an unnatural monomer possessing three freely rotating biaryl axes is pindikamine A (**55**) (for *pindika* = skew in *Swahili*) [127] (Fig. 11), which is the 6′-dimer of the dioncophyllaceous-like (i.e., (1*R*,3*R*)-configured), yet unnatural (since 6,8′-coupled) naphthylisoquinoline **53**. Starting from the known [96] trialkylstannane **51** as the naphthalene partner and the well-established isoquinoline building block **52** [242], equipped with an *O*-triflate leaving group at C-6, the key precursor **54** was easily attained by Pd-catalyzed Stille-type coupling, followed by selective cleavage of the *O*-isopropyl protective group [127]. Oxidative dimerization of this *N*-benzyl-protected biaryl **54**, with the free phenolic oxygen function specifically located next to the scheduled 6′-coupling site, led to pindikamine A (**55**) in an excellent yield (92% from **54**) after treatment with silver(I) oxide and mild hydrogenolytic reduction of the intermediate diphenoquinone, with simultaneous deprotection [127]. For comparative biotests, the free monomeric precursor **53** of pindikamine A (**55**) was also synthesized, by hydrogenolytic *N*-debenzylation of **54** [127]. Pindikamine A (**55**) displays only moderate inhibitory activities (IC_{50}, 1.63 μ*M*; NF54 strain) in vitro against *P. falciparum*, yet distinctly better than its monomeric analogue **53** (IC_{50}, 9.24 μ*M*; NF54 strain) [127].

Jozipeltine A (**57**) (Fig. 12), [129] the homo-coupled symmetric unnatural dimer of the antiplasmodial alkaloid dioncopeltine A (**23**) [204, 221], was considered as a rewarding synthesis target due to its additional benzylic oxygen functions in the side chain of the naphthalene parts, possessing three free hydroxy groups per monomeric half. After selective protection of two of the four OH and NH functionalities of dioncopeltine A (**23**) to give intermediate **56**, the coupling succeeded oxidatively, with silver(I) oxide as the reagent. Reduction of the intermediate diphenoquinone, with simultaneous cleavage of the benzyl protective groups, yielded the 6′,6″-coupled jozipeltine A (**57**) in 52% yield [129].

Jozipeltine A (**57**) is significantly less active against the malaria parasite *P. falciparum* than its monomeric precursor, dioncopeltine A (**23**). The dimer **57** exerts only moderate inhibitory effects, showing IC_{50} values of 3.41 μ*M* (for strain NF54) and 1.18 μ*M* (for strain K1) [129], whereas its monomer **23** displays excellent antiplasmodial activities, with IC_{50} values of 8.7 n*M* (for strain NF54) and 12.6 n*M* (for strain K1) [198]. The drastic decrease of the antiplasmodial activity of jozipeltine A (**57**), relative to that of the closely related dimer jozimine A_3 (**49**)

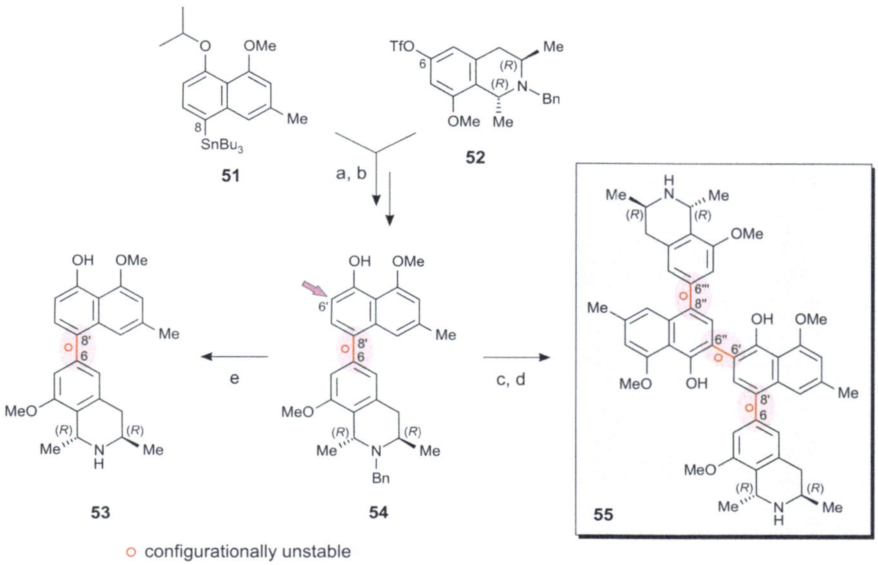

Fig. 11 Synthesis of the artificial Dioncophyllaceae-type dimer pindikamine A (**55**) with three non-stereogenic biaryl axes, via the *N*-benzyl analogue **54** of the likewise non-natural naphthylisoquinoline **53**. Reaction conditions: (a) $PdCl_2(PPh_3)_2$, PPh_3, LiCl, CuBr, DMF, 135° C, 63%; (b) BCl_3, CH_2Cl_2, −15°C, 96%; (c) Ag_2O, $CHCl_3$, Et_3N, 94%; (d) H_2, Pd/C (10%), MeOH/CH_2Cl_2, 98%; (e) H_2, Pd/C (10%), MeOH/CH_2Cl_2, 93%

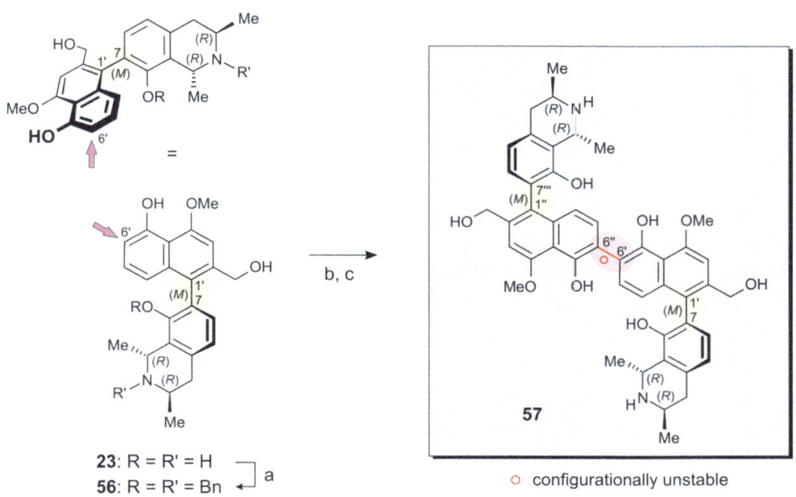

Fig. 12 Synthesis of jozipeltine A (**57**), an artificial constitutionally symmetric 6′,6″-coupled dimer of the natural Dioncophyllaceae-type naphthylisoquinoline alkaloid dioncopeltine A (**23**), by oxidative coupling of its *N*-benzyl derivative **56** with Ag_2O. Reaction conditions: (a) BnBr (2.0 equiv.), K_2CO_3, acetone, reflux, 96%; (b) Ag_2O, 0.2% Et_3N in $CHCl_3$; (c) H_2, Pd/C (10%), EtOH, r.t., 53% from **23**

[130], is apparently due to the fact that the usual methyl groups at C-2′ in the central binaphthalene core of **49** are replaced by the much more polar CH_2OH functions in **57**, leading to a higher total number of six free hydroxy groups—too high to permit good antiplasmodial activities. A similar effect was also observed for the antiplasmodial alkaloid korupensamine A (**13a**; for its structure, see Fig. 3) (IC_{50}, 0.82 μ*M*) [156], possessing three hydroxy groups, versus the corresponding dimer, michellamine A (**47a**, for its structure, see Fig. 9), with six free hydroxy functions, which was inactive against *P. falciparum* [156]. The increase from one to two free hydroxy groups triggered a distinct increase of antiplasmodial activity as shown, for example, for the dimer pindikamine A (**55**) and its monomeric precursor **53** [127, 129, 130].

Attempts to dimerize dioncophylline A (**16**) under phenol-oxidative conditions failed completely. Therefore, an alternative, reductive strategy was pursued, by Klar–Wittig coupling, which delivered the new quateraryl jozimine A (**61**) (Fig. 13) [126]. By lithiation of *O,N*-dibenzyl-protected 5-bromodioncophylline A (**59**) with *tert*-butyllithium at −95°C and subsequent transformation into the corresponding diaryl cuprate, the desired dimer **60** was obtained upon warming to room temperature. Subsequent cleavage of all four *O*- and *N*-benzyl protective groups succeeded by catalytic hydrogenation, leading to the final target molecule **61** [126]. Jozimine A (**61**) is structurally unique, since its two halves are coupled via the isoquinoline moieties and not via the naphthalene parts, thus leading to a connectivity with a sequence "naphthalene-isoquinoline-isoquinoline-naphthalene" ("N–I–I–N"), in contrast to all the other naphthylisoquinoline dimers presented here, which using the same terminology, would have to be addressed as an "I–N–N–I" sequence. Different from the 6′,6″-coupled dimers jozimine A_3 (**49**) [130], pindikamine A (**55**) [127], and jozipeltine A (**57**) [129], the inner axis of jozimine A (**61**) was stereochemically moderately stable, providing the two atropo-diastereomers, **61a** and **61b** (Fig. 12), which after their successful chromatographic resolution, undergo a slow interconversion at room temperature to give again the initial ratio of ca. 1:1 [126].

Of particular interest was the antiplasmodial activity of jozimine A (**61**). This unnatural dimer of natural dioncophylline A (**16**) displays a strong inhibitory effect in vitro against the NF54 strain of *P. falciparum* (IC_{50}, 0.10 μ*M*) [126]. With its two phenolic OH groups, it clearly exceeds the activity of its monophenolic precursor, dioncophylline A (**16**) (IC_{50}, 3.81 μ*M*) [198], being 38 times more active than its monomeric half and >200 times more active than its fully protected synthesis precursor 5-bromodioncophylline A (**58**) (IC_{50}, 20.3 μ*M*) [126].

In summary, some of the dioncophyllaceous dimers presented in this section, in particular the naturally occurring jozimine A_2 (**40a**) [122] and the non-natural jozimine A_3 (**49**) [130], showed excellent antiplasmodial activities, and can be regarded as promising lead compounds for an evaluation of their potential as novel-type antimalarial agents.

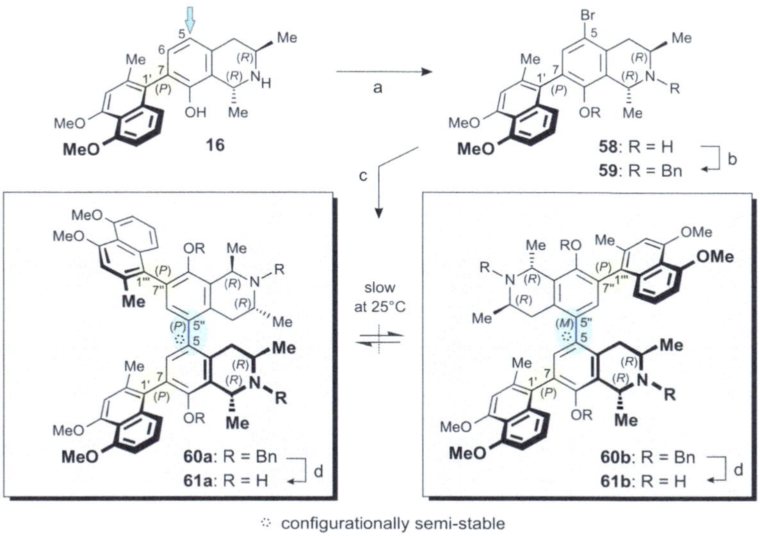

Fig. 13 Synthesis of jozimine A (**61**) by reductive dimerization of dioncophylline A (**16**), creating a unique, non-michellamine-type molecular architecture. Reaction conditions: (a) $N(nBu)_4Br_3$, $CHCl_3$, NaOAc, 0°C, 85%; (b) BnBr, K_2CO_3, acetone, reflux, 87%; (c) *t*-BuLi, 2-MeTHF, −95°C; $CuCl_2$, −95°C → r.t., 14%; (d) H_2, Pd/C (10%), EtOH, 99%

2.3 A Novel Biosynthesis Pathway in Plants: Acetogenic Naphthylisoquinoline Alkaloids

Ancistrocladine (**1a**) (Fig. 4) with its unique substitution pattern has been termed "the most unusual of all the isoquinoline alkaloids" [2, 243, 244]. Indeed, this pattern does not fit into the generally accepted biosynthesis scheme for tetrahydroisoquinolines. Specifically the methyl group at C-3, the oxygen function at C-8, and the apparently acetogenic naphthalene substituent hint at a hitherto unprecedented origin of isoquinoline alkaloids exclusively from polyketide precursors [131–134]. Approximately 2500 different isoquinoline alkaloids have so far been isolated from plants [6–12], among them structurally simple 1-alkylisoquinolines such as anhalonidine (**63**) [135], but also structurally complex compounds like the well-known benzylisoquinoline-derived, polycyclic alkaloids morphine (**64**) and codeine (**65**) [6, 10, 12, 245, 246]. All of them originate from amino-acid-derived 2-arylethylamines like, e.g., dopamine (**62**), by a Pictet–Spengler condensation with aldehydes (or α-keto acids) (Fig. 14, left). Their great structural diversity results from the variation of the aldehyde precursor and/or from a series of follow-up reactions of the initially formed tetrahydroisoquinoline [6, 7, 10, 136–140].

As already assumed by Govindachari [2], then formulated in detail [13, 243] and unambiguously proven by the authors' group [131–134], naphthylisoquinoline alkaloids like ancistrocladine (**1a**) from *A. heyneanus* (Ancistrocladaceae) or

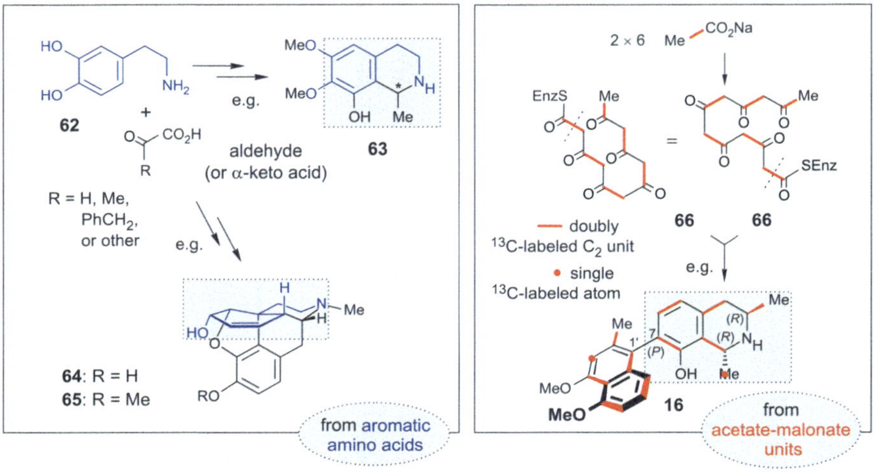

Fig. 14 Convergence in isoquinoline biosynthesis in plants—the 'usual' pathway via Pictet–Spengler condensation (left side), starting from amino acids such as dopamine (**62**), as exemplified for anhalonidine (**63**), morphine (**64**), and codeine (**65**), in comparison to the novel acetate-malonate pathway (right side) to give naphthylisoquinoline alkaloids via a joint polyketide precursor **66**, which is illustrated here for dioncophylline A (**16**)

dioncophylline A (**16**) from *T. peltatum* (Dioncophyllaceae) arise from a novel acetate-polymalonate-based biosynthesis pathway to isoquinolines (Fig. 14, right), via the open-chain hexaketide **66**. From this joint precursor, both halves of the alkaloids, the naphthalene moiety and the isoquinoline portion, are built up divergently. The two molecular portions are then joined together—convergently—at a late stage of the biosynthesis route, by phenol-oxidative biaryl coupling, thus providing the complete alkaloids [131–133]. Metabolic transformation reactions such as reduction, oxygenation, O- or N-methylation, or oxidative dimerization finally give rise to the plethora of structurally diverse naphthylisoquinoline alkaloids isolated from Ancistrocladaceae and Dioncophyllaceae plants.

As outlined in Fig. 15, the key steps of the proposed acetogenic biosynthesis pathway to naphthylisoquinoline alkaloids [131–133] comprise the formation of the highly reactive β-pentaketone **66** from six acetate units, followed by their aldol condensation and subsequent aromatization. These reactions should give rise to the monocyclic diketone **68**, from which, by further aldol cyclization, the corresponding naphthalene portion **69** should arise or, by reductive amination of the more reactive acetonyl keto function of **68** (probably by transamination), the dihydroisoquinoline **70** should result. This concept convincingly rationalizes the unusual substitution patterns in the isoquinoline and the naphthalene part. Oxidative phenolic coupling of the two molecular moieties, **69** and **70**, will then after some follow-up reactions (in particular O-methylations), directly lead to complete naphthylisoquinoline alkaloids like ancistrocladine (**1a**) or dioncophylline A (**16**).

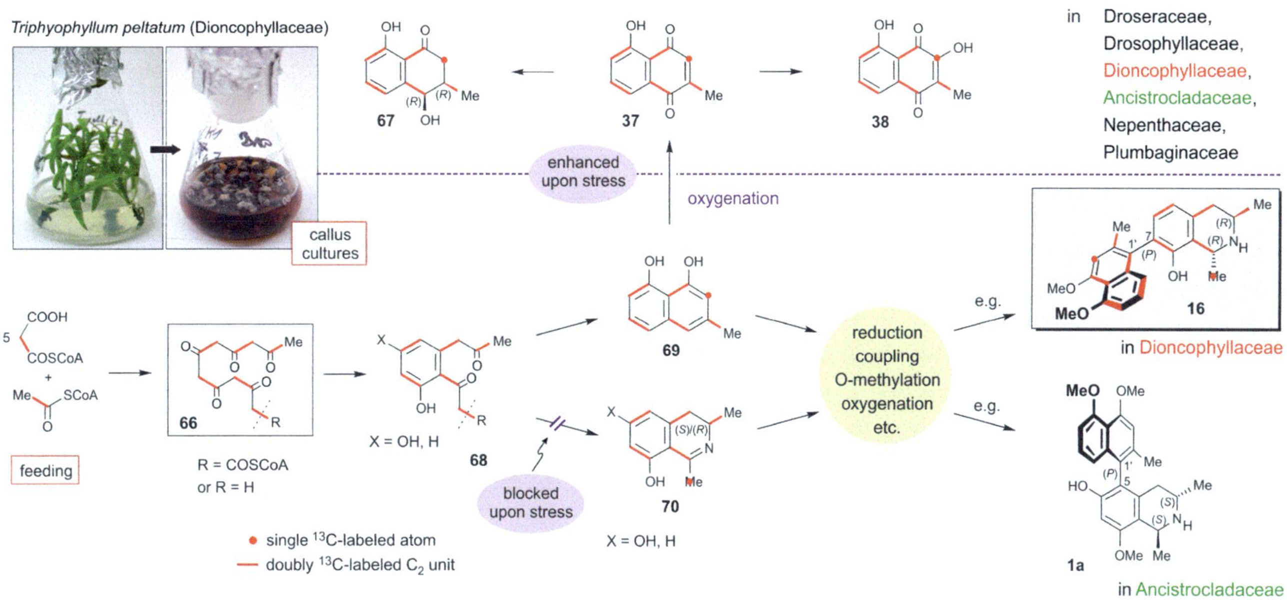

Fig. 15 Biosynthesis route to naphthylisoquinoline alkaloids like dioncophylline A (**16**) and ancistrocladine (**1a**) from acetogenic precursors via joint polyketide and diketo intermediates (e.g., **66** and **68**) giving rise to the two molecular halves **69** and **70**, and stress-induced (or constitutional) formation of the naphthalene-related compounds isoshinanolone (**67**), plumbagin (**37**), and droserone (**38**). Feeding experiments in callus cultures of *Triphyophyllum peltatum* (Dioncophyllaceae) conducted with $^{13}C_2$-labeled acetate, have revealed the entire carbon skeletons of the two molecular portions of dioncophylline A (**16**) to be derived from acetate units (Photographs: A. Irmer)

Another interesting aspect of the biosynthesis route to naphthylisoquinoline alkaloids relates to the question at which stage and by which substrate properties the configurations at the centers and at the axis are established, and which kinds of enzymes are involved in these reactions. As already described in the previous Sect. 2.1, all of the Dioncophyllaceae and some of the Ancistrocladaceae lianas (especially those from Asia) synthesize their naphthylisoquinoline alkaloids in a highly regio- and stereoselective manner. Several West and Central African plants by contrast, show a smaller degree of specificity and produce typical Ancistrocladaceae-type alkaloids such as ancistrocladine (**1a**), as chemotaxonomically expected, but also pure Dioncophyllaceae-type compounds like dioncophylline A (**16**), and even mixed hybrid-type structures with an oxygen function at C-6 and (*R*)-configuration at C-3 [122–124, 182, 184–187, 191–194]. Despite such complex mixtures of constitutionally and stereochemically diverse naphthylisoquinoline alkaloids produced by some of the African plants, the alkaloids are by no means built up in an uncontrolled or even chaotic manner. The biosynthetic formation of the compounds in an enantiomerically pure form nearly exclusively [13–18], and the likewise predominantly atroposelective construction of the biaryl linkage of the alkaloids corroborates the assumption of a strictly enzyme-catalyzed and thus efficiently controlled pathway to naphthylisoquinolines showing a high substrate specificity. This is underlined by the fact that never has any naphthyltetra- or -dihydroisoquinoline alkaloid been isolated from a certain plant species as a mixture of all imaginable stereo- or regioisomeric forms [13, 17, 18].

This unique biosynthesis pathway to acetogenic isoquinoline alkaloids is however, extremely susceptible to all sorts of chemical, physical, or biotic stress, which can easily result in an enhanced formation of naphthoquinones like plumbagin (**37**) and droserone (**38**) and tetralones such as isoshinanolone (**67**) [81, 217, 218, 226, 247–252]. The predominant production of these compounds is apparently a consequence of an impairment of the reductive incorporation of nitrogen into the assumed joint diketo precursor **68**. As a result, the plants then fail to form the dihydroisoquinoline building block **70** of naphthylisoquinoline alkaloids (Fig. 15).

1,4-Naphthoquinones like **37** and **38** and tetralones like **67** are also well known from the closest phylogenetic neighbors to the Dioncophyllaceae and Ancistrocladaceae, such as the Plumbaginaceae, Droseraceae, Drosophyllaceae, and Nepenthaceae [253–255]. These plant families however, strictly produce nitrogen-free compounds and to date, never any naphthylisoquinoline alkaloids. Thus, the capability to incorporate nitrogen into the postulated joint diketo precursor **68** to give the dihydroisoquinoline **70** is so far restricted to the Dioncophyllaceae and Ancistrocladaceae, exclusively.

2.4 The Tool Box Used for the Isolation, Structural Elucidation, and Stereochemical Assignment of Naphthylisoquinoline Alkaloids—A Short Overview

For the isolation and structural elucidation of naphthylisoquinoline alkaloids, a broad spectrum of efficient tools have been applied [13, 14, 17, 42, 73, 75–85, 191] or, where required, further developed [31, 73, 75–77, 80, 99, 101], or even newly designed [42, 75, 78–85, 87]. This section gives a short introduction into the basic principles for the preparative isolation and purification of the alkaloids and into the spectroscopic (NMR, MS, ECD) [13–15, 17, 18, 42, 73–75, 81–83], chemical (oxidative degradation) [79, 80], and computational (quantum-chemical ECD calculations) [31–33, 76–78, 83, 130, 155, 159, 160, 178, 180, 187, 256] methods used for reliable assignment of the constitutions, the relative and absolute configurations, and thus the full stereostructures of naphthylisoquinoline alkaloids.

The hyphenation of HPLC with tandem mass spectrometry (MS/MS), in combination with soft-ionization techniques such as electrospray ionization (ESI) in positive mode, usually permits a sensitive detection of naphthylisoquinoline alkaloids even in crude plant extracts [31, 42, 81, 82, 122–124, 170, 175–181, 257]. Low-energy collision-induced dissociation (CID) provides characteristic full product ion spectra by scanning a mass range from 20 to 850 amu, with mass ions resulting from typical fragmentation reactions (for a more detailed description, see Sect. 5.1.1). While monomeric naphthylisoquinolines exclusively yield monoprotonated molecules ($[M + H]^+$, *m*/*z*: 360–440), dimeric alkaloids mainly provide doubly protonated molecular ions ($[M + 2H]^{2+}$, *m*/*z*: 360–440), thus appearing in the range characteristic of monomeric compounds, besides the formation of a monoprotonated species ($[M + H]^+$, *m*/*z* >724) in the expected mass range of dimers [17, 31, 42, 81, 82, 122, 170, 175, 178, 180, 181, 257].

The typical isolation procedure for naphthylisoquinoline alkaloids starts with the exhaustive extraction (usually by repetitive maceration) of air-dried powdered plant material (root bark, twigs, stem bark, or leaves) with organic solvents (*n*-hexane, dichloromethane, methanol, either separately or as a mixture, neutral or slightly acidic), followed by consecutive fractionation of the alkaloid-containing extracts by liquid–liquid and/or solid–liquid separation. Among the common techniques are methods like high-speed countercurrent chromatography (HSCCC), fast centrifugal partition chromatography (FCPC), normal-phase silica gel column chromatography as well as ion exchange, gel permeation, and/or reversed-phase chromatography. Further resolution and purification by preparative HPLC (in most cases on C_{18} columns) finally affords pure material of the respective naphthylisoquinoline alkaloids for structural and stereochemical investigations and for biological testing.

The most important technique for the structural elucidation of natural products is NMR spectroscopy with all its manifold 1D and 2D methods. A large number of important structural features can already be derived from the one-dimensional ^{1}H NMR spectrum. Typical signals indicative of the presence of monomeric C,C-coupled naphthyltetrahydroisoquinoline alkaloids are as follows (Fig. 16):

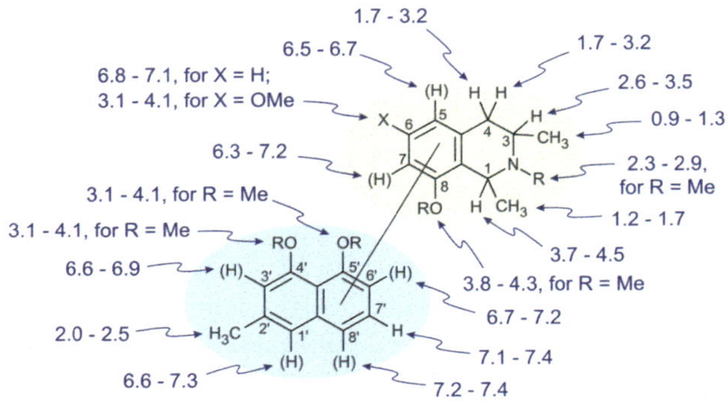

Fig. 16 Characteristic ^{1}H NMR data (δ/ppm) of monomeric C,C-coupled naphthyltetrahydroisoquinoline alkaloids

- two doublets around $\delta_H = 0.8–1.7$ ppm with a coupling constant of about 6 Hz, each corresponding to three protons (1-Me and 3-Me);
- a singlet at $\delta_H = 2.0–2.6$ ppm (2′-Me);
- signals of two aliphatic methines at $\delta_H = 3.7–4.5$ ppm (H-1) and $\delta_H = 2.6–3.5$ ppm (H-3);
- signals of two diastereotopic protons (ideally two doublets of doublets) at C-4 around $\delta_H = 1.7–3.2$ ppm;
- up to four three-proton singlets at $\delta_H = 3.1–4.1$ ppm, typical of methoxy substituents;
- a downfield-shifted three-proton singlet at $\delta_H = 2.3–2.9$ ppm, as characteristic of an *N*-methyl group;
- five (in the case of Ancistrocladaceae-type alkaloids) or six (for Dioncophyllaceae-type alkaloids) aromatic protons at $\delta_H = 6.3–7.5$ ppm: The spin patterns of the aromatic protons give a first hint at the coupling type of a respective naphthylisoquinoline. An unambiguous assignment of the coupling positions in the naphthalene and isoquinoline subunits, however, can only be achieved by specific Nuclear Overhauser Enhancement Spectroscopy (NOESY) and Heteronuclear Multiple Bond Correlation (HMBC) interactions.

Intense phytochemical studies on Asian Ancistrocladaceae lianas furnished a plethora of naphthyltetrahydroisoquinoline alkaloids like e.g., ancistectorine C_1 (**71a**) [61] showing a high structural diversity with respect to their different coupling sites and manifold variations in their O- and N-methylation patterns (see Sects. 4–8). In addition, compounds with further constitutional features have to be taken into account, as obvious from the chemical shifts in the ^{1}H and ^{13}C NMR spectra of the respective naphthylisoquinoline alkaloids (Fig. 17):

- The presence of a naphthyldihydroisoquinoline alkaloid such as ancistroheynine B (**72**) [41] is evidenced by the lack of an H-1 quartet (around $\delta_H = 4.0$ ppm in naphthyltetrahydroisoquinolines), the downfield shift of the protons of the

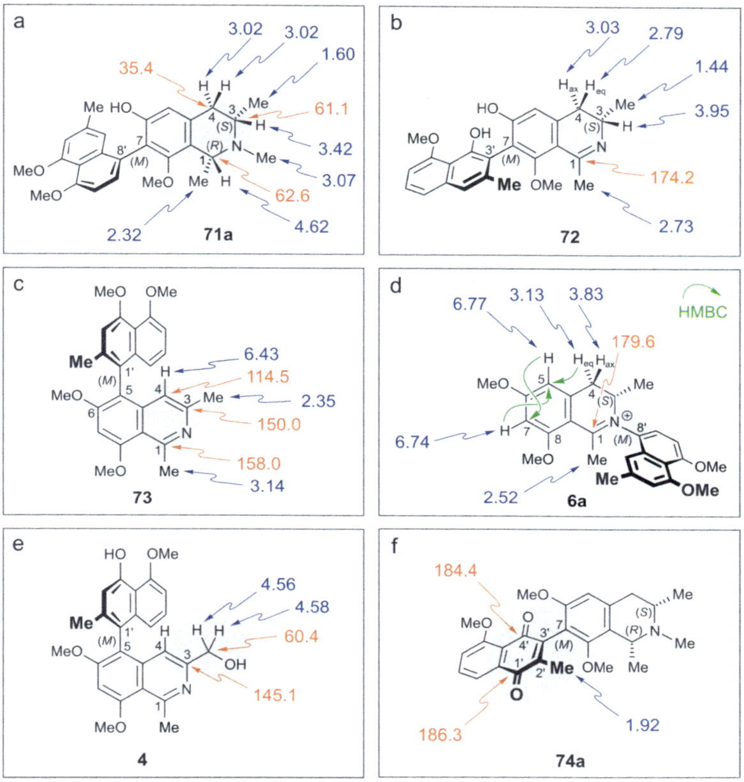

Fig. 17 ^{1}H and ^{13}C NMR chemical shifts (δ/ppm), typical of **a** naphthyltetrahydroisoquinoline alkaloids such as ancistectorine C_1 (**71a**), **b** naphthyldihydroisoquinolines like ancistroheynine B (**72**), **c** non-hydrogenated alkaloids as for 6-*O*-methylhamateine (**73**), and **d** N,C-coupled naphthylisoquinolines like ancistrocladinium A (**6a**), and alkaloids with **e** a hydroxymethylene function at C-3, as in ancistrobenomine A (**4**), or a naphthoquinone portion, as in ancistrotectoquinone A (**74a**). The latter occurs as a pair of configurationally semi-stable and thus slowly interconverting atropo-diastereomers, **74a** and **74b**. Arbitrarily, only the (*M*)-atropo-diastereomer **74a** is shown

methyl group at C-1 (δ_H = 2.2–2.7 ppm), along with the changed multiplicity (a singlet instead of a doublet around δ_H = 1.4 ppm as typical of a naphthyltetrahydroisoquinoline), and the downfield chemical shift of the ^{13}C NMR resonance of C-1, here appearing at ca. δ_C = 170–180 ppm;

- Indicative of naphthylisoquinoline alkaloids possessing a non-hydrogenated isoquinoline subunit like 6-*O*-methylhamateine (**73**) [45, 186] is an additional singlet (δ_H = 6.8–7.6 ppm), together with the lack of the diastereotopic geminal protons at C-4 and the absence of two doublets of the aliphatic methyl groups at C-1 and C-3 around δ_H = 1.1 and 1.9 ppm; further characteristic features are the missing quartet of H-1 around δ_H = 4.4 ppm, the lack of the multiplet of H-3 in the region between δ_H = 3.2 and 4.0 ppm, and the significant downfield shifts of the ^{13}C NMR resonances of C-1, C-3, and C-4;

- Some of the fully dehydrogenated naphthylisoquinoline alkaloids such as ancistrobenomine A (**4**) [30] possess a hydroxymethyl group at C-3, as obvious from the lack of proton resonances for the methyl group at C-3, replaced by the appearance of signals for the two diastereotopic protons of the CH_2OH group and the downfield-shifted ^{13}C NMR signal at δ_C = 60.4–61.4 ppm (see also Sect. 6.1);
- Alkaloids possessing a naphthoquinone entity (instead of a naphthol portion) such as ancistrotectoquinone A (**74a/74b**) [63] can readily be recognized by the appearance of two carbonyl signals at around δ_C = 180–186 ppm, the extremely upfield-shifted signal of Me-2′ resonating at δ_H = 1.7–1.9 ppm, and compared to other monomeric alkaloids, the lack of one aromatic proton (for a detailed report, see Sect. 8.5.6.2); moreover they show a typical UV spectrum with a maximum at 400 nm;
- The detection of one additional aromatic proton, as established by NOE correlations in the series of protons and substituents in the isoquinoline subunit and confirmed by specific HMBC interactions of H-5 to C-7, and by HMBC effects of both H_{eq}-4 and H-7 to C-5, supports the presence of an *N*-iminium-*C*-aryl axis in the respective naphthylisoquinoline alkaloid (see also Sects. 6.3 and 7.3), shown exemplarily for ancistrocladinium A (**6a**) in Fig. 17d; furthermore, in comparison to the UV spectra monitored for C,C-coupled naphthylisoquinolines (typically around 305 nm), the second maximum of N,C-coupled alkaloids is bathochromically shifted by 30–40 nm, to appear around 335 nm [18, 32, 47];
- Naphthylisoquinoline dimers can be detected by their (nearly) double masses in MS when using mild ionization techniques (mono-ionization), and by ^{1}H NMR, due to the lack of one aromatic proton (lost during the coupling step) [17, 31, 42, 82, 122–124, 170, 175, 182, 257].

For the elucidation of the absolute configuration of the stereogenic centers C-1 and C-3 of naphthylisoquinoline alkaloids, a smooth procedure based on the ruthenium-mediated oxidative degradation of the di- or tetrahydropyridine heterocycle in these alkaloids has been developed and miniaturized (down to ca. 0.1–0.5 mg of the analyte) [79, 80]. The method yields the simple and easy-to-analyze amino acids alanine ("Ala", formed from C-1) and 3-aminobutyric acid ("3-ABA", derived from C-3) or analogously, in the case of N-methylated naphthylisoquinolines, *N*-methylalanine ("*N*-Me-Ala") and *N*-methyl-3-aminobutyric acid ("*N*-Me-3-ABA"), usually together with their N-unsubstituted parent compounds alanine and 3-aminobutyric acid. The obtained degradation products carry the direct information about the absolute configuration at the two—now separately analyzable—stereocenters of the investigated alkaloids (Fig. 18). After their conversion with $MeOH/SOCl_2$ to give the respective methyl esters and Mosher-type derivatization with (*R*)-α-methoxy-α-trifluoromethylphenylacetyl chloride [(*R*)-MTPA-Cl, prepared from (*S*)-MTPA], the absolute configurations of the respective naphthylisoquinoline alkaloids can be assigned by gas chromatography on an achiral

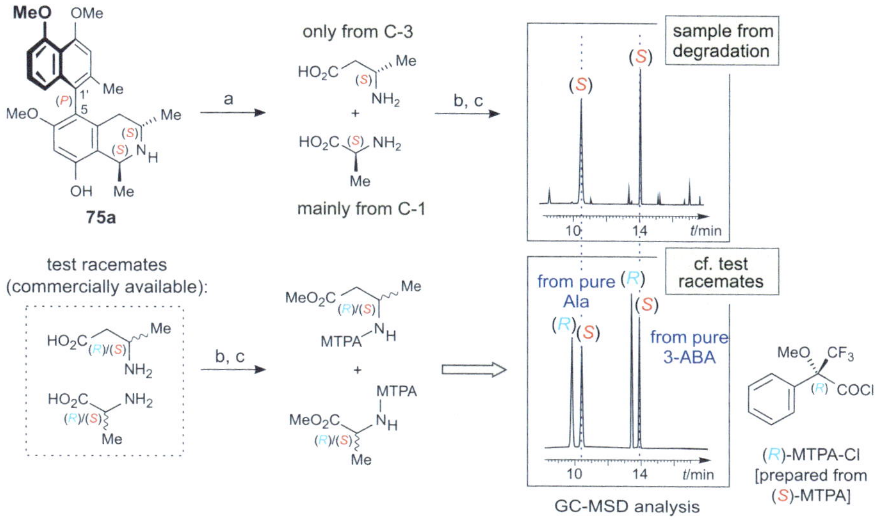

Fig. 18 Elucidation of the absolute configurations at the stereogenic centers C-1 and C-3 of naphthylisoquinoline alkaloids, as exemplified for ancistectorine A_2 (**75a**), by oxidative degradation and analysis of the resulting amino acids by esterification and gas chromatography of the Mosher derivatives on an achiral phase with mass-selective detection (GC-MSD). Reagents: (a) $RuCl_3$, $NaIO_4$; (b) $SOCl_2$, MeOH; (c) (*R*)-MTPA-Cl (= (*R*)-Mosher's chloride)

dimethylpolysiloxane-coated silica capillary column coupled to a mass-selective detector (GC-MSD). Comparison with the respective derivatives of the authentic amino acids of known configuration (from a test racemate) finally provides direct information about the absolute configuration at the chiral centers [17, 79, 80].

It must, however, be noted that stereochemical investigations at C-1 (i.e., in the case of tetrahydroisoquinolines) should always be accompanied by NOE-based attribution of the relative configuration at C-1 versus C-3 (see below). In contrast to the reliable stereochemical assignment at C-3, as established by the formation of 3-aminobutyric acid from C-3, the stereo-information about C-1 may be less unambiguous, because the 3-aminobutyric acid formed from C-3 may, to a small degree, be further degraded down to alanine, and thus the alanine peak alone does not provide reliable information about the configuration at C-1 [17, 79, 80]. Moreover, in particular *N*-unsubstituted *cis*-configured tetrahydroisoquinolines are easily oxidized at C-1 to give the respective 3,4-dihydroisoquinolines [13, 17, 242], thus losing the genuine stereo-information at C-1 totally, so that any alanine possibly formed will come from C-3 exclusively.

Within the repertoire of analytical tools required for a reliable stereochemical assignment of naphthylisoquinoline alkaloids, the oxidative degradation procedure [80] plays a key role. It is the only method that gives immediate access to the absolute configuration at the stereogenic center(s) and it can even be carried out at a sub-mg level. Otherwise, only the demanding total synthesis of the investigated molecules [84–122], or the only rarely possible X-ray structure analysis of a heavy-atom derivative [156, 258] will provide an unequivocal attribution of the stereogenic elements of the tetra- and dihydroisoquinoline portion of the alkaloids.

NOESY and ROESY spectroscopy have become reliable and widely used standard methods for the stereochemical attribution of the relative and, in combination with the oxidative degradation procedure, of the absolute configuration of naphthylisoquinoline alkaloids at centers and axes [15, 17, 18]. NMR analysis of the stereochemical relationships between the spin systems in the naphthalene subunit and protons with a defined stereo arrangement in the isoquinoline moiety (viz. single protons or whole methyl groups at C-1 and C-3, and/or C-4) provides valuable information about the relative configuration at the biaryl axis. As an example, the NOE cross-peaks of H-8′ with H-4_{eq} (which is above the isoquinoline 'plane') and of Me-2′ with H-4_{ax} (below the 'plane') in the 5,1′-coupled alkaloid ancistectorine A_2 (**75a**) (Fig. 19a), isolated from the Chinese liana *A. tectorius*, were complementary to those of its co-occurring atropo-diastereomer 5-*epi*-ancistectorine A_2 (**75b**), which displayed specific NOE interactions between H-8′ and H-4_{ax} and between Me-2′ and H-4_{eq} (Fig. 19b) [61]. The correlation between H-8′ and the *equatorial* proton at C-4 in ancistectorine A_2 (**75a**) established the two spin systems to be on the same side of the isoquinoline plane, which given the known (3*S*)-configuration (as assigned by oxidative degradation), evidenced H-4_{eq} to be on the upper side of the isoquinoline, and thus revealed the biaryl axis to be (*P*)-configured [61]. This stereochemical assignment of **75a** was further corroborated by the complementary NOE interaction between the *axial* proton at C-4 and Me-2′, both below the isoquinoline 'plane' (Fig. 19a). In the case of 5-*epi*-ancistectorine A_2 (**75b**), where the NOE correlations across the axis were opposite, an (*M*)-configuration was analogously deduced (Fig. 19b) [61].

In agreement with the thus established (1*S*,3*S*,*P*)-array for **75a** and the (1*S*,3*S*,*M*)-configuration for **75b**, the *equatorial* proton at C-4 was upfield-shifted for **75a** (δ_H = 2.26 ppm) as compared to the analogous signal in **75b** (δ_H = 2.37 ppm). Vice versa, the *axial* proton at C-4 was upfield-shifted for **75b** (δ_H = 2.06 ppm) as

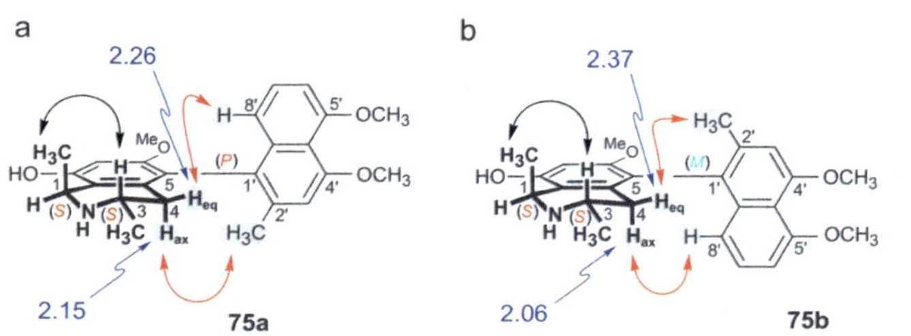

Fig. 19 Determination of the relative configuration of naphthylisoquinoline alkaloids at centers versus axes using NOESY experiments. Illustrated are specific NOESY interactions defining the relative configuration at C-3 versus C-1 within the tetrahydroisoquinoline portions (black double arrows) and the configuration at the axes (red double arrows) of the 5,1′-coupled alkaloids **a** ancistectorine A_2 (**75a**) and **b** its (*M*)-atropo-diastereomer, 5-*epi*-ancistectorine A_2 (**75b**)

compared to that of **75a** (δ_H = 2.15 ppm). In general, the protons that are closer to the larger naphthalene portion are expectedly more strongly influenced by its ring-current effect and thus show higher-shifted resonances (Fig. 19) [61].

A complementary approach for the assignment of the absolute axial configuration is based on the analysis of the chiroptical properties of naphthylisoquinoline alkaloids by electronic circular dichroism (ECD) investigations [13, 15, 17, 31, 32, 76–78, 83, 122–124, 130, 155, 159, 160, 175–182, 186, 187, 225]. The ECD behavior of naphthylisoquinolines is largely dominated by the strong contributions of their biaryl chromophore, and thus by their axial configuration, over those of the stereogenic centers in their tetrahydroisoquinoline portion [31, 122–124, 175–181]; for an exception, see jozimine A_4 (**50**), with its extended tetralone chromophore close to the stereogenic center (Fig. 10) [130]. For this reason, two atropo-diastereomeric naphthylisoquinoline alkaloids like ancistectorine A_2 (**75a**) and 5-*epi*-ancistectorine A_2 (**75b**) are usually characterized by their nearly fully opposite ECD spectra (Fig. 20) [61]. Thus, the axial configuration of a new naphthylisoquinoline may, in most cases, be easily established by comparison of its ECD spectrum with that of a structurally related known representative of the same coupling type, regardless of the configurations at the stereocenters [77, 78]. If novel-type structures are concerned for example as in the case of the already mentioned artificial dimer jozimine A_4 (**50**), with its unique naphthalene–tetralone linkage, as a consequence of its unprecedented 6′,8″-coupling type (see Fig. 10), the assignment of the absolute axial configuration may only be possible by the combination of experimental ECD spectroscopy with quantum-chemical ECD calculations [130]. The computational simulation of ECD spectra has become an efficient and reliable tool widely used for the determination of the absolute configuration of naphthylisoquinoline alkaloids, because such ECD predictions can be performed for virtually any novel alkaloid with axial and/or central chirality, even including additionally bridged naphthylisoquinoline dimers with their complex molecular architectures [31–33, 63, 76, 78, 83, 154–156, 159, 160, 165, 175–182, 216, 225].

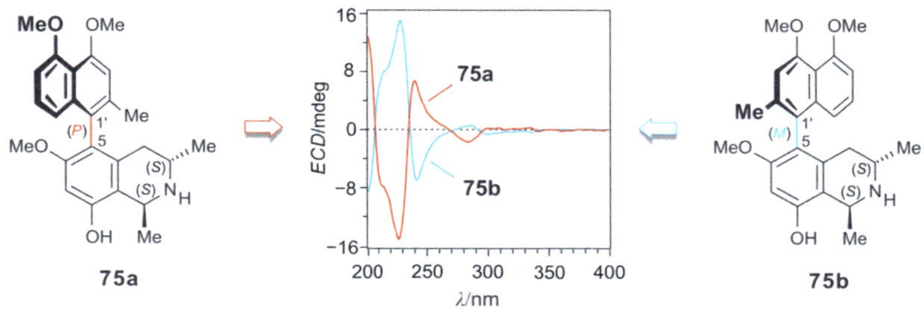

Fig. 20 Nearly opposite ECD spectra of ancistectorine A_2 (**75a**) and 5-*epi*-ancistectorine A_2 (**75b**), despite the fact that the two compounds are not enantiomeric, but diastereomeric to each other, having an opposite axial configuration, but the same configuration at the stereogenic centers, C-1 and C-3 in the tetrahydroisoquinoline subunit

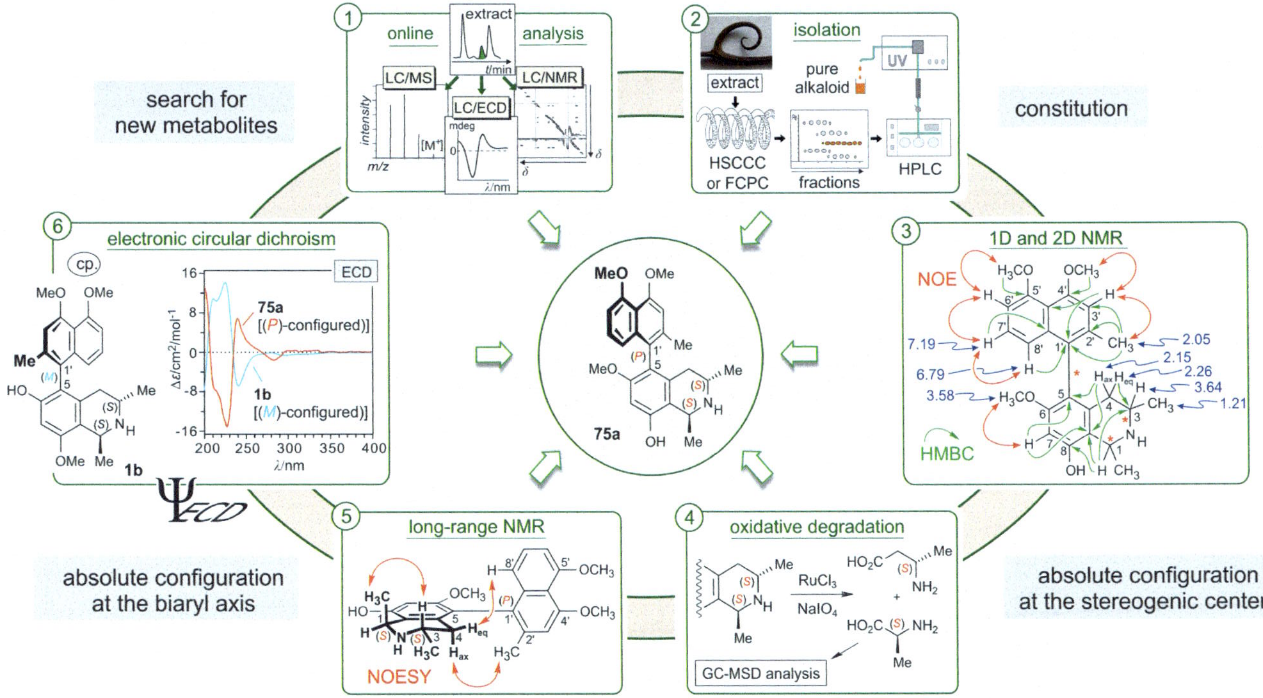

Fig. 21 As exemplified for ancistectorine A_2 (**75a**) from the Chinese liana *Ancistrocladus tectorius*, an overview on the repertoire of analytical methods used routinely for the isolation and structural elucidation of naphthylisoquinoline alkaloids. The closely related and likewise 5,1′-coupled, but (*M*)-configured alkaloid hamatine (**1b**) was used as a reference compound for the ECD investigations

As exemplified for ancistectorine A_2 (**75a**), which was isolated from the twigs of the Chinese liana *A. tectorius* [61], the most important methods routinely applied to the isolation and structural elucidation of naphthylisoquinoline alkaloids are addressed in Fig. 21:

(1) **Hyphenation of HPLC with MS, NMR, and ECD**—for a rapid screening and identification of new secondary metabolites from plant extracts that could be worth being isolated at a preparative scale (including a reliable dereplication of known structures);
(2) **Isolation**—besides conventional procedures such as preparative column chromatography (CC) on normal-phase silica gel, predominantly by the use of efficient liquid–liquid chromatographic separation techniques like FCPC or HSCCC in combination with preparative HPLC on normal- or reversed-phase silica gels;
(3) **Constitution**—by 1D NMR (^{1}H, ^{13}C) and by a series of homo- and heteronuclear 2D NMR experiments, in particular by selective HMBC and NOESY interactions;
(4) **Absolute Configuration at the Stereocenters C-1 and C-3**—by ruthenium-mediated oxidative degradation to give easy-to-analyze amino acids;
(5) **Relative Axial Versus Central Configuration**—relative configuration at the stereogenic centers, C-1 versus C-3, by specific NOE interactions, and the relative stereoarray—axis versus center—by long-range NOESY correlations;
(6) **Absolute Axial Configuration**—by experimental electronic circular dichroism (ECD) investigations, in the case of a novel-type structure (which is here not required), in combination with quantum-chemical ECD calculations.

3 *Ancistrocladus*, a Genus of Woody Lianas of the Monotypic Plant Family Ancistrocladaceae Widely Occurring in India, Sri Lanka, and Southeast Asia

The small paleotropical plant family Ancistrocladaceae comprises only one genus, *Ancistrocladus*, with presently about 20 accepted species [20, 21, 34, 162, 259] out of ca. 30 species listed in earlier plant surveys from Asia and Africa [19–21, 34, 35, 260–262]. The family is currently phylogenetically placed within the Caryophyllales, along with the closely related Dioncophyllaceae [263–267]. The presence of naphthylisoquinoline alkaloids found exclusively in these two families strongly supports this botanical classification [13–18].

The name *Ancistrocladus* is derived from the Greek words ανκιστρο, *ankistro* (= hook) and κλαδος, *klados* (= shoot) [19–21], reflecting a botanical peculiarity of the plants. All species are woody lianas or scandent shrubs equipped with a series of recurved hooked twigs, which serve as climbing devices and enable the plants to grow, through the surrounding vegetation, to reach the level of the canopy of the rainforests, attaining heights of more than 20 m [19–21].

The main centers of distribution of the Ancistrocladaceae plant family in Asia are the lowland and submontane evergreen rain forests and swamp regions in Southeast Asia, with two outliers in India and Sri Lanka. Comprehensive data about the natural diversity of the *Ancistrocladus* genus in Asia, however, are still lacking due to incomplete herbarium collections, complicated by the fact that nearly all of the type specimens of the Asian species have been collected and described before 1925, often in the vegetative state only [19, 21, 34, 35].

Based on surveys of local floras, Gereau [19] presented in 1997 an annotated checklist of *Ancistrocladus* species from Africa and Asia, with the typification of 17 Asian species names. However, in the latest comprehensive synoptic revision covering the entire Ancistrocladaceae family, only five were recognized as valid taxa by Taylor et al. [21]. These include *A. heyneanus* (with a large distribution area in the Western Ghats of India), *A. hamatus* (indigenous to Sri Lanka), *A. griffithii* (distributed from India through Myanmar, Thailand, and Cambodia to Vietnam), *A. attenuatus* (described for India, Myanmar, and the Andaman Islands), and *A. tectorius* (widely occurring all over Southeast Asia). With the botanical description of *A. benomensis*, discovered on an isolated mountain of Peninsular Malaysia, the number of taxa accepted for Asia has more recently increased to six species. This new taxon was botanically described for the first time in 2005 [34].

Ancistrocladus lianas in Asia have been known scientifically for more than 200 years. The first botanical report on an Asian *Ancistrocladus* plant was published by de Loureiro [268] in 1790 under the name of *Bembix tectoria*. This liana, discovered in a region that now belongs to Vietnam, is nowadays known as *A. tectorius*. In 1829, Wallich [269] was the first to introduce the name *Ancistrocladus* into the botanical literature, describing four species (*A. heyneanus*, *Ancistrocladus vahlii*, *Ancistrocladus extensus*, and *Ancistrocladus wallichii*) from India, Sri Lanka, Burma (today Myanmar), Malaysia, and the Andaman Islands [21]. Several years later, Planchon [270] recognized the Ancistrocladaceae as a separate family consisting of one genus with seven species from all over Southeast Asia and India (namely *A. vahlii*, *A. extensus*, *Ancistrocladus pinangianus*, *A. griffithii*, *Ancistrocladus sagittatus*, *A. wallichii*, and *A. heyneanus*).

During the following decades, most taxonomists did not accept this classification of the Ancistrocladaceae as a separate plant family, but considered the genus *Ancistrocladus* as belonging to the Dipterocarpaceae, because of the similar fruits [21]. At the turn of the nineteenth to the twentieth century, the family Ancistrocladaceae was generally acknowledged according to the reviews on the genus *Ancistrocladus* by Gilg [260, 261] and by van Tieghem [271].

A reliable typification and compilation of species names in Southeast Asia still remains a problematic issue to this day, in particular because most are only known from relatively few collections. Leaf morphology can be highly variable even within a single individual, and some of the described taxa are fairly rare, and do not flower abundantly [21]. In view of these difficulties, van Steenis [35] summarized all collections of *Ancistrocladus* known from Southeast Asia under *A. tectorius*. As

Fig. 22 **a** A young self-supporting sapling of the Vietnamese species *Ancistrocladus cochinchinensis* in its natural habitat; **b** a juvenile plant of *Ancistrocladus griffithii* from Thailand cultivated in a greenhouse of the Botanical Garden of the University of Würzburg; **c–e** *Ancistrocladus* aff. *tectorius*, showing an individual plant from Laos flowering in a greenhouse in the Botanical Garden in Würzburg; **c** lateral branch with a rosette of leaves bearing inflorescences; **d** flowers in buds at the beginning of anthesis; **e** an open flower, fully developed with five green petals and five pink-colored sepals with dark spots on their inner surface, and with ten white-yellow stamens in two alternately unequal whorls; **f** *Ancistrocladus* aff. *tectorius*, illustrating an adult woody liana in its natural habitat in Vietnam (Photographs: **a**, **b**, **f**—M. Dreyer; **c**, **d**, **e**—I. Kajahn)

Fig. 23 Flowering and fruiting specimens of *Ancistrocladus* aff. *tectorius*. Photographs **b**, **c**, and **e** taken of *Ancistrocladus* aff. *tectorius* plants cultivated in a greenhouse in the Botanical Garden of the University of Würzburg: Photographs **a** and **d** show plants in their natural habitats **a** in South-Central Vietnam (area Cao Som of the Hoa Binh Province) and **b** in South-Central Thailand (Khao Yai National Park). **a** Inflorescence emerging from a leafy hooked twig; **b** cyme branched dichotomously bearing 30–40 buds; **c** cyme with buds and young flowers with contorted petals at the beginning of anthesis; **d** lateral branch bearing immature fruits with five red-colored wings derived from enlarged sepals; **e** ripe fruit representing a mono-seeded nut with five irregularly formed wings (Photographs: **a**—M. Dreyer; **b**, **c**, **e**—A. Irmer; **d**—H. Rischer)

Fig. 24 Indian liana *Ancistrocladus heyneanus* (**a–e**, **g**), cultivated in the greenhouse of the Botanical Garden of the University of Würzburg, and *Ancistrocladus hamatus* (**f**), in its natural habitat in Sri Lanka. **a** A three-year old erect, self-supporting sapling of *A. heyneanus*; **b** inflorescence bearing flowers in buds; **c** cyme with a young flower with contorted petals at the beginning of anthesis; **d** flower at anthesis, fully developed, with a view of five pink petals; ovary fully inferior with three white styles; **e** inflorescence bearing an immature fruit with five irregularly formed 'wings' (enlarged sepals); **f** *A. hamatus* growing near Adam's Peak (the sacred mountain Śrī Pada, located in the district Ratnapura, Sri Lanka), grown to large shrubs with typical hooked lateral branchlets and rosettes of leaves; **g** seedling of *A. heyneanus* after germination from the mono-seeded nut (partly deprived of its sepals) (Photographs: **a**, **g**—K. Wolf; **b–d**—A. Irmer; **e**—M. Dreyer; **f**—H. Bringmann)

Fig. 25 *Ancistrocladus heyneanus*, endemic to the Western Ghats of India, cultivated from fruits in the greenhouse of the Botanical Garden of the University of Würzburg. **a** Mono-seeded nuts sown on soil mixtures (earth/sand/peat), keeping them constantly moist (temperature 30–33°C) for germination; **b** seedlings just after germination grown on pure river sand; **c**, **d** adaption of single plants with well-developed roots to a hydroculture substrate; **e** young prospering plants at an age of about 18 months; **f** adult woody liana, having hooked twigs and leaves arranged in terminal rosettes (Photographs: **a**, **b**—B. Wiesen; **c**–**e**—K. Wolf; **f**—A. Irmer)

Fig. 26 *Ancistrocladus benomensis* Rischer & G. Bringmann, a new plant species from Southeast Asia, discovered in 1999 on an isolated mountain in Peninsular Malaysia and cultivated in the greenhouse of the Botanical Garden of the University of Würzburg, then flowering in March 2004. **a** Young plant growing as a self-supporting sapling with widely spaced leaves; **b**, **c** lateral branch with rosette leaves and inflorescences bearing flowers and buds; **d** inflorescence emerging from a leafy hooked twig; **e**, **f** young flowers showing its reflexed sepals and petals; **g**, **h** pink-colored flower with five spots on the inner surface of the petals and ten pale-yellow stamens arranged in two whorls (Photographs: **a–h**—M. Dreyer)

a consequence, many botanists have, since then, classified most of the Southeast Asian *Ancistrocladus* collections uniformly as *A. tectorius*. Doubts about this 'one-species concept' arose from phytochemical studies on plant material of *A. tectorius* collected from geographically disjunct areas, e.g., in Thailand [51, 52, 55], Laos [49], Vietnam [32, 272], southern Peninsular Malaysia [54], and in China on the Hainan Island [31, 33, 50, 53, 56–64]. The discovery of structurally most different naphthylisoquinoline alkaloids exclusively in plant samples from distinct sites strongly indicated a species differentiation within the '*A. tectorius* complex'. Comparative phylogenetic studies [36] on *Ancistrocladus* lianas from all over Asia, based on the internal transcribed spacers (ITS) of nuclear rDNA and on the *trn*K intron of cpDNA, and on intersimple sequence repeat (ISSR) fingerprinting analyses clearly support the existence of further valid species not yet recognized.

Typical of the genus *Ancistrocladus* are its unusual dimorphic juvenile and adult habits [20, 21, 260–263, 271, 273, 274]: Young plants (Fig. 22a, b) start growing solitarily in open places as erect saplings; then they become sympodial lianas, with concomitant production of modified tendril-like shoots with a curved hook at the tip. Adult climbing plants (Fig. 22c, f) finally form stems of about 1–5 cm diameter with a brown smooth or roughened bark, bearing scattered leaves and lateral branchlets, each with one to several hooks (Fig. 22f) [20, 21]. Inflorescences usually emerge from leafy hooked twigs (Fig. 23a) and are composed of dichotomously branched cymes (Figs. 22c and 23b), which produce about 20–60 flowers (Fig. 23c). The latter are characterized by five petals (equal in size and shape), varying from pale green, white, yellow, or pink to dark red (Figs. 22d, e and 23c). They possess five oblong to obovate sepals, in most cases nearly equal in size, frequently bearing small pits and glands. The plants are equipped with ten stamens (Fig. 22e), which are arranged in one or two whorls. The ovary is partially or fully inferior, crowned by a disk in the flower. The persistent basal stylar column shows three styles [21]. The five sepals enlarge markedly during fruit development to form wings (Fig. 23d, e), which are several times larger than the body of the nut-like, mono-seeded fruit (Fig. 23e). With the five irregularly formed 'wings', the fruit is well-suited for wind dispersal. The seeds of *Ancistrocladus* species are solitary and relatively large (5–8 mm of diameter) with a starchy endosperm covered with a reddish-brown testa [20, 21]. Botanical features that characterize plant development, flowering, and fruiting of the genus *Ancistrocladus* are presented in Figs. 22, 23, 24, 25 and 26. The photos were taken from individual *Ancistrocladus* plants cultivated in the greenhouse of the Botanical Garden of the University of Würzburg or they show the lianas in their natural habitats in Southeast Asia.

Sections 3.1–3.6 describe the most interesting botanical peculiarities and refer to further important facets including phytochemical and genetic aspects of the six *Ancistrocladus* species from Asia accepted so far as valid taxa.

3.1 *Ancistrocladus heyneanus, a Large Climbing Shrub from the Indian Subcontinent*

The Indian liana *A. heyneanus* Wall. ex J. Graham (Figs. 24 and 25) has been known for a long time. In 1839, Graham [275] listed this species in his "Catalogue of the Plants Growing in Bombay and its Vicinity" and described it as "a very pretty shrub, but hardly known yet to European Botanists". The plant has been listed repeatedly in numerous Indian local plant surveys [260, 261, 269, 270, 275–288] till this day, often together with impressive illustrations of its most peculiar botanical features such as the hooked branches, inflorescences bearing buds and flowers, and the nut-like fruits [260, 261, 276, 277, 281, 283, 285] with their unusual five 'wings', derived from enlarged sepals. Interestingly, the presumably oldest illustration of *A. heyneanus* is a drawing published in 1686 in the catalogue "*Hortus Indicus Malabaricus*" [276] presenting a plant named *Modira valli* with a lateral hooked branch and a cluster of leaves. Graham [275] stated that the plant described and illustrated in this book has a great resemblance to *A. heyneanus*. Another old—and very nice—illustration of this Indian plant is presented in the handbook "Icones Plantarum Indiae Orientalis" (in English: "Figures of East Indian Plants") from 1853 [277]. Slightly modified, this illustration can also be found in subsequently published plant catalogues [260, 261, 281].

A. heyneanus is a large climbing shrub with hooked lateral branchlets widely occurring in seasonal and evergreen forests in low elevations in the Western Ghats of the Indian subcontinent. Its vast distribution area comprises the Indian states Maharashtra, Goa, Karnataka, Tamil Nadu, and Kerala [19, 21, 277, 280, 282–288]. In collaboration with the Botanical Gardens of the Universities of Münster and Würzburg, the authors' group successfully cultivated *A. heyneanus* from fruits [13, 289, 290] (Fig. 25), thus allowing to study its complete life cycle from the germination of fruits (Figs. 24g and 25a, b), giving rise to young erect saplings (Figs. 24a and 25e), and finally to adult climbing lianas (Fig. 25f). Some of them grew to a height of more than 4 m and developed the characteristic hooked branches and, at the age of three years, they even started flowering and fruiting (Fig. 24b–e).

Phytochemical investigations on aerial parts and root extracts of *A. heyneanus* revealed this Indian liana to produce a characteristic pattern of structurally unique alkaloids that is significantly different from the metabolite profiles of other *Ancistrocladus* species from Asia and Africa. A plethora of 16 naphthylisoquinoline alkaloids have so far been identified in *A. heyneanus*. Most were C,C-coupled monomeric compounds (ten representatives) exhibiting a 5,1′-, 7,1′-, 7,3′-, or 7,8′-linkage [1–5, 13, 26–29, 37–41, 70], isolated along with a series of N,6′- and N,8′-coupled alkaloids [18, 40, 70]. A comprehensive report on the secondary metabolites of *A. heyneanus* is given in Sect. 4, with a special emphasis on the structural elucidation and total synthesis of the chemical constituents. Likewise presented is a more in-depth discussion regarding the bioactivities of structurally simplified N,C-linked compounds, in particular focusing on their inhibitory effects against the pathogens causing severe tropical diseases like malaria, leishmaniasis, or trypanosomiasis.

3.2 *Ancistrocladus hamatus*, Endemic to the Lowland Evergreen Forests of Sri Lanka

A. hamatus (Vahl) Gilg. (Fig. 24f) is the only species of the genus *Ancistrocladus* that grows in Sri Lanka (formerly known as Ceylon) [19, 21, 260, 261, 263, 270, 271, 278, 279, 284, 286, 291–297]. It is a shrub indigenous to this island in the Indian Ocean, occurring mainly in lowland moist evergreen forests (100–150 m alt.), easily discernable due to its typical hooked branches and rosettes of leaves appearing on adult climbing stems. Some of the authors of handbooks of local floras from India and Sri Lanka [277, 284, 297] assumed that *A. hamatus* may be synonymous with *A. heyneanus* from the Indian subcontinent. In the comprehensive revision of the entire *Ancistrocladus* genus presented by Taylor et al. [21] in 2005 however, *A. hamatus* was acknowledged as a valid *Ancistrocladus* taxon, since it can clearly be distinguished from *A. heyneanus* by the number of the stamens (five in *A. hamatus* versus ten in *A. heyneanus*) and by the size of the fruits (with the nut portion about four times larger in *A. hamatus* than in *A. heyneanus*).

A. hamatus has been known for a long time. It is listed under various botanical names in a whole series of surveys of local floras [260, 261, 271, 278, 279, 284, 291–297]. Thwaites [293] described in detail this species (here named as *Ancistrocladus vahlii* Arn.), referring to the type collection of *Wormia hamata* Vahl [291], mainly focusing on its floral morphology. Of particular interest for present-day readers, he reported the plant to be "very abundant in some of the warmer districts of Ceylon". He further gave a commentary on this shrub regarding it as "a very troublesome weed to the cultivator", obviously due to the fact that it is "spreading so much by its roots". In 1893, Trimen [294] published a handbook on the flora of Ceylon with a really magnificent atlas of colored plates "illustrating some of the more interesting species", among them also *A. hamatus* (under the name of *A. vahlii* Arn.). Trimen [294] described the Ancistrocladaceae as an "anomalous genus as forming a separate Natural Order". He further noted that "its affinities are very doubtful, but it cannot be placed in Dipterocarpaceae" as it had been done by most authors in the nineteenth century [21, 294].

While Thwaites [293] treated the *Ancistrocladus* plants of Sri Lanka as an individual species, van Tieghem [271] even treated them as a distinct endemic genus, named *Bigamea*. He considered *Bigamea* to comprise two species, the previously described *Wormia hamata* (named *Bigamea hamata*) [279, 291, 292, 294] and his new species, named *Bigamea thwaitesii* [271]. Gilg [261] finally classified the Sri Lankan *Ancistrocladus* plants as one taxon, *A. hamatus* (Vahl) Gilg., comprising only one lectotype describing this species. For this reason, recent plant checklists [19, 21, 286, 295, 296] now consider both of the names, *A. vahlii* Arn. [270, 278, 279, 292–295] and *A. thwaitesii* Tiegh. [271] as synonyms of *A. hamatus* (Vahl) Gilg. [19, 21].

So far, *A. hamatus* is phytochemically still poorly investigated and only five naphthylisoquinoline alkaloids have as yet been identified [5, 13, 23, 24] (see Sect. 4).

3.3 Ancistrocladus griffithii, from Submontane Forests in the West and South of Southeast Asia

A. griffithii Planch. (Fig. 22b) was described for the first time in 1849 by Planchon [270], based on a type specimen collected from Burma (today Myanmar). It is a woody liana, climbing up to 4 m high, bearing some scattered leaves and hooked lateral branchlets, which show also groups of leaves [21, 270, 278, 279]. The plant is widely occurring on riversides and in submontane evergreen forests (up to 350 m above sea level) mainly all along the Tenasserim Hills and Phuket Mountain range, from Myanmar to Thailand, further extending to Cambodia and Vietnam [19, 21, 261, 278–280, 286, 298]. *A. griffithii* can easily be identified due to its very slim and elongated leaves and characteristic secondary veins on both surfaces. The identity as a separate species can reliably be determined by reproductive characters such as the sepals, which lack large abaxial glands, and by the relatively small inflorescence bracts mainly clustered at the apices of the axes near the flower buds [21]. Moreover, genetic analysis strongly supported the species status of *A. griffithii*, by the large distance of ISSR fingerprints and ITS sequences compared to the rest of the Southeast Asian *Ancistrocladus* plants [36].

Isolation work on *A. griffithii* [42, 43, 82] furnished six naphthylisoquinoline alkaloids. Four were new natural products, which have so far been identified in this Southeast Asian species exclusively, thus clearly evidencing a species-specific production of naphthylisoquinoline alkaloids and unambiguously supporting the treatment of *A. griffithii* as a valid *Ancistrocladus* taxon. A detailed report on the metabolite profile and on the identification and structure elucidation of mono- and dimeric naphthylisoquinoline alkaloids of *A. griffithii* online, directly from crude extracts, by hyphenation of HPLC with MS, NMR, and ECD, is given in Sect. 5.

3.4 Ancistrocladus attenuatus, from Lowland Evergreen Forests of Myanmar and the Indian Andaman Islands

A. attenuatus Dyer has been recognized and shortly described for the first time in 1874 in the plant catalogue 'Flora of British India' [279]. Since then this liana has been listed in numerous local plant surveys from India and Southeast Asia [19, 260, 261, 280, 285, 286, 299–302]. In the comprehensive taxonomic revision by Taylor et al. [21], covering the entire Ancistrocladaceae family, the plant was accepted as a valid species from Southeast Asia.

A. attenuatus is a woody climber with dark-colored stems bearing strong woody hooks, scattered leaves, and lateral branchlets with terminal rosettes of leaves [21, 285, 299]. The plant was found to occur in evergreen shrub and bamboo forests at 50–150 m above sea level, with an area of distribution mainly located in the Tenasserim Hills of Southeastern Myanmar, extending in the East to all groups of the Indian Andaman Islands [19, 21, 260, 261, 279, 280, 285, 286, 300–302]. In several accounts on the forest flora of the Andaman Islands, indigenous people are reported to use the very hard woody parts of *A. attenuatus* to produce arrows [286, 300, 301].

Owing to its elongated, slim leaves, the phenology of this liana is quite similar to that of *A. griffithii*, but the species is clearly distinguished by some reproductive features such as the enlarged size of both the nut portion of its seeds and the persistent sepal limbs [21, 299]. Most of the authors of recent plant checklists have regarded this taxon as quite rare [285, 300]. The conservation status of *A. attenuatus* has been discussed controversially, but no direct survival threats have been reported [285, 300, 301]. Unfortunately, no drawings of typical botanical features or photos of the plant in its natural habitat (or from plants cultivated in a greenhouse) that might permit a more concrete and reliable impression on *A. attenuatus* are available from the literature. Phytochemically, *A. attenuatus* is as yet totally unexplored.

3.5 *Ancistrocladus benomensis*, a New Submontane Liana from Peninsular Malaysia

On Gunung Benom, Pahang, an isolated mountain in Peninsular Malaysia, located in the Kerau Game Reserve in the State Pahang, the authors' group discovered a new taxon, *A. benomensis* Rischer & G. Bringmann (Fig. 26), in 1999 [34]. This submontane liana, which was found to occur from 500 to 1180 m above sea level on the slopes of Gunung Benom ('Benom' is an old Mon-Khmer word for 'mountain'), was botanically described for the first time in 2005 [34]. At the type location Gunung Benom, a large number of fallen brown nut-like fruits with five papery-coriaceous sepals forming wings as typical of the *Ancistrocladus* genus were observed as well as abundant seedlings near the adult plants [34].

The authors' group managed to cultivate some individuals of this new Malaysian liana from young seedlings to adult plants including flower development in the greenhouse of the Botanical Garden of the University of Würzburg (Fig. 26). Typical of *Ancistrocladus* plants, the saplings of *A. benomensis* were erect, unbranched with leaves arranged in terminal rosettes, while adult plants became woody lianas climbing in trees and forming sympodial branches with circinate unilateral hooks. Morphologically, *A. benomensis* clearly differs from *Ancistrocladus* aff. *tectorius*, which is the only other taxon reported to occur in Malaysia: The *Ancistrocladus* plants endemic to the rainforests of Gunung Benom are characterized by alternate, sessile, and ovate-lanceolate leaves. Inflorescences were observed to be solitary and cymes emerging from a leaf rosette, up to three- or four-times dichotomously branched, produced 50–70 buds and flowers (Fig. 26b–e). The flowers are characterized by five pale-red-colored green sepals (slightly unequal, reflexed at anthesis), and by five free bright-red petals (weakly contorted, longer than the sepals, at the tip finely toothed, reflexed) with a contrasting black spot on the inner surface [34]. Based on the short description by de Loureiro [268] and with respect to the notes on de Loureiro's specimen by Moore and Tandy [303], *A.* aff. *tectorius*, on the contrary, has opposite cuneiform leaves with a short, broad petiole, pale flowers, and a weakly branched inflorescence with only few flowers.

The spectrum of naphthylisoquinoline alkaloids produced by *A. benomensis* is substantially different from those of all other known *Ancistrocladus* species. Isolation work on the twigs and leaves led to the discovery of ten alkaloids [30, 32, 44], among them a high number of non-hydrogenated naphthylisoquinolines (seven representatives), some unexpectedly lacking an oxygen function at C-6 [44]. These compounds are therefore so far the only known naphthylisoquinoline alkaloids isolated from an Asian *Ancistrocladus* species possessing these structural features. A detailed report on the isolation and categorization of the alkaloids of *A. benomensis*, including genetic aspects regarding the taxonomic treatment of this Malaysian liana, is given in Sect. 6.

3.6 *Ancistrocladus tectorius*, Widely Distributed in the Rain Forests All Over Southeast Asia

A. tectorius (Lour.) Merr. [19, 21, 35, 36, 268, 285, 303–306] (Figs. 22c–f and 23) is a woody liana climbing up to 20 m high into the canopy of evergreen forests, with a huge distribution area, ranging from lowland to premontane regions (ca. 0–1500 m above sea level) extending from Myanmar, Thailand, Cambodia, Laos, and Vietnam to the Chinese island of Hainan. In the southern part of Southeast Asia, this liana is known to occur from Sumatra and Borneo (Indonesia) through Singapore to Malaysia including their adjacent small islands, and the Nicobar and Andaman Islands of India [19, 21, 35, 36, 268, 283, 285, 300, 303–310]. Detailed descriptions of the vegetative and reproductive morphology of *A. tectorius* including the anatomy of buds, flowers, and fruits of plants collected on Sumatra (at Bogor Botanic Garden) and in southern Malaysia (State of Johor) are presented in two publications by Keng [304, 305].

There is a notable variation of the morphological features of *A. tectorius*. In principle, plants recognized as *A.* aff. *tectorius* can be assigned to two forms: One form, mainly found in China and Vietnam, but also in Malaysia, Singapore, and Sumatra, is similar to *A. heyneanus* (India) and resembles the type specimens of *A. tectorius* (*Bembix tectoria* Lour., from Vietnam, 1790 [268]), *A. cochinchinensis* (from Vietnam), and *A. harmandii* (Laos), with three well-developed, marginate glands on the outer three sepals [21]. The other form, predominantly occurring in Malaysia, Sumatra, Singapore, and the Andaman Islands, is similar to the type specimen of *A. pinangianus* (from Malaysia), and possesses sepals without abaxial glands. These two forms are linked by a number of intermediate specimens, thus giving rise to a high number of morphological variations observed for *A. tectorius* depending on the local sites from where the plants have been collected [21].

This finding is in accordance with a high genetic variability within the *A. tectorius* complex, as evidenced by a comparative study on molecular markers such as the ITS region of nuclear rDNA and the *trn*K intron of cpDNA as well as ISSR fingerprints [36]. In addition, the occurrence of different naphthylisoquinoline alkaloids exclusively in plant samples from distinct sites in Laos [49], Vietnam

[31, 272], Thailand [51, 52, 55], Peninsular Malaysia [54], and China (on Hainan island) [31, 33, 50, 53, 56–64] gave rise to doubts regarding this one-species concept introduced by van Steenis [35].

Comprehensive isolation work on *A.* aff. *tectorius* growing on Hainan furnished by far the highest number of naphthylisoquinoline alkaloids (nearly 70 compounds) that has ever been identified in an Asian *Ancistrocladus* liana [31, 33, 50, 53, 56–64]. Among these natural products are a whole series of new and structurally unique compounds known to occur solely in this Chinese plant, for example, dimeric alkaloids [31] and new examples of monomeric naphthylisoquinolines displaying coupling types that have never been detected before (7,6′-coupling) [59, 60] or have only rarely been found (5,3′-coupling) [60] in other *Ancistrocladus* species from Asia and Africa.

Section 8 provides an overview on the alkaloids and related compounds discovered to date in samples of *A.* aff. *tectorius* collected all over Southeast Asia, along with a more detailed report on the results of a genetic study [36] on the species differentiation within the *A. tectorius* complex in Southeast Asia.

4 The Indian Liana *Ancistrocladus heyneanus* and *Ancistrocladus hamatus* from Sri Lanka: Early Studies and More Recent Discoveries

The short summary on naphthylisoquinoline alkaloids [5] published by the pioneer in this field, Govindachari, in 1977 reviewed the first six representatives of this unusual class of biaryl compounds (Fig. 27). All had been isolated from the Indian liana *A. heyneanus* and from the Sri Lankan species *A. hamatus*. Among the described naphthylisoquinoline alkaloids, elucidated with brilliance and ingenuity, involving the most modern and efficient methods of that time, were ancistrocladine (**1a**) [1–5], its 6-*O*-methyl derivative **76a** [3, 5], its dihydro analogue ancistrocladinine (**77a**) [37], and the atropo-diastereomer of **1a**, hamatine (**1b**) [23, 24]. Besides these four 5,1′-coupled naphthylisoquinoline alkaloids, two further compounds were identified, viz.

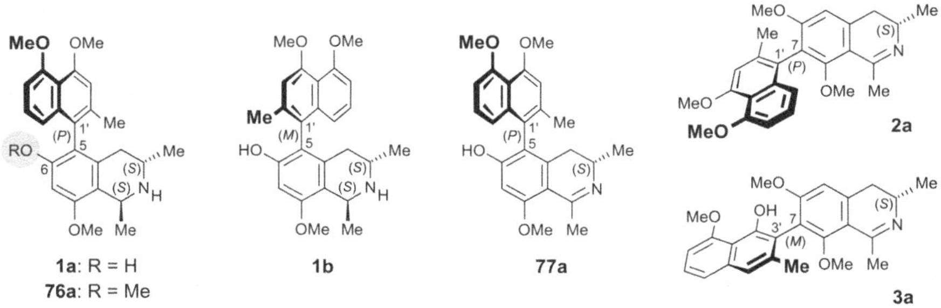

Fig. 27 Secondary metabolites of *A. heyneanus* (India) and *A. hamatus* (Sri Lanka), representing the six initially discovered naphthylisoquinoline alkaloids

ancistrocladisine (**2a**) [26–28], for which naphthalene and isoquinoline portions were linked via C-7 and C-1′, and ancistrocladidine (**3a**) [27–29], which had a 7,3′-linked biaryl axis. For some of the alkaloids, viz., **1a**, **2a**, and **3a** the constitutions and relative configurations were later confirmed by X-ray crystallography [4, 28]. The absolute configuration at the stereogenic centers in the tetrahydroisoquinoline subunit was established by exhaustive ozonolysis of the pure alkaloids (in the case of **1a**, the reaction was even performed on a 10 g preparative scale [4, 5]), to yield a degradation product of known absolute configuration, L-(+)-β-amino-*n*-butyric acid, thus proving the (*S*)-configuration at C-3. From this, in view of the relative *trans*-configuration between the methyl groups in the isoquinoline portion as assigned by NMR investigations on **1a** and related derivatives, C-1 also had to be (*S*)-configured [4, 5]. The absolute axial configurations were assigned by ECD spectroscopy, by analyzing the chiroptical behavior of the dehydrogenated product of the alkaloids, interpreted by applying the Exciton Chirality method [311–314]. Finally, the full absolute stereostructures of all of these biaryls were confirmed, in all details by total synthesis, mostly by the authors' group [84, 85, 90–94, 99, 102–104, 115, 116].

In 2003 and 2006, *A. heyneanus* again attracted attention due to the discovery of the first representatives of a structurally most peculiar new subclass of naphthylisoquinoline alkaloids [18, 32, 40], showing a high degree of structural originality. The first example of these novel plant secondary metabolites exhibiting an unprecedented *N*-iminium-aryl axis was ancisheynine (**78**) [40]. It was isolated as a minor constituent from the aerial parts of *A. heyneanus*. Ancisheynine (**78**) is a cationic naphthylisoquinoline alkaloid (natural counteranion yet unknown) based on an N,8′-coupling type. The authors, Butler et al. [40], described the compound as optically inactive ($[\alpha]_D = 0$), despite the existence of an apparently rotationally hindered heterobiaryl axis. Online LC-ECD analysis of fresh leaf extracts of *A. heyneanus* on a chiral phase by the authors' group revealed **78** to be present in the plant as a racemic mixture of its two configurationally stable atropo-enantiomers, **78a** and **78b** [83, 256] (Fig. 28).

A few years later, two similarly N,C-coupled alkaloids, ancistrocladinium A (**6**) and B (**14**) (Fig. 28), which had previously been isolated for the first time from an as yet undescribed *Ancistrocladus* species from the Central Congo Basin, were identified in the leaves of *A. heyneanus*, also discovered in the authors' group [18, 32]. In contrast to ancisheynine (**78**) [18, 40, 256], which is fully dehydrogenated, ancistrocladinium A (**6**) and B (**14**) are dihydroisoquinolinium salts. In the twigs and leaves of *A. heyneanus*, the N,8′-linked representative ancistrocladinium A (**6a**) was found to be present in the plant along with its minor *(P)*-configured atropo-diastereomer *N-epi*-ancistrocladinium A (**6b**) in a ratio of 3:1 ((*M*):(*P*)) [18, 32]. Ancistrocladinium B (**14a**) and its rotational isomer, *N-epi*-ancistrocladinium B (**14b**), are so far the only representatives of an N,6′-coupling type. They show a slow rotation about the N,C-axis at room temperature, thus occurring as a mixture of two configurationally semi-stable and slowly interconverting atropo-diastereomers [18, 32].

While ancisheynine (**78**) has so far been isolated exclusively from *A. heyneanus* [40], ancistrocladinium A (**6**) and B (**14**) are quite widespread, they have meanwhile been detected in various *Ancistrocladus* species from Asia and Africa [18, 32,

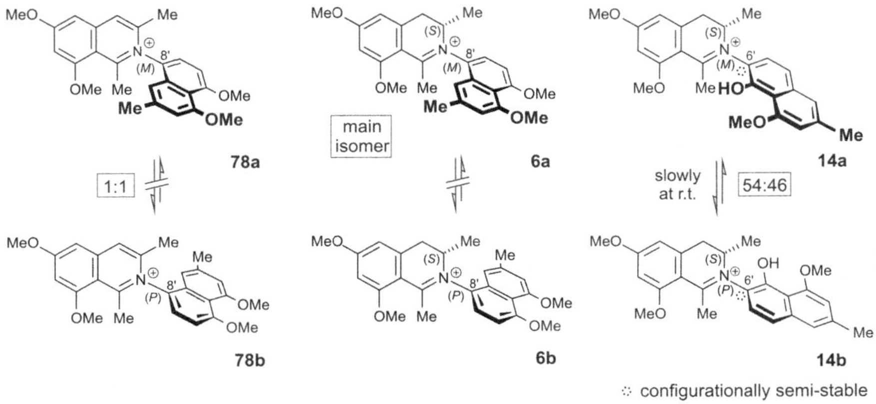

Fig. 28 N,C-Coupled naphthylisoquinoline alkaloids isolated from the leaves of *A. heyneanus*, including the racemic trace metabolite ancisheynine (**78a/78b**), ancistrocladinium A (**6**), which occurs in the plant as a pair of atropo-diastereomers, with a distinct prevalence of the (*M*)-configured isomer, **6a**, and the two configurationally semi-stable rotational isomers **14a** and **14b** of ancistrocladinium B (**14**)

33, 47, 161, 176]. All of these N,C-coupled naphthylisoquinoline alkaloids display strong inhibitory activities against *L. donovani*, the pathogen of visceral leishmaniasis [66–72]. This finding initiated intense research efforts aiming at the total synthesis [18, 69, 101, 118, 119, 256] of these unusual natural products in a stereochemically pure form and at their directed structural modification [67, 69–72, 101], with the ultimate goal to develop new promising agents.

Further isolation work on the roots, stem bark, and leaves of *A. heyneanus* furnished a plethora of C,C-linked naphthylisoquinoline alkaloids exhibiting four different coupling types (viz., 5,1′, 7,1′, 7,3′, and 7,8′) [1–5, 26–29, 37–41, 70], among them some compounds that have as yet been exclusively detected in this Indian liana, with prominent examples (Fig. 29) being the minor root metabolite isoancistrocladine (**79a**) [38], ancistroheynine A (**80**) [39], and ancistroheynine B (**72**) [41], which was predominantly isolated from the leaves of the plant. Ancistroheynine A (**80**) was found to accumulate in spherical inclusions in particular in older shoots of *A. heyneanus*. It was therefore assumed that it is formed as a stress metabolite [39], similar to the naphthoquinones plumbagin (**37**), droserone (**38**), and ancistroquinone A (**81**) [5, 315] (Fig. 29), which are also present in enhanced quantities in stem bark cells [13, 255].

While intense isolation work on *A. heyneanus* furnished a plethora of structurally unique C,C- and N,C-coupled alkaloids [1–5, 26–29, 37–41, 70], *A. hamatus* [5, 23, 24, 316] from Sri Lanka has so far only been poorly studied phytochemically. Govindachari and coworkers isolated the (*M*)-configured atropo-diastereomer **1b** of ancistrocladine (**1a**) from the roots of *A. hamatus* and named it hamatine (Fig. 30) [23, 24], according to its occurrence in *A. hamatus*. Besides hamatine (**1b**), ancistrocladine (**1a**) was also isolated from this Sri Lankan plant [5, 23]. Three further naphthylisoquinoline alkaloids, the 6-*O*-methyl analogues **76a** and **76b** of ancistrocladine (**1a**)

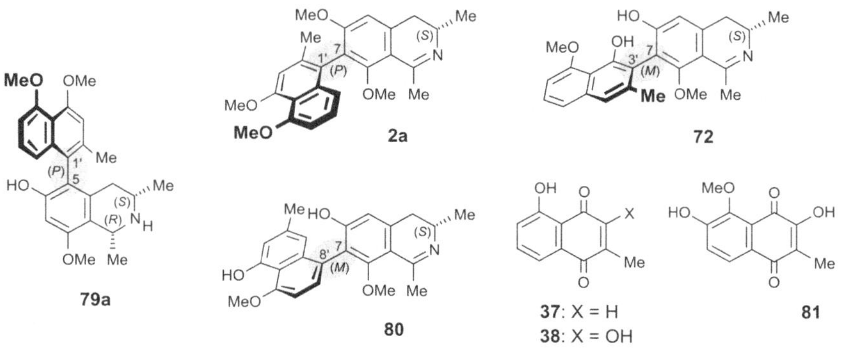

Fig. 29 Naphthylisoquinoline alkaloids from *A. heyneanus*, representing the four C,C-coupling types found in this Indian liana, namely, isoancistrocladine (**79a**) (5,1′-coupling), ancistrocladisine (**2a**) (7,1′), ancistroheynine B (**72**) (7,3′), and ancistroheynine A (**80**) (7,8′), which were isolated along with the biosynthetically related naphthoquinones plumbagin (**37**), droserone (**38**), and ancistroquinone A (**81**)

and hamatine (**1b**), respectively, and the 7,1′-coupled ancistrocladisine (**2a**) (Fig. 30) were identified in the roots, stems, and twigs of *A. hamatus* [5, 316]. Although early phytochemical studies by Govindachari and coworkers [5, 23, 24, 316] only focused on the isolation and structural elucidation of the main metabolites of this species, the results clearly support a taxonomic treatment of this Sri Lankan plant as a distinct *Ancistrocladus* species. The discovery of hamatine (**1b**) and its 6-*O*-methyl analogue **76b** as predominant constituents of *A. hamatus* [5, 23] clearly distinguishes this liana from *A. heyneanus* from India. From *A. heyneanus*, ancistrocladine (**1a**) was isolated in an atropisomerically pure form and its atropo-diastereomer hamatine (**1b**) was not detected in any plant organ of this Indian species [5, 13]. However, **1a** and **1b** have been discovered in many *Ancistrocladus* lianas from Asia and Africa, among them *A. tectorius* [50, 57, 60–62] from China, *A. griffithii* [43] from Thailand, *A. robertsoniorum* [25] from Kenya, two botanically yet undescribed Congolese species [122, 171], and *A. barteri* [13] and *A. abbreviatus* [13, 192] from Coastal West Africa. In most of the investigated *Ancistrocladus* species, a more or less preferential occurrence of the (*M*)-configured hamatine (**1b**) was detected [13, 25, 43]. In any event, more-in-depth isolation work on *A. hamatus* is required to further explore the synthesis potential of this liana that is endemic to Sri Lanka.

The unique structural features of the naphthylisoquinoline alkaloids of *A. heyneanus* and *A. hamatus* have been a strong motivation and challenge for the authors' group [18, 69, 84, 85, 90–94, 99, 101, 256] and several other researchers [102–104, 113, 115, 116, 118, 119, 317–322] to develop and apply synthesis pathways to prepare these rotationally hindered C,C- and N,C-coupled biaryl natural products possessing fascinating elements of axial and central chirality.

The lactone method [84, 85, 87, 89, 90], as developed by the authors' group, successfully gave rise to a directed, i.e., regio- and atroposelective total synthesis of enantio- and atropisomerically pure ancistrocladine (**1a**) [84, 90–92] and ancistrocladisine (**2a**) [84, 85, 90, 93, 94], or, optionally, to their atropo-

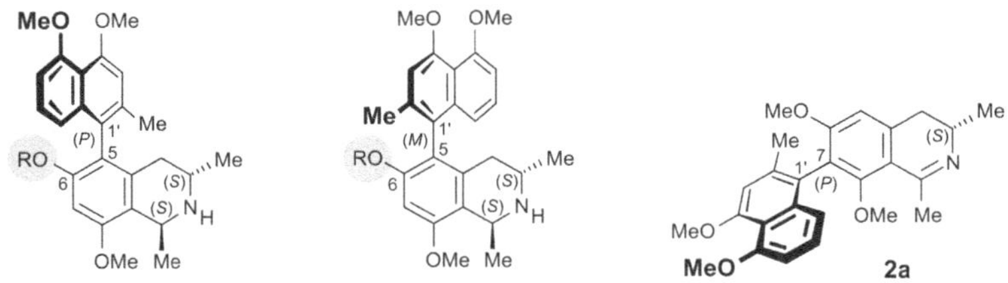

1a: R = H
76a: R = Me

1b: R = H
76b: R = Me

Fig. 30 Alkaloids of *Ancistrocladus hamatus* known thus far: the 5,1′-coupled ancistrocladine (**1a**) and its atropo-diastereomer hamatine (**1b**), their 6-*O*-methyl analogues **76a** and **76b**, and the 7,1′-linked ancistrocladisine (**2a**)

diastereomers. The stereoselective preparation of pure **1a** and **2a** confirmed the structures established by Govindachari et al. and demonstrated the efficiency of the strategy, paving the way for the preparation of further, structurally likewise challenging representatives of naphthylisoquinoline alkaloids [84, 85, 87, 89, 90, 95, 97–99]. Later, this class of natural products also came into the focus of Sargent and Rizzacasa [102–104, 317–320] and Rizzacasa and coworkers [102, 321, 322], who applied intermolecular coupling techniques as elaborated by Meyers et al. [323–326]. This synthesis approach provided access to for example, 6-*O*-methylancistrocladine (**76a**) [102, 103, 319, 321], ancistrocladinine (**77a**) [102, 104, 320], and (unnatural and racemic) dehydroancistrocladisine [102, 317, 318]. However, the specific substitution pattern of ancistrocladine (**1a**) and ancistrocladisine (**2a**), and in particular the free hydroxy group at C-6 of **1a** and the identity of the two *ortho*-substituents (2 × OMe) next to the axis, at C-6 and C-8 of the dihydroisoquinoline portion of **2a**, made a stereoselective construction of the biaryl axis by direct, intermolecular coupling techniques extremely challenging. The level of difficulty was further enhanced by the fact that **2a** and, even more so, **1a** show a very high degree of steric hindrance at the biaryl axis. For this reason, **1a** and **2a** have as yet never been prepared by any direct, intermolecular coupling strategy. Attempts by Sargent and Rizzacasa only gave modified (unnatural) analogues [102–104, 319–322] of **1a** and **2a** (often as stereoisomeric mixtures), or even entirely failed to build up the required biaryl axis at all [102, 317, 318].

In the early 2000s, Bungard and Morris [115, 116] established the first synthesis pathway to the 7,3′-coupled ancistrocladidine (**3a**) based on an *ortho*-arylation methodology for the construction of the biaryl linkage. Starting from a naphthol and an aryllead triacetate as the precursor for the isoquinoline portion, the target molecule was obtained, yet as a 1:1 mixture of **3a** and its—so far unnatural—atropo-diastereomer (see Sect. 4.4). In 2016, the authors' group reported the first regio- and stereoselective total synthesis of ancistrocladidine (**3a**) and of the likewise 7,3′-coupled ancistrotectorine (**17a**), a main metabolite of the Southeast Asian liana *A. tectorius*, by applying the lactone method [99]. This pathway permitted a highly atroposelective access to **3a** and **17a**, or, if desired, to their (unnatural) atropo-diastereomers (see Sect. 4.5).

The first total synthesis of the N,C-coupled alkaloid ancisheynine (**78**) was achieved by the authors' group, based on the cyclocondensation of a monocyclic diketone (or the respective benzopyrylium salt) with an aminonaphthalene, leading to a fully dehydrogenated isoquinolinium moiety (see Sect. 4.8) [18, 256]. Similarly in the authors' group, the first total synthesis of the N,C-linked naphthyldihydroisoquinoline alkaloids ancistrocladinium A (**6a/b**) and B (**14a/b**) succeeded, by using a highly convergent route [101]. It was based on the Buchwald–Hartwig amination of the respective naphthalene precursor with an enantiomerically pure arylisopropylamine building block (prepared from L-alanine) with subsequent Bischler–Napieralski cyclization (see Sect. 8.5.9.2). A directed synthesis of all four stereoisomers of ancistrocladinium A (**6**) in an enantio- and atropisomerically pure form was also achieved by the authors' group, based on the chromatographic resolution of the likewise rotationally hindered, and thus atropo-diastereomeric acetamide precursors, which were specifically converted into the respectively configured target molecule (see Sect. 8.5.9.3) [69].

The general applicability of the developed strategy was demonstrated by the preparation of the N,3′- and N,1′-coupled analogues of **6** and **14a/b**, which have not so far been found in Nature [18, 101].

A slightly modified pathway to ancistrocladinium A (**6**) was described by Kim and Cheon [118, 119]. Based on an asymmetric reductive amination of a respective 1-arylpropanone with the corresponding aminonaphthalene building block using a chiral phosphoric acid derived from BINOL as the catalyst, **6** was synthesized by acetylation of the resulting chiral amine, as in the previous syntheses [18, 101], and Bischler–Napieralski cyclization (see Sect. 8.5.9.5). In their first attempt, the configuration at C-3 as (*R*) was installed by employing a chiral catalyst derived from (*M*)-BINOL, yet leading to the unnatural (3*R*)-configuration, thus, giving rise to the 'wrong' enantiomer of ancistrocladinium A (*ent*-**6**), moreover as a mixture of its two atropo-diastereomers [118]. In a renewed approach, Bischler–Napieralski ring closure of the atropisomerically pure individual acetamide precursors, now equipped with the 'correct' (3*S*)-configuration (as in the natural target molecules), using triflic anhydride, provided authentic ancistrocladinium A (**6a**) and its (*P*)-configured atropo-diastereomer **6b** in a stereochemically pure form [119]. A more detailed report on the synthesis of ancistrocladinium A (**6**) and related compounds, associated with successful attempts to establish the rare, yet intriguing, phenomenon of N,C-axial chirality, is given in Sect. 8.5.9.

4.1 Ancistrocladine and Further 5,1′-Coupled Naphthylisoquinoline Alkaloids from the Roots and Leaves of *Ancistrocladus heyneanus*

Isolation work on leaf, stem bark, and root extracts of *A. heyneanus* showed that this liana, which is endemic to the Western Ghats of the Indian subcontinent,

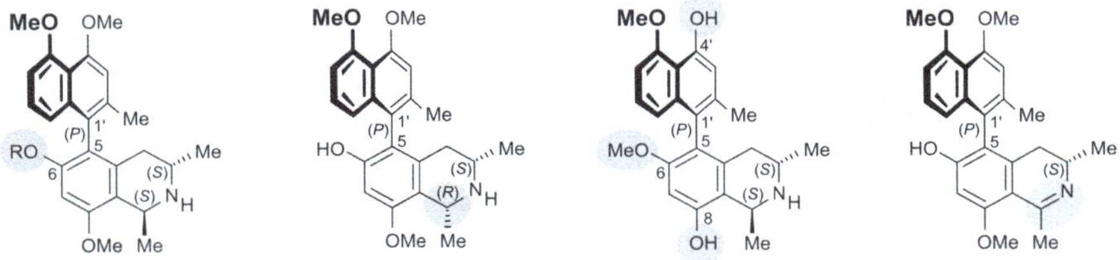

1a: R = H
76a: R = Me
79a
82
77a

Fig. 31 5,1′-Coupled naphthylisoquinoline alkaloids (all (*P*)-configured) as isolated from the Indian liana *A. heyneanus*: ancistrocladine (**1a**), its 6-*O*-methyl analogue **76a**, isoancistrocladine (**79a**), 4′-*O*-demethylancistectorine A_2 (**82**), and ancistrocladinine (**77a**). Structural modifications of the metabolites compared to **1a** are underlaid in light gray

produces a characteristic pattern of naphthylisoquinoline alkaloids [1–5, 26–29, 37–41, 70] that is significantly different from that of other species from Asia and Africa. A common phytochemical feature of *A. heyneanus* and related Asian and East African *Ancistrocladus* taxa is their exclusive production of Ancistrocladaceae-type alkaloids, i.e., bearing an oxygen function at C-6 and being (*S*)-configured at C-3. While various naphthylisoquinolines were identified in the aerial parts of *A. heyneanus* [39–41, 70], it was mainly the 5,1′-coupled ancistrocladine (**1a**) that was found to occur in the roots [1–5], along with its 6-*O*-methyl analogue **76a** [5, 316] and the related minor metabolites ancistrocladinine (**77a**) [37], and isoancistrocladine (**79a**) [38]. The latter is one of the less frequently isolated representatives of *cis*-configured naphthylisoquinoline alkaloids with a free, non-methylated NH function. In the aerial parts of *A. heyneanus*, one further 5,1′-linked compound, 4′-*O*-demethylancistectorine A_2 (**82**) [70], was identified as a major constituent. All of the five above-mentioned naphthylisoquinoline alkaloids (see Fig. 31), with their 5,1′-coupling type, were isolated from this Indian liana in an atropisomerically pure form. Neither hamatine (**1b**), i.e., the atropo-diastereomer of ancistrocladine (**1a**), nor related 5,1′-linked and (*M*)-configured compounds have so far been detected in this Indian *Ancistrocladus* species.

4.2 *Directed Total Synthesis of Ancistrocladine and Its Atropo-Diastereomer Hamatine by the Lactone Methodology*

The lactone concept [84, 85, 87, 89, 90] as developed by the authors' group, was initially designed for the regio- and stereoselective synthesis of naphthylisoquinoline alkaloids [84, 85, 90–95, 97–99], but was later applied to numerous other axially chiral natural and unnatural target molecules too [84, 85, 87, 90, 327–339]. It is an efficient method for the atropo-divergent preparation of any of the two possible

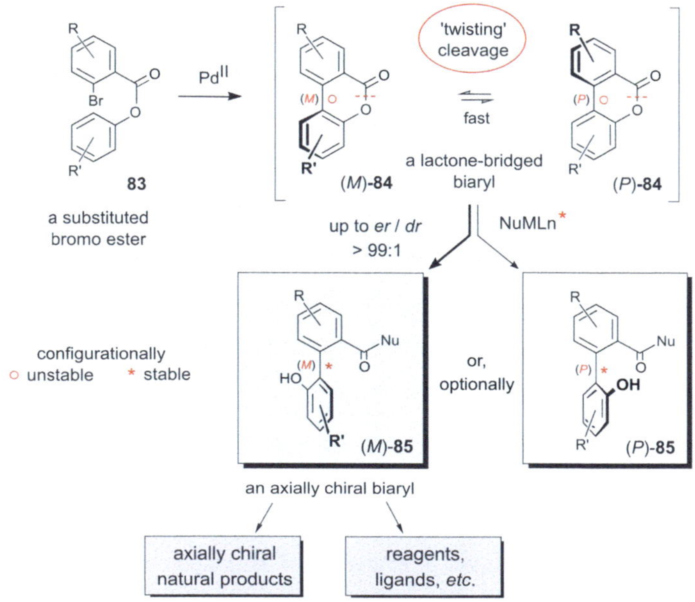

Fig. 32 Lactone methodology: efficient synthesis of configurationally not yet fixed lactone-bridged biaryls **84** by intramolecular coupling and their atroposelective cleavage, to give, atroposelectively (i.e., with directed, stereoselective "twisting"), axially chiral biaryls of any desired configuration (e.g., (*M*)-**85** is shown). The descriptors are arbitrary, without consideration of the residue R′

atropisomers in a pure form, starting from the same (late) biaryl lactone precursor. The strategy permits a recycling of any undesired minor product, literally by recyclization back to the lactone, following the principle of chiral economy. The approach has almost no restrictions concerning the substitution pattern and is even applicable to target molecules that have a very high steric hindrance at the *ortho*-position(s) of the biaryl axis [84, 85, 340–342]. The general principle of the lactone method is based on the construction of the biaryl axis in a two-step procedure (see Fig. 32):

(1) The intramolecular C–C coupling of the ester-like prefixed aromatic portions as in **83**, leads to a six-membered lactone **84**. This bridged biaryl already contains the biaryl axis, and it is chiral, helically distorted—but still configurationally unstable, existing as a mixture of rapidly interconverting atropo-enantio- or diastereomers.
(2) The coupling step is followed by the highly stereoselective lactone ring cleavage using appropriate chiral nucleophiles, thus providing the desired—now configurationally stable—chiral biaryl, here (*M*)-**85** (or, optionally (*P*)-**85**), with excellent atropo-enantio- or diastereoselectivities.

For the atroposelective total synthesis of naphthylisoquinoline alkaloids with their additional stereogenic centers in the tetra- or dihydroisoquinoline portion, the ring cleavage leads to atropo-diastereomers. It can then sometimes be performed even with achiral nucleophiles, if the internal asymmetric induction, exerted by the stereocenters, is good enough, or preferably, with chiral nucleophiles, reliably providing the desired rotational isomer by a dominating external asymmetric induction.

An instructive example evidencing the high degree of steric hindrance that can be overcome by the lactone concept is the total synthesis of the fourfold *ortho*-substituted naphthylisoquinoline alkaloid ancistrocladine (**1a**) [84, 90, 91]. Its sterically blocked biaryl linkage gives rise to rotational isomers that are configurationally stable far beyond 250°C [2, 4, 13]. The key step of the reaction sequence presented in Fig. 33 is the thermodynamically controlled induction of the axial configuration at the level of the severely distorted, here configurationally semi-stable biaryl lactone **89**. Starting from the ester **88**, as prepared from **86** and **87** by Pd(II)-catalyzed biaryl coupling, the intermediate lactone **89** was smoothly obtained in 49% yield, against strongest steric hindrance. The two helicene-like distorted diastereomeric lactones **89a** and **89b** were spectroscopically distinguishable, because they interconverted slowly at room temperature ($t_{1/2} \approx 1$ min at 299 K), with the desired ancistrocladine-like stereo-orientation prevailing. The helimeric ratio was found to be 3:1 in favor of **89a** (for R = Bn), and could even be enhanced up to "100:0" by making the isoquinoline portion more rigid (e.g., with R = $COCF_3$) [90, 91]. Out of this equilibrium mixture (e.g., of *dr* 75:25 in the case of *N*-benzyl), the lactone ring was opened with full conservation of the atropisomeric ratio by rapid reduction with lithium aluminum hydride to give the configurationally stable alcohols **90a** and **90b** in 95% yield, which were easily resolved by chromatography on silica gel.

Final deoxygenation of the hydroxymethyl group and *N*-debenzylation converted **90a** into the target molecule ancistrocladine (**1a**). The analogous transformation of **90b** afforded the likewise naturally occurring [23, 24] atropo-diastereomer of ancistrocladine (**1a**), hamatine (**1b**) [90, 91]. An even more stereo-economic synthesis of **1a** and **1b** succeeded via the respective cyclic ethers [92].

The pioneering character of these regio- and stereoselective first syntheses of ancistrocladine (**1a**)—in 1986 already!—can be seen from the fact that this goal has not been reached by any other method to date.

4.3 Directed Synthesis of Ancistrocladisine via Configurationally Unstable Lactone-Bridged Biaryl Lactones

The first, atroposelective synthesis of ancistrocladine (**1a**) (Fig. 33) [90–92], as described in the previous section, demonstrated the efficiency of the lactone approach with respect to the high steric hindrance tolerated by this methodology. The total synthesis of the 7,1′-coupled naphthylisoquinoline alkaloid ancistrocladisine (**2a**)

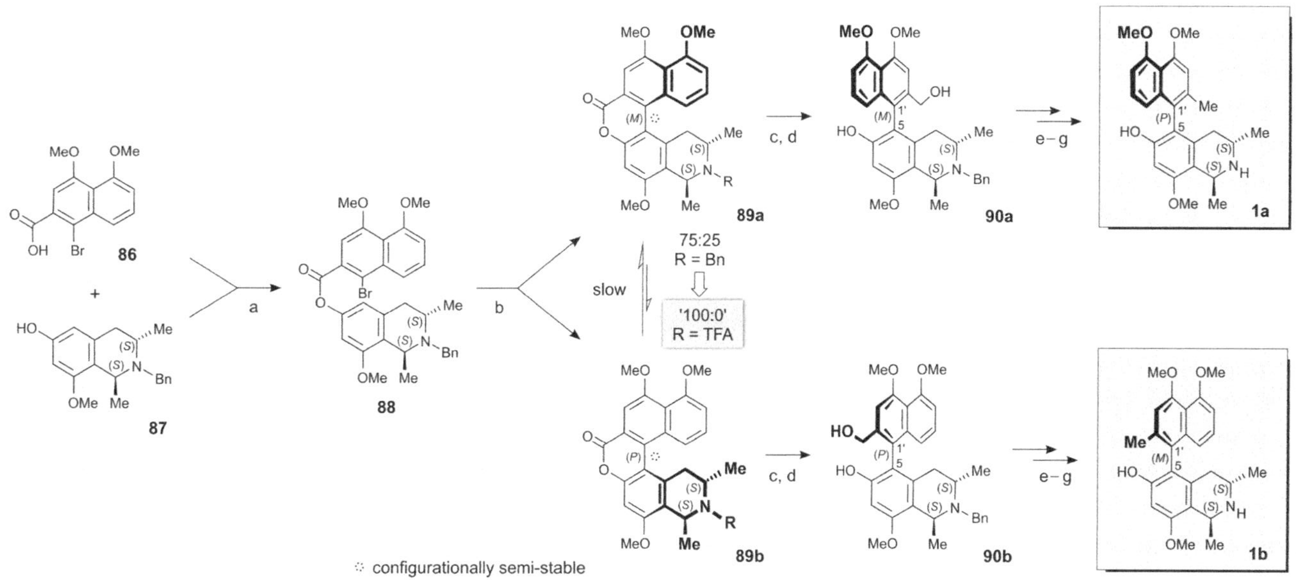

Fig. 33 Thermodynamically controlled atroposelective total synthesis of ancistrocladine (**1a**), or, optionally, hamatine (**1b**), by applying the lactone approach. Reaction conditions: (a) NEt_3, DMAP, CH_2Cl_2, 86%; (b) $Pd(PPh_3)_2Cl_2$, NaOAc, DMA, 100°C, 49%; (c) $LiAlH_4$, THF, 95%; (d) diastereomeric resolution of **90a/90b** (for R = Bn) by chromatography on silica gel; (e) $(CBrCl_2)_2$, polym. PPh_3, CH_2Cl_2 [343]; (f) $LiAlH_4$, THF; (g) Pd/C (10%), H_2, MeOH, 91% (for **90a** → **1a**) and 90% (for **90b** → **1b**)

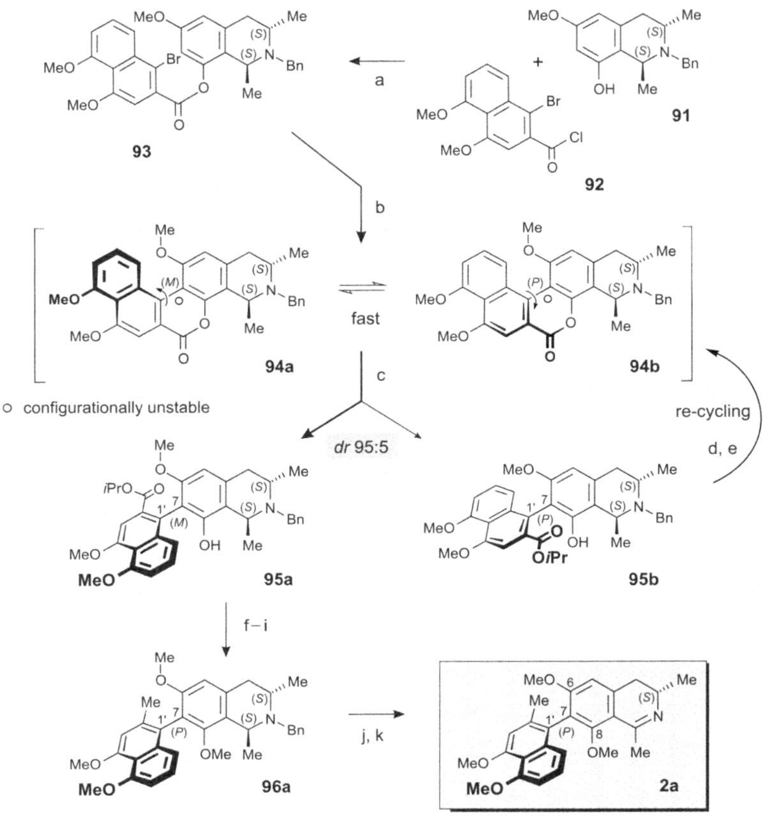

Fig. 34 Despite the high steric hindrance next to the biaryl axis and the presence of two identical *ortho*-substituents next to the axis in the dihydroisoquinoline subunit (at C-6 and C-8), the highly atropisomer-selective synthesis of the 7,1′-coupled ancistrocladisine (**2a**) can occur by cleavage of the lactone **94** with potassium isopropylate as a simple, achiral *O*-nucleophile. Reaction conditions: (a) NEt_3, DMAP, CH_2Cl_2, 96%; (b) $Pd(PPh_3)_2Cl_2$, NaOAc, DMA, 130°C, 87%; (c) KO*i*Pr, *i*PrOH, r.t., 83%; (d) 2 N NaOH, dioxane; (e) $(COCl)_2$, DMF, CH_2Cl_2, NEt_3, 71%; (f) $LiAlH_4$, THF; (g) Me_2SO_4, PTC, CH_2Cl_2, 2 N NaOH; (h) $(CBrCl_2)_2$, PPh_3, CH_2Cl_2 [343]; (i) $LiAlH_4$, THF; (j) Pd/C (10%), H_2, MeOH, 68%; (k) $KMnO_4$, THF, 14%

(Fig. 34) [90, 93, 94], achieved with complete regio- and stereocontrol, is a further impressive example that proved the usefulness, broad applicability, and the chiral economy of the lactone concept [84, 85, 87, 90]. As in the case of ancistrocladine (**1a**), the method permitted the targeted synthesis of any of the two possible atropisomers of this bulky target molecule **2a** with its four *ortho*-substituents next to the biaryl axis, from a late joint precursor, providing high chemical and optical yields. Moreover, the presence of two identical *ortho*-functionalities in the dihydroisoquinoline subunit next to the axis (6-OMe and 8-OMe), which constitutes a difficult, if not unsurmountable obstacle to any direct, intermolecular coupling method, was easily overcome by applying the lactone approach [84, 85, 90–94].

Prefixation of the two building blocks **91** and **92**, by esterification afforded **93**, which was subjected to a Pd-catalyzed intramolecular biaryl coupling, thus affording the desired 7,1′-coupled lactone **94** in an excellent chemical yield (87%) [90, 93, 94]. Different from the respective precursor **89** to ancistrocladine (**1a**) (Fig. 33), the lactone **94** presented itself as one single species in the ^{1}H NMR, occurring at room temperature as a mixture of its two rapidly interconverting atropo-diastereomers, **94a** and **94b** (Fig. 34) [90, 93, 94]. Because of the presence of stereogenic centers in **94**, the lactone bridge can be ring-opened atroposelectively even with simple, cheap achiral *O*-nucleophiles such as potassium isopropylate, to give with high internal asymmetric induction, the respective methyl ester **95**, which is configurationally stable. This provides excellent atropo-diastereoselectivities in favor of the targeted (*M*)-isomer, **95a** (*dr* = 95:5) [90, 93, 94].

Quantum-chemical investigations on related model lactones [341] showed that the observed strong asymmetric induction in the ring-opening step is a consequence of a kinetically controlled preferred—*axial*—attack of the isopropylate anion on the (*M*)-helically distorted biaryl lactone **94a** (*dr* = 95:5). The small quantities of the minor ('wrong') atropisomer **95b** were not lost, but could be re-utilized, by its recyclization back to the lactone **94**, and renewed ring opening, thus permitting a high degree of chiral economy.

A series of standard transformation reactions delivered the 2′-methyl compound **96a**, which after N-debenzylation, was oxidized with potassium permanganate. This was a most critical step, because of the required stereoelectronically unfavorable abstraction of the *equatorial* hydrogen at C-1, to give the target molecule ancistrocladisine (**2a**) [90, 93, 94]. Its identity was confirmed by comparison with authentic material isolated from *A. heyneanus* [26, 27].

4.4 Total Synthesis of Ancistrocladidine by ortho-Arylation of a Naphthol with an Aryllead Triacetate

In view of its rare 7,3′-biaryl linkage, ancistrocladidine (**3a**) [27–29] was also an attractive, but challenging target molecule for total synthesis. Two synthesis approaches to **3a** have so far been developed [99, 115, 116]. The challenging architecture of this naphthyldihydroisoquinoline, and in particular, the distinct steric bulk surrounding its biaryl bond due to the presence of four *ortho*-substituents next to the biaryl axis required the elaboration of new and efficient strategies. This was a demanding task, since well-established palladium-catalyzed cross-coupling reactions, widely used in biaryl synthesis [88, 344–346], suffer from the fact that increasing steric hindrance leads to a significant decrease of the yields of the key biaryl coupling intermediates [88, 347], or even totally fail to build up the biaryl linkage [317, 318]. The first regio- and stereoselective total synthesis of

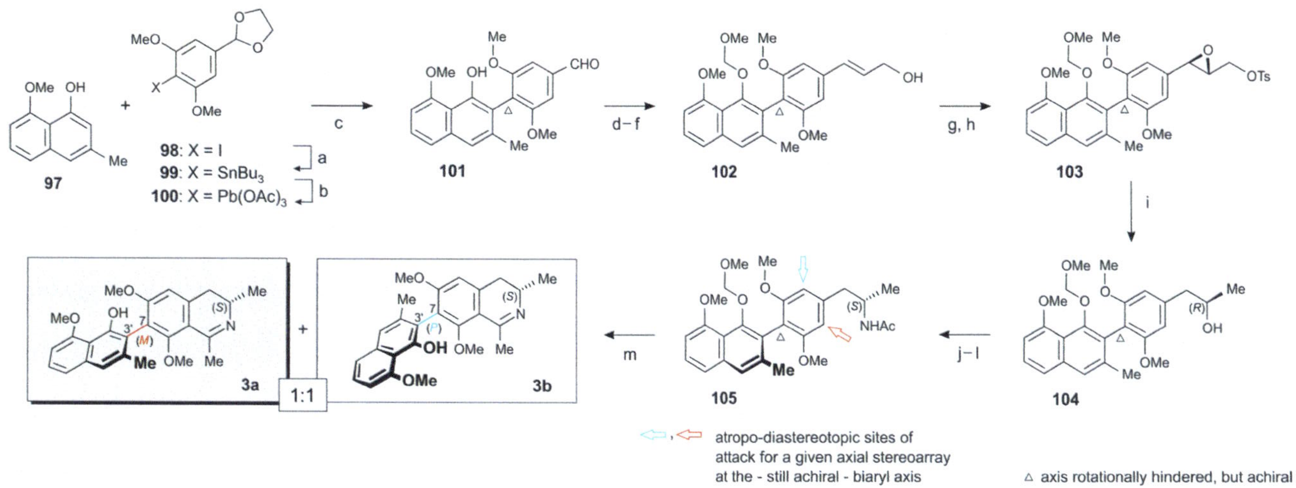

Fig. 35 Total synthesis of ancistrocladidine (**3a**) and its atropo-diastereomer **3b**: construction of the biaryl axis by Pinhey–Barton *ortho*-arylation of the naphthol **97** with the aryllead triacetate **100**. Reaction conditions: (a) *t*-BuLi, THF, −95°C, 15 min → Bu_3SnCl, r.t., 85%; (b) $Pb(OAc)_4$, cat. $Hg(OAc)_2$, CH_2Cl_2, r.t., 24 h, 93%; (c) pyridine, CH_2Cl_2, r.t., 24 h → 3% aq. H_2SO_4 (v/v), THF, r.t., 1 h, 67% (from **97**); (d) MOM-Cl, NaH, THF, r.t., 81%; (e) NaH, $(EtO)_2POCH_2CO_2Et$, C_6H_6, 0°C → r.t., 99%; (f) DIBAL-H, toluene, −78°C, 15 min, 89%; (g) 5 mol% Ti(*O*-*i*Pr)$_4$, 6 mol% D-diisopropyltartate, TBHP, CH_2Cl_2, −20°C, 5 h, 80% (90% *ee*); (h) TsCl, NEt_3, DMAP, CH_2Cl_2, 0°C, 1 h, 83%; (i) $LiAlH_4$, Et_2O, 0°C, 2 h, 94%; (j) phthalimide, DEAD, PPh_3, THF, r.t., 16 h, 82%; (k) 40% aq. $MeNH_2$, EtOH, reflux, 1 h, 99%; (l) MeCOCl, NEt_3, CH_2Cl_2, 0°C to r.t., 97%; (m) $POCl_3$, 2,4,6-collidine, MeCN, reflux, 4 h, 74% (1:1 mixture of **3a** and its atropo-diastereomer **3b**)

ancistrocladidine (**3a**) and of the related, likewise 7,3′-coupled ancistrotectorine (**17a**), was achieved by the authors' group [99] by applying the lactone method, and will be the topic of the next paragraph (see Sect. 4.5). Already earlier, based on the Pinhey–Barton *ortho*-arylation methodology for the construction of the biaryl axis and starting from the naphthol **97** and the monocyclic aryllead triacetate **100** as a symmetric precursor for the isoquinoline portion, Bungard and Morris established the first synthesis pathway to ancistrocladidine (**3a**) [115, 116]. This route provided ancistrocladidine (**3a**) and its atropo-diastereomer **3b** as a 1:1 mixture, which was separable only by crystallization (Fig. 35).

As a building block for the formation of the biaryl axis in ancistrocladidine (**3a**), the aryllead triacetate **100** was prepared by halogen-lithium exchange of the iodide **98** to give the stannane **99**, followed by tin-to-lead transmetalation using lead tetraacetate in the presence of catalytic amounts of mercuric acetate [115, 116]. Construction of the rotationally hindered—yet achiral—biaryl linkage by Pinhey–Barton methodology was readily achieved by condensation of **100** with the known [101, 348] naphthol **97**, leading to the biaryl aldehyde **101**. According to the reaction sequence in Fig. 35, the allylic alcohol **102** was prepared in three steps, by protection of the free phenolic group of **101** with (chloromethoxy)methane, and subsequent conversion of the aldehyde to the corresponding α,β-unsaturated ester by Horner–Wadsworths–Emmons reaction and reduction to give **102** (71% overall yield). Enantioselective Sharpless epoxidation to introduce the stereocenter at C-3, followed by tosylation, afforded the desired, now chiral intermediate **103** in 80% yield and 90% *ee* [115, 116]—yet still without chirality at the biaryl axis. In the next step, the alcohol **104** was obtained by concomitant cleavage of the tosyl group of **103** and ring opening of the epoxide functionality. Treatment of **104** with phthalimide under Mitsunobo conditions, followed by hydrolysis of the phthalimide group and acetylation of the resulting free amine provided the acetamide **105**.

Bischler–Napieralski cyclization of **105**—with its rotationally hindered, but still achiral, since symmetrically substituted biaryl axis—now made the axis chiral. In principle, the cyclization could occur via any of the two unsubstituted *ortho*-positions next to the isopropylacetamido side chain, which due the presence of the stereogenic center, were diastereotopic to each other. With a given stereo-orientation of the naphthalene part (e.g., as drawn for **105** in Fig. 35), the 'blue' and the 'red' positions should lead to the respective atropo-diastereomers. Actually, Bischler–Napieralski cyclization along with the in situ cleavage of the MOM protecting group occurred with complete lack of diastereoselectivity, delivering a 1:1 mixture of ancistrocladidine (**3a**) and its atropo-diastereomer **3b** in 74% overall yield [115, 116]. Resolution of the mixture succeeded only by recrystallization from toluene/petroleum ether, giving stereochemically pure ancistrocladidine (**3a**), while the non-natural atropisomer **3b** remained in the mother liquor. The synthetic material of **3a** was identical in all its physical and spectroscopic data with the isolated natural product [27–29].

4.5 Atroposelective Total Synthesis of Enantiomerically Pure Ancistrocladidine and Ancistrotectorine Using the Lactone Method

As described in the previous section, the Morris group managed to synthesize for the first time ancistrocladidine (**3a**) as a representative of the small subclass of 7,3′-coupled naphthylisoquinoline alkaloids. Their synthesis strategy was based on a Pinhey–Barton *ortho*-arylation as the decisive step for the construction of the biaryl axis—but no atroposelectivity was achieved in the Bischler–Napieralski ring closure [115, 116]. The regio- and stereoselective total synthesis of ancistrocladidine (**3a**) and ancistrotectorine (**17a**) by using the "lactone methodology" [99] is thus a further impressive proof of its efficiency and broad applicability for the C,C-coupling between the two molecular portions of a biaryl compound (see also Sects. 4.2 and 4.3). This pathway (Fig. 36) permitted a highly atroposelective access to the two alkaloids **3a** and **17a**, or if desired, to their (as yet unnatural) atropo-diastereomers **3b** and **17b**, respectively (not shown) [99]. Following the general concept of the lactone method, the main steps were the ester-type prefixation of the naphthalene unit **108** and the isoquinoline portion **107** (to give **109**) and its intramolecular biaryl coupling using the palladium-phosphinous acid POPd as a catalyst and Cs_2CO_3 as a base, thus affording the configurationally unstable 7,3′-linked lactone-bridged biaryl **110** (Fig. 36).

For the lactone formation, the two molecular portions **107** and **108** were activated with the halogen atom located in the phenolic isoquinoline moiety. This 'inverse-halogenation strategy' led to an intramolecular biaryl coupling between C-7 and the electronically unfavorable C-3′ position of the ester **109** in an overall coupling yield of 31% [99]. The at first sight moderate result compared with the outcome of the total syntheses of other naphthylisoquinoline alkaloids however, was still most satisfactory in view of the unexpectedly clear regioselectivity (virtually no 1′-coupling in the *peri*-position of the naphthalene was observed, although likewise imaginable) [99]. Previous attempts to achieve the biaryl bond formation in accordance with the 'normal' lactone concept (i.e., with the halogen substituent located on the naphthoic acid part) under standard coupling conditions (e.g., with $Pd(PPh_3)_2Cl_2$ as the catalyst and NaOAc as the base) failed completely, leading to a complex mixture, in which only traces of the desired lactone **110** were identified, while hydrodehalogenation had occurred to a large extent [99].

As anticipated, the lactone-bridged biaryl **110**, with its configurationally unstable biaryl axis, was an ideal joint precursor for both target molecules, ancistrocladidine (**3a**) and ancistrotectorine (**17a**). Reductive ring cleavage of **110** proceeded in good chemical yields, with borane activated by Corey's chiral oxazaborolidine (*R*)-**106** as the hydride transfer reagent, providing the configurationally stable, (*P*)-configured diol **111a** in a high asymmetric induction (*dr* = 94:6) (Fig. 36). Conversely, when using $BH_3{\cdot}THF$ in the presence of (*S*)-**106** as the reductant (not shown), the (*M*)-atropo-diastereomer of the diol **111b** (not shown) was selectively obtained (*dr* = 94:6) [99]. Achiral nucleophiles, by contrast, like $LiAlH_4$, gave only bad or

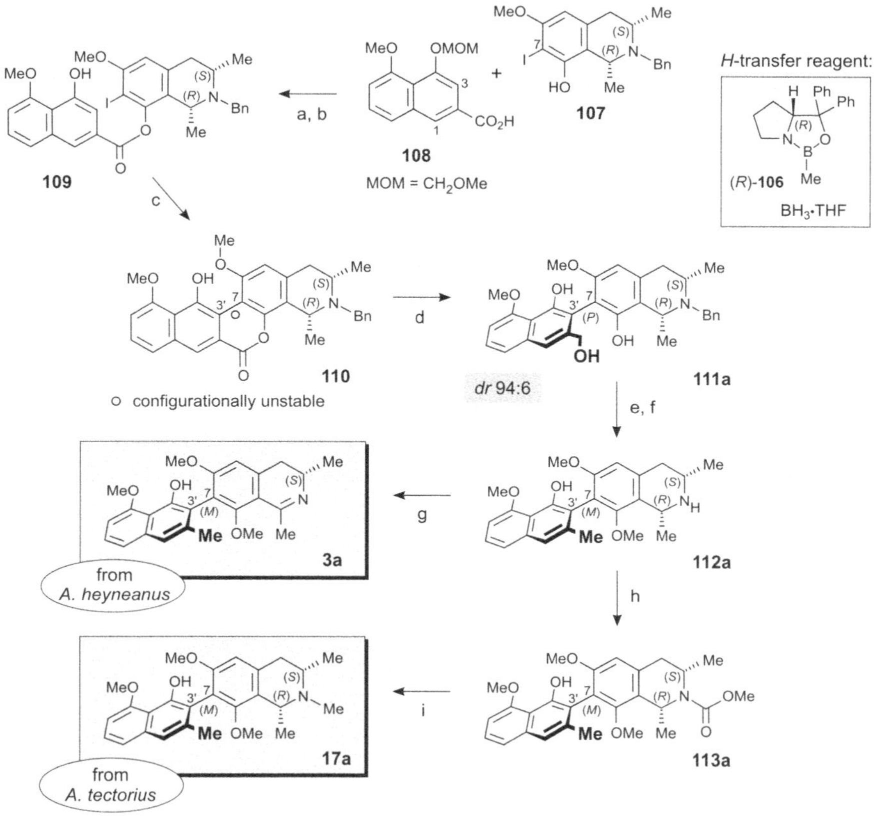

Fig. 36 Regio- and stereoselective joint total synthesis of ancistrocladidine (**3a**) and ancistrotectorine (**17a**) from identical precursors following the lactone method, by ester-type prefixation of the two molecular halves, activated in an "inverse-halogenated" form (i.e., with the halogen substituent in the isoquinoline part), followed by intramolecular coupling and reductive cleavage of the lactone auxiliary bridge with borane activated by the chiral oxazaborolidine (*R*)-**106**. Reaction conditions: (a) DCC, DMAP, CH_2Cl_2, r.t., 1 h; (b) TFA, CH_2Cl_2, r.t., 1 h, 72%; (c) POPd (1.5 equiv.), Cs_2CO_3 (5 equiv.), DMA, 110°C, 45 min, 31%; (d) (*R*)-**106**·BH_3, THF, 0°C, 30 min, 88%; (e) K_2CO_3, Me_2SO_4, acetone, r.t., 7 h, 76%; (f) Pd/C (10%), H_2, 2 N HCl, EtOH, r.t., 94%; (g) $KMnO_4$, THF, r.t., 2 h, 65%; (h) $ClCO_2Me$, $NaHCO_3$, CH_2Cl_2, r.t., 1 h, 93%; (i) $LiAlH_4$, Et_2O, r.t., 1 h, 92%

no asymmetric induction, despite the chirality of the substrate, possibly due to the relatively flat stereoarray of the *cis*-configured tetrahydroisoquinoline part [99]. Conversion of **111a** into the final joint intermediate **112a** was achieved by selective methylation of the OH functionality at C-8 with dimethyl sulfate and potassium carbonate followed by a one-step deoxygenation and *N*-deprotection by hydrogenolysis over Pd/C. The first atroposelective total synthesis of the naphthyldihydroisoquinoline ancistrocladidine (**3a**) was completed by a facile, since stereoelectronically favorable permanganate oxidation of the *cis*-configured 1,3-dimethyltetrahydroisoquinoline subunit of **112a** (Fig. 36) [99]. The synthetic product was identical in all its physical, spectroscopic, and chiroptical data with the natural material of **3a** previously isolated from *A. heyneanus* [27–29].

The second target molecule, ancistrotectorine (**17a**), was obtained by N-methylation of the joint intermediate **112a**. This was achieved by conversion of **112a** into the *N*-carbamate **113a** and subsequent reduction using $LiAlH_4$ [99] (Fig. 36). All spectroscopic data of the final product completely matched with those of authentic ancistrotectorine (**17a**) isolated from *A. tectorius* [51] and *A. guineënsis* [159].

4.6 Produced in Aging Shoots and in Cell Cultures of *Ancistrocladus heyneanus*, The 7,8′-Coupled Naphthylisoquinoline Alkaloid Ancistroheynine A and Further Stress Metabolites

Extraction of fresh stem and shoot material of *A. heyneanus* led to the isolation of a bright-yellow amorphous solid, which was chemically homogeneous. The compound attracted attention since it was found to be located in spherical inclusions in the pith and bark regions of the shoots. The solutes were easily detectable by light microscopic investigations upon irradiation with UV light (λ = 365 nm), conspicuous by their strong yellow fluorescence [39]. Elucidation of the stereostructure of the accumulated compound by spectroscopic (in particular, HRESIMS as well as 1D and 2D NMR) and chemical (oxidative degradation) methods revealed this shoot metabolite to be a new 7,8′-coupled naphthyldihydroisoquinoline alkaloid with (*S*)-configuration at C-3. It was named ancistroheynine A (**80**). Its axial configuration was unambiguously determined to be (*M*) by long-range NOE measurements between H_{ax}-4 and H-1′, despite the very long distance between these nuclei, and based on the known (see above) absolute configuration at C-3 [39] (Fig. 37a).

The occurrence of ancistroheynine A (**80**) accumulating in certain parts of the shoots of *A. heyneanus*, as observed by light and fluorescence microscopy (Fig. 37b) [39, 349], was additionally corroborated by non-destructive, since non-invasive FT-Raman microscopy [349] (Fig. 37c). Although the alkaloids of *A. heyneanus* differ only by the linkage between their molecular portions, by the OH/OMe pattern, and the degree of hydrogenation in the isoquinoline moiety, Raman spectroscopy permitted to distinguish between individual alkaloids, due to its high specificity and sensitivity. Comparison of localized Raman spectra monitored in situ in the shoot tip and the leaf midrib with the FT-Raman spectrum of the pure isolated compound clearly confirmed the presence of ancistroheynine A (**80**) in spherical yellowish inclusions [349].

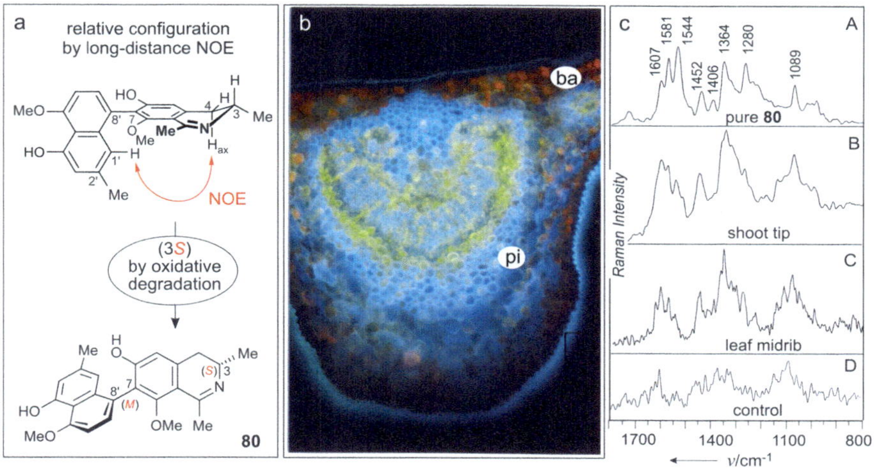

Fig. 37 Ancistroheynine A (**80**), a new 7,8′-coupled naphthyldihydroisoquinoline alkaloid isolated from spherical inclusions in the shoots of *A. heyneanus*. **a** Elucidation of the full stereostructure of **80** by oxidative degradation and long-range NMR experiments; **b** fluorescence microscopic image of the shoot, showing a very dense distribution of yellow solutes in the pith (pi) and bark (ba) regions displaying a bright-yellow fluorescence (magnification ×215); **c** in situ localization of **80** in fresh shoot material of *A. heyneanus* by a FT-Raman investigation: spectrum of pure **80** (section A), in situ spectrum of **80** monitored in the shoot tip (section B), in situ spectrum of **80** recorded in the leaf midrib (section C), and in situ spectrum taken from cells without inclusions (section D) (Photograph: H. Schneider)

The accumulation of **80** in older shoots of *A. heyneanus* [39] led to the assumption that this alkaloid was formed as a stress metabolite, presumably acting as a phytoalexin similar to the naphthoquinones plumbagin (**37**) and droserone (**38**), which are likewise produced by the plants in large quantities, as a consequence of different sorts of chemical, physical, or biotic stress [13, 250–252, 255]. Furthermore, a new, even oxygen-richer naphthoquinone, ancistroquinone A (**81**) [13, 315], was identified in aerial parts of this Indian liana, probably likewise involved in defense-mediating processes allowing the plants to cope with herbivores, with threatening microorganisms, or with neighboring plants (Fig. 38).

This stress-induced formation of naphthoquinones is also predominant in cell cultures of *A. heyneanus* and was, in turn, used for biosynthetic feeding experiments with ^{13}C-labeled acetate (Fig. 39) [134, 249]. The investigations unambiguously confirmed the acetogenic nature of plumbagin (**37**) and its reduction product, the tetralone *cis*-isoshinanolone (**67**) [249]. The ^{13}C NMR spectra of **37** and **67** isolated from cell cultures of *A. heyneanus* after administration of [$^{13}C_1$]-acetate revealed an incorporation rate of ca. 2–3% (for plumbagin, **37**) and of ca. 6% (for *cis*-isoshinanolone, **67**), with enhanced signals at the C-atoms 2-Me, C-1, C-8, C-6, C-10, and C-3. ^{13}C NMR measurements likewise showed the methylene positions alternating with the carbonyl-derived ones of the polyketide chain [249].

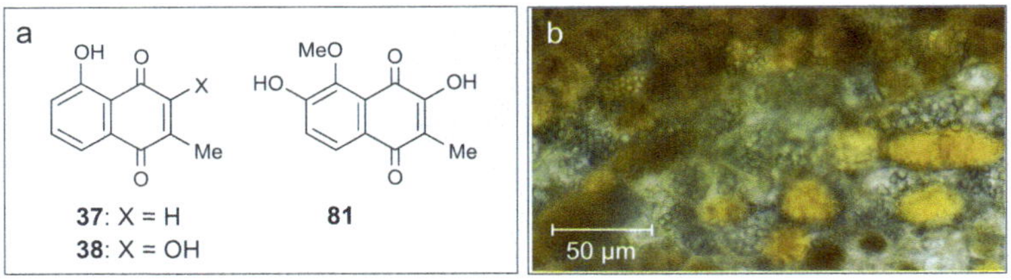

Fig. 38 **a** Oxygenated naphthalene derivatives such as plumbagin (**37**), droserone (**38**), and ancistroquinone A (**81**), isolated from *A. heyneanus*, presumably formed in the plant as phytoalexins; **b** bright-field microscopic image of the root bark showing an accumulation of plumbagin (**37**), which is deeply orange-colored after staining with ammonia vapor (Photograph: H. Schneider)

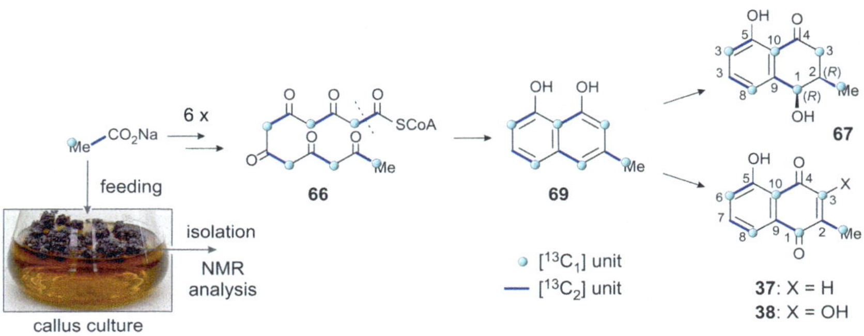

Fig. 39 Formation of the naphthalene-related plant metabolites *cis*-isoshinanolone (**67**), plumbagin (**37**), and droserone (**38**) in suspended callus cultures of *A. heyneanus* from acetate, under stress and/or under normal conditions. Feeding experiments conducted with differently ^{13}C-labeled acetate, [2-^{13}C]- or [$^{13}C_2$]-acetate, establishing the F-type folding mode of the polyketide chain (Photograph: A. Irmer)

The 2D INADEQUATE spectra of **37** and **67** obtained from bond-labeling experiments with [$^{13}C_2$]-acetate visualized the full polyketide chain and proved its folding mode [249]. These findings demonstrated an intact incorporation of C_2 units by the pairwise ^{13}C-^{13}C correlations of the carbon atoms 2-Me/C-2, C-1/C-9, C-8/C-7, C-6/C-5, and C-10/C-4, while C-3 remained isolated due to the loss of its previous neighboring carbon atom by decarboxylation [249]. This result simultaneously proved the intermediate polyketide chain to be folded according to the so-called mode F [350], here observed for the first time for plant secondary metabolites, i.e., with two intact acetate-derived C_2-units in the first ring (referred to as folding mode F, as typical of fungal producers, or, more generally, of eukaryotes [134, 249]).

4.7 A Series of 7,3′-Coupled Naphthylisoquinoline Alkaloids Isolated from Fresh Leaves of *Ancistrocladus heyneanus*

All of the phytochemical studies described in the previous sections focused on the secondary metabolites of the roots and stems of *A. heyneanus*. Isolation work on fresh leaves from individual plants of this Indian liana cultivated in the greenhouse led to the identification of a series of further new naphthylisoquinoline alkaloids by the authors' group. LC-ESI-MS/MS and 1D and 2D NMR clearly suggested the presence of three main naphthylisoquinoline alkaloids with closely related molecular architectures. All were 7,3′-coupled and had the same naphthalene portion, with a hydroxy function at C-4′ and a methoxy group at C-5′ [41].

Two of these alkaloids (peaks **B** and **C**) (Fig. 40) were found to be naphthyldihydroisoquinolines, as was obvious from their typical downfield shifts of the methyl group at C-1 ($\delta_H \sim 2.70$ ppm) and the absence of an H-1 signal. The third compound (peak **A**) (Fig. 40) was deduced to be an N-methylated naphthyltetrahydroisoquinoline, with the signals for the methyl groups at C-1 and C-3 in the isoquinoline ring being upfield-shifted ($\delta_H = 1.5$ and 1.7 ppm), and the signals of the protons H-1 and H-3 resonating at $\delta_H = 4.6$ and 3.3 ppm, respectively. Based on the results from HMQC, HMBC, and NOESY experiments, from an oxidative degradation procedure, and from ECD spectroscopy, the two less polar compounds were readily identified as known alkaloids. One of them (peak **A**) was ancistrotanzanine C (**114**), previously discovered in the leaves of the East African species *A. tanzaniensis* [119], while another (peak **C**) was ancistrocladidine (**3a**), which had previously been isolated from the roots of *A. heyneanus* [28, 29]. The third alkaloid (peak **B**) was found to be new. Its absolute axial configuration was established to be (*M*) by comparison of its ECD spectrum with that of the likewise (*M*)-configured ancistrocladidine (**3a**). Thus, the new compound had the stereostructure **72**. It was the 6-*O*-demethylated analogue of **3a**, and was named ancistroheynine B, in continuation of the names of the new alkaloids detected in the leaves of *A. heyneanus*.

Isolation work on a methanolic leaf extract of *A. heyneanus* afforded a further main metabolite, which was identified as 4′-*O*-demethylancistectorine A_2 (**82**) [70] (see Fig. 41), previously isolated as a minor metabolite from leaves and twigs of the Chinese liana *A. tectorius* [64]. A comparative study using HPLC coupled to UV and ESI-MS on the naphthylisoquinoline profiles of extracts prepared from fresh leaves (see Fig. 41a) and those obtained from air-dried material (see Fig. 41b) revealed large differences concerning the presence and the content of some of the naphthylisoquinoline alkaloids described above. Apparently, the two main metabolites of the fresh leaves, ancistrocladidine (**3a**) and ancistrotanzanine C (**114**), had vanished completely in extracts of air-dried material, whereas ancistroheynine B (**72**) and 4′-*O*-demethylancistectorine A_2 (**82**) were still present, but occurring in distinctly lower quantities than in the fresh leaves from living plants. HPLC analysis of air-dried bulk material of *A. heyneanus* collected

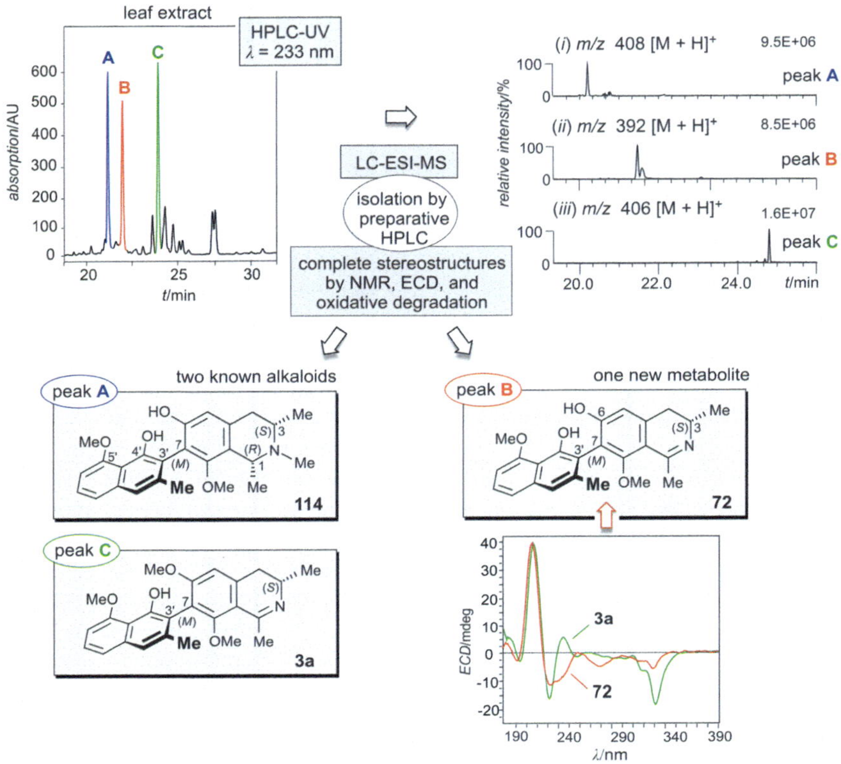

Fig. 40 Isolation of three 7,3′-coupled alkaloids from fresh leaves of *A. heyneanus*: the new naphthyldihydroisoquinoline ancistroheynine B (**72**) (= peak **B**) and the two known [28, 29, 189], structurally related compounds ancistrotanzanine C (**114**) (= peak **A**) and ancistrocladidine (**3a**) (= peak **C**). **a** HPLC-ESI-MS chromatograms of the protonated pseudomolecular ions (i) *m*/*z* 408.2 (for **114**), (ii) *m*/*z* 392.2 (for **72**), and (iii) *m*/*z* 406.2 (for **3a**); **b** absolute axial configuration of ancistroheynine B (**72**), by comparison of its ECD spectrum with that of the related known [28, 29] alkaloid ancistrocladidine (**3a**)

approximately 20 years earlier from the liana in its natural habitat in India and stored since then in the dark at constant temperature and humidity, however, led to nearly identical chromatograms (data not shown) compared to those from dried leaves from plants cultivated in the greenhouse.

LC-MS/MS analysis and co-elution experiments using authentic reference material resulted in the additional identification of the N,C-coupled naphthylisoquinoline alkaloids ancistrocladinium A (**6a**) and B (**14a/b**), in both fresh and air-dried leaves [70] (see Fig. 41). While **6a** and **14a/b** were only minor compounds in the extracts of fresh leaves, they both became the dominating metabolites in air-dried leaves, although their content had not significantly increased. Apparently, ancistrocladinium A (**6a**) and B (**14a/b**) as cationic, and thus electron-poor

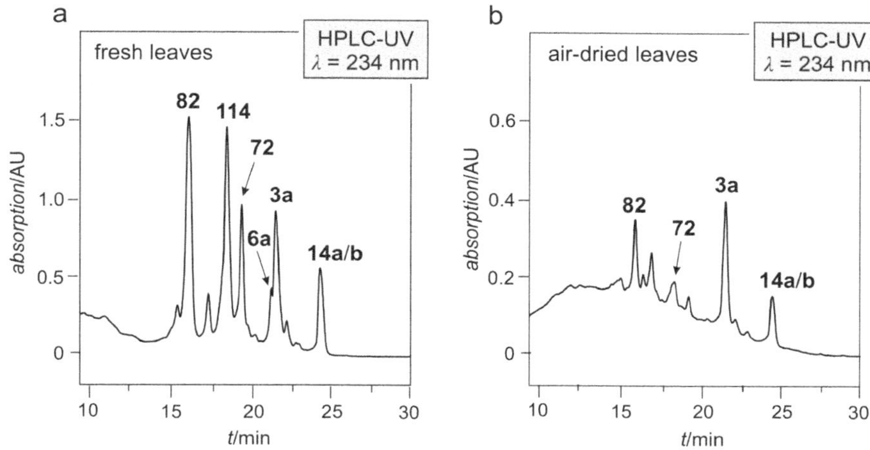

Fig. 41 Comparative HPLC analysis of extracts of **a** fresh leaves and **b** of air-dried leaves of *Ancistrocladus heyneanus*, showing different compositions and quantities of naphthylisoquinoline alkaloids. This permits the identification of the 7,3′-coupled naphthylisoquinolines ancistrotanzanine C (**114**), ancistroheynine B (**72**), and ancistrocladidine (**3a**), the 5,1′-linked 4′-*O*-demethylancistectorine A_2 (**82**), and the N,C-coupled alkaloids ancistrocladinium A (**6a**) and B (**14a/14b**)

alkaloids are more stable toward oxidative conditions, while ancistrocladidine (**3a**) and ancistrotanzanine C (**114**), due to their electron-rich ring systems, easily decompose under the influence of air and light.

4.8 Ancisheynine, a Novel-Type N,C-Coupled Naphthylisoquinolinium Alkaloid: Total Synthesis and Stereoanalysis

Ancisheynine (**78**) was discovered by Yang et al. [40] in 2003, isolated from aerial parts of *A. heyneanus*. It was the first representative of a novel-type subclass of naphthylisoquinoline alkaloids exhibiting an unprecedented iminium-aryl axis. In that original article, the stereochemical properties of ancisheynine were not investigated. The compound was reported to be optically inactive ($[\alpha]_D = 0$), despite the existence of an apparently rotationally hindered heterobiaryl axis [18, 40, 83, 256], thus suggesting that **78** was produced by the plant in a racemic form, i.e., without any enantioselectivity—or that it was not configurationally stable at the axis.

Assisted by the availability of synthetic reference material [256] in the authors' group (see below), HPLC analysis of a leaf extract revealed the presence of ancisheynine (**78**) as a trace compound of *A. heyneanus* eluting in the peak front of the main leaf constituent ancistrotanzanine C (**114**), since the two alkaloids exhibited only slightly different retention times (Fig. 42). For this reason, the

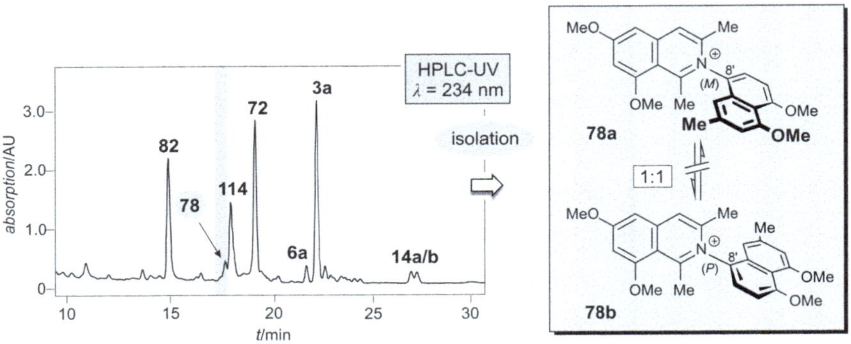

Fig. 42 Ancisheynine (**78**), the first known N,C-coupled alkaloid, discovered in the leaves of *A. heyneanus*, produced by the plant as a racemic metabolite (underlaid in gray in the chromatogram), along with 4′-*O*-demethylancistectorine A_2 (**82**), ancistrotanzanine C (**114**), ancistroheynine B (**72**), ancistrocladinium A (**6a**), ancistrocladidine (**3a**), and ancistrocladinium B (**14a/14b**)

isolation of **78** in a pure form by preparative HPLC was extremely difficult and time-consuming, in particular when trying to obtain milligram quantities of this unusual alkaloid [70, 256], e.g., for stereochemical investigations and for biological testing. Moreover, investigations on fresh leaf material obtained from plants cultivated in the greenhouse clearly showed a strong dependence of the formation of **78** with the season. During the main growth period of the plants (from September to March), **78** was clearly identified in the leaves, albeit in small quantities, while it was not detected at all during the summer period, when the plants passed through a time of rest [70].

For reliable access to ancisheynine (**78**), a convergent synthesis pathway was developed by the authors' group [256], based on the cyclocondensation of the monocyclic diketone **117** with the aminonaphthalene **120** (see Fig. 43).

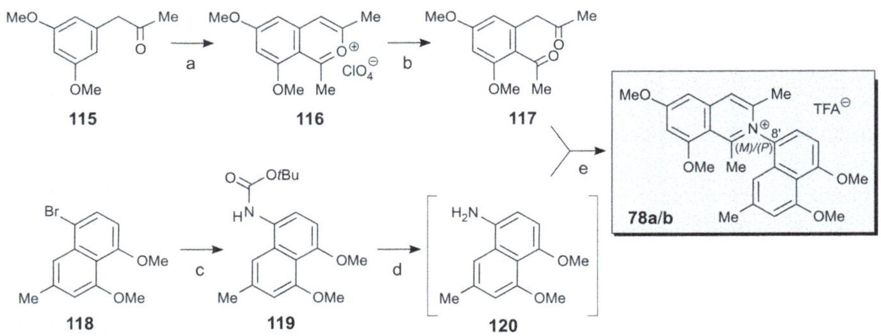

Fig. 43 Total synthesis of ancisheynine (**78a/b**), by cyclocondensation of the diketone **117** with the chemically unstable aminonaphthalene **120**. Reaction conditions: (a) $HClO_4$, Ac_2O, CH_2Cl_2, r. t., 76%; (b) 5 N K_2CO_3, H_2O/Et_2O, r.t., 78%; (c) *tert*-butyl carbamate, $Pd_2(dba)_3$, DAV-Phos, K_2CO_3, toluene, 90°C, 76%; (d) TFA, CH_2Cl_2, r.t., 5 h; (e) addition of **117**, reflux, 60 h, 48%

The required diketone **117** was prepared by treatment of the arylpropanone **115** with acetic anhydride in the presence of perchloric acid to give the benzopyrylium salt **116**, followed by ring opening with aqueous potassium carbonate. For the synthesis of the second building block, the primary amine **120**, the specifically brominated precursor **118** was transformed into the *N*-Boc-protected aminonaphthalene **119** by Buchwald–Hartwig amination, using *tert*-butyl carbamate as an ammonia equivalent. The best results for the synthesis of **119** were obtained by application of $Pd_2(dba)_3$/DAV-Phos as the catalytic system and potassium carbonate as base. Initial attempts directly leading to the free amine **120** with lithium hexamethyldisilazide (LiHMDS) as an ammonia equivalent and with $Pd_2(dba)_3$ and $P(t\text{-}Bu)_3$ as the catalytic system, were not successful, due to the rapid decomposition of the highly electron-rich free amine **120** during work-up [256].

Deprotection of the naphthalene **119** with trifluoroacetic anhydride and in situ trapping of the sensitive aminonaphthalene **120** by addition of the diketone **117**, eventually provided the target molecule ancisheynine (**78**) as its TFA salt in 48% yield [256].

Trifluoroacetate as an electron-poor counteranion was found to drastically enhance the stability of **78**, while the presence of tetrafluoroborate or perchlorate led to its complete decomposition during chromatographic purification. For the natural product ancisheynine itself, no such instability had been observed [18, 40, 83, 256], thus raising the question about its natural counteranion, which is as yet unknown.

The stable axial chirality of ancisheynine (**78**) was evidenced by the resolution of its atropo-enantiomers by HPLC separation of the synthetic material on Chiralcel® OD-RH as a chiral adsorbent, with acetonitrile/water (30:70, *v*/*v*) as the eluent, providing the two expected peaks for the enantiomeric forms of the TFA salt of **78** [18, 83, 256]. The rotational stability of **78** at room temperature was further corroborated by HPLC separation of the enantiomers at a semi-preparative level and re-injection of the enantiomerically pure or enriched samples, revealed their constant enantiomeric ratio over time (see Fig. 44, center), even at room temperature. LC-ECD online analysis demonstrated that these two LC-UV peaks indeed corresponded to the respective atropo-enantiomers of ancisheynine (**78a/b**), due to their opposite-signed single-wavelength LC-ECD signals and the mirror-imaged full LC-ECD spectra recorded in stopped-flow mode [18, 83, 256]. Since a merely empirical assignment of the absolute configuration at the heterobiaryl axis of ancisheynine was impossible due to its novel-type structure, the experimental LC-ECD spectra of the two peaks were compared with those obtained by quantum-chemical ECD calculations for the two atropo-enantiomers of **78**, by means of TDDFT (B3LYP/6-31G*/TZVP). The experimental ECD spectrum of peak **A** matched the curve calculated for the (*M*)-configured atropo-enantiomer **78a**, nearly perfectly (see Fig. 44, left), while the ECD spectrum computed for **78b**, possessing (*P*)-configuration at the biaryl axis, was in agreement with the ECD curve monitored for peak **B** (see Fig. 44, right) [18, 83, 256].

Online LC-ECD analysis of fresh plant material of *A. heyneanus* revealed the two atropo-enantiomers of ancisheynine (**78a** and **78b**) to occur in a 50:50 ratio

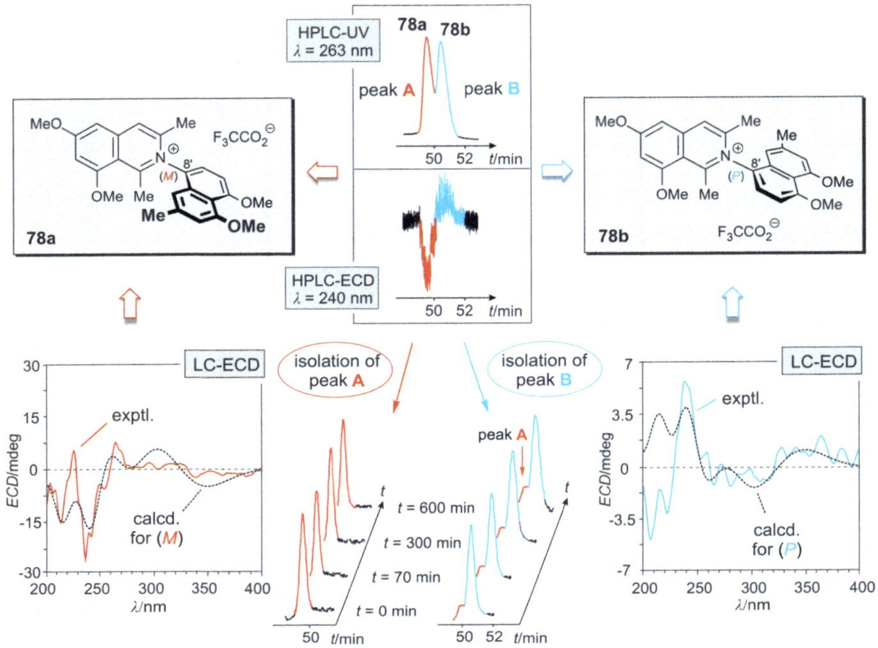

Fig. 44 Stereochemical assignment of the two enantiomers of ancisheynine, **78a** and **78b**, involving their resolution on a chiral HPLC phase (Chiralcel® OD) and comparison of the experimental LC-ECD spectra (monitored in stopped-flow mode) of peak **A** (left) and peak **B** (right) with the spectra quantum-chemically calculated for **78a** and **78b** by means of TDDFT (B3LYP/TZVP). This enables proof of the rotational stability of ancisheynine (**78**) by the constant enantiomeric ratio of enantiomerically enriched samples at r.t. over time

and, thus, in a racemic form. This finding left open whether ancisheynine (**78**) is formed in a non-enantioselective way (possibly in a non-enzymatic, merely chemical coupling step) or whether subsequent racemization of an initially enantiopure or scalemic material occurs in the biological environment [18, 83].

4.9 Simplified N,C-Coupled Arylisoquinolinium Salts as Promising Candidates for the Development of Antiparasitic Agents

Naphthylisoquinoline alkaloids, among them dioncophylline C (**27**) [97, 146, 150, 151, 198–201, 204–207, 220], 5-*epi*-ancistectorine A_2 (**75b**) [61, 191], or ealapasamine C (**121**) [17, 188] (Fig. 45), have been described as promising agents displaying strong anti-infective activities in vitro against the malaria parasite *P. falciparum*, with IC_{50} values toward the chloroquine-sensitive NF54 strain and

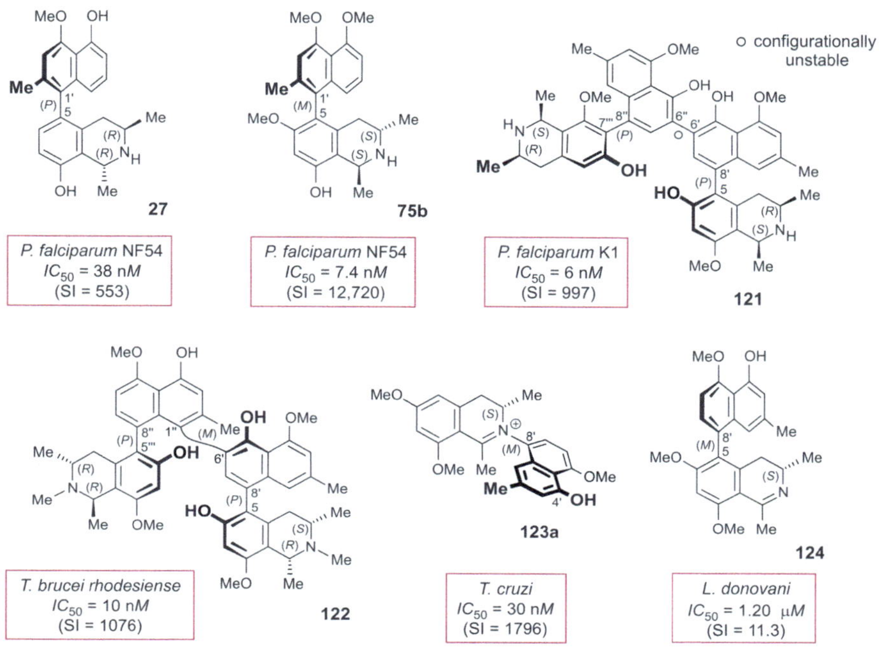

Fig. 45 A selection of naphthylisoquinoline alkaloids with promising activities against *Plasmodium falciparum*, *Trypanosoma brucei rhodesiense*, *T. cruzi*, and *Leishmania donovani*. The selectivity index (SI) is calculated as the ratio of the IC_{50} values for the cytotoxicity of the compounds against rat skeletal myoblast L6 cells relative to the IC_{50} data for the respective parasite

the chloroquine-resistant K1 one in the low nanomolar range [151]. Dioncophylline C (**27**) was found to exert potent inhibitory effects against the rodent parasite *Plasmodium berghei* also in vivo, thus even healing malaria-infected mice [204]. Potent activities of naphthylisoquinoline alkaloids against *Trypanosoma* species causing African sleeping sickness or Chagas' disease, and against *L. donovani*, responsible for visceral leishmaniasis (also known as kala-azar), have also been reported. Examples are the antitrypanosomal compounds mbandakamine B_2 (**122**) [17, 180] and 4′-*O*-demethylancistrocladinium A (**123a**) [18, 47] and the antileishmanial alkaloid ancistectorine D (**124**) [62] (Fig. 45).

These findings triggered further research efforts to explore, more thoroughly, the potency of naphthylisoquinoline alkaloids as a promising new class of antiparasitic compounds, focusing on detailed structure–activity relationship (SAR) studies including selected N,C-coupled analogues.

A more-in-depth evaluation of a broad series of structurally modified compounds related to the naphthylisoquinolinium alkaloid ancisheynine (**78**) was facilitated by the convergent, short, and flexible synthesis route leading to N,C-coupled arylisoquinolinium salts as described in the previous section, giving rise to the

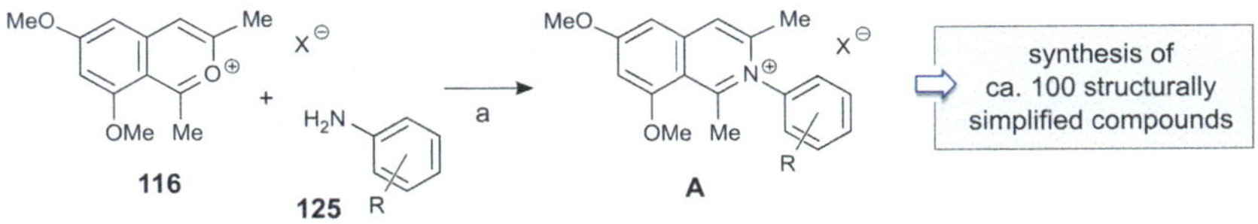

Fig. 46 Rational one-step synthesis of structurally simplified N,C-coupled arylisoquinolinium salts (basic structure **A**) related to ancisheynine (**78**), starting from the benzopyrylium salt **116** and the primary aryl amine **125**. Reagents: (a) HOAc, r.t.

first total synthesis of **78** [256]. This rational one-step synthesis pathway, starting from the benzopyrylium salt **116** and the aryl amine **125** (see Fig. 46), became the basis for the preparation of more than 100 structurally simplified ancisheynine analogues permitting a modular synthesis with a high substrate variability [66, 67, 70].

4.9.1 Inhibitory Activities of Synthetic Arylisoquinolinium Salts Against *Leishmania major*

The natural alkaloid ancisheynine (**78**) displayed only moderate to weak antiparasitic activities [66]. The IC_{50} value of **78** against promastigotes of *L. major*, which is responsible for cutaneous leishmaniasis in the Mediterranean littoral, the Middle East, the Indian Subcontinent, and Central Asia [351–354], was determined to be 30.2 μ*M* (Fig. 47). The structurally simplified, synthetically prepared naphthylisoquinolinium salt **126**, by contrast, was found to strongly inhibit the growth of *L. major* promastigotes by exerting an IC_{50} value of 2.91 μ*M* (Fig. 47), thus being effective at a similar concentration as the leishmanicidal drug amphotericin B (IC_{50}, 2.51 μ*M*), which served as a reference compound. Furthermore, **126** showed a high efficacy in decreasing the infection rate of macrophages as the host cells of *Leishmania* and was found to be effective against the intracellular amastigote form of *L. major* in the low submicromolar range, while toxicity against mammalian cells was observed at concentrations that were significantly higher than those needed to impair parasite replication [66]. Thus, the naphthylisoquinolinium salt **126** was considered to be a promising lead candidate for the development of new leishmanicidal agents. For a more-detailed insight regarding the ability of N,C-coupled compounds of type **A** (see Fig. 46) to inhibit the proliferation of *L. major* promastigotes, structure–activity relationship (SAR) studies were performed, along with investigations on the toxic effects of these structurally simplified analogues of ancisheynine (**78**) toward the macrophage cell line J774.1 [66, 67]. The test results clearly revealed that all of the compounds exhibiting strong to very good antileishmanial activities, with IC_{50} values in the submicromolar range, having lipophilic, weakly electron-donating aryl portions, equipped with either a non-substituted aromatic ring system (like in **126** and **127**) or with phenyl substituents bearing alkyl groups (as in **9** and **128–130**) (Fig. 47). Poor or even no

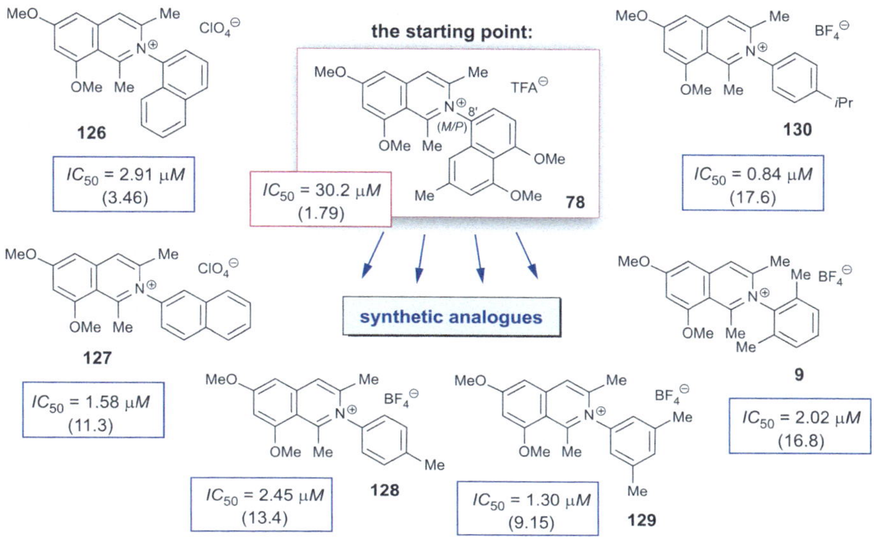

Fig. 47 A selection of synthetically prepared arylisoquinolinium salts, **9** and **126–130**, exerting promising activities against *Leishmania major* in vitro at a similar concentration to the standard, amphotericin B (IC_{50}, 2.51 μ*M*). For comparison, see the natural parent compound ancisheynine (**78**) as the much less active starting point. The indices (numbers in parentheses) were calculated as the ratio of the IC_{50} values for J774.1 macrophages relative to the IC_{50} values for *L. major*

inhibitory effects were observed for compounds possessing electron-withdrawing substituents (like, e.g., a halogen atom) or hydrophilic electron-donating ones (like, e.g., methoxy groups) [67]. Six of the most potent antileishmanial N,C-coupled arylisoquinolinium salts are presented in Fig. 47. Notably, a distinct influence of the steric demand of the aryl portion (naphthyl versus anthracenyl) or of the size of the alkyl groups (methyl versus isopropyl) on the inhibitory activity of the compounds against *L. major* was not observed, but was found to be of major importance for their toxicity. This finding was supported by the likewise strongly enhanced cytotoxicity (index <1) of compounds possessing a higher number of substituents (e.g., trimethyl) or a longer carbon side chain (e.g., *n*-butyl) at the phenyl ring compared to the more moderate toxic effects exerted by monosubstituted arylisoquinolinium salts [67, 70].

In conclusion, among the compounds tested, **9**, **128**, and **130** displayed the most potent inhibitory activities against *L. major* promastigotes, while they impaired the growth of J774.1 macrophages at concentrations that were at least sixfold higher (see Fig. 47). Their efficacy against the pathogen was similar to that of amphotericin B, although they were more toxic against the J774.1 mammalian cell line [66, 67].

Within a life cycle, *Leishmania* sp. parasites exist in two different forms, the extracellular promastigotes and the intracellular amastigotes [355]. Since compounds that show pronounced activities against *L. major* promastigotes do not have to be a priori also effective against intracellular forms of the parasite, freshly isolated peritoneal macrophages were infected with *L. major*, and then treated with increasing concentrations of **126** to analyze the activity of this arylisoquinolinum salt against the *Leishmania* host cells. Compound **126** disturbed the survival of peritoneal macrophages (Fig. 48, green curve) at a concentration (IC_{50} = 15.50 μ*M*) that was similar to that interfering with the proliferation of the J774.1 macrophage cell line (IC_{50} = 10.15 μ*M*), i.e., at a concentration that was about fivefold higher than that needed to impair the growth of the *Leishmania* parasite. The arylisoquinolinium salt **126** caused a dose-dependent decrease in the infection rate of macrophages (Fig. 48, blue curve), with a half-maximal effective concentration (EC_{50} = 0.11 μ*M*) for their survival that was 25-fold lower than that needed to inhibit the growth of *L. major* promastigotes by 50%. The efficacy of **126** in decreasing the infection rate of peritoneal macrophages was similar to that of amphotericin B (EC_{50} = 0.08 μ*M*) [67]. Simultaneous exposure of **126** and the macrophage activator interferon-γ (IFN-γ) to peritoneal macrophages infected with *L. major* hinted at a synergistic effect of the two compounds, lowering the concentration of **126** required for a 50% decrease of the infection rate to 0.03 μ*M* (Fig. 48, red curve) [67]. The reason that low concentrations of **126** strongly decreased the infection rate of the macrophages—but did not kill promastigotes—may be due to a higher sensitivity of amastigotes toward **126** or an enhanced selectivity of this arylisoquinolinium salt against amastigotes. It is also conceivable that **126** enters the host cells by a selective uptake mechanism via the plasma membrane of the macrophages, thus leading to enriched concentrations of the compound in the parasitophorous vacuole within the host cell [67, 68].

The significant decrease in the macrophage infection rate induced even at low concentrations of the naphthylisoquinolinium salt **126** was obviously based on an inhibitory effect of **126** directly targeted toward the intracellular parasites, since the leishmanicidal activity was not associated with the production of nitric oxide or

Fig. 48 Dose–response curves showing the effect of the naphthylisoquinolinium salt **126** on the survival and infection rate of peritoneal macrophages, with the macrophage infected with *L. major* promastigotes for 24 h. After removal of the extracellular parasites, the cells were incubated for a further 48 h in the absence of **126** and in the presence of increasing concentrations of **126**. Macrophage survival and infection rates were normalized to 100%

with the stimulation of cytokine secretion by infected macrophages [67]. Investigations regarding the influence of **126** on tumor necrosis factor alpha (TNF-α) and on pro-inflammatory cytokines like interleukins IL-1β and IL-6, which are secreted in early stages of infections with *L. major* [356–358], likewise showed that the arylisoquinolinium salt did not exert any effect on cytokine regulation. Moreover, no significant changes in the levels of IL-1β, IL-6, and TNF-α were found in the macrophages [67]. However, **126** caused a significant reduction of the release of IL-6 and TNF-α in response to an additional treatment of the infected macrophages with potent stimulators such as INF-γ or bacterial lipopolysaccharide (LPS).

Furthermore, the host defense against *L. major* infections induced by the IL-12-driven expansion of T-helper cells 1 (Th1) was strongly inhibited by **126** [67]. The results indicated that this arylisoquinolinium salt did not induce cytokine secretion by macrophages but rather interfered with their cytokine response to potent activators, thus suggesting that **126** may reinforce processes in the host cells to limit the extent of inflammatory response and to restore homeostasis.

One possible mode of action of N,C-coupled arylisoquinolinium salts may be an apoptosis-like death pathway, similar to the antileishmanicidal agents amphotericin B and miltefosine [67]. Double staining experiments using annexin V and propidium iodide to monitor the incidence of apoptosis revealed a significant increase of stained cells after incubation of *Leishmania*-infected macrophages with **126**, thus corroborating that this arylisoquinolinium salt acted against the intracellular parasites by inducing apoptosis [67].

Morphological changes of *L. major* promastigotes treated with **126** were examined by light and electron microscopy (Fig. 49). They became evident after 2 h, reaching their maximum after 24 h of incubation [68]. While light microscopic images of non-treated cells showed the characteristic promastigote cigar-shaped cell body, with a flagellum longer than the cellular body (Fig. 49a, b), images after treatment with 3 μM of the *N*-arylisoquinolinium salt **126** (i.e., the half-maximum inhibitory concentration) revealed drastic morphological alterations in the parasite structure, varying from discrete changes to total destruction of the promastigotes. The total body and flagellum lengths decreased, the cells became round or ovoid and lost their motility (Fig. 49c, d), and they were distorted by the appearance of a huge vacuole [68].

Electron microscopy on the toxic effects of **126** on the ultrastructure of the parasite demonstrated that the arylisoquinolinium salt dramatically affected the cell compartments (Fig. 49f, g). Untreated promastigotes showed the typical elongated cell body with the anterior flagellum and flagellar pocket, the nucleus, and a single mitochondrion containing the kinetoplast (Fig. 49e), while parasites treated with **126** changed to an ovoid shape, accompanied by the condensation of the nucleus, shrinkage of the kinetoplast, and the appearance of a large cytoplasmic vacuole [68]. The formation of such huge vacuoles that eventually led to the disruption of the promastigotes hinted at an autophage-like or necrotic mechanism of cell death [68].

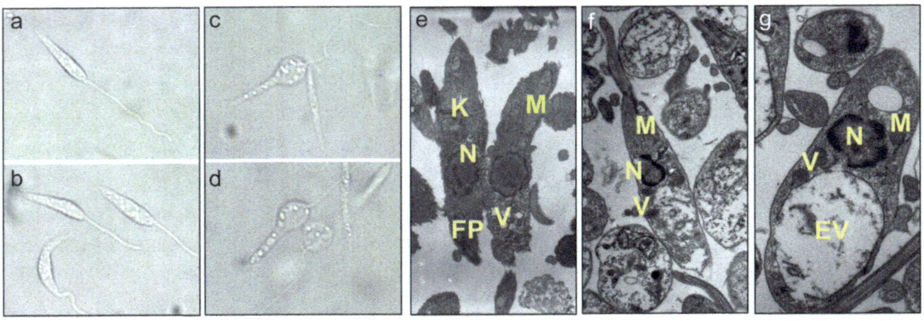

Fig. 49 Dramatic alterations of morphology and size of *L. major* parasites induced by the naphthylisoquinolinium salt **126**. **a–d** Light microscope images of **a**, **b** untreated *L. major* promastigotes and **c**, **d** those after 2 h of incubation with **126**; **e–g** electron microscope images of *L. major* parasites after osmium tetroxide staining (magnification ×10,000), with photograph **e** showing a non-treated parasite, and images **f** and **g** taken of the promastigotes after 6 h of treatment with 3 μ*M* of **126**. Legend: K, kinetoplast; M, mitochondria; N, nucleus; FP, flagellar pocket; V, vacuoles; EV, empty vacuoles

Fluorescence microscopy of *L. major* promastigotes incubated with the autofluorescent arylisoquinolinium salt **126** (3 μ*M*) revealed that the above-described morphological changes strongly correlated with a continuous accumulation and precipitation of **126** in intracellular organelles. Simultaneously, a drastic decrease of concentration of the compound in the culture supernatant was observed, by approximately 80% after only 1 h of application of **126**, essentially remaining at this level until 24 h of incubation [68]. The compound rapidly entered the cells and was already detectable inside *L. major* after 5 min of incubation (Fig. 50). Labeling experiments using Lyso Tracker red as a specific marker for acidic compartments suggested that **126** might specifically accumulate in acidocalciosomes, which are fundamental for parasite nutrition and survival, and might thus constitute the cellular target of **126** [68].

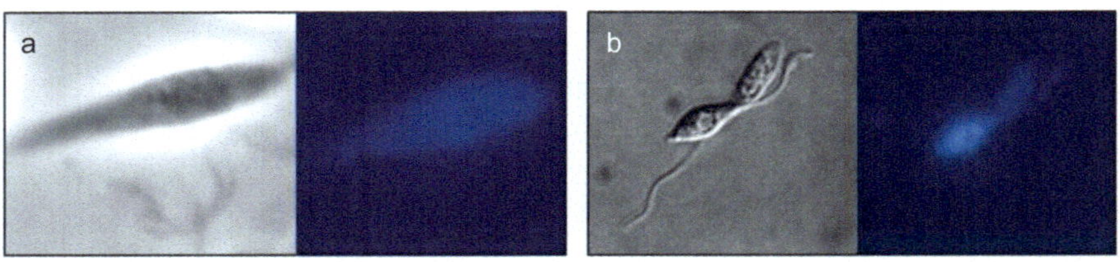

Fig. 50 Light (right) and fluorescence (left) microscope images of *L. major* promastigotes after **a** 5 min and **b** 3 h of incubation with the naphthylisoquinolinium salt **126**. The (blue) autofluorescent compound easily enters the parasite and rapidly accumulates in intracellular compartments

4.9.2 Inhibitory Activities of Synthetic Arylisoquinolinium Salts Against *Trypanosoma brucei brucei*

Trypanosoma brucei spp. are protozoan parasites comprising the animal-infectious form *Trypanosoma brucei brucei* and the two human-pathogenic subspecies *Trypanosoma brucei gambiense* (*T.b.g.*) and *T. brucei rhodesiense* (*T.b.r.*) causing African sleeping sickness [359–361]. Currently, according to the World Health Organization (WHO), about 70 million people in sub-Saharan Africa are at risk of infection [362, 363]. Because of the severe toxic side effects of existing antitrypanosomal drugs and the rapid emergence of chemoresistance [359, 364], the search for a new generation of therapeutic agents remains an urgent need.

Isoquinolinium salts with aryl substituents possessing hydrophilic, electron-donating groups displayed pronounced and selective inhibitory activities against *Trypanosoma brucei brucei* [70]. The sensitivity of the *Trypanosoma* parasites compared to that of the promastigotes of *L. major* was found to be enhanced by a factor of up to 25 toward the N,C-coupled arylisoquinolinium salts evaluated in the screening program. The compounds presented in Fig. 51 exhibited pronounced activities against *T. brucei brucei*, with half-maximum inhibitory concentrations (IC_{50}) in the submicromolar range (down to 0.04 μ*M*, see below). Although the tested compounds were less active than pentamidine (IC_{50}, 0.02 μ*M*), which was used as a positive control, the antitrypanosomal potential of the arylisoquinolinium salts **8** and **131**–**134** (Fig. 51) was high [70]. Derivatives with a methoxy or a trifluoromethyl group at the phenyl ring, *ortho* or *meta* to the heterobiaryl axis, displayed very good antitrypanosomal activities, whereas the corresponding *para*-substituted compounds were only moderately active (not shown, IC_{50}, ~2 μ*M*). Additionally, the monomeric arylisoquinolinium salts **8**, **131**, and **132** (IC_{50}, <0.4 μ*M*), the 'double', bis-cationic analogues **133** (IC_{50}, 0.04 μ*M*) and **134** (IC_{50}, 0.24 μ*M*) were likewise promising drug candidates, with potent inhibitory activities against *T. brucei brucei*, simultaneously showing excellent therapeutic indices (Fig. 51).

Investigations on the mechanism of action revealed remarkable morphological and ultrastructural alterations in *T. brucei brucei* upon treatment with **131** (Fig. 52), varying in intensity—from small modifications to total disruption of the parasite structure [70]. As observed in *L. major* promastigotes (see previous section), the antitrypanosomal activity of arylisoquinolinium salts seems to involve the formation of a huge vacuole, which eventually occupies the whole cell (Fig. 52d), finally promoting cell lysis. Treatment of *T. brucei brucei* with the standard drug amphotericin B led to an impairment of parasite replication and resulted in an induction of dramatic morphological changes (Fig. 52b) in a similar manner as observed for **131** (Fig. 52c) [70]. Thus, an apoptosis-like pathway as known for amphotericin B [365, 366] may be the mechanism of action of arylisoquinolinium salts [70].

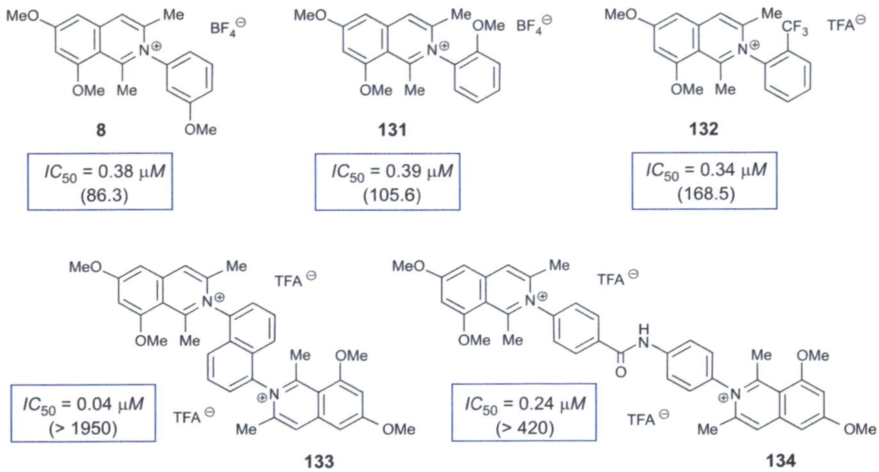

Fig. 51 Synthetically prepared "simple" and "double" arylisoquinolinium salts, **8** and **131–134**, displaying pronounced inhibitory activities against *Trypanosoma brucei brucei* in vitro. The indices (numbers in parentheses) were calculated as the ratio of the IC_{50} values for J774.1 macrophages divided by the IC_{50} values for *T. brucei* brucei. For comparison, pentamidine was used as a standard (IC_{50}, 0.02 μM)

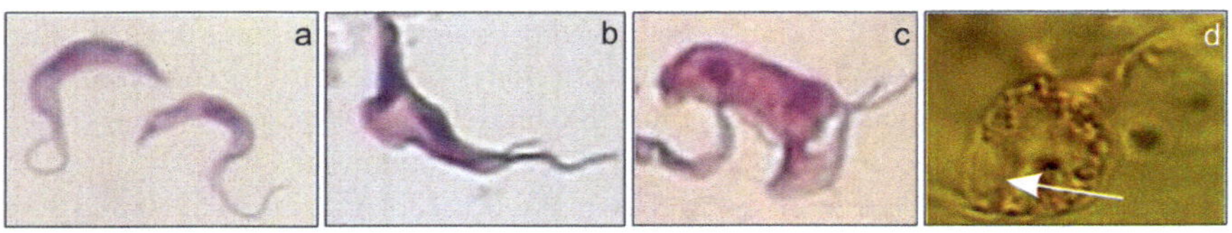

Fig. 52 Morphological changes of *Trypanosoma brucei brucei* induced by **b** amphoterin B and **c** the arylisoquinolinium salt **131**, monitored 2 h after application of the agents, in comparison to **a** untreated parasites (control), and **d** the formation of swollen large vacuoles inside the parasite (marked by a white arrow) after incubation of virulent *T. brucei brucei* isolates with **131**

4.9.3 Antiplasmodial Activities of Synthetic Arylisoquinolinium Salts and the Formation of Complexes with Hemin

Malaria remains one of the most lethal human tropical disorders, with about half of the world's population being at risk of getting infected with *Plasmodium* parasites, the causative agent of this protozoan disease [367–369]. As reported by the World Health Organization (WHO) in 2020 [370], an estimated 1.5 billion malaria cases and 7.6 million malaria deaths occurred worldwide in the period 2000–2019. Over the past two decades, 29 countries accounted for 95% of malaria cases globally, with a share of disease of 51% recorded in West and Central Africa, namely, in Nigeria (27%), the Democratic Republic of the Congo (12%), Uganda (5%),

Mozambique (4%), and Niger (3%) [368, 370]. Malaria is not only a burden to human health, but also strongly affects the socio-economic development of many countries, especially in Africa and Asia. Despite significant progress regarding diagnosis, prevention and treatment, there are still about 110 countries where malaria is endemic, and most have ongoing malaria transmission challenges [371, 372]. Investments in vector control (e.g., in long-lasting insecticidal nets, LLINs, against mosquitoes) and diagnostic products, basic research and drug development to combat malaria are urgently required, in particular with respect to the rapid emergence of resistance of *Plasmodium* parasites against frequently used agents such as chloroquine or artemisinin [373–375].

N,C-Coupled arylisoquinolinium salts equipped with aryl moieties bearing electron-withdrawing substituents such as halogens or amide functions exhibited strong and selective inhibitory effects against the chloroquine-resistant K1 strain of *P. falciparum* (Fig. 53) [70]. The concentrations needed to achieve a half-maximum inhibition of the growth of the parasites (IC_{50} value) were in the low submicromolar range and comparable to those of chloroquine (IC_{50} = 0.19 μ*M*) or even lower by up to one order of magnitude. The most potent antiplasmodial compounds had an aryl portion with substituents that were located *para* or *ortho* to the heterobiaryl axis, e.g., the 4′-fluoro- and 3′-bromophenyl derivatives **135** and **137** (Fig. 53) [70]. Also remarkable were the pronounced inhibitory activities of the acetamide **141** and the 6'-benzothiazole analogue **10** (Fig. 54), which displayed excellent IC_{50} values of 0.21 μ*M* (for **141**) and even 0.04 μ*M* (for **10**), while exerting only weak cytotoxic effects against rat skeletal myoblast (L6) mouse cells [70]. Interestingly, the regioisomeric 2′-benzothiazole derivative (not shown), like most of the other

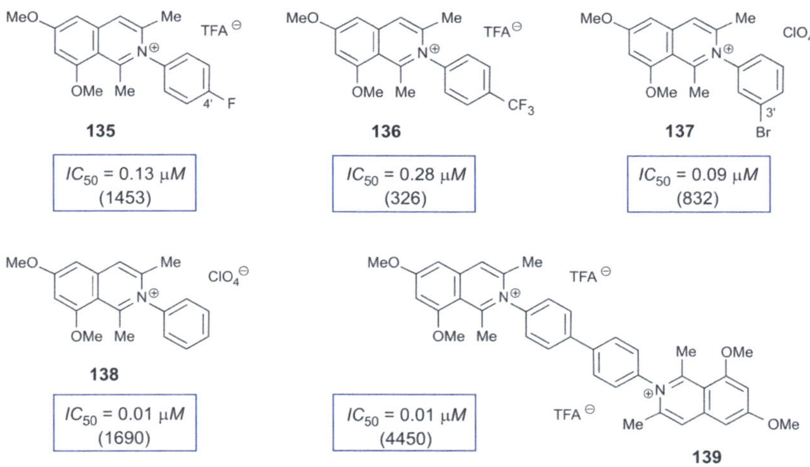

Fig. 53 A series of arylisoquinolinium salts, **135–139**, showing promising inhibitory activities in the submicromolar range against *Plasmodium falciparum* (K1 strain) in vitro. The indices (numbers in parentheses) were calculated as the ratio of the IC_{50} values for mammalian host cells (rat skeletal myoblast L6 cells) divided by the IC_{50} values for *P. falciparum*

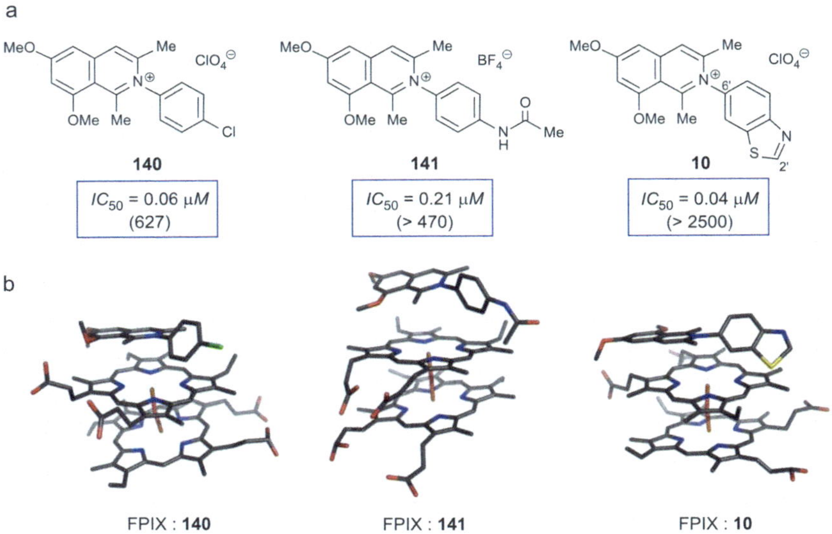

Fig. 54 **a** Three synthetic arylisoquinolinium salts, **140**, **141**, and **10**, displaying pronounced inhibitory activities against *Plasmodium falciparum* (K1 strain) in vitro. The indices (numbers in parentheses) were calculated as the IC_{50} values for mammalian host cells (rat skeletal myoblast L6 cells) divided by the IC_{50} values for *P. falciparum*; **b** calculated structural models of the complexes of **140**, **141**, and **10** with the presumed target ferriprotoporphyrin IX heme (FPIX)

compounds tested that were equipped with a heteroaromatic portion, displayed no inhibitory activity at all, thereby suggesting that the linkage of the isoquinoline subunit to the phenyl part and not directly to the heteroaromatic ring system is important for the antiplasmodial effects exerted by N,C-coupled arylisoquinolinium salts [70]. In addition to the monomeric unsubstituted *N*-phenylisoquinolinium salt **138**, its dimer **139** also displayed potent inhibitory activities against *P. falciparum*, simultaneously exerting excellent therapeutic indices (Fig. 53) [70]. The fact that compounds like **10** (SI >2500) and dimer **139** (SI 4450) possessing large but more hydrophilic aryl moieties were not cytotoxic, indicated that the lipophilicity, more than the bulkiness of the aryl portion, seems to be of particular importance for the (undesired) toxicity of *N*-arylisoquinolinium salts [70].

During the blood stage of the malaria infection, *Plasmodium* parasites, living in erythrocytes, digest hemoglobin inside their vacuoles, which gives rise to the production of a soluble Fe(II) complex, named ferriprotoporphyrin IX heme (FPIX) [376–378]. In order to protect themselves against this toxic waste product, the parasites convert FPIX into its crystalline form by polymerization to give hemozoin, which is non-toxic to the pathogens. This compound appears as dark black spots ('malaria pigment') in the red blood cells of infected patients and is toxic for the human body, causing severe fever when it gets released from ruptured erythrocytes [378, 379].

In aqueous solution, FPIX predominantly forms a μ-oxo dimer, which consists of two FPIX monomers bridged by a Fe–O–Fe bond [380]. Most commonly used antimalarial drugs such as the aminoquinolines chloroquine, mefloquine, or primaquine easily enter the digestive vacuoles of the parasites, and inhibit hemozoin crystallization by forming another dimeric heme complex (FPIX-μ-oxo dimer: drug molecule), which is highly toxic to the parasites. The increased load of FPIX and heme complex eventually leads to autocatalytic death of the parasites [381, 382].

N,C-Coupled arylisoquinolinium salts likewise interfere with hemozoin formation inside the digestive vacuole of *Plasmodium* parasites in a manner comparable to the inhibitory mode of action displayed by the aminoquinolines [70]. The distribution coefficients log*D* (*n*-octanol/aqueous phosphate buffer) determined at the pH values 7.4 (as prevalent in cytosol) and 5.2 (as existing in digestive vacuoles of the parasites) confirmed that **141** and **10** (Fig. 54) accumulate in the vacuoles of *P. falciparum*. Because of their different lipophilicity under neutral (**141**: log*D* = 1.29; **10**: log*D* = 2.47) and acidic conditions (**141**: log*D* = −1.06; **10**: log*D* = −0.250), the compounds are rapidly enriched as a consequence of the pH gradient differences between the red blood cells and the digestive vacuoles of the parasite—similar to chloroquine [log*D* (pH 7.4) = 2.95, log*D* (pH 5.2) = −1.20] [70]. In addition, the *N*-arylisoquinolinium salts, like most antimalarial drugs, inhibit the formation of hemozoin, with IC_{50} values (**140**: 2.49 μ*M*, **141**: 1.67 μ*M*, and **10**: 2.14 μ*M*) comparable to that of chloroquine (IC_{50} = 1.191 μ*M*). This effect was only found for antiplasmodial compounds, while inactive arylisoquinolines (IC_{50} >10 μ*M*) did not induce any inhibition [70].

Complex formation between FPIX and the N,C-coupled compounds **140**, **141**, and **10** was investigated more thoroughly by UV spectroscopy. Job's plot analysis of the changes in the absorbance at 402 nm showed that a complex was formed with a molar ratio of two FPIX molecules per one *N*-arylisoquinoline molecule (Fig. 54) [70]. For each of the compounds, such a 2:1 stoichiometry of the FPIX: arylisoquinoline complex was found for all of the three pH levels investigated (pH 5.6, 6.0, and 7.4). The complex structures were calculated with molecular dynamics (MD) simulations using intermolecular distance restraints between the respective *N*-arylisoquinoline and the iron center in FPIX, determined from NMR paramagnetic relaxation [70]. The structural models thus obtained showed a high similarity to complexes formed by FPIX and antimalarial agents of the quinolone family [70, 205]. As presented in Fig. 54, the calculated *N*-arylisoquinoline:FPIX-μ-oxo dimer complexes were located above the central iron atom in FPIX, but did not interact directly with the iron, and the isoquinoline portions were tilted with respect to the FPIX tetrapyrrole moieties. No covalent interactions or hydrogen bonding effects were observed [70, 205]. The structures gave no evidence of an electrostatic interaction of the positively charged nitrogen in the N,C-coupled compounds with the negatively charged propionic acid side chains in FPIX. This hinted at a non-covalent stabilization of the complex, probably through π–π interactions between the isoquinoline portions of **140**, **141**, and **10**, which were almost parallel to one another, and the FPIX tetrapyrrole units [70, 205].

4.9.4 Antibiotic Effects of Dimeric Isoquinolinium Salts on *Staphylococcus aureus*, *Staphylococcus epidermidis* and Inhibitory Effects Against *Candida albicans*

The progression of multi-resistance to antibiotic treatment of infectious diseases has become a global health threat [383, 384], warranting intense studies aiming at the identification of new antibacterial lead compounds to enable efficient medical therapies.

Staphylococcus epidermidis can cause severe wound infections and inflammations. It is a most common bacterium associated with biofilm formation on medical devices. Bacteria residing inside the biofilm can survive antibiotic therapy, thus massively contributing to the spread of resistance and to treatment failure [385, 386]. Similar to *S. epidermidis*, *Staphylococcus aureus* is also a highly dangerous—and common—staphylococcal bacterium [387, 388]. These pathogens often cause

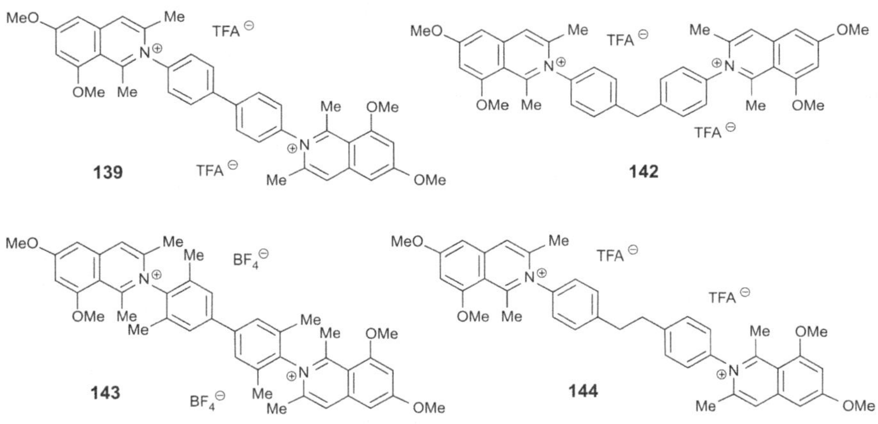

139 **142** **143** **144**

	MIC/μ*M*			
	139	**142**	**143**	**144**
Staphylococcus aureus 325	0.63	2.50	0.31	0.63
Staphylococcus epidermidis RP62A	0.63	2.50	0.16	0.31
Candida albicans	1.25	2.50	5.00	1.25

Fig. 55 Antibiotic effects of the dimeric arylisoquinolinium salts **139** and **142–144** against the bacteria *Staphylococcus aureus* and *S. epidermidis* and against the opportunistic fungal pathogen *Candida albicans*. The antimicrobial activities are given as minimal inhibitory concentrations (*MIC*s), i.e., the lowest concentrations that inhibit the visible in vitro growth of the respective microorganism

skin infections but are also responsible for pneumonia and heart valve inflammation [386, 388]. Systemic fungal infections, in particular those caused by the yeast *Candida albicans*, have emerged as an important cause of morbidity and mortality in immune-compromised patients after having undergone a surgery, cancer chemotherapy, or organ or stem cell transplantation [389, 390]. The ability of *C. albicans* to form biofilms on implanted medical devices contributes to its virulent character [391, 392].

Dimeric arylisoquinolinium salts such as the benzidine-derived bis-cationic compound **139** and related dimers, among them **142–144** (Fig. 55), showed excellent growth-inhibitory activities against pathogenic Gram-positive *Staphylococci* bacteria, and against the opportunistic yeast *C. albicans* [65, 70]. Monomeric analogues like the phenylisoquinolinium salts **138** (Fig. 53) and **129** (Fig. 47), by contrast, did not exert any antimicrobial activity at all. The antibacterial effects of the dimeric arylisoquinolinium salts were particularly pronounced against the two biofilm-producing strains *S. aureus* 325 (clinical isolate from blood culture, wild type) and *S. epidermidis* RP62A (multi-resistant reference strain). All tested compounds showed little or no inhibitory activities against enterobacteria such as *Escherichia coli* and against pseudomonads such as *Pseudomonas aeruginosa* [65, 70].

In addition to its strong growth-retarding effect, the benzidine-derived bis-cation **139** significantly inhibited the ability of *S. epidermidis* to form biofilms. In the presence of 0.63 μ*M* of **139**, which corresponded exactly to its minimal inhibitory concentration (*MIC*), a drastic reduction of biofilm formation, by approximately 90%, was observed as compared to untreated control cultures. This effect decreased in a dose-dependent manner (see Fig. 56) [70]. Interestingly, biofilm-inhibiting activity also occurred at concentrations that were still too low to influence the growth of the bacteria. Thus, in addition to the growth-inhibiting effect described above, the dimeric isoquinolinium salt **139** showed an anti-biofilm effect on *S. epidermidis* independent from the mere antibiotic effect [70].

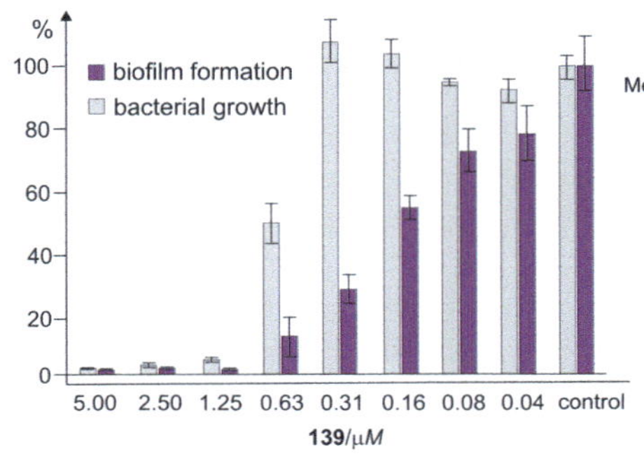

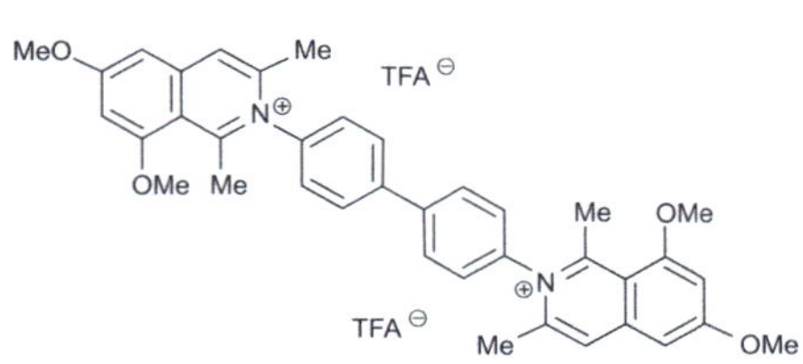

139

Fig. 56 Inhibition of biofilm formation for the methicillin-resistant human pathogen *Staphylococcus epidermidis* (strain RP62A) in response to the presence of a subminimal (sub-*MIC* <0.31 μ*M*) and minimal inhibitor

Investigations on the metabolic activity in combination with gene expression data analysis and in silico pathway modeling permitted to study more closely the effects of **139** on *S. aureus*, *S. epidermidis*, and human cell lines [65]. The clinical *S. epidermidis* strain RP62A and the cell line *S. aureus* USA 300 were grown in the presence of **139** in two different concentrations, at 0.16 μ*M* (i.e., a quarter of the *MIC* value of 0.63 μ*M*) and at 1.25 μ*M* (i.e., twice the *MIC* value). At the lower concentration, **139** was bacteriostatic, while it was bactericidal at the higher concentration. The result suggested that **139** targeted the nucleotide and energy metabolism of *S. epidermidis* and *S. aureus* [65]. The data revealed a complex pattern of cell alterations upon administration of **139**. The arylisoquinolinium salt significantly disturbed the NAD(H) pool and the multi-enzyme complexes of oxidative phosphorylation by directly inhibiting NADH oxidation by complex I of the respiratory chain. Pathway modeling also indicated down-regulation of the purine metabolism and revealed a strong impact of **139** on pyrimidine metabolism and on other nucleotide-producing enzymes [65]. The expression of genes encoding proteins involved in the transport of macromolecules was significantly affected by **139**, among them the ATP-binding cassette (ABC) transporter, peptide and choline transporters, and metabolic enzymes of carbohydrate pathways [65]. Experiments in human cell lines (J774.1 macrophages, kidney epithelial 293 T cells) supported the conclusions from bioinformatic pathway modeling revealing that **139** exerted a low cytotoxicity [65], thus suggesting that this *N*-arylisoquinolinium salt might be a promising new lead for the development of agents against infections and inflammation caused by *Staphylococcus* bacteria.

5 Full Absolute Stereostructures of Naphthylisoquinoline Alkaloids Directly from Crude Extracts: Characterization of New Metabolites from *Ancistrocladus griffithii* by the HPLC-MS/MS-NMR-ECD Triad

Ancistrocladus griffithii Planch. widely occurs on river banks and in submontane evergreen forests (up to 350 m above sea level) mainly along the Tenasserim Hills and Phuket mountain range, from Myanmar to Thailand, further extending to Cambodia and Vietnam [21, 270, 278–280, 286, 298]. The taxonomic classification of the Asian *Ancistrocladus* plants presented by Taylor et al. [21] in 2005 accepted *A. griffithii* as a valid species. Investigations by the authors' group on the naphthylisoquinoline pattern of this previously phytochemically unexplored liana clearly evidenced a species-specific production of typical Ancistrocladaceae-type alkaloids, i.e., possessing an oxygen function at C-6 and displaying the (*S*)-configuration at C-3. Among them were three new monomeric compounds, ancistrogriffines A (**145**), B (**146**) and C (**147**), which have as yet only been detected in *A. griffithii* [43]. The main metabolite of leaf and twig extracts was found to be an unprecedented twofold 7,8′-coupled naphthylisoquinoline dimer, named ancistrogriffithine

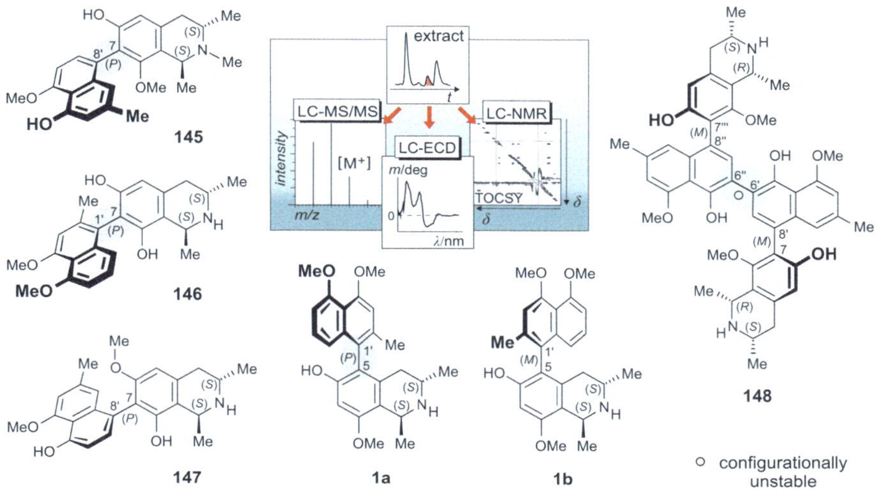

Fig. 57 Ancistrogriffines A (**145**), B (**146**), and C (**147**), and the dimer ancistrogriffithine A (**148**), four new naphthylisoquinoline alkaloids from *Ancistrocladus griffithii* that have been structurally elucidated and stereochemically fully assigned directly from crude plant extracts, by LC hyphenation with MS/MS, NMR, and ECD, along with two further, known [1–5, 23–25] atropo-diastereomeric alkaloids, ancistrocladine (**1a**) and hamatine (**1b**). The different biaryl coupling types found in the plant are underlaid in gray, and for the central axis of **148** in light blue

A (**148**) [42, 82]. This unique alkaloid profile thus strongly supports the treatment of *A. griffithii* as an individual *Ancistrocladus* taxon, which is in agreement with results from genetic studies [36] revealing a large distance of intersimple sequence repeat (ISSR) fingerprints and internal transcribed spacers (ITS) compared to the other *Ancistrocladus* plants from all over Southeast Asia.

Discovery and structure elucidation of the monomeric alkaloids ancistrogriffines A–C (**145–147**) and of the dimer ancistrogriffithine A (**148**) (see Fig. 57) was achieved online directly from crude plant extracts, without previous isolation of the alkaloids. Using the analytical triad HPLC-MS/MS-NMR-ECD [42, 43, 82] in combination with the microscale ruthenium-mediated oxidative degradation procedure [79, 80], both developed by the authors' group, the configuration(s) at the stereogenic center(s) in the isoquinoline portion and at the stereoarray at the biaryl axis were determined.

5.1 Hyphenation of HPLC with UV, MS, NMR, and ECD

Hyphenation of efficient separation techniques like high-performance liquid chromatography (HPLC) with ultraviolet-visible spectroscopy (HPLC-UV-vis), tandem mass spectrometry (HPLC-MS/MS) [17, 18, 42, 75, 81, 82, 159, 175–181, 257], and nuclear magnetic resonance spectroscopy (HPLC-NMR) [42, 74, 75, 81, 82] is a useful strategy for a structure-based metabolic screening aiming at a rapid characterization of the naphthylisoquinoline pattern of an *Ancistrocladus* species [41–43, 47, 74, 75, 81, 82, 122–124, 175–181, 257]. It is a successful general approach for the facile acquisition of plant secondary metabolite profiles and a robust basis for a reliable decision of which extracts should have the highest priority to be analyzed [82]. Coupling HPLC with MS/MS and NMR is not only a useful method for fast dereplication of already known alkaloids and related compounds [41, 42, 47, 81, 82, 122, 159, 257], but also a valuable analytical tool for the elucidation of the constitutions and relative configurations of as yet unknown natural products [41–43, 47, 75, 81, 82]. A further important aspect underlining the promising potential of this online approach is due to the fact that only small quantities of the (sometimes rare) biological material are needed to search for new compounds directly in crude extracts.

5.1.1 Characterization of Mono- and Dimeric Naphthylisoquinoline Alkaloids by HPLC-ESI-MS/MS

The hyphenation of HPLC with tandem mass spectrometry (MS/MS) in combination with soft-ionization techniques such as electrospray ionization (ESI) is a versatile analytical approach for the sensitive detection of naphthylisoquinoline alkaloids in crude plant extracts, even down to ng quantities. Low-energy collision-induced dissociation (CID) yields characteristic full ion spectra by scanning a mass range from 20 to 850 amu giving rise to the protonated molecules of mono- and dimeric naphthylisoquinoline alkaloids and to mass ions resulting from characteristic fragmentation reactions [17, 18, 31, 41, 122, 123, 175–181, 257].

Initial experiments revealed the electrospray process in the positive mode to ionize effectively monomeric naphthylisoquinoline alkaloids such as ancistrocladine (**1a**) (see Fig. 58a), affording mass spectra corresponding to monoprotonated molecules $[M + H]^+$ as the most abundant ions [17, 18, 42, 75, 81, 257]. Furthermore, product ions apparently originating from the loss of ammonia (−17 amu) can be observed, besides the formation of the most characteristic product ions resulting from the neutral loss of a C_2H_5N unit (−43 amu) due to retro-Diels–Alder fragmentation [42, 43, 75, 257]. The loss of another neutral retro-Diels–Alder fragment C_3H_7N (−57 amu), i.e., larger by a CH_2 unit, was likewise found to be the main cleavage pathway for N-methylated tetrahydroisoquinolines e.g., ancistrogriffine A (**145**) (see Fig. 66) [43].

HPLC-ESI-MS/MS experiments on naphthyldihydroisoquinolines such as ancistrocladisine (**2a**) (see Fig. 58b) deliver substantially different mass spectra as compared to those that are obtained from naphthyltetrahydroisoquinolines under the same CID conditions. Obviously, due to the more stable C=N double bond, dihydroisoquinolines show a reduced tendency for fragmentation reactions. Besides the monoprotonated molecules $[M + H]^+$ as the most abundant ions, product ions such as $[M + H - CH_3]^+$, resulting from the cleavage of a vinylic methyl group (−15 amu), and $[M + H - CH_3OH]^+$, indicating the loss of methanol (−32 amu), were monitored.

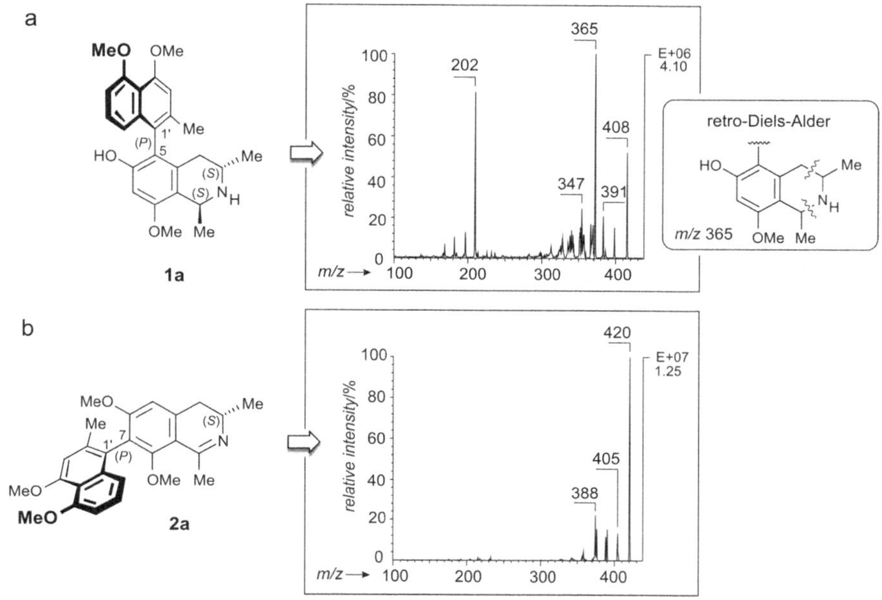

Fig. 58 HPLC-MS/MS spectra obtained by electrospray ionization (ESI, 30 eV) with fragmentation ion products typical of naphthyltetraisoquinolines and dihydroisoquinolines, as, for example, monitored for **a** ancistrocladine (**1a**) and **b** ancistrocladisine (**2a**); the precursor ions are *m/z* 408.3 for **1a** and *m/z* 420.3 for **2a**. Further characteristic ions are the retro-Diels–Alder fragment *m/z* 365.1 $[M + H - C_2H_5N]^+$ and the naphthalene fragment *m/z* 201.9 $[C_{13}H_{13}O_2 + H]^+$ in the case of ancistrocladine (**1a**), and the respective ions *m/z* 405.8 $[M + H - CH_3]^+$ and *m/z* 388.2 $[M + H - CH_3OH]^+$ for ancistrocladisine (**2a**)

A further characteristic fragment of naphthyltetrahydroisoquinoline alkaloids is the product ion *m/z* 202 (see Fig. 58a), which results from the cleavage of the central C–C bond between the naphthalene and the isoquinoline part and the simultaneous transfer of a positive charge to the naphthalene moiety. Thus, the occurrence of the product ion *m/z* 202 is typical of naphthylisoquinoline alkaloids with a naphthalene portion possessing two methoxy groups, e.g., ancistrocladine (**1a**) (Fig. 58a), and a fragment ion of *m/z* 188 hints at the presence of a tetrahydroisoquinoline with a naphthalene substituent displaying only one methoxy function at C-4′ or at C-5′ (while the second oxygen function is a free hydroxy group).

Monomeric alkaloids exclusively yield protonated molecules $[M + H]^+$. Naphthylisoquinoline dimers, by contrast, such as ancistrogriffithine A (**148**) (see Fig. 63) mainly exhibit doubly protonated, and thus bis-cationic, molecular ions $[M + 2H]^{2+}$, consequently appearing in the range typical of monomeric compounds, besides the formation of the monoprotonated species $[M + H]^+$ in the expected mass range of dimers [17, 31, 122–124, 175–181, 257]. The mass peak $[M + 2H]^{2+}$, hinting at a dimer, can be identified based on the characteristic difference of 0.5 amu between the monoisotopic ions $[[^{12}C]dimer + 2H]^{2+}$ and $[[^{13}C]dimer + 2H]^{2+}$ [17, 257].

For the doubly protonated precursor ions of naphthylisoquinoline dimers, two major reaction pathways can be observed. The loss of a fragment with formally 8.5 amu, which can be explained by the neutral cleavage of one ammonium unit per doubly charged molecule, gives the product ion $[M + 2H - NH_3]^{2+}$. Cleavage of 21.5 amu arising from a retro-Diels–Alder fragmentation leads to the formation of the doubly protonated ion $[M + 2H - C_2H_5N]^{2+}$ [17, 257].

As discussed in more detail in the following sections, coupling of HPLC with electrospray ionization mass spectrometry can give valuable insight into the presence of structurally unique mono- and dimeric naphthylisoquinoline alkaloids in twig and leaf extracts of *A. griffithii*, with the most prominent example being the discovery of the dimer ancistrogriffithine A (**148**) (see Sect. 5.3).

5.1.2 Constitution and Relative Configuration of Naphthylisoquinoline Alkaloids by HPLC-NMR

NMR methods are well suited for the reliable elucidation of the constitution and the relative configuration at the stereogenic centers and the biaryl axis (relative to the centers) of a new naphthylisoquinoline alkaloid. All spin systems are usually well separated, and the chemical shifts already give a rapid first insight informing on the coupling position, the O- and N-methylation pattern, and the saturation degree of the isoquinoline moiety. It can even be possible to distinguish between atropo-diastereomeric naphthylisoquinoline alkaloids. Thus, combining well-established strategies of modern NMR spectroscopy, such as ^{1}H measurements and 2D pulse sequences, e.g., total-correlation spectroscopy (TOCSY), nuclear Overhauser enhancement spectroscopy (NOESY), or rotating-frame Overhauser enhancement spectroscopy (ROESY) with the very good chromatographic resolution of the HPLC-NMR experiment, high-quality spectral data can be obtained. The potential of HPLC-NMR for the investigation and structural elucidation of novel-type natural products has strongly been extended by the development of powerful solvent-suppression schemes, e.g., WET (Water Suppression Enhanced by T_1 Relaxation) [393] or double presaturation [394]. Thus, extracts can routinely be analyzed in the on-flow mode for a continuous acquisition of ^{1}H NMR spectral data during the screening process. The stopped-flow mode can be applied for 1D and 2D experiments, in order to identify spatial proximities of proton spins, and to therefore establish the constitutions and relative configurations in the isoquinoline moieties, online, right from the peak in the chromatogram.

The first successful adaptation of a 2D ROESY sequence to stopped-flow HPLC-NMR experiments for the online analysis of leaf extracts of *A. guineënsis* (from Nigeria) by the authors' group provided the full constitutions of two new 5,8′-coupled naphthylisoquinoline alkaloids, ancistroguineines A (**149**) and B (**150**). In addition, HPLC-NMR-WET-ROESY permitted a reliable attribution of the relative configurations of **149** and **150** at the stereogenic centers C-1 and C-3 in the tetrahydroisoquinoline portions (see Fig. 59), without the necessity of first

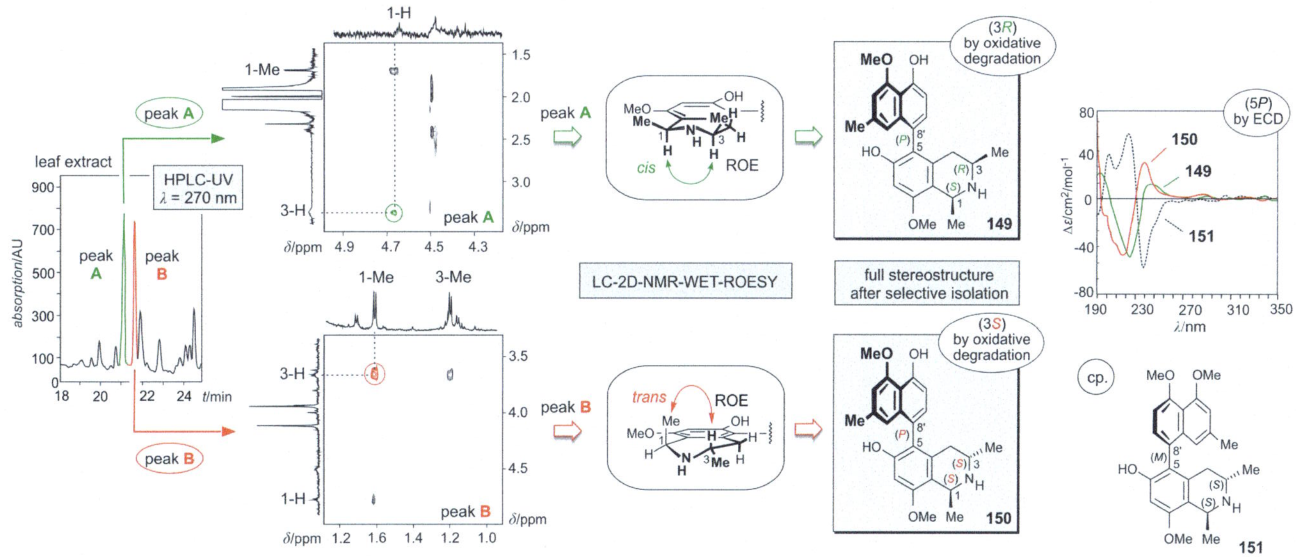

Fig. 59 First application of LC-ROESY coupling in natural product analysis: online elucidation of the relative *cis*- and *trans*-arrays of the methyl groups in the isoquinoline moieties of the diastereomeric alkaloids ancistroguineine A (**150**) (= peak **B**) and B (**149**) (= peak **A**), directly from the leaf extract of *A. guineënsis*. The absolute configurations were determined offline, after isolation of **149** and **150** in a pure form, by oxidative degradation and by ECD spectroscopy and comparison with the ECD spectrum of the structurally related, but (*M*)-configured alkaloid ancistrobrevine B (**151**) [192]

isolating the compounds [75]. This was, at the same time, the very first application of HPLC-ROESY coupling in natural product analysis.

600-MHz-HPLC-^{1}H-NMR measurements and 2D TOCSY correlation experiments in the stopped-flow mode on a crude leaf extract of *A. guineënsis* revealed the presence of two predominantly occurring diastereomeric naphthylisoquinoline alkaloids, possessing the same constitution, but differing in the relative configuration at C-1 versus C-3. HPLC-MS/MS coupling investigations on the two compounds delivered nearly identical spectra and showed the two isomers to have protonated molecular ions $[M + H]^+$ of *m/z* 394.4 [75].

2D HPLC-NMR-ROESY experiments, applying a conventional ROESY sequence with a double-presaturation block during the relaxation delay, evidenced the more rapidly eluting diasteromer (= peak **A**) to be *cis*-configured at C-1 versus C-3 in the isoquinoline part, as obvious from a ROESY correlation between H-1 and H-3. The relative configuration of the two stereogenic centers in the tetrahydroisoquinoline portion of the second alkaloid (= peak **B**) was established to be *trans*, from a ROESY interaction between H-3 and Me-1 (see Fig. 59) [75].

The results of the facile monitoring of the metabolite profile of *A. guineënsis* in combination with the acquisition of important structural information paved the way for a selective isolation of these two new naphthylisoquinoline alkaloids. With the pure compounds in hands, the absolute configurations of the two diastereomers at C-1 and C-3 and at the biaryl axis were determined. Ruthenium-mediated oxidative degradation of the first alkaloid (= peak **A**) afforded the (*R*)-enantiomer of 3-aminobutyric acid (derived from C-3). This revealed C-3 to be (*R*)-configured and, given the aforementioned relative *cis*-configuration, C-1 had to be (*S*) [75, 159]. The absolute configurations of the two stereogenic centers in the isoquinoline portion of the more slowly eluting alkaloid (= peak **B**), again deduced from the oxidative degradation procedure, evidenced (*S*), both for C-1 and C-3 [75, 159].

The absolute axial configuration of these two new naphthylisoquinolines was established to be (*P*) due to their virtually opposite ECD spectra compared to that of the known [192], structurally closely related and likewise 5,8′-coupled, but (*M*)-configured, ancistrobrevine B (**151**) from the West African liana *A. abbreviatus*. The two new alkaloids thus had the full absolute stereostructures **149** (= peak **A**) and **150** (= peak **B**) as shown in Fig. 59. In view of their occurrence as the main metabolites of the Nigerian liana *A. guineënsis*, they were named ancistroguineine A (**150**) and ancistroguineine B (**149**) [75, 159].

5.1.3 Online Assignment of Absolute Configurations of Naphthylisoquinoline Alkaloids by HPLC-ECD

Hyphenation of HPLC not only to MS/MS and NMR, but also to ECD spectroscopy as a third online-analytical tool, finally led to the introduction of the "analytical triad", LC-MS/MS-NMR-ECD [42, 43, 47, 81, 82] into natural product analysis by the authors' group. The LC triad permits the attribution of full absolute stereostructures of chiral molecules directly from the peaks in crude mixtures, without the necessity of first isolating the compounds. ECD spectroscopy can distinguish between stereoisomers that show sufficient chiroptical differences, i.e., usually between enantiomers, because their ECD curves are fully opposite to each other, but sometimes, in favorable cases, also between diastereomers (see below) [395–398]. It is a particularly valuable tool for the determination of the absolute configuration at the biaryl axis of naphthylisoquinoline alkaloids [76–78, 83]. Since the shape of the ECD spectrum of a compound with stereogenic axes and centers is usually dominated by the chiral biaryl chromophore (in the presence of which the ECD contributions of the stereogenic centers are hardly to be seen), the ECD spectra of atropo-diastereomeric naphthylisoquinoline alkaloids will give, in most cases, near-opposite ECD curves [31, 32, 124, 160, 170, 175, 179, 186, 191, 225]. Thus, the axial configuration of a new naphthylisoquinoline may usually be established by comparison with the spectrum of a structurally related known one of the same coupling type, regardless of the configurations at the stereocenters. If the compounds to be compared do not resemble each other enough and/or the chiroptical similarity is not convincing, or if one is even dealing with a totally new, unprecedented structural type, the quantum-chemical calculation of ECD spectra is an elegant (and sometimes the only) way to address this problem, not only for naphthylisoquinoline alkaloids [31–33, 101, 123, 155, 159, 160, 178, 180, 187, 225], but also for other—natural or synthetic—chiral products [76–78, 83, 399–410].

The online discovery of four new naphthylisoquinoline alkaloids from the as yet phytochemically unexplored Thai liana *A. griffithii* convincingly proved the efficiency of the LC triad for a rapid chemical metabolic screening of plant extracts. It permitted the fast dereplication of two already known alkaloids, ancistrocladine (**1a**) and hamatine (**1b**), in situ directly from the peaks in the chromatogram and, in particular, enabled the assignment of the complete stereostructures of three new monomeric alkaloids, ancistrogriffines A (**145**), B (**146**), and C (**147**), and of the novel-type dimer ancistrogriffithine A (**148**).

5.2 A Photometric Chemical Screening Method to Trace Up Michellamine-Type Naphthylisoquinoline Dimers

Ancistrogriffithine A (**148**) was discovered in the leaves of *A. griffithii* by applying a photometric chemical screening method [42] developed for michellamine-type naphthylisoquinoline dimers. The procedure, as elaborated in the authors' group, is based on the lead tetraacetate oxidation of the central naphthalene core (substructure **A**) of the alkaloids giving rise to the formation of dinaphthoquinones (substructure **B**) showing an intense violet-blue color (Fig. 60). Addition of the oxidant to leaf extracts of *A. griffithii* resulted in the appearance of a characteristic long-wavelength absorption in the UV spectrum of the sample, with the maximum determined to be at λ = 525 nm. In this region, only very few plant pigments with interfering absorbance maxima occur in Ancistrocladaceae plants [42].

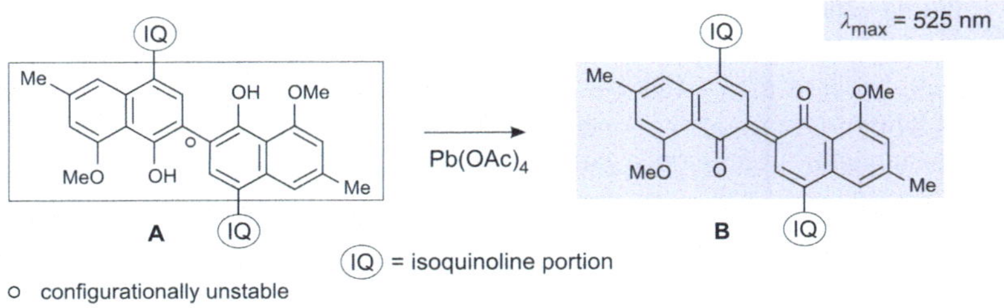

Fig. 60 Key reaction of the photometric test for naphthylisoquinoline dimers: formation of the diquinoid products **B** from binaphthols of type **A**, by oxidation with lead tetraacetate [42, 86, 125]

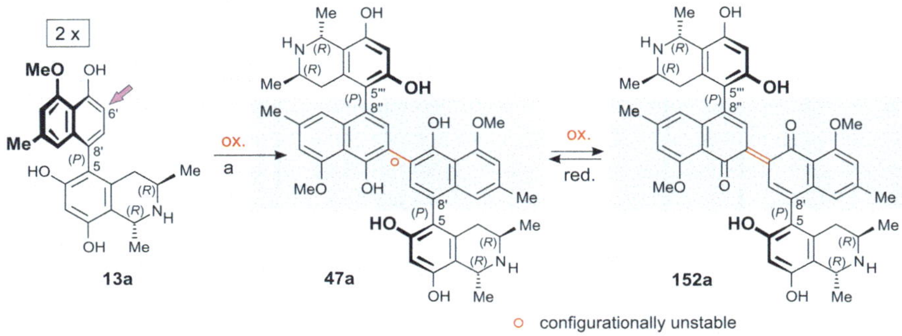

Fig. 61 One-step biomimetic synthesis of michellamine A (**47a**), by oxidative dimerization of korupensamine A (**13a**) using $Pb(OAc)_4$ as the oxidizing agent, with reversible overoxidation to give the diphenoquinone **152a**. Reaction conditions: (a) $Pb(OAc)_4$, $BF_3{\cdot}Et_2O$, CH_2Cl_2, 0°C, 89%

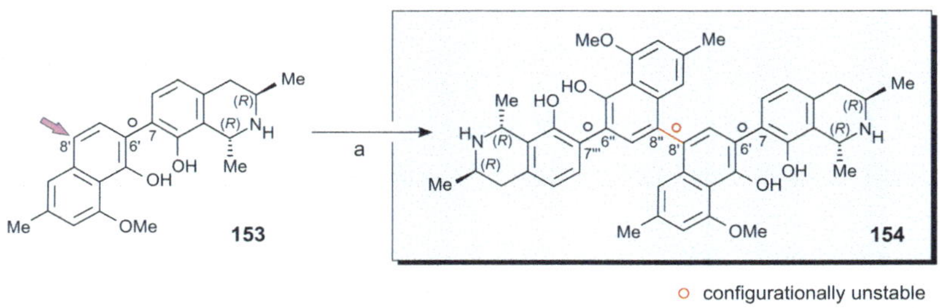

Fig. 62 Phenol-oxidative coupling of dioncophylline B (**153**) with $Pb(OAc)_4$, giving rise to jozimine D (**154**), an unnatural dimer with three configurationally unstable biaryl axes. Reaction conditions: (a) $Pb(OAc)_4$, $BF_3{\cdot}Et_2O$, CH_2Cl_2, 0°C, 81%

The fundamental principle of this photometric assay was discovered in the course of the first total synthesis of michellamine-type dimers, when constructing the central axis biomimetically, by oxidative dimerization of their unprotected monomeric naphthylisoquinoline precursors. The smooth, regioselective phenolic coupling of korupensamine A (**13a**) gave rise to the first non-enzymatic direct preparation of michellamine A (**47a**) from its monomeric half (Fig. 61). The same principle was applied to the synthesis of entirely new quateraryls like jozimine D (**154**) (Fig. 62) [128], the unnatural dimer of the 7,6′-coupled naphthylisoquinoline alkaloid dioncophylline B (**153**). The latter monomer is easily available in sufficient amounts by isolation [14, 219] from the Dioncophyllaceae liana *T. peltatum*, but also by total synthesis [411, 412].

Upon treatment of korupensamine A (**13a**) with excessive amounts of lead tetraacetate (1.2 equiv.), immediate coupling occurred—with overoxidation—to give the deeply violet-colored diketone **152a** [86, 128]. The reversal of this over-reaction, luckily, did not require an additional reaction step because, in accordance with previous experience with such diphenoquinones [86, 96, 413], reduction occurred already by filtration of the deep-violet-colored reaction mixture over Celite with dichloromethane/methanol (8:2) as the eluent. This biomimetic approach provided michellamine A (**47a**) in a highly pure, virtually colorless form, and in an excellent coupling yield of 89%. The dimer **47a** was obtained stereochemically homogeneous, and no hints were found at the formation of any regioisomeric coupling products [86, 128].

A further interesting dimer is jozimine D (**154**) [128] (Fig. 62). Structurally, it strongly differs from its natural analogues, the michellamines, which are widely present in various Central African *Ancistrocladus* species [163–165, 167, 168, 170, 173, 176], possessing a most different molecular shape. Using the lead tetraacetate methodology [86], it was synthesized by direct phenol-oxidative coupling of free, unprotected dioncophylline B (**153**) [128]. Again, dimerization occurred in the naphthalene part exclusively, thus giving rise to a novel-type bis-naphthylisoquinoline, this time linked via a *peri*-position (via C-8′) of the naphthalene moiety; no coupling was observed in the monocyclic aromatic part of the isoquinoline [128]. The intermediate diphenoquinone (whose formation was evident from its typical violet color, see above) was reduced to the colorless target molecule **154** during work-up in methanol/dichloromethane (without any additional required synthesis efforts), leading to **154** in an overall yield of 81% from **153**. Besides its novel 8′,8″-coupling type at the central biaryl axis, jozimine D (**154**) was the first unnatural dimer of a natural naphthylisoquinoline monomer with three configurationally unstable biaryl axes [128].

The photometric assay developed by the authors' group for a simplified screening of a number of *Ancistrocladus* lianas from Asia and Africa, is based on the observation that even the biphenolic aromatic target dimers themselves like michellamine A (**47a**) can be re-oxidized to give the colored diquinoid intermediate (like e.g., **152a**) using lead tetraacetate (see Fig. 61). Stated succinctly, the assay is positive for dimers with a low steric load at the central axis—but also for their monomeric precursors, which will dimerize under the same conditions and then give the color reaction. In the screening of a broad series of Ancistrocladaceae plants for the potential occurrence of michellamine-type dimers, significantly positive results were obtained for only very few extracts, among them, as expected, a leaf extract of *A. korupensis*, which exhibited the by far highest value (ΔE_{525} = 1.48) of the whole screening series. From this Cameroonian liana, seven such 6′,6″-coupled dimeric naphthylisoquinoline alkaloids had already been isolated earlier [163–165, 195]. The michellamines and their monomeric halves, the korupensamines [156], are even the main secondary metabolites of *A. korupensis*, to be found mainly in the leaves of the plant, in particular in withering leaves. This Central African *Ancistrocladus* species occupies a special position within the Ancistrocladaceae, due to its predominant production of mono- and dimeric naphthylisoquinoline alkaloids belonging to the 5,8′-coupling type. With these phytochemical characteristics, *A. korupensis* closely resembles *A. congolensis* [172, 173] and *A. likoko* [166–169] from the northwestern Congo Basin. In the photometric test, distinct, but much weaker signals were monitored for these two *Ancistrocladus* species from Central Africa, namely ΔE_{525} = 0.35 (for *A. congolensis*) and ΔE_{525} = 0.28 (for *A. likoko*) [42]. This was in accordance with the results of the ensuing isolation work on these two plant species. New michellamine-type dimers with structures closely related to those of their parent compounds, michellamines A (**47a**) and B (**47b**) (for the structure of **47b**, see Fig. 65), which were also detected in the plants, were discovered in *A. congolensis* (seven representatives) [173] and in *A. likoko* (six compounds) [167, 168], yet being produced only as minor or even trace metabolites. In this context, it is noteworthy that the likewise related Congolese species *A. ealaensis* only showed a weak signal (ΔE_{525} = 0.14) in the photometric assay [42]. This can probably be attributed to the fact that only four, minor dimeric naphthylisoquinoline alkaloids with a central 6′,6″-coupled binaphthalene core were detected in *A. ealaensis* [176, 188], being present in the plant mostly in trace quantities. One of these dimers was a conventional michellamine-type alkaloid [176], built up from two 5,8′-coupled monomers, whereas the three other compounds were unusual heterodimers [188], consisting of a 5,8′-coupled naphthylisoquinoline and a 7,8′-linked molecular half. All the other African *Ancistrocladus* species investigated so far, such as *A. barteri* and *A. abbreviatus* from Coastal West Africa, the Gabonese liana *A. letestui*, and the two East African species *A. tanzaniensis* and *A. robertsoniorum*, did not reveal any significant dimer concentrations in the photometric assay [42]. These results were in accordance with the fact that no michellamine-type dimers had ever been identified in the above-mentioned *Ancistrocladus* species. Samples of the Indian liana *A. heyneanus* and the Sri Lankan species *A. hamatus*, but also several plant extracts of *A. tectorius* from different locations in Thailand, Laos, Vietnam, and Malaysia, gave no positive signals in the photometric assay [42], which was in agreement with the fact that no dimers were found in these species. Only *A. griffithii* from Thailand showed a strongly positive reaction (ΔE_{525} = 0.63). It was the second-highest value ever monitored in the photometric survey of *Ancistrocladus* plants from Africa and Asia. This made the liana particularly interesting for detailed phytochemical studies of its alkaloid pattern.

5.3 *Online Discovery of the First Ancistrocladaceae-Type Dimer, Ancistrogriffithine A, in Crude Leaf Extracts of* **Ancistrocladus griffithii**, *by the LC-MS/MS/LC-NMR/LC-ECD Triad*

Following the positive results for *A. griffithii* in the photometric assay (see above) [42], the leaf extract of this Thai liana was analyzed for the presence of dimeric naphthylisoquinoline alkaloids by HPLC-MS/MS. The main compound (= peak **A**) gave signals at both, *m/z* 785.4 and 393.4, which were in agreement with the mono- and deprotonated molecular ions $[M + H]^+$ and $[M + 2H]^{2+}$ of a possible dimer with a molecular formula of $C_{48}H_{52}N_2O_8$ (see Fig. 63a) [42, 43]. This was supported by an MS/MS experiment on $[M + 2H]^{2+}$ at *m/z* 393.4, which provided a mass spectrum with fragment ions up to 742.4 amu (see Fig. 63b), corresponding to a monoprotonated ion $[M + H - C_2H_5N]^+$ from a retro-Diels–Alder reaction of one tetrahydroisoquinoline subunit of a 6,8-dioxygenated naphthylisoquinoline dimer. The appearance of typical diprotonated mass fragments such as $[M + 2H - NH_3]^+$ with *m/z* 384.8 and $[M + 2H - C_2H_5N]^+$ with *m/z* 371.7 further evidenced the main metabolite of the leaves of *A. griffithii* to be a new dimeric naphthylisoquinoline alkaloid (see Fig. 63b).

After successful verification of the result from the photometric chemical screening assay using HPLC-ESI/MS/MS [42], the constitution and the relative configuration of the new plant metabolite were elucidated online, directly from the leaf extract, by HPLC-NMR [42, 82]. ^{1}H NMR measurements in the stopped-flow mode showed only one single set of twelve proton signals, not two sets, as imaginable for the 'double' molecular mass, indicative of a constitutionally symmetric dimer (see Fig. 63c), consisting of two equivalent monomeric molecular halves, which were thus homo- or enantiomorphous to each other, of which the latter (i.e., the presence of a *meso*-compound) was excluded due to the optical activity ($[\alpha]_D = 73.5$) of the compound [42, 43, 82].

The elucidation of the complete constitution and relative configuration of the new dimeric naphthylisoquinoline alkaloid, directly from the leaf extract, was achieved by LC-NMR-ROESY experiments in combination with the efficient WET sequence [393] as the solvent-suppression scheme. This experimental setup led to a significant improvement in sensitivity and spectra quality of the online 2D NMR measurements and permitted assignment of the constitution, the coupling type, and the relative configuration at the biaryl axis and at the stereogenic centers C-1 and C-3 in the tetrahydroisoquinoline part of the naphthylisoquinoline portions of the new compound [42, 82].

Comprehensive analysis of the online acquired ^{1}H NMR and TOCSY data [42] revealed the presence of a naphthyltetrahydroisoquinoline subunit possessing two methoxy groups, one located at C-8 in the isoquinoline part and one at C-4′ in the naphthalene portion. The position of the biaryl axis between the isoquinoline and the naphthalene moieties was deduced from ROE correlations from the two naphthalene protons H-1′ and H-7′, both to MeO-8, and from the two correlation

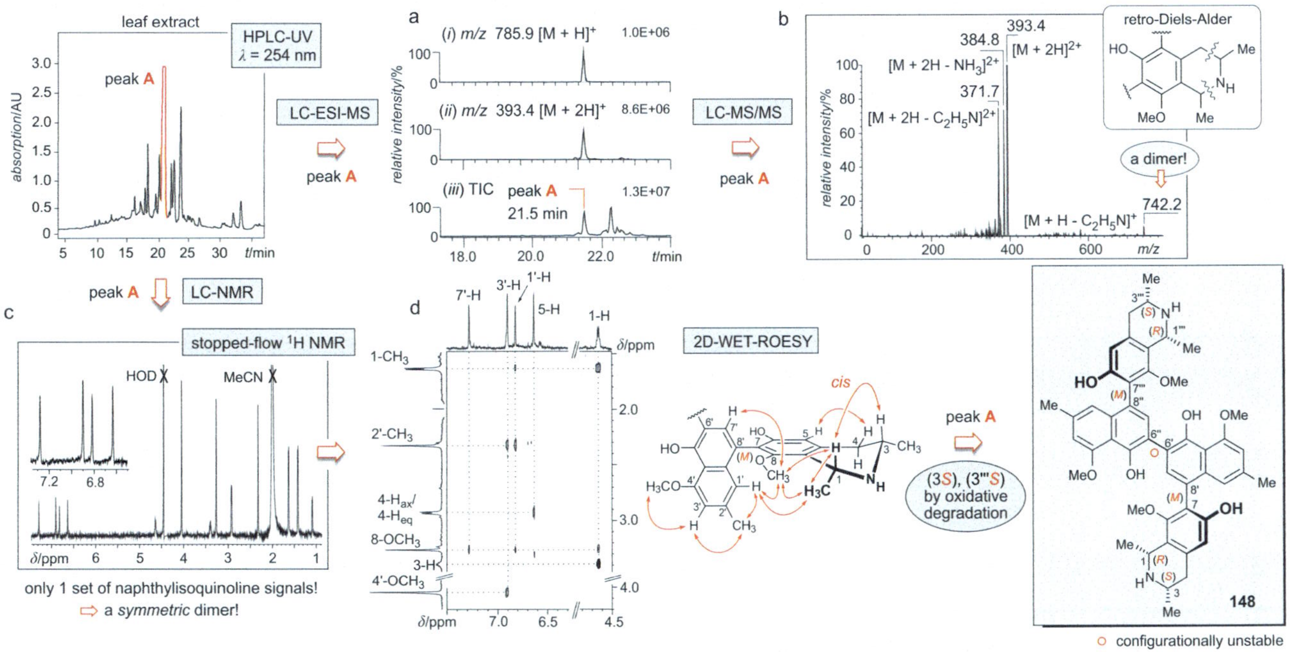

Fig. 63 Elucidation of the full absolute stereostructure of ancistrogriffithine A (**148**) directly from a crude leaf extract of *A. griffithii*: **a** LC-ESI-MS chromatogram with reconstructed ion chromatograms of (i) *m*/*z* 785.4 and (ii) *m*/*z* 393.4, and (iii) reconstructed total ion chromatogram (TIC); **b** product ion spectrum of the precursor ion *m*/*z* 393.4 of peak **A**; **c** stopped-flow NMR spectrum of peak **A**; **d** 2D-WET-ROESY measurements and the resulting assignment of the relative configurations, and, for the absolute configuration of **148**, further complemented by applying the ruthenium-catalyzed oxidative degradation procedure to the entire leaf extract

series {H-1′ ↔ Me-2′ ↔ H-3′ ↔ MeO-4′} and {H_{eq}-4 ↔ H-5}. These data gave conclusive evidence of a 7,8′-coupling site in the two monomeric halves, while the only position left for the central biaryl linkage was C-6′ in the naphthalene subunits (see Fig. 63d). Owing to the presence of only two *ortho*-substituents next to this 6′,6″-linkage, the central biaryl axis was configurationally unstable, and thus not an additional element of chirality [42, 82].

From a ROESY interaction between H-1 and H-3, the relative configuration of the two methyl groups at C-1 and C-3 was deduced to be *cis* (Fig. 63d) [42]. Ruthenium-mediated oxidative degradation of a portion of the total leaf extract of *A. griffithii* followed by GC-MSD analysis of the resulting chiral amino acids clearly evidenced the formation (>99:1) of 3-aminobutyric acid exclusively as its (*S*)-enantiomer, thus corroborating that only tetra- or dihydroisoquinoline substructures with an (*S*)-configuration at C-3 were present in the crude extract, which is typical of an *Ancistrocladus* species from Asia. Given the relative *cis*-configuration at C-1 versus C-3 in the two tetrahydroisoquinoline subunits of the new dimer, the absolute configuration at C-1 and C-1‴ had to be (*R*) [42]. The configuration at the two outer biaryl axes relative to the stereogenic centers became evident from a specific long-range ROESY interaction between Me-1 and H-1′ (Fig. 63d). This, in combination with the absolute configuration at the two stereogenic centers, as determined by oxidative degradation, clearly showed that the two naphthalene-isoquinoline linkages had to be (*M*)-configured [42, 43]. It was for the first time that such an ROE effect across a biaryl axis was observed online, by HPLC-NMR coupling—and here even in a dimer. Thus, the new naphthylisoquinoline dimer possessed the full absolute stereostructure **148**, with the (1*R*,3*S*,7*M*)-configuration in both monomeric halves as presented in Fig. 63. The new Ancistrocladaceae-type homodimer was named ancistrogriffithine A [42, 43, 82].

Preparative isolation of **148** from leaf and twig extracts of *A. griffithii* for biological testing and for further structural investigations offline, by ROESY, HMQC, and HMBC measurements, fully confirmed the previous online assignment [43]. Ancistrogriffithine A (**148**) was found to strongly inhibit the growth of the K1 strain (resistant to chloroquine and pyrimethamine) of the malaria parasite *P. falciparum* (IC_{50} = 0.045 μ*M*) comparable to the activity of the antiplasmodial agent chloroquine. It exerted only weak inhibitory effects against the causative pathogens of Chagas' disease and African sleeping sickness, *T. cruzi* and *T. brucei rhodesiense*, respectively [43].

Although the knowledge about naphthylisoquinoline dimers has undergone a tremendous development during the past decade in terms of their increasing numbers (ca. 50 representatives are known so far), their structural diversity, and their fascinating molecular architectures [17], ancistrogriffithine A (**148**) is still the only example of a 6′-homo-coupled dimer of a 7,8′-linked monomeric naphthylisoquinoline alkaloid ever found in Nature. Within the group of 6′,6″-coupled naphthylisoquinoline dimers (23 compounds), **148** is one of only four known C_2-symmetric representatives (see Fig. 64). The other examples are the michellamines A (**47a**) and C (**47c**) from the Cameroonian liana *A. korupensis* [163, 164], and michellamine A_6 (**155**) from an as yet botanically unidentified species [170].

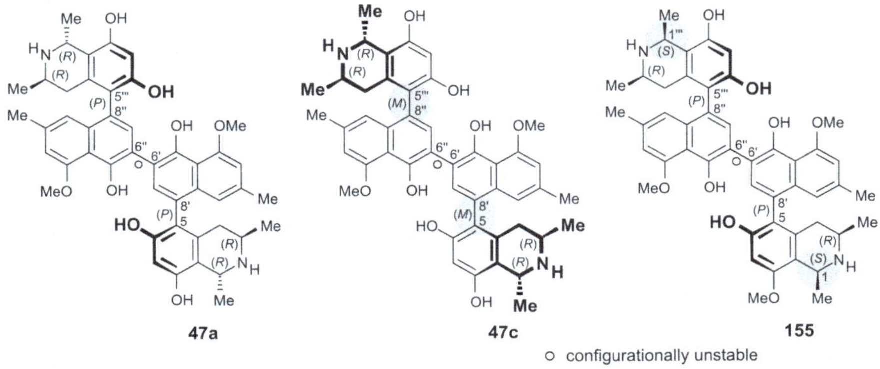

Fig. 64 Michellamines A (**47a**) and C (**47c**) from *A. korupensis* [163, 164] and michellamine A_6 (**155**) from an unknown Congolese *Ancistrocladus* species [170], the only other known 6′,6″-coupled naphthylisoquinoline dimers (besides ancistrogriffithine A, namely, structure **148**, see Fig. 63) that consist of two fully identical, i.e., homomorphous naphthylisoquinoline halves, thus being C_2-symmetric. Differences of **47c** and **155** as compared to **47a** are highlighted in gray

Michellamine A_6 (**155**) is structurally closely related to michellamine A (**47a**), from which it differs only by the absolute configuration at C-1 and C-1‴, thus being the 1,1‴-bis-epimer of **47a**. These compounds have attracted attention due to their unique molecular architectures and because of their interesting bioactivities. The michellamines A (**47a**) and C (**47c**) display pronounced antiviral effects against HIV-1 (RF strain) and HIV-2 (CBL-20 strain) [163, 164], whereas michellamine A_6 (**155**) exhibits good-to-moderate inhibitory activities on HeLa and PANC-1 pancreatic cancer cells [170].

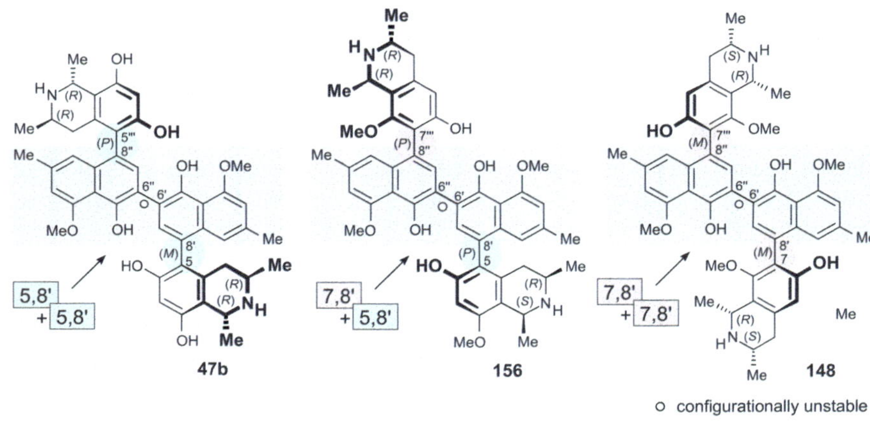

Fig. 65 A unique series of dimeric naphthylisoquinoline alkaloids found in Nature possessing a 6′,6″-linked central biaryl axis, which is not sterically hindered and thus, can freely rotate: michellamine B (**47b**) (5,8′–5,8′), ealapasamine A (**156**) (5,8′–7,8′), and ancistrogriffithine A (**148**) (7,8′–7,8′)

With the discovery of ancistrogriffithine A (**148**), the series of dimers with a 6′,6″-linked central biaryl axis, which is thus not rotationally hindered, has become complete regarding the 5,8′- and 7,8′-coupling types of their two outer biaryl axes. The series now comprises michellamine-type dimers (18 representatives) built up from two 5,8′-coupled monomers (5,8′–5,8′) like michellamine B (**47b**) [163, 164], heterodimers (four compounds) such as ealapasamine A (**156**) [188] consisting of a 5,8′-coupled naphthylisoquinoline and a 7,8′-linked molecular half (5,8′–7,8′), and the twofold 7,8′-coupled ancistrogriffithine A (**148**) (7,8′–7,8′) [42]. Among all these dimers, ancistrogriffithine A (**148**) is the only known example representing a 6′-homocoupling product of a 7,8′-linked monomeric alkaloid [42, 43, 82] (Fig. 65).

5.4 *Complete Structural Elucidation of Ancistrogriffines A–C, Three New Monomeric Naphthylisoquinoline Alkaloids Online, Directly from the Extract of Ancistrocladus griffithii*

Extract screening by HPLC coupled to UV spectroscopy, MS-MS, NMR, and ECD led to the discovery of three new monomeric naphthylisoquinolines, named ancistrogriffines A (**145**), B (**146**), and C (**147**), along with two related, but known compounds, ancistrocladine (**1a**) and its atropo-diastereomer hamatine (**1b**), directly from extracts of the leaves and twigs of *A. griffithii* [43].

By the combination of HPLC-MS/MS and HPLC-NMR, the constitution and the relative configuration of the main constituent of the leaf extract (= peak **A**) were assigned. As described in the previous section, this plant metabolite was identified as a new naphthylisoquinoline dimer, ancistrogriffithine A (**148**) (Fig. 63) [42, 43, 82]. The second major constituent (= peak **B**) (see Fig. 66), detected by HPLC-UV and HPLC-ESI-MS in the leaf extract of *A. griffithii*, displayed a mass peak of a monoprotonated ion $[M + H]^+$ with *m/z* 408. Daughter ion experiments, which produced fragments of *m/z* 351 $[M + H - C_2H_4NCH_3]^+$ (see Fig. 66a), suggested this metabolite to be a monomeric *N*-methylated naphthyltetrahydroisoquinoline alkaloid with two methoxy substituents [43]. The presence of an *N*-methyl group was confirmed by HPLC-NMR experiments, by online ROE interactions of *N*-Me both with H-1 and Me-3. From a ROESY correlation between Me-1 and H-3, the relative configuration at C-1 versus C-3 in the tetrahydroisoquinoline portion was established to be *trans* (see Fig. 66c) [43]. The substitution pattern in the naphthalene subunit was deduced from two ROESY correlation sequences {H-3′ ↔ Me-2′ ↔ H-1′} and {H-7′ ↔ H-6′ ↔ OMe-5′}, leaving C-8′ as the only remaining coupling position. The axis-bearing carbon atom in the isoquinoline portion was assigned as C-7 due to ROE effects between H-1 and the protons of MeO-8 and between H-4_{eq} and H-5. Additional evidence of a 7,8′-coupling type was obtained from an ROE interaction across the biaryl axis, between H-7′ and the methyl protons of MeO-8 (see Fig. 66c) [43].

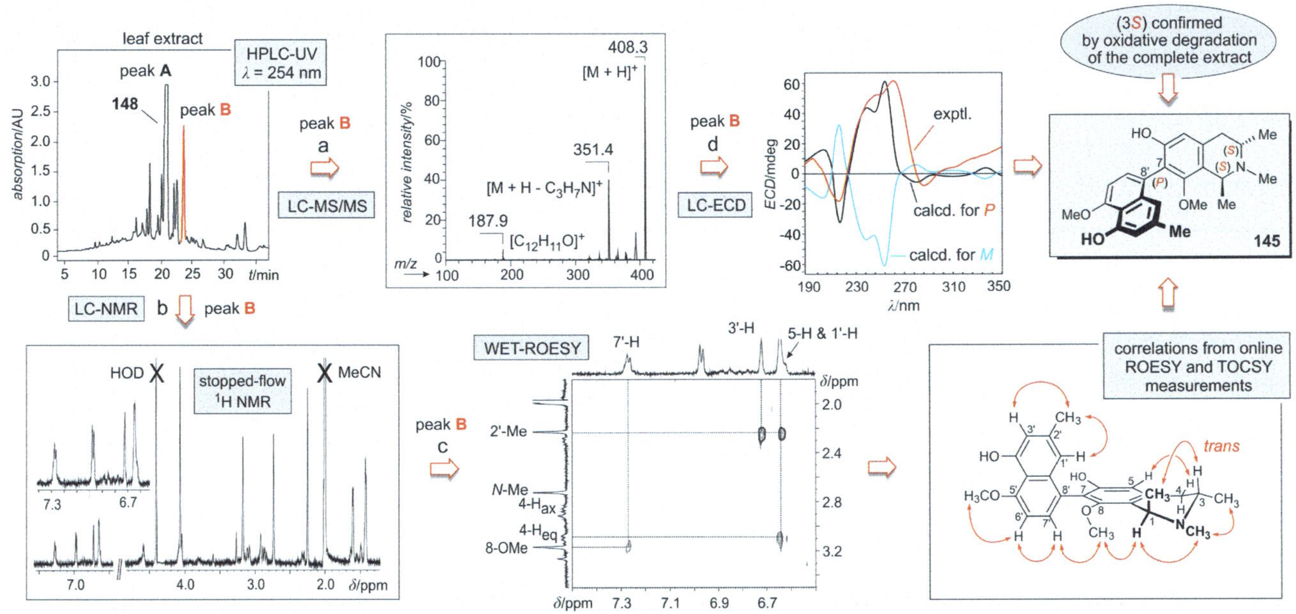

Fig. 66 Online elucidation of the full stereostructure of ancistrogriffine A (**145**) (= peak **B**) from *A. griffithii*, by HPLC-MS/MS-NMR-ECD: **a** product ion spectrum of the precursor ion *m*/*z* 408.3 $[M + H]^+$; **b** stopped-flow ^{1}H NMR spectrum of peak **B**; **c** 2D-WET-ROESY measurements supporting the basic structure of **145** and likewise revealing a relative *trans*-configuration between 1-Me and 3-Me and the axial configuration relative to the centers; **d** assignment of the absolute configuration of **145** as (1*S*,3*S*,7*P*), by comparison of the LC-ECD curve of peak **B** with the ECD spectra predicted for (*P*)-**145** and (*M*)-**145** by quantum-chemical ECD calculations (AM1, CNDO/S), complemented, offline, by oxidative degradation, thus confirming the absolute configuration at the stereocenters; peak **A**: the dimer ancistrogriffithine A (**148**)

This ROE effect also has a diagnostic value for the relative axial configuration, as follows: as evidenced by a significant ROE effect of the protons of MeO-8 with H-1, the *O*-methyl group at C-8, which is in a rapid up-down movement, seems to occupy a preferential position *syn* to H-1 and, given the above-mentioned NOE of MeO-8 with H-7′, also *syn* to that aromatic proton, thus establishing a relative axial configuration in which H-7′ and H-1 are on the same side of the molecule [43].

Quantum-chemical ECD calculations (AM1, CNDO/S) gave rise to a theoretical ECD spectrum of the (*P*)-configured atropo-diastereomer, which showed excellent agreement with that obtained experimentally for the new alkaloid by online HPLC-ECD measurements in stopped-flow mode. This permitted a clear attribution of the absolute axial configuration as (*P*), in accordance with the above-described, tentative assignment (see Fig. 66d). Considering the exclusive detection of the (*S*)-enantiomers of 3-aminobutyric acid and its *N*-methyl derivative after the oxidative degradation of leaf and twig extracts of *A. griffithii* as mentioned above, and with respect to the given relative *trans*-array of Me-1 versus Me-3, the entire absolute stereostructure of the new alkaloid **145**, named ancistrogriffine A, was established as (1*S*,3*S*,7*P*) [43]. Offline 1D- and 2D-NMR investigations on pure isolated material of ancistrogriffine A (**145**) fully confirmed the 3D structure, as previously deduced online, directly from the plant extract [43].

Similar to the dimer ancistrogriffithine A (**148**), the monomeric alkaloid ancistrogriffine A (**145**) also showed strong inhibitory effects against the chloroquine-resistant K1 strain of *P. falciparum*, comparable to those exerted by chloroquine [43]. Furthermore, **145** exhibited a distinct activity against *L. donovani*, the pathogen causing visceral leishmaniasis. Ancistrogriffine A (**145**) showed an enhanced selectivity against the parasite in comparison to their major host cell, the macrophage. The antitrypanosomal activities of **145**, as evaluated against the pathogens of Chagas' disease and African sleeping sickness, were only moderate to low [43].

An impressive example demonstrating the efficiency and sensitivity of hyphenated methods in phytochemical analysis is the elucidation of the full absolute stereostructure of the trace alkaloid ancistrogriffine B (**146**) (= peak **C**) (see Fig. 67) from a rather small subfraction of a twig extract of *A. griffithii*, as resulting from chromatography over silica gel [43]. HPLC-ESI-MS investigations of the new metabolite delivered a mass peak *m*/*z* 392 for its monoprotonated ion $[M + H]^+$. In combination with 1D and 2D HPLC-NMR experiments in the stopped-flow mode (see Fig. 67), this finding suggested the presence of a 6,8-dihydroxylated tetrahydroisoquinoline with two methoxy groups in the naphthalene subunit [43]. As in the case of ancistrogriffine A (**145**) (Fig. 66), the relative configuration at C-1 versus C-3 was evidenced to be *trans* from ROESY interactions between Me-1 and H-3, and the coupling position in the isoquinoline part was similarly attributed to be located at C-7 from a correlation between H-4_{eq} and H-5. Dissimilar from the 7,8′-coupled alkaloid **145**, however, correlation sequences in the series {Me-2′ ↔ H-3′ ↔ OMe-4′} and {OMe-5′ ↔ H-6′ ↔ H-7′ ↔ H-8′} established C-1′ as the axis-bearing carbon atom in the naphthalene subunit. Thus, the new minor metabolite ancistrogriffine B (**146**) was a 7,1′-coupled naphthylisoquinoline [43].

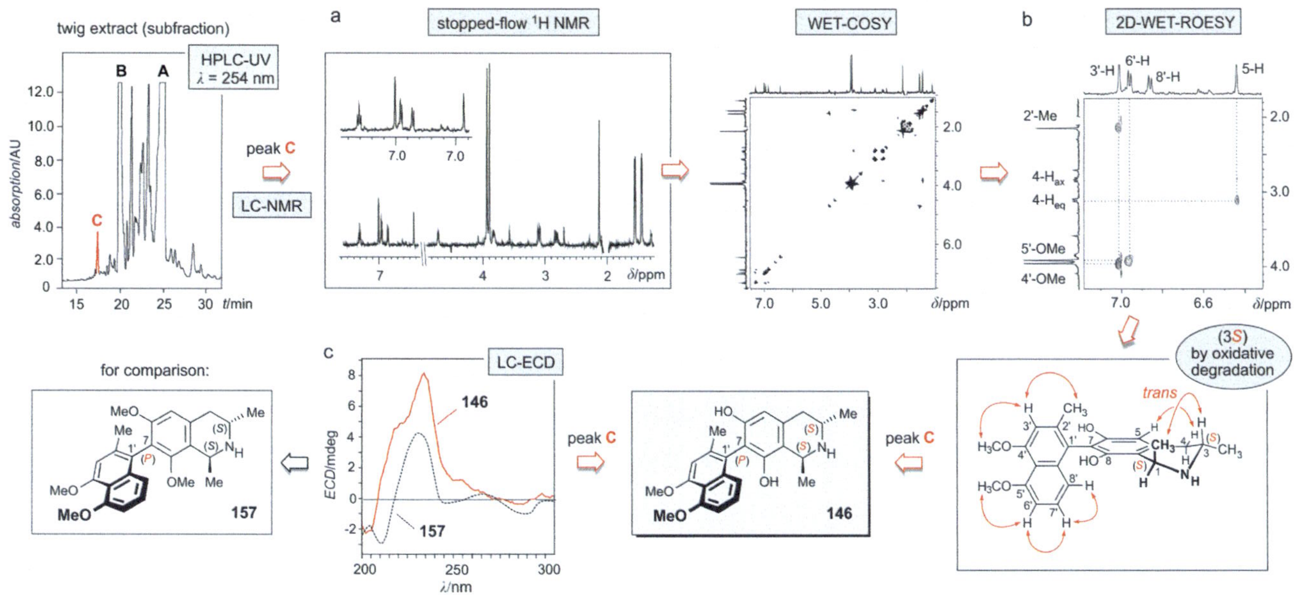

Fig. 67 Online assignment of the full stereostructure of the 7,1′-coupled minor naphthylisoquinoline alkaloid ancistrogriffine B (**146**) (= peak **C**), obtained directly from a polar subfraction of a twig extract of *A. griffithii*: **a** stopped-flow ^{1}H NMR and WET-COSY spectra; **b** ROESY correlations detected by HPLC-NMR experiments; **c** comparison of the online LC-ECD spectrum of **146** with that of the offline ECD curve of 6-*O*-methylancistectorine B_1 (**157**), available from earlier synthetic work [190], and more recently identified in the Chinese liana *A. tectorius* as a true natural product [62]. Further peaks in the chromatogram are ancistrogriffithine A (**148**) (= peak **A**) and ancistrogriffine A (**145**) (= peak **B**)

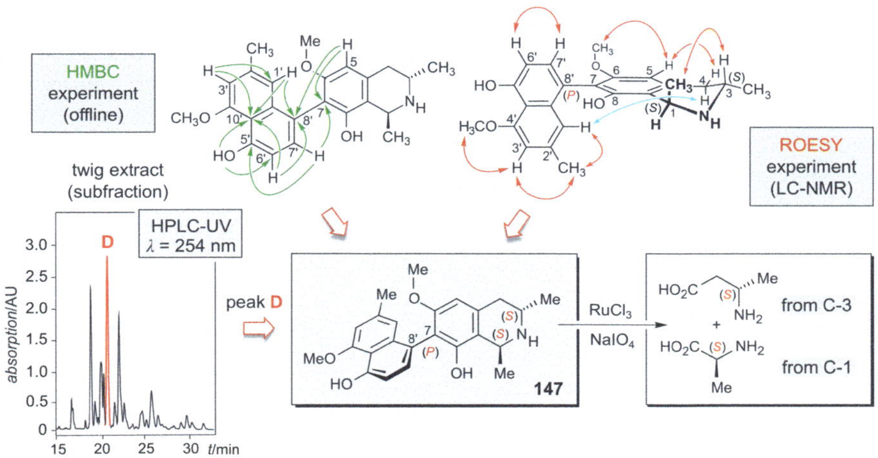

Fig. 68 Structure elucidation (online and offline) of ancistrogriffine C (**147**) (= peak **D**), a new naphthylisoquinoline alkaloid from the twigs of *A. griffithii*, by HMBC interactions (green single arrows, offline) supporting a 7,8′-coupling type, and by ROESY correlations (red double arrows), monitored by HPLC-NMR and by offline NMR (blue double arrow: correlation observed with the isolated material)

No ROE effects between the two molecular portions of this alkaloid that might have been indicative of the relative axial configuration were observed online, by LC-WET-ROESY measurements. Nonetheless, a clear assignment of the absolute configuration at the biaryl axis succeeded as (*P*), by HPLC-ECD, by comparison of the ECD spectrum of ancistrogriffine B (**146**) with that of the structurally related, likewise 7,1′-coupled and (*P*)-configured 6-*O*-methylancistectorine B_1 (**157**), well known from isolation work on the Chinese liana *A. tectorius* [62] and from semi-synthesis, obtained by *trans*-selective reduction of the related dihydroisoquinoline ancistrocladisine (**2a**) [190]. As outlined in Fig. 67, the two ECD curves were in good agreement with each other. Oxidative degradation revealed the new alkaloid to be (*S*)-configured at C-3. Thus, from the relative *trans*-configuration in the isoquinoline part as deduced by LC-NMR, ancistrogriffine B was established to possess the absolute stereostructure **146**, with (1*S*,3*S*,7*P*)-configuration (see Fig. 67) [43]. HPLC-MS and HPLC-NMR investigations on an alkaloid-enriched subfraction of the twig extract of *A. griffithii* revealed the presence of a third new monomeric naphthyltetrahydroisoquinoline alkaloid (= peak **D**) (see Fig. 68), possessing two methoxy groups, one at C-6 in the isoquinoline moiety, the second at C-4′ in the naphthalene part [43]. In the online ROESY experiment, a correlation observed between Me-1 and H-3 evidenced a relative *trans*-configuration at the stereocenters C-1 and C-3 in the isoquinoline part. Furthermore, online ROESY measurements suggested two possible coupling positions, 7,8′ or 7,6′, of which the latter seemed less probable because of the absence of any ROESY interactions of the *peri*-proton at C-1′ with a hypothetical additional *peri*-proton at C-8′ using LC-NMR [43].

For a reliable verification of the therefore assumed 7,8′-linkage between the two molecular portions, additional ROESY, HMQC, and HMBC experiments were performed on pure material of the compound after its isolation by preparative HPLC. HMBC measurements finally confirmed the 7,8′-coupling site of the new metabolite, due to two ^{3}J correlations of the hydrogen of the hydroxy group at C-5′ to the adjacent positions C-10′ and C-6′ (see Fig. 68, left), thus clearly confirming C-8′ as the only possible coupling position in the naphthalene part. The 7,8′-coupling type was further corroborated by a series of C–H correlations as illustrated in Fig. 68 (right), and by a long-range correlation between H-1′ and H_{ax}-4 clearly observed in offline ROESY experiments. The latter ROE effect simultaneously revealed a *syn*-array of these two protons, in agreement with only a (*P*)-configuration at the biaryl axis [43]. Ruthenium-mediated oxidative degradation yielded 3-aminobutyric acid exclusively as its (*S*)-enantiomer, thus indicating that the stereocenter C-3 of the new alkaloid was (*S*)-configured and hence, due to the above-mentioned *trans*-arrangement of the methyl groups at the chiral centers, C-1 had to be (*S*)-configured, too. The new alkaloid, named ancistrogriffine C, thus had the absolute stereostructure **147** [43], as presented in Fig. 68.

Furthermore, the two known alkaloids ancistrocladine (**1a**) (= peak **E** in Fig. 69) and hamatine (**1b**) (= peak **F** in Fig. 69) were identified in the twigs of *A. griffithii*, by comparison of the spectroscopic data obtained online, directly from their peaks in the extract, with those from the pure authentic alkaloids [5, 25, 43]. HPLC-NMR experiments on an alkaloid-enriched subfraction gave ^{1}H and ^{13}C NMR spectra of two 5,1′-coupled tris-O-methylated tetrahydroisoquinoline alkaloids revealing signal patterns similar to those of **1a** and **1b**. Their identity was further proven by HPLC-MS analysis by the protonated molecule ion $[M + H]^{+}$ giving *m*/*z* 408. The elution order of the atropo-diastereomeric alkaloids **1a** and **1b** was determined by spiking the extract with reference solutions and testing for co-elution, thus permitting attribution the more rapidly eluting peak **E** to hamatine (**1b**) and the slower peak **F** to ancistrocladine (**1a**). This result was in accordance with the ECD ellipticities of the

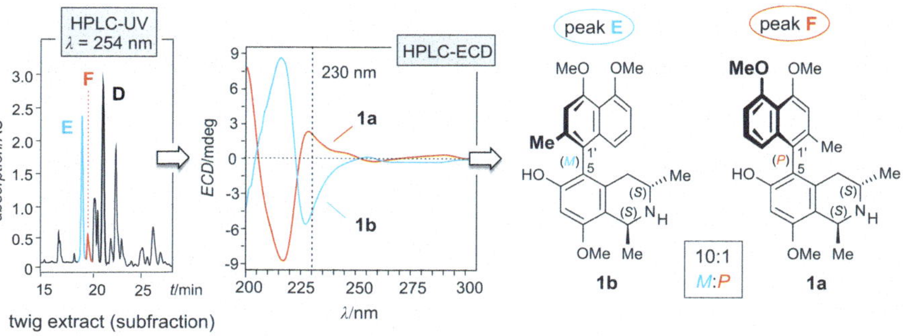

Fig. 69 Hamatine (**1b**) (= peak **E**) and its atropo-diastereomer ancistrocladine (**1a**) (= peak **F**), identified online in an alkaloid-enriched subfraction of a twig extract of *A. griffithii*, occurring in the plant in a ratio of ca. 10:1 (**1b**: **1a**) as deduced from LC-ECD. Peak **D** represents the new alkaloid ancistrogriffine C (**147**)

two alkaloids, which, as expected, exhibited a mirror-image like behavior. HPLC-ECD analysis at a single wavelength (λ = 230 nm) showed a negative signal for peak **E** and a positive one for peak **F** revealing **1b** and **1a** to occur in the plant in a ratio of ca. 10:1, in favor of the (*M*)-configured hamatine (**1b**).

In summary, phytochemical investigations on the previously unexplored Thai liana *A. griffithii* using the hyphenation triad LC-MS/MS, LC-NMR, and LC-ECD led to the identification of four new naphthylisoquinoline alkaloids, among them the first dimeric representative discovered in a Southeast Asian *Ancistrocladus* species, ancistrogriffithine A (**148**) [17, 42, 43], which was found to be present in the plant as a major constituent. Stereochemically and from its Asian source, **148** was found to be entirely unique. Meanwhile five further Ancistrocladaceae-type dimers have been detected in *A. tectorius* [17, 31] (see Sect. 8.5.11), produced by this Chinese liana as minor metabolites. Ancistrogriffithine A (**148**) and ancistrogriffines A (**145**) and B (**146**) were structurally elucidated solely online, without isolation of the compounds [42, 43]. Only in the case of ancistrogriffine C (**147**), the necessity of further C–H correlations required its additional preparative isolation [43]. Tests against the malaria parasite *P. falciparum* [43] revealed promising antiplasmodial activities of **145**, **147**, and **148** against the K1 strain (resistant to chloroquine and pyrimethamine), comparable to that of the standard drug chloroquine.

6 *Ancistrocladus benomensis*, a New Species from Malaysia, with an Unprecedented Series of Non-hydrogenated Ancistrocladaceae- and Dioncophyllaceae-Type Naphthylisoquinoline Alkaloids

A. benomensis Rischer & G. Bringmann is an intriguing plant species from Peninsular Malaysia [34]. It was discovered only in 1999, growing from 500 to 1180 m above sea level on the slopes of an isolated mountain called Gunung Benom ('Benom' is an old Mon-Khmer word for 'mountain') located in the Kerau Game Reserve in the state of Pahang. Based on field studies in the natural habitat of the plants and on observations on some individuals cultivated in the greenhouse of the Botanical Garden of the University of Würzburg, this new *Ancistrocladus* taxon was then botanically described in 2005 by the authors' group in collaboration with G. Heubl's group (LMU Munich), including illustrations of its typical morphological features [34]. A detailed portrayal of this Malaysian liana, focusing on its development from a young seedling to an adult plant, which produces hooked lateral branchlets with leaf rosettes, has already been given in Sect. 3, including a report on the formation of inflorescences bearing buds and bright-red flowers (see also Fig. 26).

The status of *A. benomensis* as a discrete *Ancistrocladus* species was corroborated by comparative genetic studies on the internal transcribed spacer (ITS) regions of the nuclear rDNA and of the *trn*K intron of cpDNA, and by an intersimple

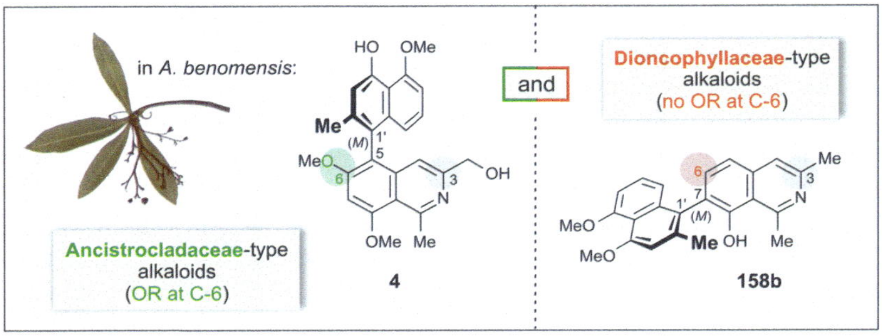

Fig. 70 Naphthylisoquinoline alkaloids with a non-hydrogenated isoquinoline portion isolated from the new Malaysian taxon *A. benomensis* and their tentative classification as Ancistrocladaceae- (green) and Dioncophyllaceae-type (red) compounds, as exemplified for ancistrobenomine A (**4**) and *ent*-dioncophylline A (**158b**). The presence or absence of a 6-oxygen function as a criterion for classification is given priority over the absolute configuration at C-3 (not applicable here, and hence underlaid in gray)

sequence repeat (ISSR) fingerprint analysis of this Malaysian plant [34, 36], but also by phytochemical investigations on its naphthylisoquinoline pattern [30, 32, 44]. Isolation work on stem bark and leaf extracts of *A. benomensis* furnished several new alkaloids, among them no less than six non-hydrogenated naphthylisoquinoline alkaloids like, e.g., ancistrobenomine A (**4**) [30] and *ent*-dioncophylleine A (**158b**) (Fig. 70). The discovery of **158b** was a great surprise due to the lack of an oxygen function at C-6 [44]—unique among all Asian-derived naphthylisoquinoline alkaloids, all of which had so far been 6-OR substituted [13, 15, 17, 18]. Within the classification of fully dehydrogenated naphthylisoquinolines, categorizing them as Dioncophyllaceae- or Ancistrocladaceae-type alkaloids, the presence or absence of the 6-oxygen function is given priority as a criterion over the —here not applicable—absolute configuration at C-3. Thus, *ent*-dioncophylleine A (**158b**) was tentatively categorized as a Dioncophyllaceae-type naphthylisoquinoline, although, due to its lack of a stereocenter at C-3, it is not known whether this compound is really derived from a precursor with the (*R*)-configuration at C-3 or, possibly, with the (*S*)-configuration (which would make it an extremely rare representative of inverse hybrid-type compounds, see Sect. 2) [44, 186]. In a similar way, fully dehydrogenated naphthylisoquinoline alkaloids possessing an oxygen function at C-6 such as ancistrobenomine A (**4**), likewise isolated from *A. benomensis* [30], were consequently classified as Ancistrocladaceae-type compounds (Fig. 70) [30, 47, 64, 167, 168, 186]. As another unique structural feature, **4** is equipped with a hydroxymethylene group at C-3 [30, 47, 64]. This finding clearly demonstrated that the metabolic spectrum of *A. benomensis* [30, 32, 44] is completely different from that of all the other known *Ancistrocladus* species, and it does not match at all with the naphthylisoquinoline profile of *A.* aff. *tectorius* [31, 33, 49–64] (see Sect. 8), which is widely occurring all over Southeast Asia.

Fully dehydrogenated naphthylisoquinoline alkaloids, in which the biaryl axis is the only element of chirality, have as yet been found in Nature quite rarely [30, 40, 44, 45, 47, 64, 152, 167, 168, 186]. While some of them occur in the plants in a completely racemic form [40, 44, 152], others were found to be optically active, even enantiomerically pure [30, 44, 45, 47, 64, 167, 168, 186], except for very few examples that are present in the plants as a scalemic mixture of the respective atropo-enantiomers [44, 186].

6.1 Ancistrobenomine A, the First Naphthylisoquinoline Alkaloid with a Hydroxymethylene Function at C-3, and Related 5,1′-Coupled Compounds

Liquid–liquid extraction and preparative HPLC of a stem bark extract of *A. benomensis* furnished three naphthylisoquinoline alkaloids [30]. The molecular formula of the least polar compound was $C_{25}H_{25}NO_5$, as evidenced by HRESIMS. Its ^{1}H NMR spectrum exhibited signals typical of a non-hydrogenated Ancistrocladaceae-type alkaloid, with the absence of two doublets of the aliphatic methyl groups at C-1 and C-3 around $\delta_H = 1.10$ and 1.80 ppm, the lacking of a quartet of H-1 around $\delta_H = 4.40$ ppm, and the missing multiplet of H-3 in the region between $\delta = 3.20$ and 4.00 ppm. The absence of a signal for the diastereotopic geminal protons at C-4 was in agreement with the appearance of an additional aromatic singlet at $\delta_H = 6.81$ ppm, and by a significant downfield shift of the ^{13}C NMR resonances for C-1, C-3, and C-4 (Fig. 71) [30]. Moreover, the new alkaloid possessed another unprecedented functionality, viz. a hydroxymethylene group connected to C-3 of the isoquinoline portion. Such a molecular entity has previously only been known from the naphthalene-devoid isoquinoline alkaloids 6-*O*-methylanciscochine (**160**) from *A.* aff. *tectorius* from Malaysia [54] and its 6-O-demethylated analogue anciscochine (**161**) identified in the Vietnamese liana *A. cochinchinensis* [48] (Fig. 72). The coupling site of the biaryl axis of the new metabolite of *A. benomensis* was determined to be at C-1′ in the naphthalene portion, based on a continuous NOESY correlation sequence in the series {Me-2′ ↔ H-3′ ↔ OH-4′ ↔ CH_3O-5′ ↔ H-6′ ↔ H-7′ ↔ H-8′} and due to HMBC interactions from H-8′ and H-3′ to the quaternary carbon atom C-1′. From the upfield-shifted signal of the methoxy group of C-6 ($\delta_H = 3.91$ ppm), the biaryl axis was assigned to be located at C-5 in the isoquinoline moiety, which agreed with the NOESY interaction of H-7 with the protons of both, OMe-6 and OMe-8 (Fig. 71). In conclusion, the new alkaloid was established to be 5,1′-coupled and to possess the constitution **4** shown in Fig. 71, with a fully dehydrogenated isoquinoline moiety and a hydroxymethylene function at C-3. It was named ancistrobenomine A [30].

The absolute configuration of **4** at the biaryl axis as the only stereogenic element of the compound was attributed to be (*M*) due to its nearly perfectly mirror-imaged ECD curve compared to that of ancistrocladeine (**162**), a known [4, 414], likewise

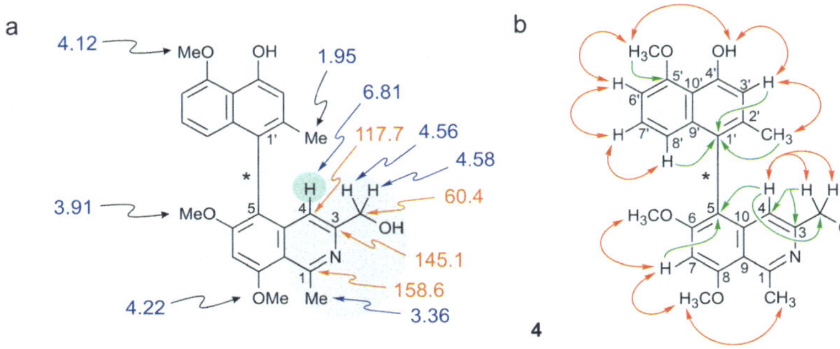

Fig. 71 Structure elucidation of the non-hydrogenated alkaloid ancistrobenomine A (**4**) with a novel hydroxymethylene group at C-3: **a** selected ^{1}H (in blue) and ^{13}C NMR (in orange) data (δ/ppm), and **b** HMBC (green single arrows) and NOESY (red double arrows) interactions relevant for the constitution of **4**. The characteristic structural features are highlighted in gray, and the aromatic singlet at C-4, typical of a non-hydrogenated alkaloid, is underlaid in green

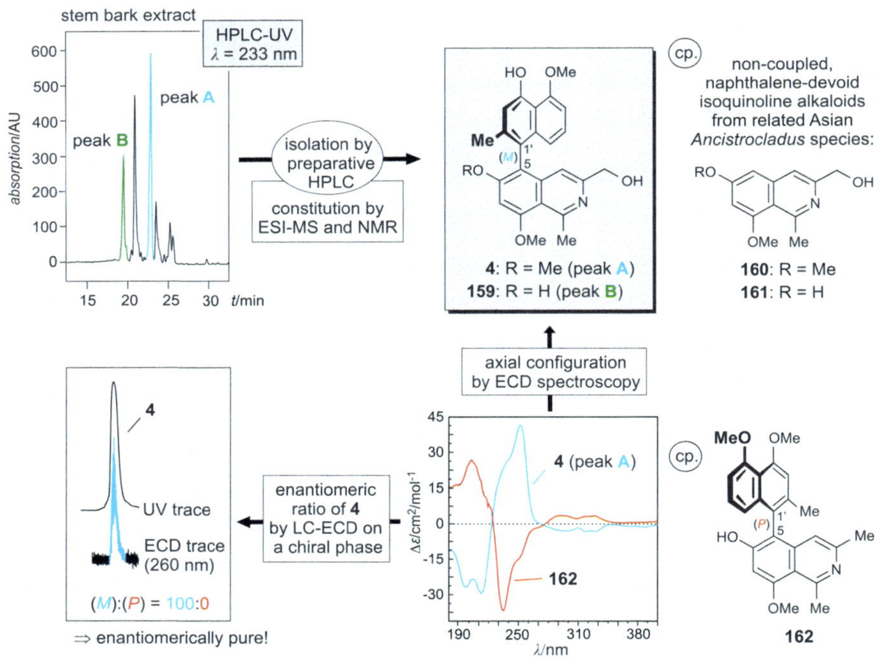

Fig. 72 Assignment of the absolute axial configuration of ancistrobenomine A (**4**) (= peak **A**) by comparison of its ECD spectrum with that of the known [4, 414], likewise 5,1′-coupled and fully dehydrogenated, but (*P*)-configured ancistrocladeine (**162**). HPLC analysis on a chiral phase (Chiralcel® OD-H) evidenced **4** to be produced by *A. benomensis* in an enantiomerically pure form, with peak **B** in the chromatogram representing the similarly new 6-*O*-demethyl analogue **159** of ancistrobenomine A (**4**)

fully dehydrogenated and 5,1′-coupled, but (*P*)-configured (and 6-O-demethylated) naphthylisoquinoline, previously obtained semi-synthetically, by dehydrogenation of ancistrocladine [414] (Fig. 72). Ancistrobenomine A (**4**) was found to be present in the plant in an enantiomerically pure form, as deduced by HPLC-ECD analysis on a chiral phase (Chiralcel® OD-H), resulting in only one sharp peak (Fig. 72). The online ECD spectra monitored from different positions of the peak were all identical [30].

The second and most polar alkaloid (peak **B** in the chromatogram, see Fig. 72) isolated from the stem bark extracts of *A. benomensis* possessed the structure **159**, and it was a new 6-*O*-demethyl derivative of ancistrobenomine A (**4**) (peak **A** in the chromatogram, see Fig. 72). The absolute axial configuration of **159** was determined to be (*M*) by the close resemblance of its ECD spectrum with that of its parent compound **4** [30].

The third compound (peak **C** in the chromatogram, see Fig. 73), possessing a polarity between that of the two ancistrobenomines **4** and **159**, was assigned to be a *N*-methylated naphthyltetrahydroisoquinoline alkaloid, likewise 5,1′-coupled [30]. Except for the presence of a free hydroxy group at C-5′ in the naphthalene part, this new plant metabolite was structurally quite similar to the known [53] alkaloid ancistrocline (**163**) from the Chinese liana *A. tectorius*. The absolute configuration at C-3 was determined to be (*S*) by ruthenium-catalyzed oxidative degradation. The relative configuration at C-1 versus C-3 was deduced to be *cis* from a NOESY cross-

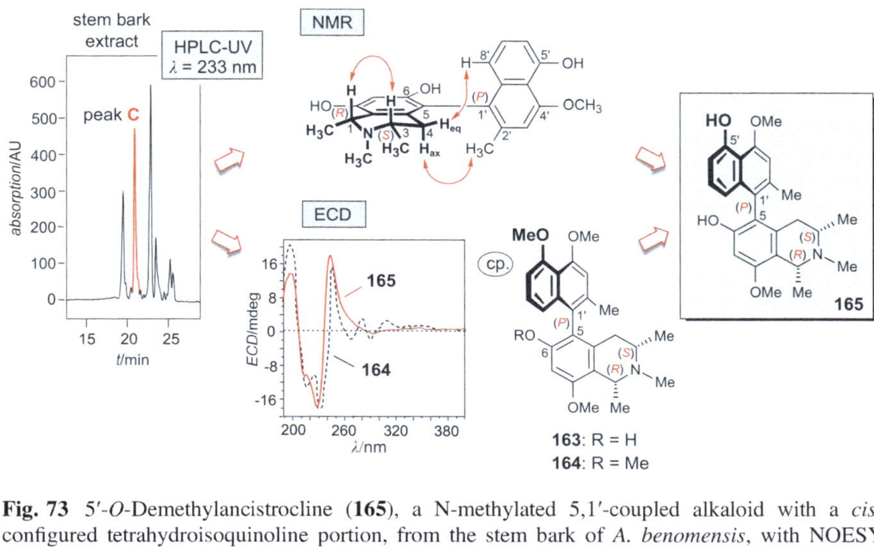

Fig. 73 5′-*O*-Demethylancistrocline (**165**), a N-methylated 5,1′-coupled alkaloid with a *cis*-configured tetrahydroisoquinoline portion, from the stem bark of *A. benomensis*, with NOESY interactions (red double arrows) relevant for the constitution and relative configuration of **165** at the biaryl axis and at C-1 and C-3 in the isoquinoline part (center, top) shown. Also presented is the assignment of the absolute axial configuration of **165**, by comparison of its ECD spectrum with that of the known [322], structurally closely related, and similarly (*P*)-configured compound 6-*O*-methylancistrocline (**164**) (center, bottom)

peak between H-1 and H-3, which, in conjunction with the absolute (*S*)-configuration at C-3, established C-1 to be (*R*)-configured [30]. This, in combination with specific NOESY interactions across the biaryl axis, between H-4_{ax} and Me-2′ and between H-4_{eq} and H-8′ (Fig. 73), showed that the biaryl axis had to be (*P*)-configured. The assignment of the absolute axial configuration was confirmed by the fact that the ECD spectrum of the new metabolite was virtually identical to that of the likewise (*P*)-configured, but O,N-permethylated 6-*O*-methylancistrocline (**164**), well known from total synthesis [322]. The new alkaloid thus had the full stereostructure **165** (see Fig. 73) and was hence named 5′-*O*-demethylancistrocline [30].

6.2 Discovery of the First Dioncophyllaceae-Type Naphthylisoquinoline Alkaloids in the Leaves of an Asian Ancistrocladus Species, A. benomensis

By using liquid–liquid fast centrifugal partition chromatography (FCPC) in combination with preparative HPLC, three further new alkaloids were isolated from the leaf extract of *A. benomensis* [44]. The UV spectra of all compounds displayed a third maximum at around 370 nm, which is characteristic of fully dehydrogenated naphthylisoquinoline alkaloids [30, 64, 167, 168, 186]. Moreover, the molecular weight, as deduced from HREISIMS, was unusually low for alkaloids isolated from an Asian *Ancistrocladus* plant, hinting at the presence of only three oxygen atoms, which, in turn, indicated that these naphthylisoquinolines belong to the subclass of Dioncophyllaceae-type alkaloids, i.e., possessing no oxygen function at C-6 in the isoquinoline subunit.

The most polar compound (= peak **A** in Fig. 74) had a molecular formula of $C_{24}H_{23}NO_3$, as deduced from its monoprotonated molecular ion at *m*/*z* 373.1676, recorded by HRESIMS. The ^{1}H NMR spectrum revealed the presence of seven aromatic protons, two more than in ‘usual’ naphthyltetraisoquinolines and dihydroisoquinolines alkaloids [44]. This was in accordance with the appearance of a spin system of three adjacent protons in the naphthalene portion, H-6′ (doublet at δ_H = 6.73 ppm), H-7′ (pseudo-triplet at δ_H = 7.21 ppm), and H-8′ (doublet at δ_H = 6.82 ppm), along with two doublets at δ_H 7.48 and 7.62 ppm, assigned to be protons located at C-5 and C-6 in the isoquinoline subunit. These data hinted at the presence of a fully dehydrogenated naphthylisoquinoline alkaloid, which was further corroborated by the lack of resonances for protons at C-1 and C-3, and by the occurrence of an additional aromatic proton at δ_H = 7.64 ppm, instead of the signals of the two diastereotopic geminal protons at C-4 typical of alkaloids with a tetra- or dihydroisoquinoline moiety [44]. The ^{1}H and ^{13}C NMR spectroscopic data and specific HMBC and ROESY interactions (see Fig. 74) of the new metabolite evidenced the presence of a 7,1′-coupled naphthylisoquinoline alkaloid with a constitution identical to that of dioncophylleine A (**158a**), which had previously been obtained by semi-synthesis from dioncophylline A (**16**), but yet not found to

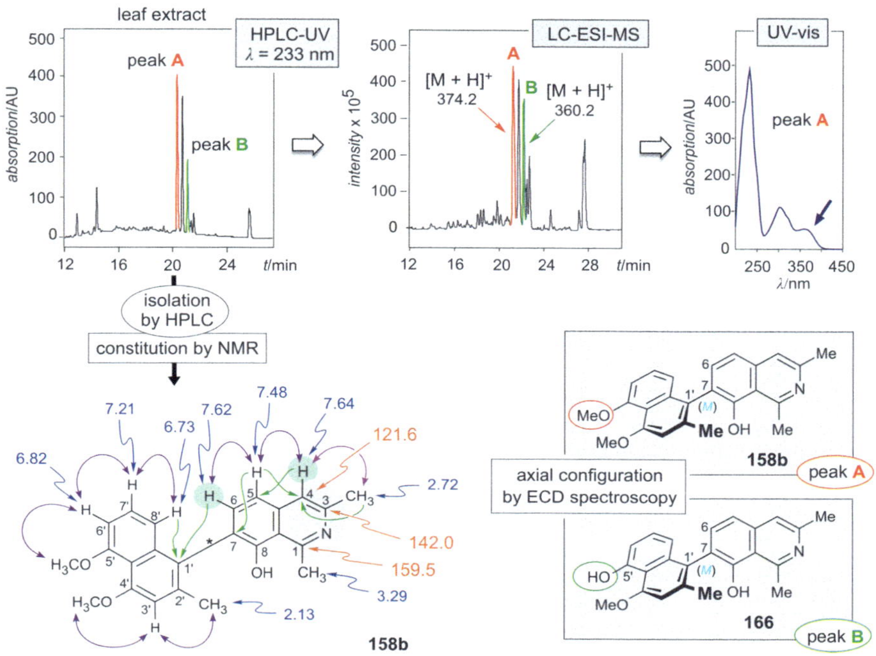

Fig. 74 *ent*-Dioncophylleine A (**158b**) (= peak **A**) and its 5′-*O*-demethyl analogue **166** (= peak **B**), two fully dehydrogenated 7,1′-coupled Dioncophyllaceae-type naphthylisoquinoline alkaloids from *A. benomensis*. HPLC-ESI-MS analysis of the leaf extract, UV spectrum of peak **A**, and ^{1}H and ^{13}C NMR data (δ/ppm), HMBC (green single arrows), and NOESY (purple double arrows) interactions relevant for the constitution of **158b**. The two aromatic protons typical of a non-hydrogenated Dioncophyllaceae-type alkaloid are highlighted in green

occur in Nature [13, 414]. The absolute axial configuration of the isolated plant metabolite was assigned to be (*M*) due to its nearly perfectly mirror-imaged ECD spectrum compared to that of the (*P*)-configured dioncophylleine A (**158a**) (see Fig. 75). Thus, the new alkaloid had the full absolute stereostructure **158b** and was, hence, the enantiomer of **158a**. It was therefore named *ent*-dioncophylleine A [44].

HPLC analysis of **158b** on a chiral phase (Chiralcel® OD-H) resulted in two peaks with a ratio of 7:93, with the more rapidly eluting small peak revealing an ECD spectrum being identical with that of **158a**, while the major peak was clearly identified to represent *ent*-dioncophylleine A (**158b**) (see Fig. 75). Obviously, this fully dehydrogenated naphthylisoquinoline alkaloid was produced by *A. benomensis* in a scalemic form [44]. Its 5′-*O*-demethyl analogue **166**, by contrast, which was likewise found to occur in the leaves of *A. benomensis* as a major metabolite (= peak **B** in Fig. 74), was determined to be enantiomerically pure. HPLC analysis of **166** on a chiral phase (Chiralcel® OD-H) resulted in only one peak [44].

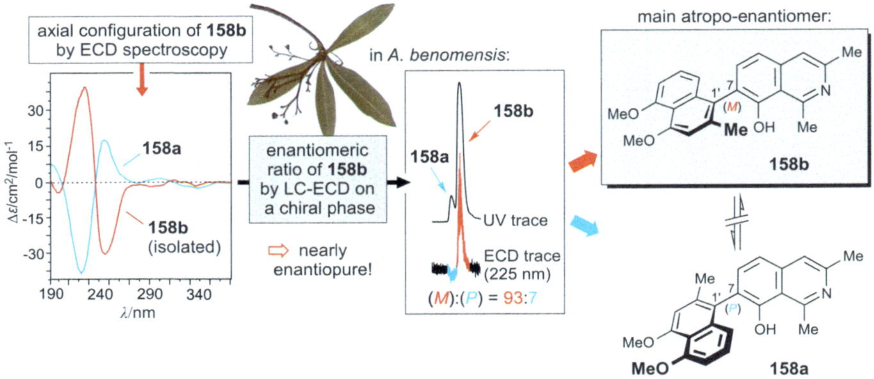

Fig. 75 Assignment of the absolute axial configuration of *ent*-dioncophylleine A (**158b**) by comparison of its ECD spectrum with that of its known [13, 414] (*P*)-enantiomer, dioncophylleine A (**158a**), and HPLC-ECD analysis on a chiral phase (Chiralcel® OD-H) showing **158b** to occur in the plant in a nearly enantiopure form, together with its enantiomer **158a**, in a ratio of 93:7

A further new alkaloid (= peak **C** in Fig. 76), with a polarity between those of *ent*-dioncophylleine A (**158b**) and its 5′-*O*-demethyl analogue **166**, was discovered in the leaf extract of *A. benomensis*. It showed the same molecular weight (*m/z* 373) as **158b**, and displayed a molecular formula of $C_{24}H_{23}NO_3$, which was identical to that of **158b**, as deduced by HRESIMS [44]. Its ^{1}H NMR data again revealed the presence of a non-hydrogenated Dioncophyllaceae-type naphthylisoquinoline alkaloid. HMBC interactions from H-1′ and H-6 to C-8′ and from H-7′ and H-5 to C-7 established the position of the biaryl axis at C-8′ in the naphthalene subunit and at C-7 in the isoquinoline portion [44]. This assignment was corroborated by the normal, not high-field-shifted signal of Me-2′ (δ_H = 2.30 ppm), and by NOESY correlation sequences in the series {H-1′ ↔ Me-2′ ↔ H-3′ ↔ OMe-4′} and {OMe-5′ ↔ H-6′ ↔ H-7′} in the naphthalene part, and {H-6 ↔ H-5 ↔ H-4 ↔ Me-3} in the isoquinoline moiety. Thus, the new alkaloid was a 7,8′-linked isomer of *ent*-dioncophylleine A (**158b**), with the constitution **167** as presented in Fig. 76. It represents the "D-type" coupling within the subclass of Dioncophyllaceae-type naphthylisoquinoline alkaloids and was henceforth named dioncophylleine D [44].

Unlike *ent*-dioncophylleine A (**158b**), however, dioncophylleine D (**167**) was optically inactive ($[\alpha]_D$ = 0), and it likewise did not show any Cotton effect in its ECD spectrum. HPLC analysis of **167** on a chiral phase (Chiralcel® OD-H) gave only one peak, even under different chromatographic conditions, thus corroborating **167** to be present in the plant as a racemic mixture of its two atropo-enantiomers, **167a** and **167b** [44]. The reason for the low rotational barrier in **167** and thus, being different from *ent*-dioncophylleine A (**158b**), is the occurrence of atropo-enantiomerization at room temperature, due to the lack of a methyl group next to the biaryl axis (as in **158b**).

The only other known fully dehydrogenated alkaloid with a 7,8′-coupling, possessing a similar molecular framework, ancistrobreveine C (**168**) (Fig. 76), an

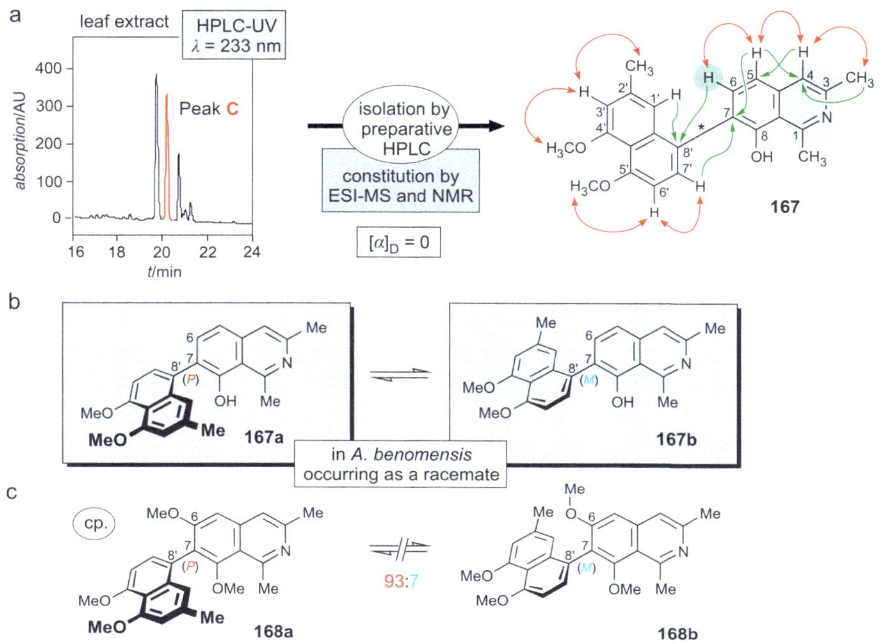

Fig. 76 **a** Dioncophylleine D (**167**), a 7,8′-linked fully dehydrogenated Dioncophyllaceae-type naphthylisoquinoline alkaloid from the leaves of *A. benomensis*: **a** selected HMBC (green single arrows) and NOE interactions (red double arrows) relevant for the constitution; **b** occurrence of the new alkaloid in the plant as a racemic mixture of its two atropo-enantiomers, **167a** and **167b**; **c** comparison of **167** with the structurally closely related, similarly 7,8′-coupled ancistrocladaceous alkaloid ancistrobreveine C (**168**) [186], which is configurationally stable at the biaryl axis, and is present in its natural producer *A. abbreviatus* scalemic, as an atropo-enantiomeric mixture, with the (*P*)-enantiomer **168a** being the by far major one in the plant

Ancistrocladaceae-type compound identified in the roots of *A. abbreviatus* from Coastal West Africa, is however, optically active. It occurs in the plant as a scalemic, yet nearly enantiopure, mixture, the ratio of the two configurationally stable enantiomers, **168a** and **168b**, being ca. 93:7 ((*P*):(*M*)) [186].

6.3 *Online Structural Assignment and Observation of the Atropisomerization of the N,6′-Coupled Alkaloid Ancistrocladinium B*

One less polar compound detected in leaf and stem bark extracts of *A. benomensis* was easily identified as the known N,C-coupled naphthyldihydroisoquinoline ancistrocladinium B (**14a/14b**) (Fig. 77). This structurally unprecedented alkaloid had for the first time been discovered in air-dried leaves of a botanically yet undescribed Congolese *Ancistrocladus* species [18, 32], and had later also been isolated from the aerial parts of the Indian liana *A. heyneanus* [32] (Sect. 4), from the Southeast Asian species *A.* aff. *tectorius* [32, 33] (Sect. 8), and from another, so far unidentified *Ancistrocladus* liana endemic to the north-central region of the Democratic Republic of the Congo [161]. Ancistrocladinium B (**14a**) and its rotational isomer, *N-epi*-ancistrocladinium B (**14b**), are the as yet only representatives of the unprecedented N,6′-coupling type, interconverting by a slow rotation about the hetero biaryl axis at room temperature, thus occurring as a 54:46 mixture of two configurationally semi-stable atropo-diastereomers [18, 32]. The complete stereostructures of the two atropo-diastereomers **14a** and **14b** were established by ruthenium-mediated oxidative degradation, specific ROESY correlations across the N,C-axis [32], and by experimental ECD spectroscopy in combination with quantum-chemical ECD calculations [18, 32, 76, 78, 83]. More recently, ancistrocladinium B (**14a/b**) has been synthesized via a short sequence of eight linear steps [101] (see Sect. 8.5.9.2).

Ancistrocladinium B (**14a/b**) was discovered in the aerial parts of *A. benomensis* by analytical HPLC of the alkaloid-enriched extract on a C_{18} column, revealing the presence of two baseline-separated peaks, **A** and **B**, in the less polar region of the chromatogram. Preparative isolation of peak **A** in a pure form again resulted in a slow interconversion, leading back to a mixture of the two peaks **A** and **B**. In the same way, pure fractions of peak **B** underwent a renewed equilibration to give mixtures of peaks **A** and **B**. Even immediate HPLC analysis of the freshly resolved single peaks showed a gradual interconversion of the two atropo-diastereomers already at room temperature [18, 32, 78]. By HPLC-MS, a mass of *m/z* 406 was determined for both peaks **A** and **B**, and HRESIMS evidenced that the two alkaloids were cationic, both corresponding to a molecular formula of $C_{25}H_{28}NO_4^+$. They exhibited nearly the same NMR data, just slightly shifted for some of the signals, with virtually identical coupling patterns and constants in the ^{1}H NMR spectrum, suggesting the presence of two slowly interconverting atropo-diastereomers [18, 32]. By ruthenium-mediated oxidative degradation performed directly on a mixture of peaks **A** and **B**, the absolute configuration at C-3 was established to be (*S*) for both atropisomers, exclusively providing the (*S*)-enantiomer of 3-aminobutyric acid [32]. The absolute configuration at the N,C-axis of the two epimers was determined by atropisomer-specific ROESY correlations in combination with the known (*S*)-configuration at C-3. Long-range ROESY cross-peaks between H-3 and H-7′ permitted assignment of the absolute axial configuration of the more rapidly eluting isomer **14b** as (*P*) (peak **A**). For the more slowly eluting atropisomer **14a** (peak **B**), a likewise specific, yet complementary ROESY interaction was observed, now between the protons of Me-3 and H-7′, thus revealing an absolute (*M*)-configuration at the iminium-aryl axis (Fig. 77, top) [18, 32, 78]. The ECD spectra of the two alkaloids **14a** and **14b** monitored by HPLC-ECD in the stopped-flow mode were found to be virtually opposite, thus additionally corroborating ancistrocladinium B (**14**) to be a mixture of two atropo-diastereomers. In view of the unprecedented N,6′-coupling type, quantum-chemical ECD calculations at the TDDFT (B3LYP/TZVP) level were performed for the configurational assignment of the two atropisomers.

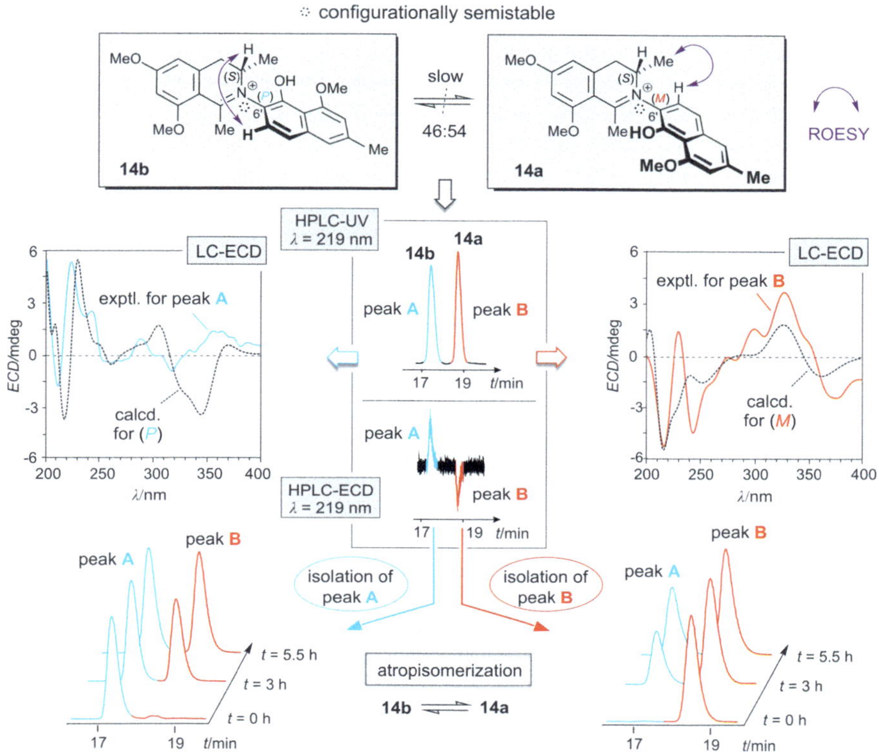

Fig. 77 Resolution and online assignment of the stereostructures of the two rotational isomers of ancistrocladinium B, **14a** and **14b**, by specific NOESY correlations and by experimental ECD investigations in combination with quantum-chemical calculations. Experimental LC-ECD curves were measured on the nearly pure peak **A** (left) and pure peak **B** (right), monitored in the stopped-flow mode, and comparison conducted with the ECD spectra calculated for the (*M*)- and the (*P*)-configured atropo-diastereomers **14a** and **14b**, respectively, using the TDDFT (B3LYP/TZVP) approach. The determination of the atropisomerization rates of **14a** and **14b**, in terms of their separate thermal equilibration over time in MeOH/H_2O (1:1) at 65°C, was analyzed by HPLC on an achiral C_{18} phase

The ECD spectrum theoretically predicted for the (*P*,*S*)-diastereomer **14b** was in accordance with the experimental curve of peak **A**, while the spectrum calculated for the (*M*,*S*)-diastereomer **14a** agreed well with that measured for peak **B** (Fig. 77, center) [18, 32, 76, 78, 83]. The results fully confirmed the attribution of the absolute axial configuration of **14a** and **14b** as deduced from the above-presented ROESY measurements.

The configurational semi-stability of ancistrocladinium B (**14a**) and its epimer **14b** was investigated more closely by HPLC-UV to determine the atropisomerization barrier by monitoring over time, the decrease of the diastereomeric purity of freshly resolved samples enriched with the (*P*)- or the (*M*)-atropo-diastereomer (Fig. 77, bottom) [18, 32, 78]. The diastereomerization rates for both directions were determined at different temperatures (between 55 and 85°C). The Gibbs free energies of activation were determined to be $\Delta G^{\#}_1 = 105.7$ kJ mol^{-1} for the atropisomerization from **14a** to **14b**, and $\Delta G^{\#}_2 = 105.8$ kJ mol^{-1} for the opposite direction; they were in agreement with the slow interconversion of the two atropo-diastereomers at room temperature [18, 32, 78]. The results also fitted well with those of DFT calculations (B3LYP/6-31G*) carried out in vacuo (i.e., neglecting any solvent effects), providing a value of $\Delta G^{\#}_1 = 113.9$ kJ mol^{-1} for the axial rotation from **14a** to **14b**, and of $\Delta G^{\#}_2 = 112.6$ kJ mol^{-1} for the atropisomerization process back from **14b** to **14a**. The calculated stability difference of 1.3 kJ mol^{-1} in favor of the (*M*)-atropo-diastereomer **14a** was in agreement with the experimentally monitored equilibrium ratio of 54:46 ((*M*):(*P*)) for **14a**:**14b** [18, 32, 78].

6.4 The New Species *Ancistrocladus benomensis* and Its Extraordinary Chemotaxomic Position Within the Ancistrocladaceae Family

Despite the broad structural variety of naphthylisoquinoline alkaloids isolated from African and Asian lianas belonging to the Ancistrocladaceae [13, 16–18], the plants obviously follow strict biosynthesis principles, which are, thus, a useful criterion for the taxonomic classification of different plant species. The botanically related Dioncophyllaceae lianas, like, e.g., *T. peltatum* [13, 14, 16] from the Ivory Coast, solely produce alkaloids with an (*R*)-configuration at C-3 and lacking an oxygen function at C-6 (the so-called "Dioncophyllaceae-type" compounds). *Ancistrocladus* species from East Africa and Asia, in contrast, exclusively produce typical Ancistrocladaceae-type alkaloids, i.e., with an oxygen function at C-6 and the (*S*)-configuration at C-3 in their di- or tetrahydroisoquinoline subunit [1–5, 23–31, 33, 37–64, 155, 189, 190]. Besides the presence of Ancistrocladaceae-type compounds, the metabolic profiles of Central African *Ancistrocladus* species are characterized by a remarkable number of so-called hybrid-type naphthylisoquinoline alkaloids, i.e., also oxygenated at C-6, but with the (*R*)-configuration at C-3 [154, 156, 160, 163–170, 172–181, 185, 188, 195]. Besides these Ancistrocladaceae-, Dioncophyllaceae-, and hybrid-type compounds, only one single example of an inverse hybrid-type alkaloid (i.e., 6-H, (3*S*)) has been identified in *A. abbreviatus* from West Africa [124].

The alkaloid pattern (Fig. 78) of the newly discovered highland liana *A. benomensis* [30, 44] from Southwest-Malaysia attracted special attention since it was found to be substantially different from those of all the other known Asian and African *Ancistrocladus* species [13, 16–18]. In particular, it strongly differs from that of *A.* aff. *tectorius*, the only other taxon reported to occur in Malaysia [54]. The plant predominantly contains naphthylisoquinoline alkaloids with a non-hydrogenated isoquinoline portion as the main metabolites [30, 44]. Six representatives have so far been identified, among them the very first naphthylisoquinoline alkaloids discovered in an Asian *Ancistrocladus* taxon that lack an oxygen function at C-6 such as *ent*-dioncophylleine A (**158b**) and dioncophylleine D (**167**) (Fig. 78), thus being classified as Dioncophyllaceae-type compounds [44, 186]. They may also be considered as Ancistrocladaceae-Dioncophyllaceae inverse hybrid-type naphthylisoquinoline alkaloids, but there is no stereocenter at C-3 as the second required criterion for this classification. Moreover, the 7,1′-coupled *ent*-dioncophylleine A (**158b**) was the first fully dehydrogenated Dioncophyllaceae-type naphthylisoquinoline alkaloid with the biaryl axis being the only element of chirality that was optically active, whereas the 7,8′-linked dioncophylleine D (**167**) is configurationally unstable at the biaryl axis, thus occurring in the plant as a racemate [44]. It is therefore one of the very rare, de facto achiral naphthylisoquinoline alkaloids, besides the 7,6′-linked dioncophyllacine B (**169**) (Fig. 78), which is so far only known to occur in the leaves of the West African liana *T. peltatum* (Dioncophyllaceae) [13, 14].

Further unique secondary metabolites discovered in the aerial parts of *A. benomensis* are the similarly non-hydrogenated, but Ancistrocladaceae-type alkaloid ancistrobenomine A (**4**) (Fig. 78), for which the methyl group at C-3 is oxygenated, giving an unprecedented hydroxymethyl substituent [30], and ancistrocladinium B (**14a**) and its rotational isomer **14b** [18, 32, 78] (Fig. 78). The latter two alkaloids are based on the very rare N,6′-coupling type, showing a slow rotation about the central N,C-axis, thus occurring as a mixture of two configurationally semi-stable atropo-diastereomers.

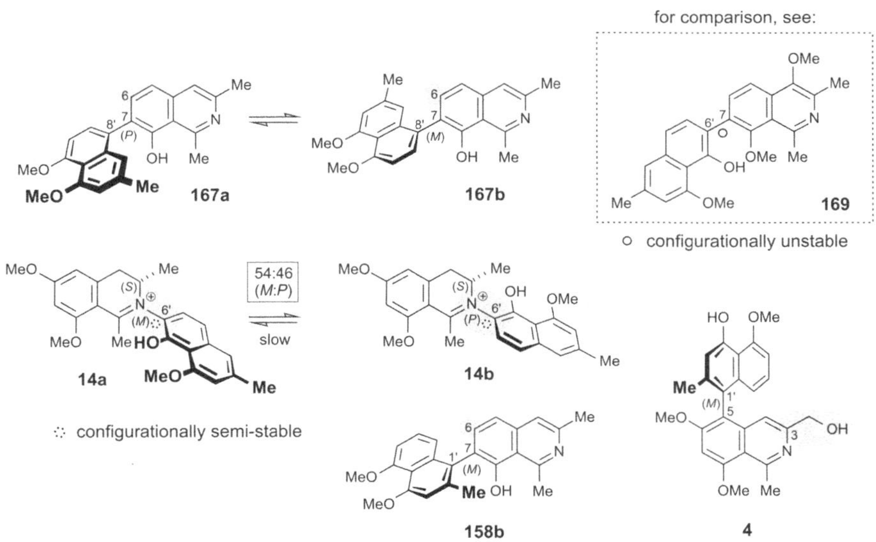

Fig. 78 Examples of selected C,C- and N,C-coupled biaryls with unusual structural features (underlaid in gray) discovered in the Malaysian species *A. benomensis*. Among these are the first Dioncophyllaceae-type naphthylisoquinolines such as *ent*-dioncophylleine A (**158b**) and dioncophylleine D (**167**) ever found in an Asian *Ancistrocladus* species, i.e., lacking an oxygen function at C-6. See also the structure of dioncophyllacine B (**169**) as the only other known example of an achiral naphthylisoquinoline alkaloid, besides dioncophylleine D (**167**)

Summarizing, the structures of the alkaloids fully support the exceptional position of this liana within the Ancistrocladaceae plant family, and—in accordance with the results obtained by morphological [34] and genetic [36] studies—clearly confirm *A. benomensis* to be a valid, distinct *Ancistrocladus* taxon.

Ancistrocladinium B (**14**) is considered to be a promising candidate as a lead compound for leishmanicidal drug therapy [32, 66, 70] (see Sect. 8). All the other naphthylisoquinoline alkaloids isolated from *A. benomensis* exhibit only moderate to weak inhibitory effects against *P. falciparum* and the pathogen of African sleeping sickness, *T. brucei rhodesiense* [30, 44]. They are even completely inactive against *Leishmania* parasites (*L. donovani* and *L. major*) and *T. cruzi* (Chagas' disease) [30, 44].

7 *Ancistrocladus cochinchinensis* from Central Vietnam, a Distinct *Ancistrocladus* Taxon? Metabolite Pattern and Phylogenetic Relationship to *Ancistrocladus* aff. *tectorius* from China

A. cochinchinensis is a woody liana indigenous to the rainforests of Central Vietnam [19, 21, 303, 415–417]. In folk medicine, the plant is used as a diuretic, antifebrile, and antiphlogistic remedy [46, 417–419]. The revisions of the *Ancistrocladus* genus undertaken by van Steenis [35] and by Taylor et al. [21] excluded *A. cochinchinensis* as a valid taxon, reducing it to taxonomic synonymy under *A. tectorius*. A molecular taxonomic study [36] on *A.* aff. *tectorius* (75 samples collected from 24 locations all over Southeast Asia, see Fig. 79) using comparative sequencing of the ITS region of nuclear rDNA and the *trn*K intron of cpDNA in combination with intersimple sequence repeat (ISSR) fingerprinting showed a high *intra*specific genetic variability (see Sect. 8.1), comparable to the *inter*specific differences among the African taxa [36]. These investigations clearly suggest that *A. tectorius* can be split into at least two species [36] represented by two morphological forms described by Taylor et al. in 2005 [21]. One group was similar to *A. heyneanus* from the Western Ghats of India and closely resembled the type specimens of *A. tectorius* (from Vietnam), *A. cochinchinensis* (from Bin Dinh, Vietnam), and *A. harmandii* (from Koukou, Mun River, Laos) [21, 36]. Most of the plant samples attributed to this morphological group were collected in the northern distribution range of *Ancistrocladus* lianas, from China through Vietnam and many were from Malaysia, Singapore, and Sumatra [21]. The other group of plants found in Malaysia, Sumatra, and Singapore resembled the type of *A. pinangianus* (from Pilau Pinang, Malaysia) [21, 36]. The type specimens investigated for their ITS sequences were positioned in different phylogenetic clades. Those of *A. pinangianus* were placed together with samples from *Ancistrocladus* plants from southern Thailand and Malaysia, while *A. cochinchinensis* formed an own clade together with most of the samples of *Ancistrocladus* lianas from eastern Thailand, Laos, and China, thus indicating that *A.*

Fig. 79 Map of the area investigated in a molecular taxonomic study [36] on *A.* aff. *tectorius* (75 samples collected from 24 locations all over Southeast Asia), with the marks indicating the sample sites (▲) and loci classici (●) of type and isotype specimens; the main centers of distribution of Ancistrocladaceae species in Southeast Asia are underlaid in green

pinangianus and *A. cochinchinensis* might indeed deserve the status of being distinct species [36]. Furthermore, the genetic investigations revealed a significant similarity of the ITS data of the isotype specimen of *A. cochinchinensis* (from its *locus classicus* in Binh Dinh, Central Coastal Vietnam) with those of *A.* aff. *tectorius* collected in the rainforests of the Chinese Island Hainan [36].

With respect to the botanical background of *A. cochinchinensis*, suggesting a close phylogenetic relationship to *A.* aff. *tectorius* from China [21, 36], more in-depth phytochemical studies on the naphthylisoquinoline pattern of this Vietnamese liana were of particular interest. *A. cochinchinensis* was found to be a rich source of predominantly 5,1′-coupled naphthylisoquinoline alkaloids (nine compounds), among them 4′-*O*-demethylancistrocladine (**170a**) [47], earlier found

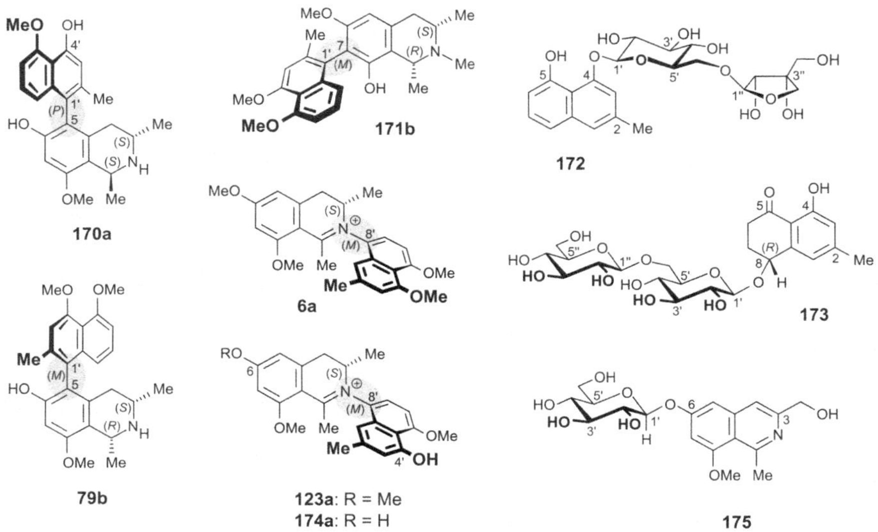

Fig. 80 Selected naphthylisoquinoline alkaloids representing the three biaryl coupling types detected in the Vietnamese liana *A. cochinchinensis*: 4′-*O*-demethylancistrocladine (**170a**) and isohamatine (**79b**) (both 5,1′-coupled), 7-*epi*-ancistrobrevine D (**171b**) (7,1′-linked), and ancistrocladinium A (**6a**) and its mono- and bisphenolic analogues **123a** and **174a** (all N,8′-coupled). Also discovered in the plant were the biosynthetically related, yet not coupled, metabolites ancistronaphthosides A (**172**) and B (**173**) as well as anciscochine-6-*O*-*β*-D-glucopyranoside (**175**)

in *A. tectorius* [54], and the newly discovered isohamatine (**79b**) [48] (Fig. 80). These compounds were isolated from leaf and stem bark extracts, along with three 7,1′-linked alkaloids such as 7-*epi*-ancistrobrevine D (**171b**) (Fig. 80) [45, 46] and the first representatives of N,8′-coupled naphthyldihydroisoquinolines possessing free phenolic hydroxy groups, 4′-*O*-demethylancistrocladinium A (**123a**) and 6,4′-*O*,*O*-didemethylancistrocladinium A (**174a**) (Fig. 80). They were identified together with their known, non-phenolic parent compound ancistrocladinium A (**6a**) and the corresponding (*P*)-configured atropo-diastereomers **6b** (Fig. 84), **123b**, and **174b** (Fig. 85) [47].

This Vietnamese liana strictly follows the chemotaxonomic rule for Asian *Ancistrocladus* species by producing exclusively typical Ancistrocladaceae-type naphthylisoquinoline alkaloids, i.e., with an oxygen function at C-6 and the (*S*)-configuration at the stereocenter at C-3. Moreover, an unusually high number (eleven compounds) of biosynthetically related, yet non-coupled naphthoquinones and tetralones [46, 48, 418, 419] and, vice versa, naphthalene-devoid isoquinolines [48] were detected in the plant, among them the ancistronaphthosides A (**172**) and B (**173**) as well as anciscochine-6-*O*-*β*-D-glucopyranoside (**175**) (Fig. 80), so far only known to occur in this *Ancistrocladus* liana from Central Vietnam [48].

7.1 In the Leaves of Ancistrocladus cochinchinensis*: Predominant Occurrence of Naphthylisoquinoline Alkaloids with a 5,1′-Coupling Site*

A characteristic feature of the alkaloid pattern of *A. cochinchinensis* is the predominant occurrence of 5,1′-linked naphthylisoquinoline alkaloids (nine compounds) as the main secondary metabolites [45–48]. The isolation of ancistrocladinine (**77a**) and its atropo-diastereomer hamatinine (**77b**) [45, 46], which were detected in the plant in a ratio of ca. 1:1, along with two congeners, 5′-*O*-demethylhamatinine (**177**) [47] and the permethylated 6-*O*-methylhamatinine (**176**) [45, 46] (Fig. 81), strongly supported a close relationship to *A.* aff. *tectorius* from the Chinese Province of Hainan, since this liana was likewise found to produce the same compounds, along with a plethora of further 5,1′-coupled alkaloids as the major constituents. A striking fact, however, in comparison to other *Ancistrocladus* species, is the remarkably high number of naphthyl*dihydro*isoquinolines (four compounds) and non-hydrogenated alkaloids (three compounds) identified in the leaves of *A. cochinchinensis*, whereas naphthyl*tetrahydro*isoquinoline alkaloids, which are the most common metabolites in many Asian and African *Ancistrocladus* species, were significantly less represented. Only two

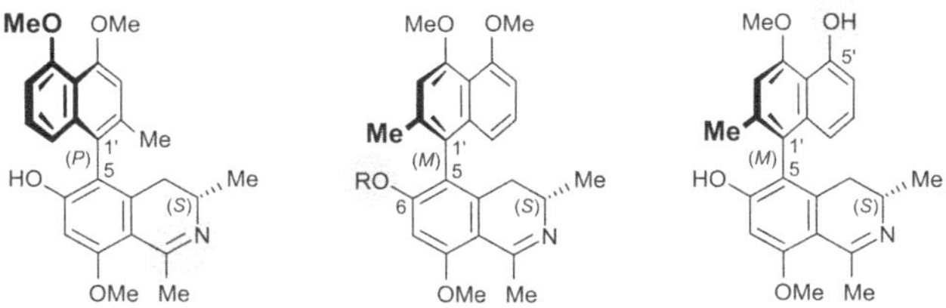

77a **77b**: R = H **176**: R = Me **177**

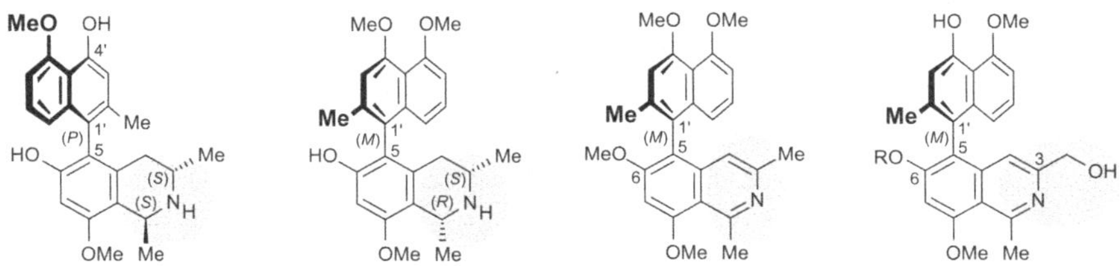

170a **79b** **73** **4**: R = Me **159**: R = H

Fig. 81 A series of nine 5,1′-linked naphthylisoquinoline alkaloids detected in the leaves of *A. cochinchinensis* constituted by the two atropo-diastereomeric alkaloids ancistrocladinine (**77a**) and hamatinine (**77b**), 6-*O*-methylhamatinine (**176**), 5′-*O*-demethylhamatinine (**177**), 4′-*O*-demethylancistrocladine (**170a**), isohamatine (**79b**), the fully dehydrogenated 6-*O*-methylhamateine (**73**), and ancistrobenomine A (**4**) and its 6-*O*-demethyl analogue **159**. The structural landmarks of the compounds are underlaid in gray

5,1′-coupled alkaloids with a tetrahydroisoquinoline subunit were discovered in *A. cochinchinensis*, namely, the *trans*-configured 4′-*O*-demethylancistrocladine (**170a**) [47], earlier found in the Chinese liana *A. tectorius* [64], and the new *cis*-configured isohamatine (**79b**) [48], possessing a free, unmethylated NH function. The latter compound has so far only been isolated from this Vietnamese liana. The plant apparently produces **79b** in an atropisomerically pure form. Complementary to *A. cochinchinensis*, its atropo-diastereomer, isoancistrocladine (**79a**, see Fig. 29), was found to occur in the Indian species *A. heyneanus* in a stereochemically homogeneous form, i.e., without isohamatine (**79b**) being detectable [38].

Isolation work on the leaves of *A. cochinchinensis* also afforded the non-hydrogenated and, thus, fully aromatic alkaloid 6-*O*-methylhamateine (**73**) (Fig. 81) as a major metabolite [45, 46]. This compound displayed a virtually opposite ECD curve compared to the spectrum measured for the likewise 5,1′-linked ancistrocladeine (**162**) (see Fig. 72), a structurally similar, but (*P*)-configured (and 6-*O*-demethylated), as yet merely artificial, semi-synthetic naphthylisoquinoline [4, 414], thus assigning **73** to be (*M*)-configured [45]. For a long time, 6-*O*-methylhamateine (**73**) had been found in *A. cochinchinensis*, exclusively. Only recently has this compound been detected in the roots of *A. abbreviatus* from Coastal West Africa, yet as a minor metabolite [186].

Furthermore, two fully dehydrogenated 5,1′-coupled naphthylisoquinoline alkaloids equipped with a hydroxymethylene function at C-3 were found to be produced by *A. cochinchinensis*, ancistrobenomine A (**4**) and 6-*O*-demethylancistrobenomine A (**159**) [47] (Fig. 81), which had firstly been discovered in the Malaysian species *A. benomensis* [30] (see Sect. 6). They were identified online directly in the crude dichloromethane extracts of the leaves of *A. cochinchinensis* by analytical HPLC on a reversed-phase column. A combination of LC-MS, LC-NMR, and LC-ECD coupling permitted a rapid and straightforward dereplication of these two already known [30, 47] naphthylisoquinoline alkaloids, including their absolute configuration, without the necessity of their isolation. Their spectroscopic and chromatographic data agreed with those reported in the literature [30, 47].

7.2 7-epi-Ancistrobrevine D and Related 7,1′-Coupled Naphthyltetrahydroisoquinoline Alkaloids

From the leaves of *A. cochinchinensis*, three 7,1′-coupled naphthylisoquinoline alkaloids have so far been identified, namely, 7-*epi*-ancistrobrevine D (**171b**) and two related analogues, 6-*O*-demethyl-7-*epi*-ancistrobrevine D (**178**) and 6-*O*-demethyl-8-*O*-methyl-7-*epi*-ancistrobrevine D (**179**) (Fig. 82), all possessing a 1,3-*cis*-configured and *N*-methylated tetrahydroisoquinoline portion [45, 46]. In *A. cochinchinensis*, 7-*epi*-ancistrobrevine D (**171b**) is present in the leaves and stems as the main secondary metabolite, produced by the plant in an atropisomerically pure form [45, 46], i.e., without any co-occurring ancistrobrevine D (**171a**), which in turn was also identified even as the major constituent of the West African liana *A.*

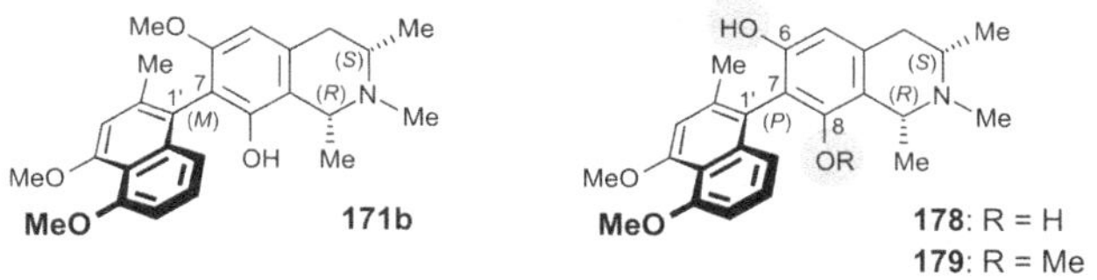

Fig. 82 7-*epi*-Ancistrobrevine D (**171b**), a 7,1′-coupled naphthyltetrahydroisoquinoline alkaloid, which occurs in *A. cochinchinensis* in an atropo-diastereomerically pure form, along with two related stereochemically pure analogues, **178** and **179**

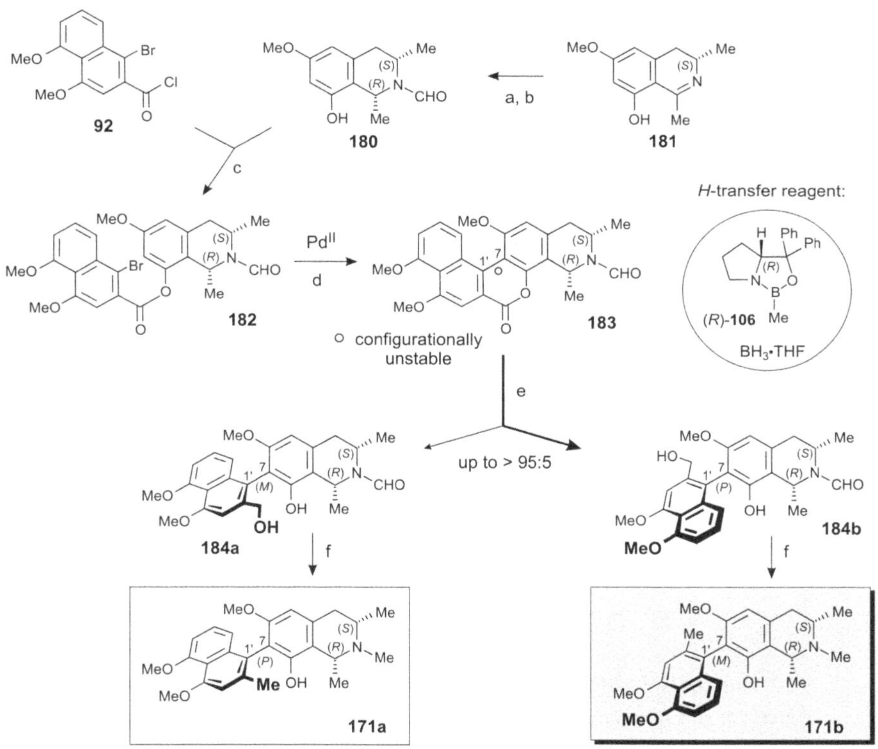

Fig. 83 Atropo-divergent first total synthesis of 7-*epi*-ancistrobrevine D (**171b**) from *A. cochinchinensis* and its likewise naturally occurring atropo-diastereomer, ancistrobrevine D (**171a**), produced by *A. abbreviatus* from Coastal West Africa [13, 191], by using the lactone approach. Reaction conditions: (a) $NaBH_4$, MeOH, 0°C, 80%; (b) *p*-nitrophenyl formate, Hünig's base, DMF, 0–25°C, 89%; (c) oxalyl chloride, DMF, NEt_3, CH_2Cl_2, 0°C, 84%; (d) $(PPh_3)_2PdCl_2$, NaOAc, DMA, 84%; (e) (*R*)-**106**·BH_3, THF, 0°C, 29%; (f) $(CBrCl_2)_2$, PPh_3, CH_2Cl_2 [343] → $LiAlH_4$, THF, reflux, 72% (for **171b**) and 83% (for **171a**)

abbreviatus, again in an atropo-diastereomerically pure form [13, 191]. Elucidation of the complete stereostructures of **171a** and **171b** was achieved by a fruitful interplay of NMR measurements (in particular by analysis of atropisomer-specific long-range NOE interactions) with ECD spectroscopy, oxidative degradation [13, 45, 191], and the atropo-divergent total synthesis of all four possible *cis*-configured stereoisomers [420].

The atroposelective preparation of **171b**, or, if desired, **171a** (Fig. 83), and their unnatural enantiomers, *ent*-**171b** (1*S*,3*R*,7*P*), or, optionally, *ent*-**171a** (1*S*,3*R*,7*M*), was achieved by applying the lactone method, a strategy developed in the authors' group for the regio- and stereoselective construction of biaryl axes (see Sect. 4.2). In view of the fact that only 7-*epi*-ancistrobrevine A (**171b**) and its 7-epimer **171a** are genuine natural products and since all of the decisive steps within the synthesis route were basically identical, only the atropo-divergent synthesis of **171b** and **171a** will be presented here, exemplarily, in more detail.

cis-Selective reduction of the known [242] deoxygenated monophenolic dihydroisoquinoline **181** with $NaBH_4$ followed by N-formylation provided the (1*R*,3*S*)-configured tetrahydroisoquinoline building block **180** with good diastereoselectivity (*dr* = 93:7) [13, 420]. Prefixation of the known [91, 93] bromonaphthoyl chloride **92** and the *N*-formyl-protected tetrahydroisoquinoline part **180** gave the bromo ester **182**, which was subjected to a Pd-catalyzed intramolecular coupling, thus affording the desired 7,1′-coupled lactone **183** in excellent yield (84%) [13, 420].

As anticipated, this bridged biaryl **183** was configurationally unstable at the biaryl axis and consequently an ideal joint precursor for both atropisomeric target molecules, by reductive cleavage of **183** using borane activated by the chiral oxazaborolidine catalyst **106**, which provided the ring-opened diol **184** in high asymmetric inductions. Thus, by the use of the CBS catalyst (–)-(*R*)-2-methyl-oxazaborolidine, (*R*)-**106**, the configurationally stable (*M*)-configured alcohol **184b** was obtained almost exclusively (*dr* = 97:3), while ring opening of **183** in the presence of (*S*)-**106** as the catalyst led to an excellent atropisomeric ratio in favor of the (*P*)-configured alcohol **184a** (>95:5) (Fig. 83). Hydroxy-bromine exchange and reduction by standard procedures finally yielded the target molecule, 7-*epi*-ancistrobrevine D (**171b**) or, optionally, its atropisomer ancistrobrevine D (**171a**) [13, 420].

7.3 N,8′-Coupled Naphthylisoquinoline Alkaloids Directly Analyzed Online, by LC-MS/MS-NMR-ECD: Ancistrocladinium A and Two New Phenolic Analogues

The N,8′-coupled naphthylisoquinoline alkaloid ancistrocladinium A (**6a/6b**), for the first time discovered in a botanically undescribed Congolese *Ancistrocladus* plant [32], was found to be a main metabolite in the leaf and the bark extracts of *A. cochinchinensis* (Fig. 84) [47]. LC-ESI-MS on the respective peak gave a signal at *m/z* 420.2, which was in accordance with the molecular ion $[M]^+$ of **6a/6b** and its cationic molecular formula of $C_{26}H_{30}NO_4^+$. The typical downfield shift of the methyl group at C-1 (δ_H = 2.52 ppm), the chemical shift of the ^{13}C NMR signal at C-1 (δ_C = 179.6 ppm), together with the lack of an H-1 quartet, revealed the presence of a naphthyl*dihydro*isoquinoline subunit. The aromatic region of the ^1H

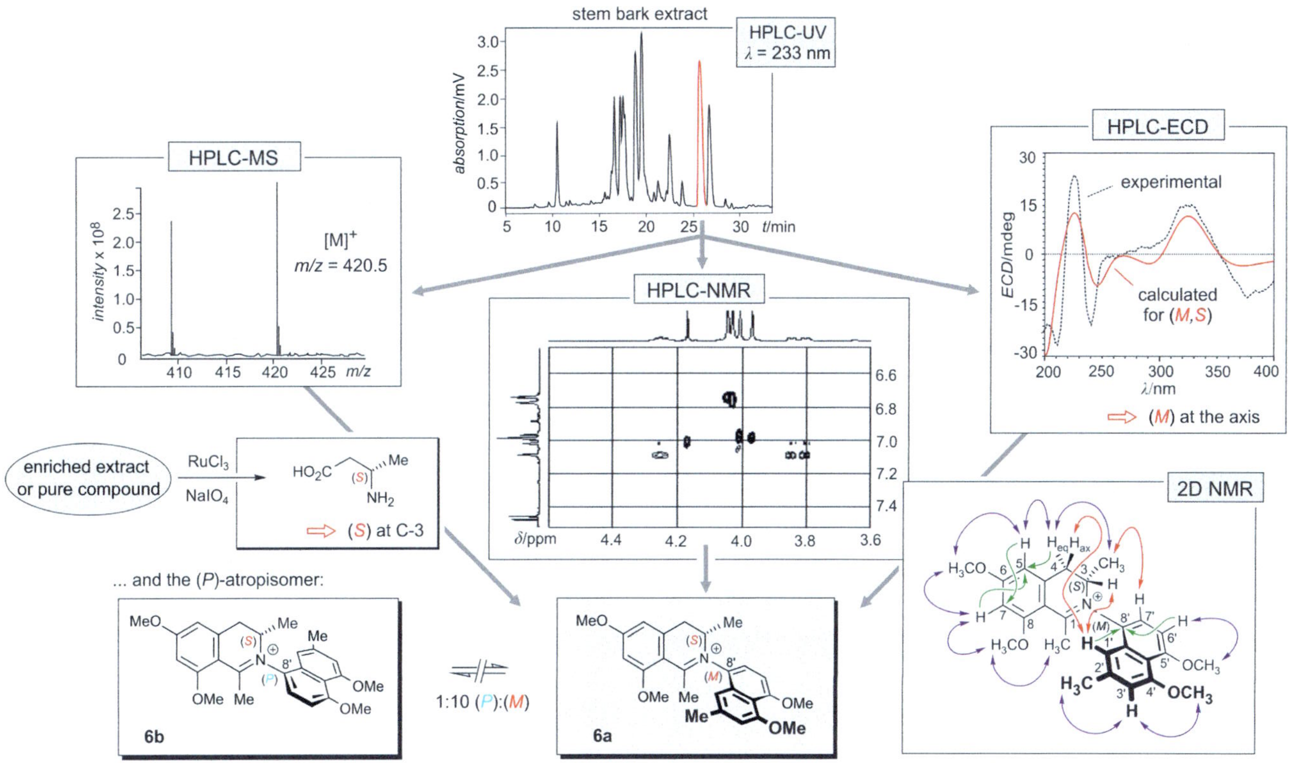

Fig. 84 Online elucidation of the complete stereostructure of ancistrocladinium A (**6a**), directly from a crude stem bark extract of *A. cochinchinensis*, by HPLC-ESI-MS, HPLC-NMR, shown with its most diagnostic HMBC (green single arrows) and ROESY interactions (purple double arrows, with those relevant for the relative axial configuration are drawn in red), and by LC-ECD investigations in combination with quantum-chemical ECD calculations, and by ruthenium-mediated oxidative degradation

NMR spectrum showed six protons, i.e., one more than found in the case of C,C-linked Ancistrocladaceae-type naphthyldihydroisoquinoline alkaloids. Two of these protons were *ortho*-coupled, and the other four ones were singlets, thus suggesting this leaf metabolite of *A. cochinchinensis* to possess an *N*-iminium C-aryl axis. By an HPLC-2D-WET-ROESY experiment, the constitution of **6a/6b**, including its coupling type, and its relative configuration were determined online, directly from the crude leaf extract. ROE correlations between H-6′ and the protons of the methoxy group at C-5′ revealed C-8′ to be the axis-bearing carbon atom, which was confirmed by cross-peaks in the series {H-1′ ↔ Me-2′ ↔ H-3′ ↔ OMe'-4} and by HMBC interactions from both H-1′ and H-6′ to C-8′ (Fig. 84) [18, 32, 47]. The ROE correlation sequence {Me-1 ↔ OMe-8 ↔ H-7 ↔ OMe-6 ↔ H-5 ↔ H_{eq}-4 ↔ Me-3} clearly revealed the coupling position in the isoquinoline part to be at N-2, which agreed with HMBC interactions from H-5 to C-7 and both from H_{eq}-4 and H-7 to C-5.

By ruthenium-mediated oxidative degradation, the absolute configuration at C-3 was deduced to be (*S*). This, together with ROE correlations between H-1′ and both H-3 and H_{ax}-4, and between H-7′ and Me-3, showed the axis to be (*M*)-configured [32, 47]. This assignment had previously also been evidenced by quantum-chemical ECD calculations (Fig. 84) [32]. For this purpose, the two possible atropo-diastereomers, ancistrocladinium A (**6a**) and the (*P*)-configured N,8′-*epi*-ancistrocladinium A (**6b**), were independently submitted to a conformational analysis using the semi-empirical PM3 approach, and the results were further optimized by DFT calculations at the B3LYP/6-31G* level, leading to four minimum structures. ECD computations by means of TDDFT (B3LYP/TZVP) [421, 422], followed by a weighted summation of the calculated ECD curves according to the Boltzmann statistics and the usual UV correction [76–78], finally provided the overall theoretical ECD spectra for **6a** and **6b**. Comparison of the spectrum calculated for **6a** with the experimental ECD curve of the natural product showed a good agreement (Fig. 84), whereas the spectrum predicted for the (*P*)-diastereomer **6b** was virtually opposite [18, 32, 47]. Thus, the combination of MS, NMR, and ECD methods with HPLC permitted a rapid and straightforward identification of the known [32] compound ancistrocladinium A (**6a/b**), including the assignment of its full 3D structure, directly in extracts of *A. cochinchinensis*.

Simultaneously, by applying the same strategy, two further, new N,8′-linked naphthylisoquinoline alkaloids were discovered in the leaves and stem bark of this Vietnamese liana, 4′-*O*-demethylancistrocladinium A (**123a**) and 6,4′-*O*,*O*-didemethylancistrocladinium A (**174a**). These compounds were the first N,C-coupled alkaloids with free phenolic hydroxy groups (Fig. 85) [47]. Their UV profiles already hinted at the presence of N,C-linked naphthylisoquinolines, because the second maximum of each of the compounds (335 nm) was shifted bathochromically by 30–40 nm in comparison to those previously reported for C,C-coupled alkaloids (typically ca. 306 nm) [18, 32, 47] (Fig. 85). By the same methods as outlined in Fig. 84 for the assignment of the complete stereostructure of ancistrocladinium A (**6a**) using ruthenium-mediated oxidative degradation, specific ROESY correlations across the N,C-axis, and ECD spectroscopy, the new mono- and bisphenolic alkaloids **123a** and

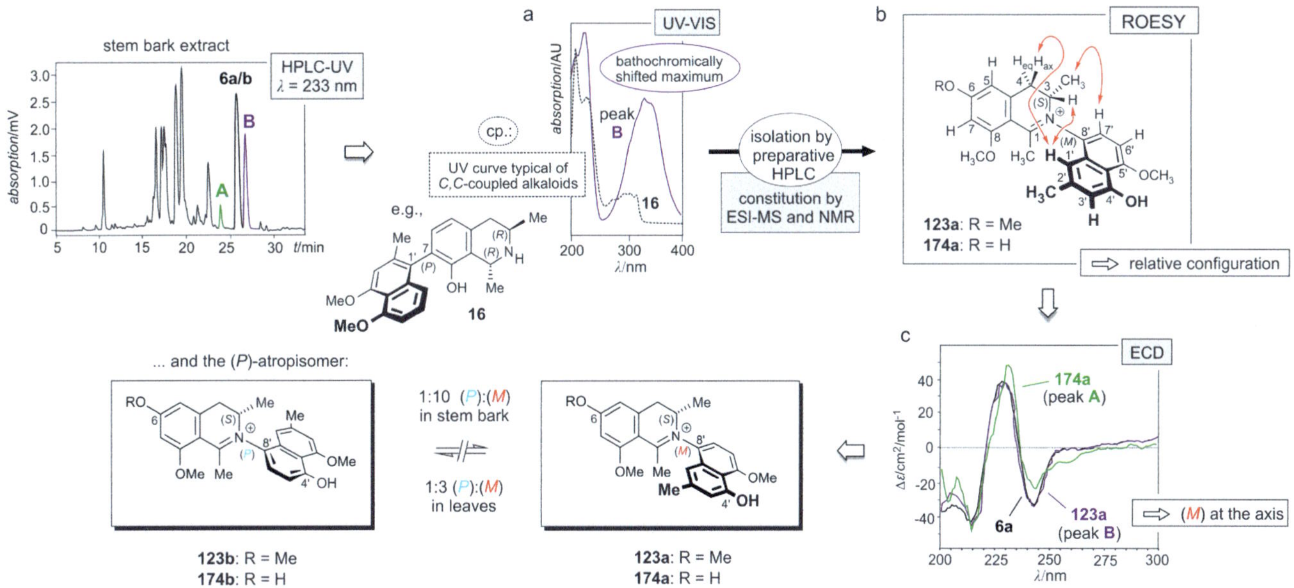

Fig. 85 Two new N,8′-linked naphthylisoquinoline alkaloids with free phenolic hydroxy groups, 4′-*O*-demethylancistrocladinium A (**123a/b**) (= peak **B**) and 6,4′-*O*,*O*-didemethylancistrocladinium A (**174a/b**) (= peak **A**), isolated along with their parent compound ancistrocladinium A (**6a/b**) (see Fig. 84) from stem bark and leaves of *A. cochinchinensis*. Identification was conducted using **a** their bathochromically shifted (by 30–40 nm) second maxima (for comparison, see the UV spectrum of the C,C-coupled naphthylisoquinoline alkaloid dioncophylline A, **16**), **b** their specific ROESY correlations across the N,C-axis, and **c** their experimental ECD spectra compared to that of the parent compound, **6a/6b**

174a were both attributed to be (*S*)-configured at C-3 and (*M*)-configured at the iminium-aryl axis (Fig. 85) [47]. In the NMR spectra of **123a**, also the other, (*P*)-configured atropo-diastereomer *N-epi*-4′-*O*-demethylancistrocladinium A (**123b**) was detectable in the bark in low quantities in a ratio of 1:10 ((*P*):(*M*)), and in the leaves in a ratio of 1:3 ((*P*):(*M*)). These atropisomeric ratios were comparable to those for ancistrocladinium A (**6a**) and its minor atropisomer **6b**. Similar results were obtained for the bisphenolic analogue **174a**, for which the respective (*P*)-atropo-diastereomer **174b** was detectable in traces only. Again, as it has previously also been the case for ancistrocladinium A (**6a/6b**), all attempts to resolve the two rotational diastereomers of **123** and **174** failed, even on chiral HPLC phases [47].

The free hydroxy groups of **123a** and **174a** exerted a strong influence on the anti-infective properties of these new alkaloids. The monophenolic analogue **123a** of ancistrocladinium A (**6a**) showed excellent activities against *T. cruzi*, the pathogen of Chagas' disease, and against *L. donovani* causing visceral leishmaniasis (also known as kala-azar). Its inhibitory effects were distinctly better than those of the parent compound **6a** [32], and more potent than those exhibited by the standard drugs benznidazole (for trypanosomiasis) and miltefosine (for leishmaniasis) [47]. 4′-*O*-Demethylancistrocladinium A (**123a**) was also active against the insect form of *L. major* suggesting a non-stage specific mode of action [47]. The bisphenolic analogue **174a**, in contrast, was much less active, as it displayed only moderate to weak antitrypanosomal effects against the pathogens causing African sleeping sickness (*T. brucei rhodesiense*) and Chagas' disease (*T. cruzi*), and it likewise showed only low activities against the related animal pathogen *T. brucei brucei* [47], even lower than the non-phenolic parent compound ancistrocladinium A (**6a**) [32]. Furthermore, **174a** was only poorly active against *L. donovani* amastigotes and *L. major* promastigotes [47].

In view of the promising antiparasitic activities of ancistrocladinium A (**6a/6b**) and its monophenolic analogue **123a/123b**, these unusual alkaloids have become attractive synthesis targets. Ancistrocladinium A (**6a/6b**) was prepared synthetically (see Sect. 8.5.9.2), by Buchwald–Hartwig amination followed by N-acetylation and Bischler–Napieralski cyclization, preferentially leading to the naturally predominant (*M*)-atropo-diastereomer **6a** [101]. Furthermore, a directed route to all the four possible stereoisomers in an atropisomerically pure form was developed by the authors' group (see Sect. 8.5.9.3) [69] and, later, also Kim and Cheon [118, 119] published a similar approach (see Sect. 8.5.9.5), yet building up the stereocenter at C-3 by reductive amination, efficiently leading to ancistrocladinium A (**6a/b**) and its stereoisomers.

7.4 *Biosynthetically Related, Mostly Non-coupled Naphthoquinones, Tetralones, and Free Isoquinolines from Ancistrocladus cochinchinensis*

Isolation work on the *n*-hexane and ethyl acetate extracts obtained from the leaves of *A. cochinchinensis* using column chromatography on silica gel yielded a series of

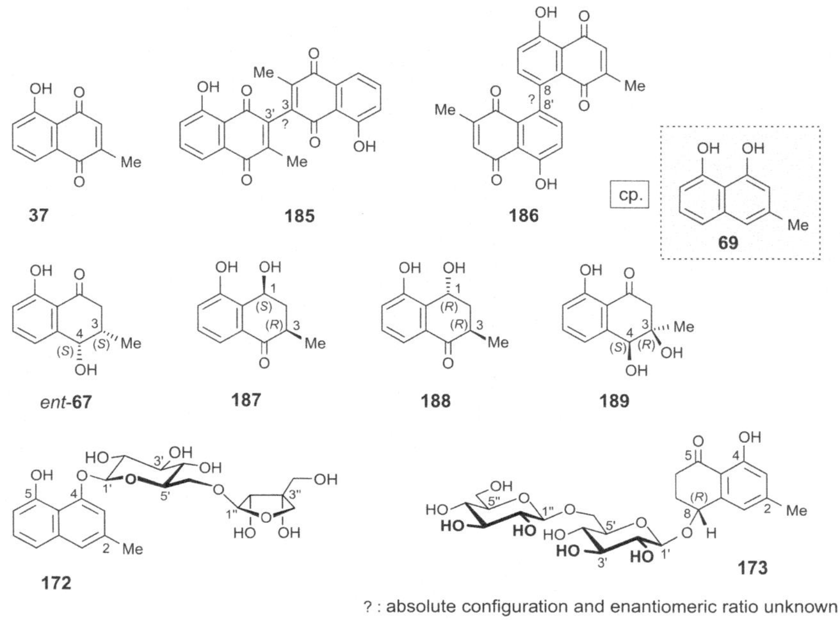

Fig. 86 Naphthalene derivatives from the leaves of *A. cochinchinensis* biosynthetically related to the presumable isocyclic precursor **69** of naphthylisoquinoline alkaloids; among them known compounds like plumbagin (**37**) [46, 418], the biplumbagins **185** and **186** [46, 418], and (3*S*,4*S*)-*cis*-isoshinanolone (*ent*-**67**) [46, 419], but also five new natural products such as the tetralones **187**, **188**, and **189** [46, 419], and the ancistronaphthosides A (**172**) and B (**173**) [48]

naphthoquinones and tetralones (nine compounds) [46, 48, 418, 419] (Fig. 86), which are apparently derived from the highly oxidation-sensitive, since electron-rich dihydroxynaphthalene **69**. In the plants, this biosynthesis intermediate can either be oxidatively coupled with an isoquinoline portion to give a naphthylisoquinoline alkaloid, or it can be oxygenated (e.g., as a consequence of chemical, physical, or biotic stress) to provide naphthoquinones like plumbagin (**37**), which then can also be reduced to tetralones like (3*S*,4*S*)-*cis*-isoshinanolone (*ent*-**67**, absolute configuration not verbally expressed, but graphically shown in the literature [419]) or the shinanolone derivative ancistronaphthoside B (**173**) [48].

In early phytochemical studies on *A. cochinchinensis* in 1996 by Adam et al. [418], plumbagin (**37**) was identified together with two related, known dimers, 3,3′-biplumbagin (**185**) [253, 423–428] and 8,8′-biplumbagin (**186**, also known as maritinone) [253, 426–431] (Fig. 86). The two latter compounds had previously been detected in various species of the Plumbaginaceae [254, 423, 424, 430] and Ebenaceae [254, 426–429, 431] plant families, but it was the first time that these two dimeric naphthoquinones were now found in an *Ancistrocladus* liana. No stereochemical investigations on these apparently chiral bisnaphthoquinones were pursued.

Plumbagin (**37**) and its dimers **185** and **186** have recently attracted attention for their high antibacterial activities against drug-sensitive and -resistant strains of *Mycobacterium tuberculosis* [432, 433]. Especially the dimers **185** and **186** emerged as candidates for further drug development as novel antituberculosis agents since they displayed no measurable cytotoxicity toward mammalian cells [433].

The great interest in the pharmacological potential of plumbagin (**37**) and related naphthoquinones becomes obvious from the considerable number of publications, including several reviews [227–233, 254, 255] that appeared in the course of the past decade, dealing with the chemical, biotechnological, and pharmaceutical aspects of plumbagin-related compounds. The main research efforts focused on the strong anticancer potential of plumbagin (**37**) [228, 230–232] since this naphthoquinone significantly inhibited the proliferation of different types of leukemia and malignant tumor cells in vitro and in vivo, in particular lung, ovarian, prostate, and breast cancer cells [227, 228, 231]. The antineoplastic effects of **37** were studied intensely and were found to be associated with the interruption of the cell cycle, the inhibition of invasion and metastasis of cancer cells, and with the induction of oxidative stress, apoptosis, and autophagy [232].

Plumbagin (**37**) also showed a promising antibacterial potential against a panel of sensitive and multidrug-resistant Gram-negative and Gram-positive bacteria, among them *E. coli*, *Klebsiella pneumoniae*, and MRSA strains of *S. aureus* [227, 229, 434, 435]. Moreover, **37** is capable of exerting pronounced effects against biofilm-associated infections caused by the yeast *C. albicans* [227, 229, 434, 436]. Furthermore, inhibitory activities of plumbagin (**37**) against the human-pathogenic protozoan parasites *L. donovani* [229, 437, 438], *T. cruzi* [229, 439], and *P. falciparum* [229, 440] were reported. Recent investigations also revealed **37** to possess neuroprotective [229, 441] and anti-inflammatory [229, 442] properties combined with the ability to improve memory performance [229, 443], thus suggesting plumbagin (**37**) to be a potential candidate for the management of neurodegenerative disorders [229, 443, 444].

Phytochemical studies by Adam et al. in 1997 [419] on the ethyl acetate extract of the leaves of *A. cochinchinensis* furnished three new tetralones (**187**–**189**), along with the known (3*S*,4*S*)-*cis*-isoshinanolone (*ent*-**67**) (Fig. 86). The structures of these novel natural products **187**, **188**, and **189** were elucidated in particular by NMR spectroscopy, on the basis of $^{1}H,^{1}H$ coupling constants, in combination with HMBC and NOE interactions. The NMR data of the tetralone **187** and its diastereomer **188** were in agreement with a relative configuration assuming a half-chair conformation with the 3-methyl group occupying an *equatorial* position and the 4-OH group an *axial* one [46, 419]. Furthermore, NOESY measurements on **188** showed that this tetralone occurred as a mixture of two possible half-chair conformations, with about 85% in favor of the arrangement with the methyl group adopting an *equatorial* position [419]. 1D and 2D NMR investigations on the third new metabolite **189** (Fig. 86) demonstrated the presence of a threefold oxygenated 3-methyltetralone, occupying a half-chair conformation and displaying a *cis*-array between the methyl group at C-3 and the *axial* proton at C-4.

Phytochemical studies on the leaves of *A. cochinchinensis* by Lien et al. [48] in 2016, additionally furnished two new naphthalene-derived compounds, named ancistronaphthosides A (**172**) and B (**173**) (Fig. 86). NOESY and HMBC experiments revealed ancistronaphthoside A (**172**) to be a 2-methylnaphthalene derivative possessing a phenolic hydroxy group at C-5 and an oxygen function at C-4, which is linked to a *β*-D-apiofuranosyl-(1 → 6)-*O*-*β*-D-glucopyranosyl disaccharide unit. Treatment of **172** with methanolic HCl at 60°C led to complete deglucosidation, delivering D-glucose and D-apiose, which were identical with authentic reference samples [48]. The NMR data of ancistronaphthoside B (**173**) hinted at the presence of a 2-methylnaphthalene-derived disaccharide with a close structural resemblance to **172**, but the appearance of a carbonyl function at C-5 and the deficiency of aromatic protons in the quinone part of the aglycon portion strongly suggested the presence of a tetralone derivative [48]. HMBC interactions revealed the new metabolite to be equipped with a phenolic hydroxy group at C-4 and a further oxygen function at C-8 that was linked to a *β*-D-glucopyranosyl-(1 → 6)-*O*-*β*-D-glucopyranosyl disaccharide moiety. The absolute configuration at C-8 was assigned to be (*S*) due to a positive Cotton effect observed in the ECD spectrum of **173**, which was in accordance with the ECD data of related, known [247] tetralone-type compounds. No antiproliferative activities against several human cancer cell lines including promyelocytic leukemia (HL-60), lung carcinoma (LU-1), and melanoma cells (SK-MEL-2) were found for ancistronaphthosides A (**172**) and B (**173**) [48].

Furthermore, two naphthalene-devoid isoquinoline alkaloids, anciscochine (**161**) and its 6-*O*-*β*-glucopyranoside **175** (Fig. 87), with a hydroxymethylene group at C-3, were discovered in the leaves of *A. cochinchinensis* [48]. They represent the isoquinoline half of the two co-occurring non-hydrogenated naphthylisoquinoline alkaloids ancistrobenomine A (**4**) and its 6-*O*-demethyl analogue **159**. Similar to ancistronaphthosides A (**172**) and B (**173**), anciscochine (**161**) and its 6-*O*-*β*-glucopyranoside **175** showed no significant cytotoxic effects against the above-mentioned human cancer cell lines HL-60, LU-1, and SK-MEL [48].

Free, non-coupled isoquinoline alkaloids have as yet only rarely been identified in *Ancistrocladus* species from Asia and Africa [48, 54, 156, 168, 176, 194, 445]. Remarkably, there was one *Ancistrocladus* liana collected in the rainforests at the East Coast of the Malaysian State of Johor, assigned as *A.* aff. *tectorius*, which was reported by Montagnac et al. [54] to be even dominated by the presence of three naphthalene-devoid isoquinoline alkaloids (see Sect. 8.4), among them 6-*O*-methylanciscochine (**160**), which has been prepared by total synthesis only recently [446]. Furthermore, the 5,1′-linked naphthylisoquinoline alkaloid 4′-*O*-demethylancistrocladine (**170a**, see Fig. 81) was isolated from this Malaysian *Ancistrocladus* liana, yet only as a minor metabolite [54]. The latter alkaloid was also identified in the leaves of *A. cochinchinensis* [47]. Since attempts of Montagnac et al. [54] to isolate and purify further naphthylisoquinoline alkaloids detected in the Malaysian *Ancistrocladus* plant had been unsuccessful, it will remain unanswered whether there is any further similarity regarding the constituents of *A. cochinchinensis* and *A.* aff. *tectorius* from Malaysia. The findings clearly underline the necessity for further, more in-depth phytochemical and phylogenetic investigations on Southeast

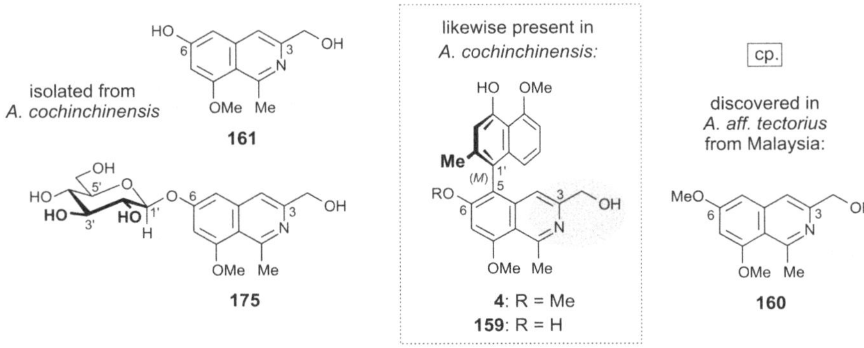

Fig. 87 Anciscochine (**161**) and its 6-*O*-*β*-glucopyranoside **175** from the leaves of *A. cochinchinensis*, two newly discovered naphthalene-free isoquinoline derivatives for which the methyl group at C-3 is oxygenated like in the co-occurring ancistrobenomine A (**4**) and its 6-*O*-demethyl analogue **159**. In comparison, 6-*O*-methylanciscochine (**160**) is a major metabolite in *A.* aff. *tectorius* from Malaysia [54]

Asian *Ancistrocladus* lianas belonging to the *A. tectorius* complex to overcome the uncertainty in the delimitation of *Ancistrocladus* species and aiming at an unequivocal identification of valid taxa. For a more detailed report, see also Sect. 8.1.

7.5 Phytochemical Relationship of *Ancistrocladus cochinchinensis* and *Ancistrocladus aff. tectorius* from the Chinese Island of Hainan

In summary, 30 secondary metabolites were isolated from *A. cochinchinensis* [45–48, 418, 419], with the most prominent ones being ancistrocladinium A (**6a/b**) (Fig. 84) and its mono- and bisphenolic analogues **123a/123b** and **174a/174b** (Fig. 85) [47], with their unusual N,8′-biaryl linkage. A total of 19 naphthylisoquinoline alkaloids comprising three different coupling types have so far been discovered in the leaves and twigs of this Vietnamese *Ancistrocladus* liana [45–48]. The isolation of a quite high number of nine biosynthetically related nitrogen-free metabolites—four naphthoquinones and five tetralones [48, 418, 419]—along with two new naphthalene-devoid isoquinolines [48], hints at a most unusually reduced capacity of phenol-oxidative cross-coupling of the plant. It is the highest number of such compounds found so far in a certain *Ancistrocladus* plant. While *A. cochinchinensis* significantly differs from other Asian and African *Ancistrocladus* species regarding the production of naphthalene-free isoquinolines or non-coupled naphthalenes [54, 156, 168, 174, 176, 194, 445], the naphthylisoquinoline pattern of this Vietnamese liana showed a huge overlap with that of *A.* aff. *tectorius* from the Chinese Hainan Island, as can be seen from the fact that 15 out of the 19 naphthylisoquinolines identified in the leaves and twigs of *A. cochinchinensis* were also detected in *A.* aff. *tectorius* [33, 54, 57–62, 64]. This finding is in accordance with results from studies on the internal transcribed spacers (ITS) of nuclear rDNA revealing as strong phylogenetic relationship between the isotype specimen of *A. cochinchinensis* (i.e., from its *locus classicus* in Binh Dinh, Central Coastal Vietnam) and *A.* aff. *tectorius* from China [31, 33, 50, 53, 56–64]. The taxonomic status of these two *Ancistrocladus* lianas from Vietnam and China still is a matter of controversy. On the one hand, there is a significant similarity regarding the molecular markers mentioned above and a pronounced consistency with respect to the occurrence of structurally identical naphthylisoquinoline alkaloids; on the other hand, the metabolite profiles of the two plants show clear differences. That is true, for example, for the discovery of naphthylisoquinoline dimers [31] and Ancistrocladaceae-type 7,6′-coupled monomers [59, 60] only in *A.* aff. *tectorius* from China, and the abundant formation of non-coupled isoquinolines and naphthalenes only found in *A. cochinchinensis* [48, 418, 419], whereas no free molecular halves of naphthylisoquinoline alkaloids have as yet been identified in *A.* aff. *tectorius* from Hainan Island.

In the latest comprehensive revision by Taylor et al. [21] covering the entire *Ancistrocladus* genus, *A. cochinchinensis* was reduced to taxonomic synonymy under *A. tectorius*. In contrast to this assumption, a comparative genetic study by Meimberg et al. [36] on samples from different locations from all over Southeast Asia using sequence data of the *trn*K intron of cpDNA, the ITS region of molecular rDNA, and ISSR fingerprints indicated that *A. cochinchinensis* might possibly deserve the status of an own species (for a more detailed discussion, see Sect. 8). Further phylogenetic investigations by Lien et al. [447] focusing on samples of *Ancistrocladus* aff. *tectorius* collected throughout Vietnam likewise showed a high genetic diversity of the *A. tectorius* population, thus corroborating the assumption that there might be further, yet unrecognized *Ancistrocladus* species in this region of Southeast Asia.

A precise morphology-based species delineation within the *A. tectorius* complex is extremely difficult or even nearly impossible where only sterile material is available. The use of multilocus microsatellite markers as described in a study to delimit Congolese *Ancistrocladus* lianas [448] might also facilitate a reliable identification of hitherto unrecognized taxa in Southeast Asia. The development of advanced approaches for further-in-depth genetic and phytochemical investigations on Asian *Ancistrocladus* lianas remains a rewarding task for the future.

8 Widespread Throughout Southeast Asia: *Ancistrocladus tectorius*, a Rich Source of Unique, Structurally Most Diverse Mono- and Dimeric Naphthylisoquinoline Alkaloids

For a long time, *A. tectorius* (Lour.) Merr. had been considered to be the only true species of the genus widely occurring in the rain forests all over Southeast Asia according to the 'one-species concept' established by van Steenis [35]. The comprehensive taxonomic revision by Taylor et al. [21] covering the entire *Ancistrocladus* genus additionally recognized *A. griffithii* and *A. attenuatus* as valid taxa, with an area of distribution ranging from Myanmar to Thailand, extending in the east to Cambodia and Vietnam, and in the west to the Indian Andaman Islands. The newly discovered Malaysian liana *A. benomensis* [34] has also been accepted as an own species [21], whereas six species have been summarized (i.e., reduced to taxonomic synonymy) under *A. tectorius*, and two species are considered to be of dubious identity [21] (see Table 1).

Phytochemical investigations on plant material of *A.* aff. *tectorius* collected from geographically disjunct areas in Laos [49] (Sect. 8.2), Vietnam [32, 272], Central and South Thailand [51, 52, 55] (Sect. 8.3), southern Peninsular Malaysia [54] (Sect. 8.4), and PR China (Province Hainan) [31, 33, 50, 53, 56–64] (Sect. 8.5) gave rise to doubts on this 'one-species concept'. *Ancistrocladus* aff. *tectorius* seems to be extremely diverse with respect to the occurrence of structurally most different naphthylisoquinoline alkaloids. The metabolic profiles of plants from different habitats in Southeast Asia vary strongly, containing a series of compounds that were exclusively detected in *Ancistrocladus* lianas from distinct sites [31, 33, 49–64, 272]. This indicates that *A.* aff. *tectorius* might be a taxon consisting of several morphologically similar subspecies or even species [34, 36]. The huge distribution area of this taxon with climatic conditions ranging from humid to semi-humid regions with drought periods further corroborated the assumption of the existence of a so-called *A. tectorius* complex [34, 36] exhibiting a wide range of variations regarding phytochemical features.

The high structural variability of naphthylisoquinoline alkaloids described for *A.* aff. *tectorius* [31, 33, 49–64] was found to be paralleled by a high genetic diversity of the plant material assigned to this taxon. The results of genetic investigations [36, 447] focusing on the internal transcribed spacers (ITS) of nuclear rDNA and on the *trn*K intron of cpDNA, in combination with intersimple sequence repeat (ISSR) fingerprints clearly suggested a species differentiation within the *A. tectorius* complex, strongly supporting the presence of further valid taxa not yet recognized botanically.

In the following paragraphs, the outcome of a comparative study [36] on the genetic relationships of plant material assigned as *A.* aff. *tectorius* collected from different sites in Southeast Asia and the results of phytochemical investigations on plant material from Laos, Thailand, Malaysia, and the Island Hainan (PR China) are presented. The findings clearly demonstrate the broad structural diversity of

Table 1 Taxonomic treatment of *Ancistrocladus* species reported to occur in Southeast Asia

Ancistrocladus species [first report in the literature]	Year of the first report	Area of distribution	Species concept according to Taylor et al. [21]
Ancistrocladus tectorius (Lour.) Merr. [268, 306]	1790	Lowland to premontane evergreen forests in Myanmar, Thailand, Cambodia, Laos, PR China (Hainan Island), Vietnam, Malaysia, Singapore, Sumatra and Borneo of Indonesia, Indian Andaman and Nicobar Islands	*Ancistrocladus tectorius*
Ancistrocladus extensus Wall. ex Planch. [270]	1849	Myanmar	Summarized under *A. tectorius*
Ancistrocladus pinangianus Wall. ex Planch. [270]	1849	Malaysia	Summarized under *A. tectorius*
Ancistrocladus cochinchinensis Gagnep. [415]	1909	Vietnam	Summarized under *A. tectorius*
Ancistrocladus harmandii Gagnep. [415]	1909	Laos	Summarized under *A. tectorius*
Ancistrocladus hainanensis Hayata [449]	1913	South-Central China, Hainan Island (PR China)	Summarized under *A. tectorius*
Ancistrocladus carallioides Craib [450]	1925	Northern Thailand	Summarized under *A. tectorius*
Ancistrocladus stelligerus Wall. ex DC. [278]	1864	Myanmar	Dubious identity
Ancistrocladus wallichii Planch. [270]	1849	Bangladesh, Assam State of India	Dubious identity
Ancistrocladus attenuatus Dyer [279]	1874	Myanmar, Andaman Islands (India)	*Ancistrocladus attenuatus*
Ancistrocladus griffithii Planch. [270]	1849	Myanmar, Thailand, Cambodia, Vietnam, Andaman Islands (India)	*Ancistrocladus griffithii*
Ancistrocladus benomensis Rischer & G. Bringmann [34]	2005	Peninsular Malaysia	*Ancistrocladus benomensis*

naphthylisoquinoline alkaloids of *A.* aff. *tectorius* and indicate a considerable number of hybridization events during the evolution of this taxon. The results emphasize the necessity for further intense research efforts regarding the taxonomy of the *Ancistrocladus* genus in Southeast Asia.

8.1 Evidence for Species Differentiation Within the *Ancistrocladus tectorius* Complex in Southeast Asia

Collections of *Ancistrocladus* lianas were examined with an investigated area extending from Hainan and South China, via Laos, East and South Thailand to the Malay Peninsula and Borneo. The study focused on the question whether the high variability of naphthylisoquinoline alkaloids in *A.* aff. *tectorius* is reflected by a significant genetic variability [36].

Using sequence data from the cpDNA *trnk*K intron and the nuclear internal transcribed spacers (ITS), including intersimple sequence repeat (ISSR) fingerprints, the sequence divergence of *Ancistrocladus* plants from Southeast Asia was analyzed in comparison to well-characterized African species [36]. Also included was the analysis of DNA isolates prepared from type or isotype specimens of *A. cochinchinensis*, *A. attenuatus*, *A. stelligerus*, *A. pinangianus*, and *A. harmandii*. The genetic data obtained from 75 samples of *Ancistrocladus* lianas, collected from 24 different locations all over Southeast Asia (for a map showing the areas of distribution of Southeast Asian *Ancistrocladus* plants, see Fig. 79 in Sect. 7), clearly indicated that the current species concepts by van Steenis [35] and Taylor et al. [21] underestimated the diversity of the *Ancistrocladus* genus in Southeast Asia.

Remarkably, the diversity of the DNA sequences of all Asian *Ancistrocladus* plants investigated (i.e., including *A. heyneanus* from India and *A. hamatus* from Sri Lanka) was similarly high in comparison to all African species [36]. The variability of the DNA data sets within the samples from Southeast Asia had a similar value as the variability in a data matrix comprising the African and Indian taxa. The variability was also very high if *A.* aff. *tectorius* was circumscribed according to Taylor et al. [21], i.e., excluding *A. griffithii*, *A. attenuatus*, and *A. benomensis*. The intraspecific variability of *A.* aff. *tectorius* was comparable to the interspecific variability among the African taxa, as obvious in particular from the results of the ISSR fingerprints, which were distinctly more diverse between the Asian samples than between the African ones [36].

The ITS and the *trn*K intron phylogenies both provided two more or less resolved clades formed from the individuals of *A.* aff. *tectorius* from all over Southeast Asia [36]. The first group was predominantly built up from samples from East and Central Thailand, Laos, and the Island Hainan, while the second group mainly comprised samples from South Thailand and Peninsular Malaysia.

The type specimen of *A. cochinchinensis* formed a clade together with most of the samples from East Thailand, Laos, and the Chinese Hainan Province, while *A. pinangianus* was positioned in another clade together with *Ancistrocladus* samples from South Thailand and Malaysia, and *A. attenuatus* was placed together with samples from southern Thailand [36].

In addition, the variation within one geographical population was high. In several local sites, highly divergent ITS and *trn*K intron sequences were found, some of which were highly similar to sequences of individuals from other sites.

For example, sequences of *A. griffithii* were detected in three locations together with individuals possessing sequences that had been assigned to a different clade in the phylogenetic analysis [36].

The strongest evidence that *A.* aff. *tectorius* consists of several differentiated taxa became obvious from the highly deviating ISSR fingerprint patterns monitored for some samples of nearby locations, thus hinting at a restricted gene flow between the individuals within one sampling site, and suggesting the occurrence of different species in a distinct region. This assumption was further corroborated by the fact that several samples from distant sites clustered together, showing nearly identical or very similar ITS, *trn*K, and ISSR data sets. In addition, comparison of the molecular data with HPLC fingerprints of alkaloid-containing extracts obtained from some of the samples collected all over Southeast Asia revealed a high correlation to the genetic distances calculated from ISSR fingerprints [36].

The ISSR sequences suggested the presence of up to four taxa occurring in the Thai part of the Malay Peninsula. One of the species was identified to be *A. griffithii*, which is morphologically characterized by its very slim, elongated leaves [21]. The species status of this Thai liana was strongly supported by the large distance of ISSR fingerprints and ITS sequences compared to all other *Ancistrocladus* samples from Southeast Asia [36]. Furthermore, the ITS sequence of the isotype of *A. attenuatus* showing a phenology similar to that of *A. griffithii*, but morphologically distinguishable by its reproductive characters [21, 279, 285, 299], was also considerably different from the material investigated and assigned as *A. griffithii* [21, 270, 278, 279].

Phylogenetic analysis of ITS sequences, supported by ISSR fingerprints, indicated that *A.* aff. *tectorius* can be split at least into two species represented by two morphological forms [21, 36]. One form, mainly found in China and Vietnam, but also in Malaysia, Singapore, and Sumatra, is similar to *A. heyneanus* from West India and resembles the type specimens of *A. tectorius*, *A. cochinchinensis*, and *A. harmandii* with three well-developed marginate glands on the outer three sepals [21]. The other form, predominantly occurring in Malaysia, Sumatra, Singapore, and the Andaman Islands, is similar to the type specimen of *A. pinangianus* and possesses sepals without abaxial glands [21]. Based on molecular markers such as ITS sequences and ISSR data sets [36], the two morphological forms of *A.* aff. *tectorius* were found to correspond quite well to the two major clades. One of the clades comprised most of the samples from the northern part of the range of distribution of *Ancistrocladus* lianas in Southeast Asia, and the second one included most of the samples collected from habitats in the southern part of Southeast Asia. The sympatric occurrence of groups of individuals showing divergent ISSR fingerprints (i.e., they are assumed to be formed from an initially interbreeding population that split into two or more distinct species sharing a common range) was a further strong hint at the presence of additional full taxa in the two major clades of *Ancistrocladus* plants from Southeast Asia [36].

ITS sequence and ISSR analysis revealed *A. benomensis* from Central Malaysia to be a true, distinct *Ancistrocladus* species [34, 36]. The status of this liana as a valid taxon was also confirmed by the high similarity of *A. benomensis* compared to the individuals collected in the Trat Province in South Thailand near the Cambodian border. These plants were significantly different from other *Ancistrocladus* lianas that grew in their closest vicinity. *Ancistrocladus* plants from the Trat Province can thus be assumed to be conspecific with *A. benomensis*, or these plants might even represent an own species [36].

Potential, yet unexplored *Ancistrocladus* taxa might also occur in South and Central Thailand and in Laos. Several individuals were identified that possessed a higher genetic similarity to populations from locations at a larger geographical distance than to *A. pinangianus* or *A. griffithii*, which were found to be present in the close vicinity [36]. Plant material collected in the northern part of Southeast Asia formed a highly supported monophyletic group with respect to their ITS data sets (i.e., representing a group of taxa composed only of a common ancestor, or, more precisely, of an ancestral population, and all of its lineal descendants). The group comprised the sequences of the specimens of *A. cochinchinensis* and *A. harmandii*, but also samples from *Ancistrocladus* lianas collected in the Hainan Province in China, presumably representing *A. hainanensis*. In the two most eastern populations in Thailand, sequences belonging to the *A. harmandii* group, but also those distinctly different from *A. harmandii*, were monitored. The latter showed a high similarity to the ITS sequences of *A. cochinchinensis* and *A. hainanensis* [36]. Some individuals even contained both of these sequence types. Thus, the respective populations of *Ancistrocladus* lianas in Southeast Thailand at the border areas to Laos and Cambodia might constitute a hybrid zone between differentiated species represented by the types of *A. harmandii* and *A. cochinchinensis* [36].

Summarizing, the molecular data sets of this comparative study clearly support the hypothesis of a species differentiation within the *A. tectorius* complex, hinting at the occurrence of distinguishable taxa not yet recognized. The outcome of the genetic investigations demonstrates that the current species concepts by van Steenis [35] and Taylor et al. [21] do not correctly reflect the natural diversity of the *Ancistrocladus* genus in Southeast Asia. This molecular survey based on nuclear and chloroplast DNA markers and DNA fingerprints represents a first step in the development of a more precise approach to a morphology-independent safe species delineation aiming at both a differentiation between taxonomically recognized plant species and at the identification and classification of botanically unknown taxa.

The broad structural variability of the alkaloid patterns reported for *A.* aff. *tectorius* depending on the place of origin of the plants investigated underlines the necessity to gain reliable knowledge on the abundance and distribution of well-defined taxa. This is of great importance for the reproducibility of phytochemical investigations on *Ancistrocladus* lianas. Vice versa, the results from natural product research will help to further support the taxonomical classification of *Ancistrocladus* specimens.

8.2 *Ancistrocladus aff. tectorius from Laos*

First isolation work by Foucher et al. [49] on *Ancistrocladus* aff. *tectorius* from Laos revealed the presence of ancistrocladine (**1a**) in root extracts. Besides this well-known [1–5] naphthylisoquinoline alkaloid, a second, likewise 5,1′-coupled, but fully dehydrogenated compound was isolated by this French group. Its reported structure, however, cannot be considered as fully established due to the missing recognition of its axial chirality and, thus, lacking the assignment of its absolute configuration.

More recently, ancistrocladinium A (**6a/6b**) and its 4′-*O*-demethyl analogue **123a/123b** were identified as the main metabolites of leaf and twig extracts obtained from a Laotian *Ancistrocladus* species. This plant, also assigned as *A.* aff. *tectorius*, had been cultivated by the authors' group in the greenhouse of the Botanical Garden of the University of Würzburg. Similar to the results obtained for *A. cochinchinensis* (see Sect. 7.3), the (*M*)-configured atropo-diastereomers **6a** and **123a** were found to be the dominating rotational isomers, while the (*P*)-configured diastereomers **6b** and **123b** were only detectable in low quantities, occurring in a ratio of ca. 1:10 ((*P*):(*M*)) (Fig. 88).

Furthermore, two less polar compounds detected in leaf extracts were easily identified as the known [32] N,C-coupled naphthyldihydroisoquinolinium alkaloids ancistrocladinium B (**14a**) and its rotational isomer **14b**. The two configurationally semi-stable atropo-diastereomers **14a** and **14b** had previously been isolated from two botanically yet undescribed Congolese *Ancistrocladus* species [32, 161] and from the Malaysian liana *A. benomensis* (see Sect. 6.3).

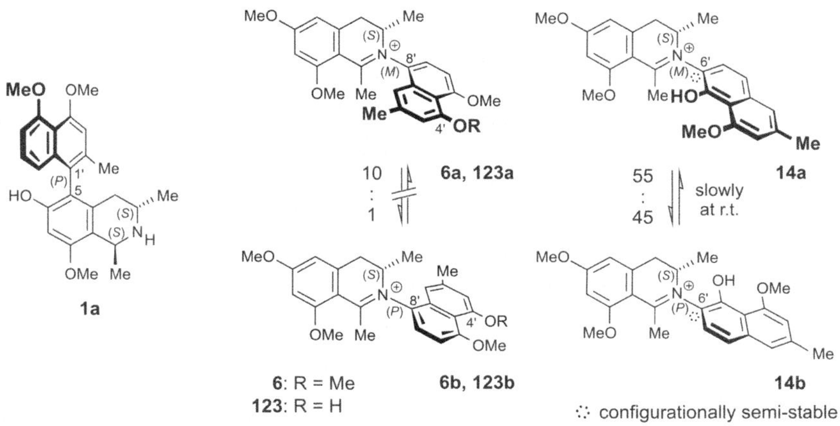

Fig. 88 Ancistrocladine (**1a**) and the N,C-coupled naphthylisoquinoline alkaloids ancistrocladinium A (**6a/6b**), 4′-*O*-demethylancistrocladinium A (**123a/b**), and ancistrocladinium B (**14a/14b**) identified from a Laotian *Ancistrocladus* taxon assigned as *A.* aff. *tectorius*

8.3 Ancistrocladus aff. tectorius from Thailand

Ancistrocladus aff. *tectorius* is widely distributed in lowland to premontane, rather dry forests all over Thailand [19, 21, 35, 36, 309]. Initial isolation work on leaf extracts of *A.* aff. *tectorius* collected from the Nakhon Ratchasima Province in South-Central Thailand by Cordell and coworkers [51, 52] furnished two naphthylisoquinoline alkaloids (see Fig. 89) with the otherwise quite rare 7,3′-biaryl linkage, ancistrotectorine (**17a**) [51] and ancistrocladidine (**3a**) [52]. The complete stereostructure of **3a** was already known from earlier isolation work by Govindachari et al. on the Indian liana *A. heyneanus* [5, 27, 28]. The constitution and the relative configuration of ancistrotectorine (**17a**) were deduced by using spectroscopic methods (in particular 1D and 2D NMR) in combination with an X-ray structure analysis [51]. In view of the fact that the absolute configuration at C-1 and C-3 was not determined directly, and that the absolute axial configuration was assigned by comparison of the ECD data with those of an electronically different naphthyldihydroisoquinoline, **3a**, ancistrotectorine (**17a**) was re-investigated by the authors' group in the course of its isolation from *A. guineënsis* from Nigeria [159]. The alkaloid has a relative *cis*-array of the methyl functions at C-1 and C-3, with a (1*R*,3*S*)-configuration as established by oxidative degradation. The absolute axial configuration of **17a** was attributed to be (*M*), by comparison of the experimental ECD spectrum with the ECD curve quantum-chemically calculated for the (*M*)-configured ancistrotectorine (**17a**) and that computed for its (*P*)-diastereomer **17b**, with the latter exhibiting a near-opposite ECD curve [159] (for a more detailed report on the assignment of the complete stereostructure of **17a**, see Sect. 8.5.6.1).

More recently, ancistrocladidine (**3a**) has been synthesized via two different routes: firstly, by Bungard and Morris [115, 116] with the Pinhey–Barton *ortho*-arylation of a naphthol with an aryllead triacetate being the key step and, secondly, atroposelectively, by the authors' group, using the lactone approach [99]. In a similar way, again following the lactone method, ancistrotectorine (**3a**) was synthesized stereoselectively by the authors' group [99] (for a more detailed report on these two synthesis pathways giving rise to **17a** and **3a**, see Sect. 4.5).

A further phytochemical study by Manfredi et al. [55] on extracts obtained from leaves and twigs of *A.* aff. *tectorius* collected from the Khao Sok National Park in southern Thailand (Province Surat Thani) led to the identification of four

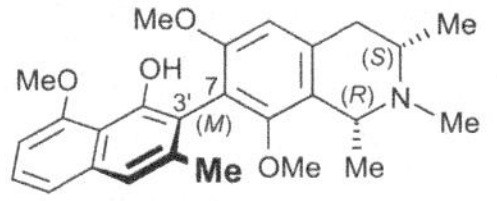

17a

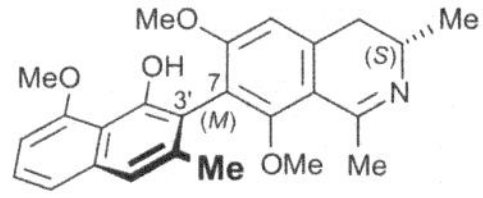

3a

Fig. 89 Ancistrotectorine (**17a**) and ancistrocladidine (**3a**), two 7,3′-coupled naphthylisoquinoline alkaloids isolated as the main secondary metabolites from the leaves of *A.* aff. *tectorius* from South-Central Thailand

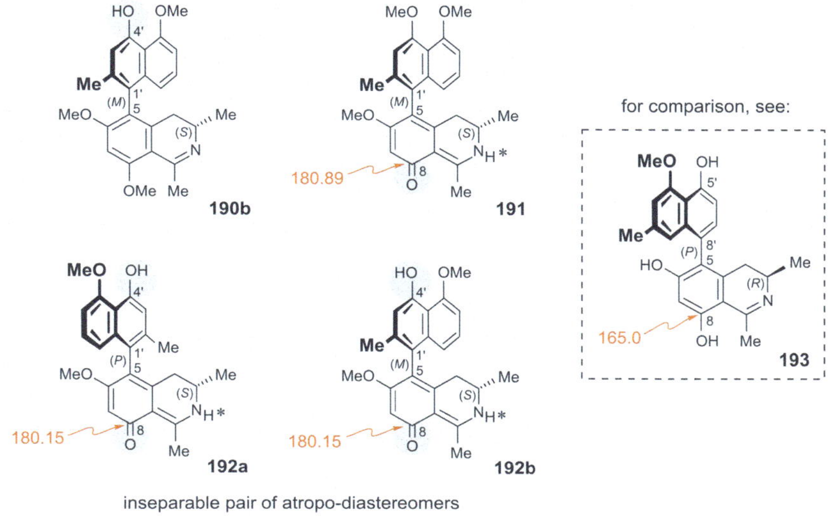

Fig. 90 Four most unusual 5,1′-coupled naphthyldihydroisoquinoline alkaloids (with three being **191**, **192a**, and **192b**, and described to exist predominantly in their keto forms) dominating the metabolite pattern of the leaves of *A.* aff. *tectorius* from southern Thailand. For comparison, see the structurally closely related 5,8′-linked alkaloid ancistrolikokine E (**193**) revealing, in contrast to **191** and **192a/b**, no intense downfield shift of the carbon atom C-8; *H atom not drawn in the published report [55], but plausible from the MS data

5,1′-coupled naphthyldihydroisoquinoline alkaloids, **190b**, **191**, **192a**, and **192b**, all possess a methoxy group at C-6 in the isoquinoline part (Fig. 90). The constitutions and relative configurations of these new compounds, which differ from each other by the presence of a methoxy or a hydroxy group at C-4′ in the naphthalene portion and at C-8 in the isoquinoline moiety, were assigned by 1D and 2D NMR spectroscopy and by difference NOE experiments [55]. The intense downfield shift of the carbon atom C-8 to a value around δ_C = 181.0 ppm, as compared to the chemical shifts of structurally similar 5,1′- or 5,8′-coupled naphthylisoquinolines bearing a hydroxy group at C-8 (around δ_C = 155.0–165.0 ppm [61, 156, 164, 167, 168, 182, 191]), suggested that **191**, **192a**, and **192b** exist predominantly in the keto form at C-8 of the isoquinoline subunit [55], as depicted in Fig. 90. This was also in accordance with the appearance of a major absorption band at λ = 384 nm, which has not been reported for related alkaloids, and was observed to disappear on addition of acid, again detectable after neutralization [55].

Ruthenium-mediated oxidative degradation exclusively delivered the (*S*)-enantiomer of β-aminobutyric acid, thus establishing the isoquinoline moieties to be (*S*)-configured at C-3 [55]. The absolute axial configurations of **190b** and **191** were determined to be (*M*) by ECD investigations, by comparison of the signs of the Cotton effects with those of structurally related and likewise 5,1′-coupled naphthylisoquinoline alkaloids [37, 45, 192]. While **190b** and **191** were isolated from the plant in an atropisomerically pure form, the third new alkaloid **192a** was obtained along with its atropo-diastereomer **192b**, as an inseparable mixture, in a ratio of ca. 1:1 ((*P*):(*M*)), as evidenced from NMR analysis showing a 'doubling' of almost all peaks. Remarkably, this series of four naphthyldihydroisoquinoline alkaloids, **190b**, **191**, **192a**, and **192b**, were reported so far to occur only in this Thai liana assigned as *A.* aff. *tectorius* [55]. Three of them (viz. **191**, **192a**, and **192b**) were suggested to be present in the plant exclusively in a tautomeric form (see Fig. 90). A reason for this unusual behavior of **191** and **192a/192b** was not given by the authors [55]. Such an unprecedented occurrence of alkaloids in the keto form at C-8 in the dihydroisoquinoline part had never been observed before, not even in the case of the structurally closely related 5,8′-coupled naphthyldihydroisoquinoline alkaloid ancistrolikokine E (**193**) (see Fig. 90) isolated from the Congolese species *A. likoko* [167]. Ancistrolikokine E (**193**) likewise possesses a free phenolic hydroxy function at C-8, and—also similar to **192a/192b**—a naphthalene portion equipped with a free hydroxy function *para* to the biaryl axis. Unlike for **191** and **192a/192b**, which showed an intense downfield shift in ^{13}C NMR for C-8 (see Fig. 90) [55], no such effect was monitored in the case of ancistrolikokine E (**193**) [167].

8.4 Ancistrocladus aff. tectorius from Malaysia

Using column chromatography on silica gel, Païs et al. [54] isolated three non-coupled, and thus naphthalene-devoid isoquinoline alkaloids, **194**, **195**, and **160**, as the major constituents of stem bark extracts of *A.* aff. *tectorius* collected in rainforests located at the East Coast of the Malaysian State of Johor. The 5,1′-linked naphthylisoquinoline alkaloid 4′-*O*-demethylancistrocladine (**170a**) was found to be present in the plant only as a minor metabolite (Fig. 91). Remarkably, the production of naphthalene-free isoquinolines has not been reported from any of the phytochemical studies [31, 33, 49–53, 55–64] on *A.* aff. *tectorius* collected from other regions in Southeast Asia.

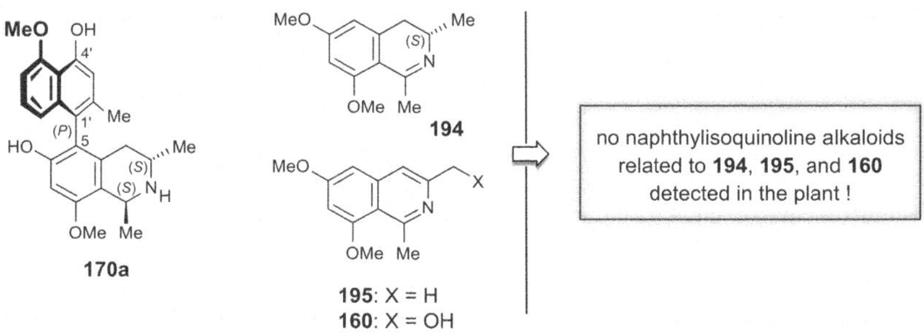

Fig. 91 Isolation of 4′-*O*-demethylancistrocladine (**170a**) from the Malaysian liana *Ancistrocladus* aff. *tectorius* as a minor metabolite, along with the naphthalene-free isoquinoline alkaloids **194**, **195**, and **160**, being present in the plant as major constituents

The metabolic pattern of this Malaysian liana was substantially dominated by the fully dehydrogenated isoquinoline **160** [54] bearing a hydroxymethylene function at C-3. This compound constitutes the isoquinoline half of ancistrobenomine A (**4**), a main alkaloid in *A. benomensis* [30] (likewise from Malaysia) and in the Vietnamese liana *A. cochinchinensis* [47], and a minor metabolite in *A.* aff. *tectorius* from Hainan [64]. The O-permethylated dihydroisoquinoline **194** [54] is the molecular portion of a distinct number of alkaloids, among them the quite widespread 5,1′-coupled compound 6-*O*-methylhamatinine (**176**) (Fig. 98), the more rarely occurring 5,3′-linked ancistrotanzanine A (**12a**) (Fig. 95), and the two 5,8′-coupled atropo-diastereomeric naphthylisoquinoline alkaloids ancistrotanzanine B (**218a**) and ancistroealaine A (**218b**) (Fig. 102). These four compounds from the twigs and leaves of the Chinese liana *A.* aff. *tectorius* [60, 62, 64] are also known to be produced by some African species [45, 46, 155, 167, 171, 174, 191].

The formation of the usual, complete naphthylisoquinoline alkaloids in only low quantities in this Malaysian liana hints at a reduced capability to perform the phenol-oxidative coupling in the biosynthesis of these biaryls, possibly due to an excessive O-methylation of the respective phenolic biosynthetic intermediates. As a result of the diminished production of naphthylisoquinoline alkaloids, the metabolite pattern of this Malaysian *A.* aff. *tectorius* plant is exceptional compared to those of other Asian and African *Ancistrocladus* lianas investigated so far. Most were found to be rich sources of structurally diverse naphthylisoquinoline alkaloids, whereas their non-coupled molecular halves have only been isolated rarely [46, 48, 54, 156, 168, 174, 176, 194, 445], usually being present in the plants only as minor or even trace compounds. The metabolite profile of *A. cochinchinensis* is an impressive example that underlines this finding. This Vietnamese liana produces a considerable number of structurally diverse naphthylisoquinoline alkaloids as main constituents [45–48], along with a series of naphthoquinones and naphthalene-devoid isoquinolines [46, 48, 418], which were isolated from the plant only in minor or even trace quantities (see Sect. 7.4). Among these compounds were anciscochine (**161**) and its glycoside **175** [48] (Fig. 87). Both likewise feature an unusual 3-hydroxymethylene unit—similar to the isoquinoline **160** (Fig. 91) from that Malaysian liana *A. tectorius* described in this section.

More recently, a convenient synthesis pathway to various 6,8-dioxygenated 1,3-dimethylisoquinolines has been reported by Kaufman et al. [451, 452], employing a strategy based on a 6π-azaelectrocyclization. This synthesis sequence is a novel and efficient alternative to already existing methods [453, 454]—it is short (only four steps needed), based on simple and readily available starting materials, and does not require any protecting groups. The route to 6,8-dimethoxy-1,3-dimethylisoquinoline (**195**) (Fig. 92) [451, 452] involves a cerium(III)-mediated oximation of 2,4-dimethoxyacetophenone (**196**) with methoxylamine hydrochloride as the reagent, giving rise to **197** as a chromatographically separable 93:7-mixture of (*E*)/(*Z*) geometric isomers. Direct allylation of the pure oxime (*E*)-**197** using allyl acetate as a convenient donor of the required three-carbon unit was performed in the presence of $[RuCl_2(p\text{-cymene})]_2$ as the catalyst and $AgSbF_6$ as an additive for C–H activation [452]. This ruthenium-mediated reaction provided the

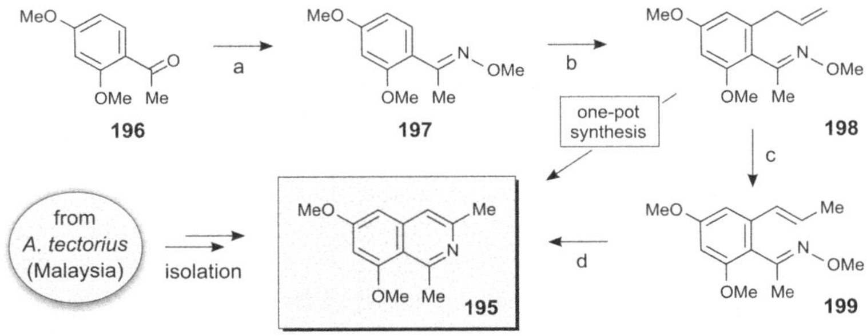

Fig. 92 Total synthesis of 6,8-dimethoxy-1,3-dimethylisoquinoline (**195**) via the 6π-azaelectrocyclization approach. Reaction conditions: (a) $MeONH_2 \cdot HCl$, $CeCl_3 \cdot 7H_2O$, NaOAc, EtOH, 50° C, 30 min, 94% ((*E*)/(*Z*) = 93:7; 87% of pure (*E*)-isomer); (b) allyl acetate (2 equiv.), $(RuCl_2(p\text{-cymene}))_2$ (5 mol%) in toluene, $AgSbF_6$ (20 mol%), CH_2Cl_2, r.t., 16 h, 73%; (c) RuHCl(CO)$(PPh_3)_3$ (3 mol%) in toluene, 90°C, 24 h, 95% ((*E*)/(*Z*) = 90:10); (d) AcOH, $PhCF_3$, microwave reactor (three cycles, 1 h each), 40%. Reaction conditions for the sequential one-pot allyl group isomerization → 6π-azaelectrocyclization procedure starting from **198**: RuHCl(CO)$(PPh_3)_3$ (3 mol%), toluene, 90°C, 24 h, in a closed vessel; further heating for 22 h at 150–160°C using a silicone oil bath, 43%

desired product **198** exclusively, in a good yield of 73%. Ruthenium-promoted allyl to propenyl isomerization in the presence of RuHCl(CO)$(PPh_3)_3$ then afforded the key intermediate **199**. The isomerized methoxime **199** was eventually subjected to a microwave-assisted 6π-azaelectrocyclization to give the target molecule **195** in an overall yield of 27%. The latter two steps can also conveniently be carried out as a one-pot reaction, thus increasing the efficiency of this synthesis approach [452]. The isoquinoline **195** exhibited only moderate to weak cytotoxic activities against H1299 lung cancer cells and against the two MDA-MB-468 and MCF7 breast carcinoma cell lines [451].

Following the above-described tandem C–C/C–N formation approach, the bark metabolite **160** of *A. tectorius*, which is the 6-*O*-methyl analogue of anciscochine (**161**) from *A. cochinchinensis* (see Sect. 7.4), was synthesized [446]. The route comprised a Ce(III)-mediated oximation of the ketone **196** to yield **200**, followed by oxime activation through pivaloylation giving rise to **201**. Introduction of the three-carbon chain via a rhodium-catalyzed pivaloxime-directed alkenylation with methyl acrylate using $(Cp^*RhCl_2)_2$ as the catalyst, followed by C–N bond formation with concomitant Ag(I)-promoted aromatization in situ delivered the targeted isoquinoline-3-carboxylate ester **202**. Reduction of **202** promoted by $NaBH_4/CaCl_2$ completed the sequence and afforded 6-*O*-methylanciscochine (**160**) in an overall yield of 43% [446] (Fig. 93).

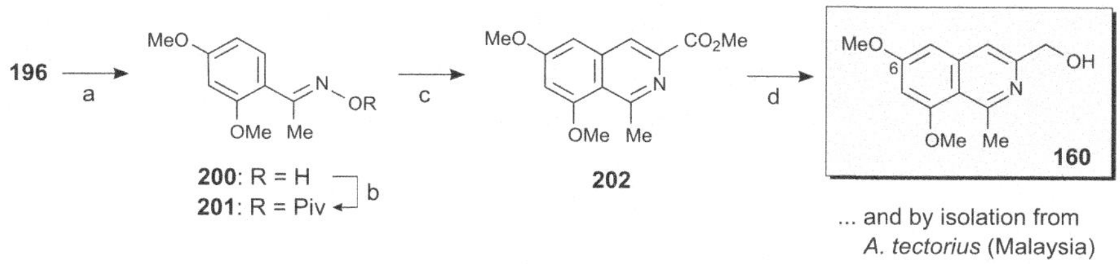

Fig. 93 A rhodium-catalyzed alkenylation/cyclization-based approach as a key step in the total synthesis of 6-*O*-methylanciscochine (**160**). Reaction conditions: (a) $HONH_2{\cdot}HCl$, $CeCl_3{\cdot}7H_2O$ (5 mol%), NaOAc, EtOH, 50°C, 1 h, 98%; (b) Piv_2O (= pivalic anhydride, $[(CH_3)_3CCO]_2O$), CH_2Cl_2, r.t., 16 h, 83%; (c) $(Cp^*RhCl_2)_2$ (Cp* = pentamethylcyclopentadienyl) [455] (2.5 mol%), $AgSbF_6$ (15 mol%), PivOH (3 equiv.), AgOAc (2 equiv.), 1,2-dichloroethane, 90°C, 20 h, 73%; (d) $NaBH_4$, $CaCl_2$, THF-EtOH (1:1, *v*/*v*), 0°C to r.t., 16 h, 72%

8.5 *Ancistrocladus aff. tectorius from the Chinese Island of Hainan*

The area of distribution of *Ancistrocladus* aff. *tectorius* in China is mainly restricted to rainforests near the town Ledong on Hainan Island [19, 21, 306–308, 449]. In traditional Chinese medicine, a decoction of leaves and stems of this plant is used for the treatment of malaria and parasite infections, in particular by people of the Li ethnicity [60, 456, 457].

Comprehensive isolation work on leaf, twig, and stem extracts of *A.* aff. *tectorius* by various Chinese scientists [50, 56–60] and by the authors' group [31, 33, 53, 61–64] led to the isolation and structure elucidation of nearly 70 different naphthylisoquinoline alkaloids. Thus—if considered as one taxon—this Chinese liana is one of the phytochemically most intensely investigated plant species within the entire *Ancistrocladus* genus. Again, similar to (nearly) all of the other Asian and also East African taxa, but in contrast to most of the West and Central African *Ancistrocladus* species, *A.* aff. *tectorius* from Hainan strictly follows the biosynthesis criterion of exclusively producing naphthylisoquinolines with the typical characteristics of a genuine Ancistrocladaceae-type alkaloid [13, 17], since all of the alkaloids from this Chinese liana are (*S*)-configured at C-3 and bear an oxygen function at C-6 [31, 33, 49–64].

Ancistrocladus aff. *tectorius* produces a plethora of naphthylisoquinoline alkaloids (Fig. 94), displaying high structural diversity, among them typical monomeric compounds, e.g., the 5,1′-linked ancistectorines A_1 (**203**) and A_3 (**204**) [61], but also novel-type alkaloids with unique structural features. Most remarkable was the isolation of ancistrotectoquinone A (**74a/b**) with its quinoid molecular portion [63], and the discovery of the pentacyclic N,C-coupled ancistrocyclinone A (**7**), along with its assumed precursor 4′-*O*-demethylancistrocladinium A (**123a**) [33]. Furthermore, the first 7,6′-coupled Ancistrocladaceae-type alkaloids, e.g., ancistrotectoriline C (**206**) [59] were identified, and new examples belonging to the very rare group of naphthylisoquinolines with a 5,3′-biaryl linkage such as

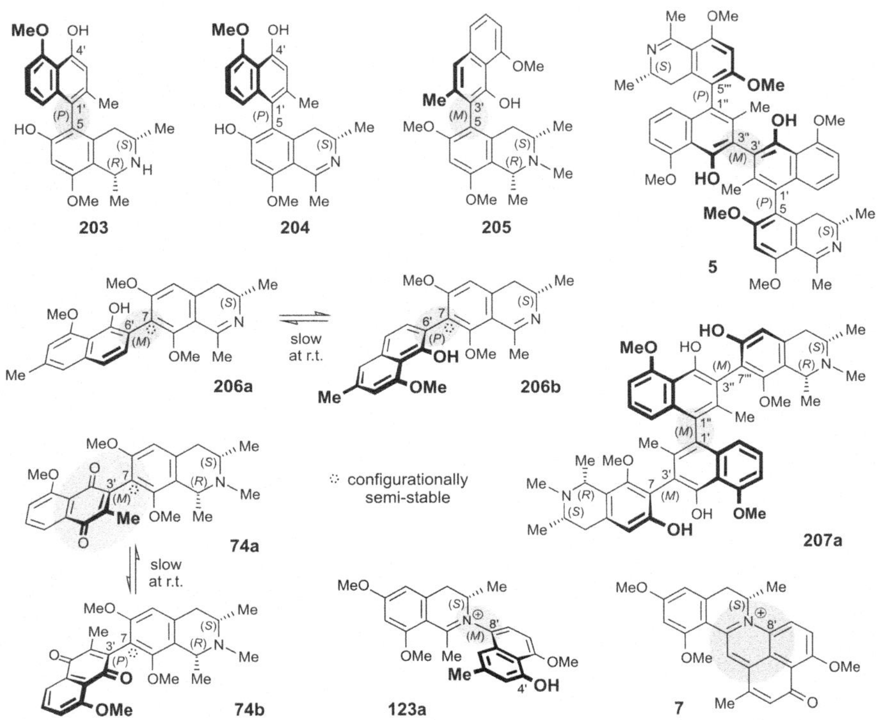

Fig. 94 Structural diversity of examples of selected naphthylisoquinoline alkaloids from *A.* aff. *tectorius* from the Chinese Hainan Province, among them the monomers ancistectorines A_1 (**203**) and A_3 (**204**), ancistrotectorine C (**205**) and ancistrotectoriline C (**206a/b**), which were isolated along with the dimers shuangancistrotectorines C (**5**) and D (**207a**), the quinoid alkaloid ancistrotectoquinone A (**74a/b**), 4′-*O*-demethylancistrocladinium A (**123a**), and the pentacyclic N,C-coupled ancistrocyclinone A (**7**). – The structural landmarks of the compounds are underlaid in gray

ancistrotectorine C (**205**) [60]. Likewise unprecedented was the discovery of a series of five new naphthylisoquinoline dimers, among them shuangancistrotectorines C (**5**) and D (**207a**) [17, 31]. These dimers are constitutionally symmetric, with their monomeric halves being linked via the sterically more constrained 3′,3″- or 1′,1″-positions in the naphthalene moieties, so that even the central axis is rotationally hindered. As a consequence, these molecules have three consecutive chiral axes.

The Chinese liana *A.* aff. *tectorius* from Hainan is the only plant species of the entire *Ancistrocladus* genus that is known to produce monomeric naphthylisoquinoline alkaloids exhibiting all of the seven C,C-coupling types (5,1′, 5,3′, 5,8′, 7,1′, 7,3′, 7,6′, and 7,8′) that have so far been discovered in Nature (Fig. 95) [31, 49–64]. Furthermore, representatives of the known two coupling types of N,C-linked naphthyldihydroisoquinoline alkaloids, N,6′ and N,8′, were identified in

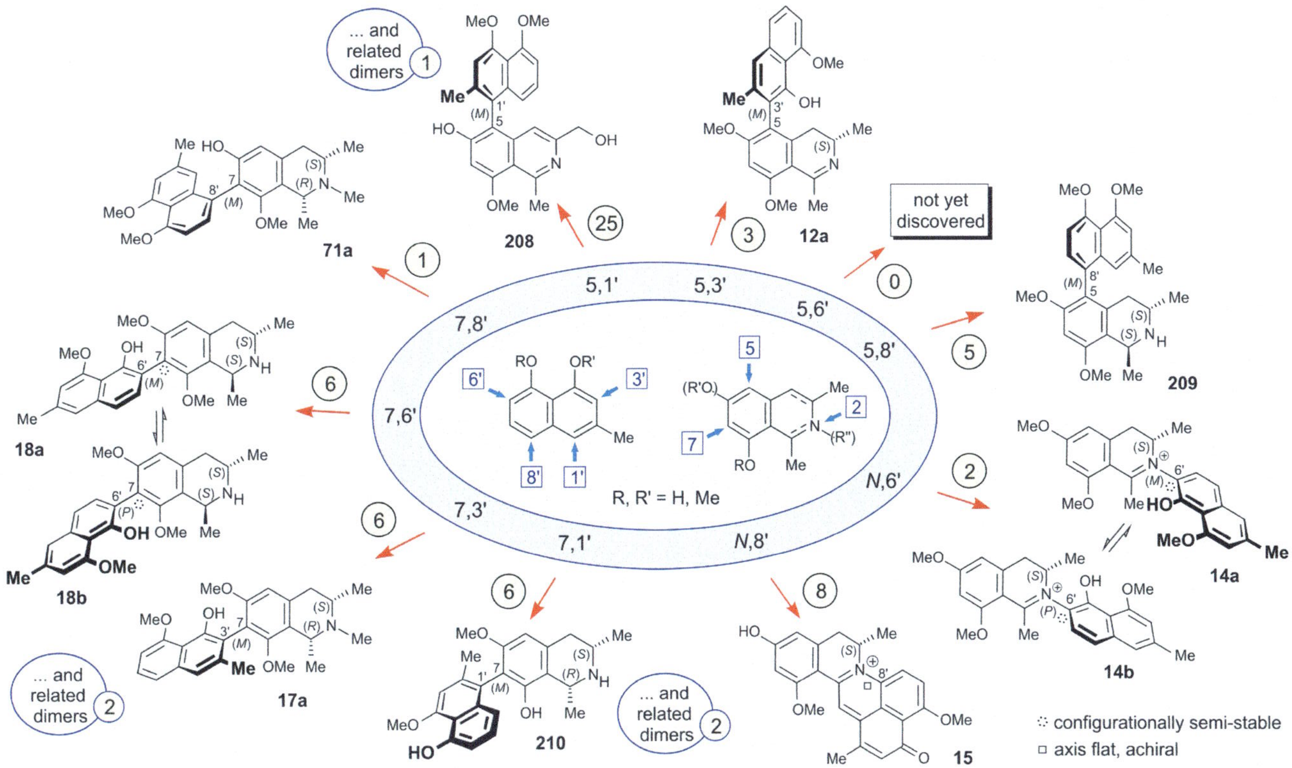

Fig. 95 Structural variability of examples of selected naphthylisoquinoline alkaloids isolated from *Ancistrocladus* aff. *tectorius* from the Chinese Hainan Province. In most cases, the biaryl axis is configurationally stable and constitutes an additional element of chirality, with exceptions being, e.g., ancistrotectorine B (**18a/b**) and ancistrocladinium B (**14a/14b**), which occur as a pair of configurationally semi-stable, and thus slowly interconverting atropo-diastereomers

the plant, among them ancistrocladinium B (**14a/14b**) (Fig. 95) and 4′-*O*-demethylancistrocladinium A (**123a**) (Fig. 94) and the further cyclized pentacyclic analogue ancistrocyclinone B (**15**) [18, 33] (Fig. 95). By far the highest number of naphthylisoquinoline alkaloids isolated from this Chinese liana are 5,1′-coupled compounds (25 representatives) [31, 49, 50, 53–61, 64] comprising examples with a tetrahydro- or a dihydroisoquinoline portion such as in ancistectorines A_1 (**203**) and A_3 (**204**) (Fig. 94) [61], but also far less frequently occurring fully dehydrogenated alkaloids, e.g., ancistrobenomine C (**208**) [64] (Fig. 95). In contrast to the impressive plethora of 5,1′-linked compounds found in *A. tectorius*, only one example of a naphthylisoquinoline alkaloid of the 7,8′-coupling type, namely ancistectorine C_1 (**71a**) [61] (Fig. 95), has so far been identified in the plant.

Compounds displaying the 5,3′-coupling site are extremely rare in Nature. So far, only three such examples have been discovered, and were found in just two *Ancistrocladus* species. For a couple of years, ancistrotanzanine A (**12a**) (Fig. 95) from the East African species *A. tanzaniensis* [155] had been the only naphthylisoquinoline alkaloid showing this unusual 5,3′-biaryl linkage, until, in 2013, Jiang et al. [60] reported the identification of **12a** and two further representatives in the leaves and twigs of *A. tectorius*.

Remarkably, from a chemotaxonomic point of view is the fact that only five examples of 5,8′-linked naphthylisoquinoline alkaloids [56, 60, 62], among them ancistrotectoriline A (**209**) [56] (Fig. 95), have so far been discovered in *A. tectorius*. They are the only representatives of the 5,8′-coupling type ever found in any Asian Ancistrocladaceae plant. Thus, *A. tectorius* and all the other *Ancistrocladus* species from Asia clearly differ from many of the African taxa, which are characterized by the presence of a high number of 5,8′-coupled naphthylisoquinoline alkaloids [156, 159, 161, 163–181] (see Sect. 2).

The majority of the 7,1′-linked compounds identified in *A. tectorius* from Hainan Island—six out of a total of eight alkaloids—are known to occur only in this Chinese liana, as minor metabolites such as ancistectorine B_2 (**210**) (Fig. 95) [62]. The alkaloid patterns of West African *Ancistrocladus* [13, 124, 184, 186, 187, 191, 194] and Dioncophyllaceae species [13, 14, 74, 81, 82, 157, 158, 221–224] by contrast are characterized by the presence of a significant number of 7,1′-coupled naphthylisoquinolines, which in some cases are the main constituents of the respective plants. Most of these alkaloids are typical Dioncophyllaceae-type compounds, i.e., devoid of an oxygen function at C-6 and with an (*R*)-configuration at C-3 (see Sect. 2).

So far, only 14 naphthylisoquinoline alkaloids with a 7,3′-biaryl linkage have been identified in Ancistrocladaceae and Dioncophyllaceae species in total [18, 27–29, 31, 41, 51, 52, 63, 189, 225]. The Chinese liana *A. tectorius* is by far the richest source of 7,3′-linked alkaloids, with no less than eight representatives, among them six monomeric compounds such as ancistrotectorine (**17a**) (Fig. 95) [58, 60, 63] and two unusual dimers such as shuangancistrotectorine D (**207a**) [31] (Fig. 94). The molecular halves of **207a** are closely related to those of ancistrotectorine (**17a**), connected through an unprecedented 1′,1″-coupling in the central binaphthalene core. The discovery of the so far only examples of quinoid naphthylisoquinoline alkaloids consisting of a 1,4-naphthoquinone portion coupled to an N-methylated tetrahydroisoquinoline subunit, such as ancistrotectoquinone A (**74a/74b**) (Fig. 94) [63], is unprecedented. It further underlines the outstanding biosynthesis potential of *A. tectorius*. In contrast to ancistrotectorine (**17a**), which shows complete configurational axial stability [51, 159], ancistrotectoquinone A (**74**) occurs as a pair of configurationally semi-stable—and thus slowly interconverting—atropo-diastereomers, **74a** and **74b** [63].

Six Ancistrocladaceae-type naphthylisoquinoline alkaloids of the less frequently occurring 7,6′-coupling type [59, 60], as present, e.g., in ancistrotectorine B (**18a/18b**) (Fig. 95) [60], have been isolated from *A. tectorius*. Their biaryl axes exert only a medium degree of steric hindrance, therefore occurring in the plant as a mixture of two configurationally semi-stable atropo-diastereomers. Only six further representatives of the 7,6′-coupling type are known from Nature; all are, however, typical Dioncophyllaceae-type alkaloids (6-H, (3*R*)) and have all been isolated from *T. peltatum* exclusively [13, 14, 153, 219]. In comparison to the Ancistrocladaceae-type representatives of this coupling site such as **18a/18b**, the axial configuration of 7,6′-coupled Dioncophyllaceae-type alkaloids is unstable due to the absence of an oxygen function at C-6 [13–15, 153, 219] (see Sect. 8.5.7).

In the following sections, an overview on the structural peculiarities of the naphthylisoquinoline alkaloids isolated from this Chinese *Ancistrocladus* species will be given. The compounds are categorized according to the position of their biaryl linkage.

8.5.1 Ancistrocline from Early Isolation Work and Its Total Synthesis

Ancistrocladus aff. *tectorius* from Hainan was phytochemically investigated in 1981 by Chen et al. [50], leading to the isolation of three naphthylisoquinoline alkaloids from the stems and twigs of the plant. Already earlier, two of the compounds, ancistrocladine (**1a**) and hamatine (**1b**), had been found in *A. heyneanus* (from India) and *A. hamatus* (from Sri Lanka), and their structures had been fully elucidated by the pioneering work of Govindachari and his group [1–5, 23, 24]. The third compound, ancistrocline (**163**), a similarly 5,1′-coupled alkaloid, possessing a *cis*-configured tetrahydroisoquinoline portion, was a new natural product. It was the (chemically more stable) *N*-methyl homologue of isoancistrocladine (**79a**), which had previously been identified in the roots of *A. heyneanus* (see Sect. 4) [38]. The full absolute stereostructure of ancistrocline (**163**) was established in the authors' group [53], based on 1D and 2D NMR spectroscopy in combination with the microanalytical ruthenium-mediated chemical degradation procedure [79, 80], and was finally confirmed by semi and total synthesis (Fig. 96), by applying the lactone concept [53, 84]. One key step was the Pd-catalyzed intramolecular aryl coupling of the ester-type prefixed molecular moieties as established in **211**, leading to the helicene-like distorted biaryl lactones **212a** and **212b**. This configurationally semi-stable bridged biaryl was converted into the configurationally stable alcohols

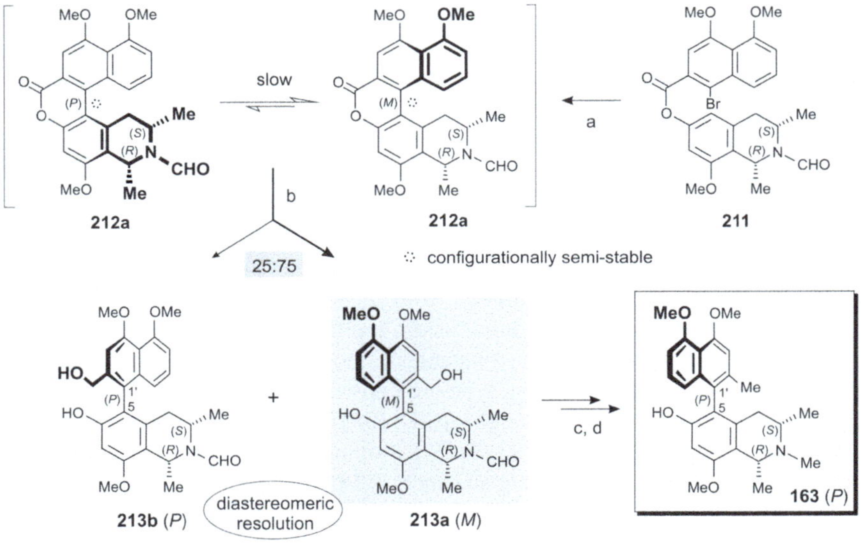

Fig. 96 Total synthesis of ancistrocline (**163**) by the lactone methodology [84–87, 89, 90]. Reaction conditions: (a) $Pd(PPh_3)_2Cl_2$, NaOAc, DMA, 100°C, 56%; (b) $LiBH_4$, THF, 42%; (c) $(CBrCl_2)_2$, polym. PPh_3, CH_2Cl_2 → (d) $LiAlH_4$, THF, 50%

213a (*M*) and **213b** (*P*) using lithium borohydride as the reductant, giving rise to an atropisomeric ratio of 75:25 in favor of the targeted isomer **213a**. Hydroxy/bromine exchange and reduction eventually yielded the target molecule ancistrocline (**163**) [53].

8.5.2 Further Naphthylisoquinoline Alkaloids with a 5,1′-Coupling Site and Their Broad Structural Diversity

By far the highest number of naphthylisoquinoline alkaloids discovered in the twigs, stems, and leaves of *A.* aff. *tectorius* from the Chinese Province Hainan are 5,1′-linked naphthylisoquinoline alkaloids, like already ancistrocline (**163**) above, showing a great structural variety in their oxygenation and O-methylation patterns compared to their parent compounds ancistrocladine (**1a**) and hamatine (**1b**), which were likewise identified in all parts of the plant. Intense isolation work by various research groups [31, 49, 50, 53–61, 64], based on liquid–liquid extraction or fast centrifugal partition chromatography (FCPC), followed by column chromatography on silica gel or aluminum oxide, and finally preparative HPLC furnished no less than 25 pure naphthylisoquinoline alkaloids displaying the 5,1′-coupling type. Within this impressive series of compounds, seven representatives were detected only in this Chinese liana [59, 61, 64]. Interestingly, nearly all of the *trans*-configured naphthyltetrahydroisoquinoline alkaloids (twelve compounds) were isolated

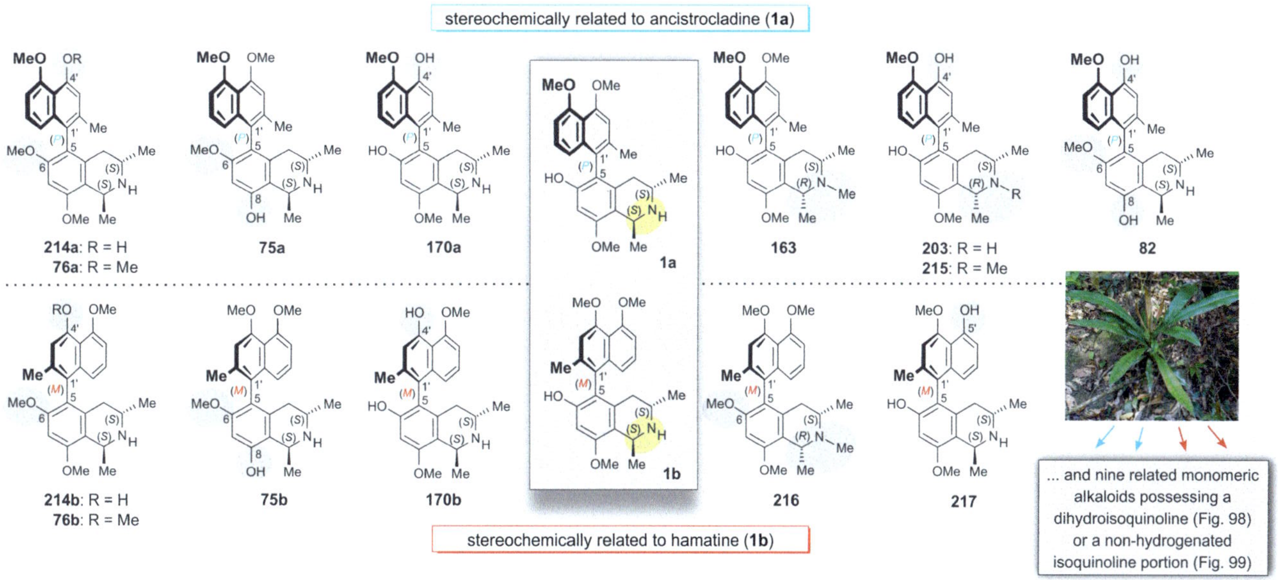

Fig. 97 A series of 14 naphthyltetrahydroisoquinoline alkaloids from the Chinese liana *A.* aff. *tectorius*, all 5,1′-coupled, yet varying in the degree of O-methylation, isolated along with their parent compounds ancistrocladine (**1a**) and hamatine (**1b**). – The structural modifications in these compounds compared to **1a** and **1b** are underlaid in light gray; the photograph shows *A.* aff. *tectorius* in its natural habitat in Hainan Province, China (Photograph: H. Rischer)

as atropo-diastereomeric mixtures (Fig. 97) [50, 56–61, 64]. All of the naphthyldihydroisoquinolines identified so far (six compounds [58, 60, 61, 64], see Fig. 98), however, with the exception of ancistrocladinine (**77a**) and hamatinine (**77b**), seem to occur atropisomerically pure in the plant. Likewise stereochemically homogeneous at the axis are the four *cis*-configured naphthyltetrahydroisoquinolines identified in *A. tectorius* from Hainan Island [53, 58, 60, 61], among them ancistrocline (**163**) [53] and the new alkaloid ancistectorine A_1 (**203**) [61] (Fig. 97). From the same plant material, three further representatives of the 5,1′-coupling type were isolated belonging to the rare group of non-hydrogenated alkaloids possessing a hydroxymethylene entity at C-3, like the newly discovered ancistrobenomine B (**11**) [64] (Fig. 99). They were found to be optically active despite the absence of stereogenic centers, with the only stereoelement being the rotationally hindered and, thus, chiral biaryl axis.

6-*O*-Methylancistrocladine (**76a**) and its atropo-diastereomer 6-*O*-methylhamatine (**76b**) occur widely in Asian and African *Ancistrocladus* species [5, 13, 64, 161, 171, 173, 191, 316, 458]. They were isolated from twig and stem bark extracts of *A.* aff. *tectorius* [64], along with their 4′-*O*-demethyl analogues **214a** and **214b** [56], which had likewise been identified in the roots of *A. abbreviatus* from Coastal West Africa [191, 459]. 4′-*O*-Demethylancistrocladine (**170a**) [64] had earlier been reported to be a minor metabolite in *Ancistrocladus* plants from Malaysia assigned as *A.* aff. *tectorius* (see Sect. 8.4) [54]. This alkaloid had also been detected in extracts of the Vietnamese liana *A. cochinchinensis*, analyzed online, using LC-NMR, LC-MS, and LC-ECD coupling [47]. Its atropo-diastereomer, 4′-*O*-demethylhamatine (**170b**), however, has so far only been discovered in *A.* aff. *tectorius* from the Chinese Island Hainan [59]. The regioisomeric 5′-*O*-demethylhamatine (**217**) [64] had earlier been identified as a minor constituent in *A. congolensis* [173] and in one other, botanically yet unknown *Ancistrocladus* liana from the Central Congo Basin [171], while the O,N-permethylated and *cis*-configured ancistrobertsonine B (**216**) [58, 60] had previously been known to occur only in the East African taxon *A. robertsoniorum* [190].

From the twigs and stems of *A. tectorius* a series of four new 5,1′-linked naphthyltetrahydroisoquinoline alkaloids [61] were isolated, differing from their parent compounds ancistrocladine (**1a**) and hamatine (**1b**) by the presence of a free hydroxy function (instead of a methoxy group) at C-4′ or at C-8. A fifth compound detected in the aerial parts of the plant, 4′-*O*-demethylancistectorine A_2 (**82**) [64], possessed even two phenolic hydroxy substituents at C-4 and C-8. This alkaloid was likewise found to occur as a major metabolite in fresh leaves of the Indian liana *A. heyneanus* (see Sect. 4.7). Among the new compounds, ancistectorine A_1 (**203**) [61] attracted attention since it is one of the quite rare examples of *cis*-configured naphthyltetrahydroisoquinolines with a free, unmethylated NH function. Within the subgroup of 5,1′-coupled alkaloids, only two further such compounds are known to date from Nature, isoancistrocladine (**79a**, see Fig. 29) [38] from the Indian liana *A. heyneanus*, and its atropo-diastereomer, isohamatine (**79b**, see Fig. 81) [48] from the Vietnamese plant *A. cochinchinensis*. Among the nearly 70 naphthylisoquinolines discovered in *A. tectorius* from China, the 7,1′-linked ancistectorine B_2 (**210**,

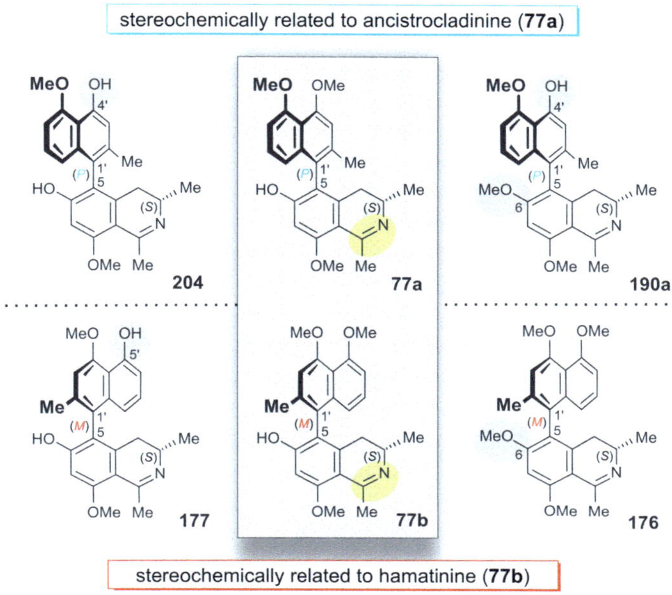

Fig. 98 Six 5,1′-coupled naphthyl*dihydro*isoquinoline alkaloids isolated from the Chinese liana *A.* aff. *tectorius*, among them the two new compounds ancistectorine A_3 (**204**) and its 6-*O*-methyl analogue **190a**, discovered along with 5′-*O*-demethylhamatinine (**177**) and 6-*O*-methylhamatinine (**176**), and the parent compounds ancistrocladinine (**77a**) and hamatinine (**77b**)

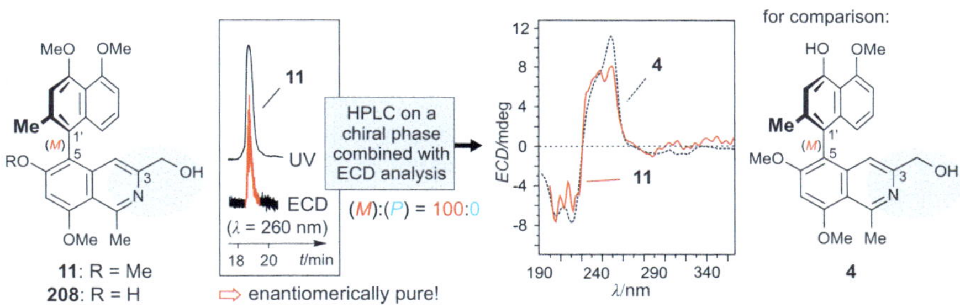

Fig. 99 Ancistrobenomines B (**11**) and C (**208**), new 5,1′-coupled non-hydrogenated naphthylisoquinoline alkaloids isolated as minor metabolites from twigs and stems of the Chinese liana *A.* aff. *tectorius*. Assignment of the absolute axial configuration of **11**, by comparison of its ECD spectrum with that of the known [30] ancistrobenomine A (**4**), which occurs also in *A.* aff. *tectorius*, and HPLC-ECD analysis on a chiral phase (Chiralcel® OD-H), giving only one peak, revealing that **11** occurs enantiomerically pure in the plant. The hydroxymethylene unit at C-3 as the structural pecularity of **11**, **208**, and **4** is underlaid in gray

Fig. 95) [62] is, besides the above-mentioned ancistectorine A_1 (**203**) [61], the only other compound that shows this structural element of an unmethylated NH function.

Three of the new 5,1′-linked compounds, *N*-methylancistectorine A_1 (**215**), ancistectorine A_2 (**75a**), and its atropo-diastereomer 5-*epi*-ancistectorine A_2 (**75b**), showed pronounced effects against the chloroquine-resistant K1 strain of *P. falciparum*, with half-maximum inhibitory concentration (IC_{50}) values of 0.08 μM (for **215**), 0.07 μM (for **75a**), and 0.03 μM (for **75b**). They were more active than the standard chloroquine (IC_{50}, 0.26 μM) by a factor of 3–7 and, given their extremely weak cytotoxicities against rat skeletal myoblast (L6) cells [61], the alkaloids showed excellent selectivity indexes of >3000. In addition, 5-*epi*-ancistectorine A_2 (**75b**) revealed a strong inhibitory activity against the chloroquine-sensitive *P. falciparum* strain NF54 (IC_{50}, 0.0074 μM), and it was found to be about 8.5 times more active than the standard chloroquine (IC_{50}, 0.065 μM) [191]. Hence, according to the TDR/WHO guidelines [238], ancistectorine A_2 (**75a**) and 5-*epi*-ancistectorine A_2 (**75b**) can be considered a promising novel antiplasmodial leads [61, 151]. The (*M*)-configured 5-*epi*-ancistectorine A_2 (**75b**) was even more potent, exerting a much better antiplasmodial activity than the (*P*)-configured ancistectorine A_2 (**75a**). Simultaneously, **75b** had a lower cytotoxicity than **75a**, thus impressively evidencing the substantial impact of axial chirality on the biological properties of naphthylisoquinoline alkaloids. The specificity of the antiplasmodial activity of **215**, **75a**, and **75b** was supported by the fact that these three alkaloids did not exert any inhibitory activities against bacterial organisms, among them *S. aureus*, *S. epidermidis*, *Enterococcus faecalis*, *Enterococcus faecium*, *E. coli*, *P. aeruginosa*, *Yersinia pseudotuberculosis*, and *Yersinia pestis* [61].

Furthermore, isolation work on the twigs and stems of *A. tectorius* led to the identification of six 5,1′-coupled naphthyl*dihydro*isoquinoline alkaloids (Fig. 98) [61, 64]. This series of minor metabolites comprised two new alkaloids, ancistectorine A_3 (**204**) [61] and its 6-*O*-methyl analogue **190a** [64], and the known naphthylisoquinolines 6-*O*-methylhamatinine (**176**) and 5′-*O*-demethylhamatinine (**177**), which were found to be present in the plant along with their parent compounds ancistrocladinine (**77a**) and hamatinine (**77b**) [58, 60, 64]. 6-*O*-Methylhamatinine (**176**) and 5′-*O*-demethylhamatinine (**177**) had previously been isolated as the main constituents from the Vietnamese liana *A. cochinchinensis* [45–47] and from a botanically as yet undescribed Central Congolese *Ancistrocladus* species [171]. Ancistectorine A_3 (**204**) had so far exclusively been detected in *A. tectorius*, differing from **177** only by its (*P*)-configuration at the biaryl axis and by a free phenolic hydroxy function at C-4′ on the naphthalene portion (and not at C-5′ as in **177**). All of these 5,1′-linked naphthyldihydroisoquinoline alkaloids displayed only moderate to weak inhibitory activities against pathogens belonging to the genera *Plasmodium*, *Leishmania*, and *Trypanosoma* [61, 64, 171].

Compared to the large number of naphthyltetraisoquinolines and naphthadihydroisoquinolines alkaloids (ca. 240 representatives) identified in *Ancistrocladus* lianas from Asia and Africa, compounds possessing a non-hydrogenated isoquinoline portion had been discovered only rarely in these plants (ca. 20 examples are known today) [30, 44, 45, 64, 167, 168, 186]. From the twigs and stems of

A. tectorius collected in the region of Ledong of the Chinese Hainan Province, three 5,1′-coupled fully dehydrogenated naphthylisoquinoline alkaloids were isolated, the new ancistrobenomines B (**11**) and C (**208**) (Fig. 99), along with the known alkaloid ancistrobenomine A (**4**) [64]. The latter had earlier been identified together with its 6-*O*-demethyl analogue in the Malaysian species *A. benomensis* [30] (Sect. 6.1) and in the Vietnamese liana *A. cochinchinensis* [47] (Sect. 7.1). The ancistrobenomines have a hydroxymethylene unit at C-3; this structural peculiarity is otherwise only known from three simple, naphthalene-devoid isoquinolines isolated from the Malaysian liana *A. tectorius* (one example) [54] (Sect. 8.4) and from *A. cochinchinensis* (two examples) [48] (Sect. 7.4). Despite the absence of stereogenic centers, the four ancistrobenomines A (**4**), B (**11**), C (**208**), and 6-*O*-demethylancistrobenomine A (**159**; for its structure, see Fig. 81 in Sect. 7.1) are optically active—due to the presence of the rotationally hindered biaryl axis as the only element of chirality. The absolute axial configuration of the new *O*-permethylated ancistrobenomine B (**11**) was attributed to be (*M*), due to the fact that **11** and its known [30, 47] parent compound ancistrobenomine A (**4**) displayed nearly identical ECD spectra (Fig. 99). Similar as in the case of **4**, HPLC analysis on a chiral phase (Chiralcel® OD-H) coupled to ECD spectroscopy revealed **11** to occur enantiomerically pure in *A. tectorius* [64]. The second new metabolite, ancistrobenomine C (**208**) (Fig. 99), differing from **11** only by the presence of a free phenolic hydroxy function at C-6 on the isoquinoline portion, was likewise attributed to be (*M*)-configured at the biaryl axis. HPLC-ECD analysis on a chiral phase resulted in only one sharp peak, and the ECD spectra measured in intervals at different sections of the peak were all identical. In the plant, ancistrobenomine C (**208**) was thus present in an enantiopure form [64].

Ancistrobenomines B (**11**) and C (**208**) exhibited only moderate effects in vitro toward *P. falciparum* and *T. brucei rhodesiense* [64]. The inhibitory activities of **11** (IC_{50}, 4.75 μ*M*) and **208** (IC_{50}, 5.08 μ*M*) against the K1 strain of *P. falciparum* were distinctly weaker than those of the parent compound ancistrobenomine A (**4**) (IC_{50}, 1.81 μ*M*), by a factor of 2.6 and 2.8, respectively, and similar to those of 6-*O*-demethylancistrobenomine A (IC_{50} = 5.38 μ*M*) [30]. Furthermore, ancistrobenomine B (**11**) showed strong antiproliferative activities against drug-sensitive acute lymphoblastic CCRF-CEM leukemia cells (IC_{50}, 3.5 μ*M*) and their multidrug-resistant subline, CEM/ADR5000 (IC_{50}, 21.4 μ*M*) [64]. The latter leukemia cells, which are highly resistant to the standard drug doxorubicin [460, 461] (ca. 1770-fold compared to CCRF-CEM), exerted only a low degree of cross-resistance to **11** (6.1-fold) [64], thus indicating that this non-hydrogenated and fully O-permethylated naphthylisoquinoline might be an efficient inhibitor also for other drug-sensitive and -resistant tumor cells. Ancistrobenomine C (**208**), by contrast, was nearly one order of magnitude less active toward the CCRF-CEM tumor cells (IC_{50}, 26.5 μ*M*), exerting only very weak effects on the cell viability of these two leukemia cell lines [64].

8.5.3 Ancistrotanzanine A and Two Further New Examples of the Rare Subclass of 5,3′-Coupled Naphthylisoquinoline Alkaloids

Phytochemical studies by the Hua group [60] on air-dried and powdered stems and leaves of *A.* aff. *tectorius* led to the identification of three naphthylisoquinoline alkaloids exhibiting the extremely rare 5,3′-coupling type, among them ancistrotanzanine A (**12a**) (Fig. 100), which had previously been isolated from the East African species *A. tanzaniensis* [155]. Prior to the discovery of **12a** and two similar alkaloids (Fig. 101) in *A.* aff. *tectorius*, ancistrotanzanine A (**12a**) had been the only known representative of this unusual subclass of 5,3′-coupled naphthylisoquinoline alkaloids. The structural elucidation of **12a** from *A. tanzaniensis* in the authors' group had been achieved by HRESIMS, 1D and 2D NMR, oxidative degradation, and by experimental and quantum-chemical ECD investigations [155]. In $CDCl_3$, the aromatic protons displayed the coupling pattern of a triplet, two doublets, and two singlets, which evidenced that the biaryl axis could not be located at C-6′ or C-8′, leaving only C-3′ or C-1′ for the coupling position in the naphthalene part. Of these, the latter was excluded by NOESY interactions of H-1′ with H-8′ and Me-2′. HMBC cross-peaks from H-1′ and H-6′ to C-8′, from H-1′ to Me-2′, and vice versa from the protons of the aromatic methyl group to C-1′ confirmed C-3′ to be the axis-bearing carbon atom [60, 155]. The high-field shifts of Me-2′ and OMe-6 finally determined the position of the biaryl axis to correspond to the previously unknown 5,3′-coupling type (Fig. 100a). The absolute configuration at C-3 of ancistrotanzanine A (**12a**) was established to be (*S*) by ruthenium-mediated oxidative degradation [155]. Attempts to assign the configuration at the biaryl axis relative to that of the stereogenic center at C-3 by specific NOESY interactions between H_{eq}-4 and Me-2′ failed because of the overlap of the signals of the two diastereotopic, but largely isochronous protons at C-4. Therefore, the absolute axial configuration of **12a** had to be attributed by a combination of experimental and computational ECD investigations [155]. For this purpose, the molecule was submitted to a conformational analysis using the Tripos force field and the structure was further refined using the semiempirical AM1 method. For each of the 48 conformers thus identified, the corresponding ECD spectrum was calculated by applying the CNDO/2S approach. The single ECD spectra were then added up Boltzmann-weighted, with respect to the different occurrence of the conformers according to their energies, to give the overall calculated ECD spectrum. The ECD curve calculated for the (*M*,*S*)-isomer of **12a** revealed a good agreement with that measured (Fig. 100b), thus permitting assignment of the absolute configuration of the chiral axis of ancistrotanzanine A (**12a**) as (*M*) [155].

Two new examples of 5,3′-coupled naphthylisoquinoline alkaloids now discovered in *A.* aff. *tectorius* were found to be the atropo-diastereomer of ancistrotanzanine A (**12a**), named ancistrotectorine D (**12b**), and a structurally closely related N-methylated naphthyltetrahydroisoquinoline, named ancistrotectorine C (**205**) [60]. The absolute configuration at C-3 in the two compounds was only tentatively assumed to be (*S*), based on the fact that so far exclusively

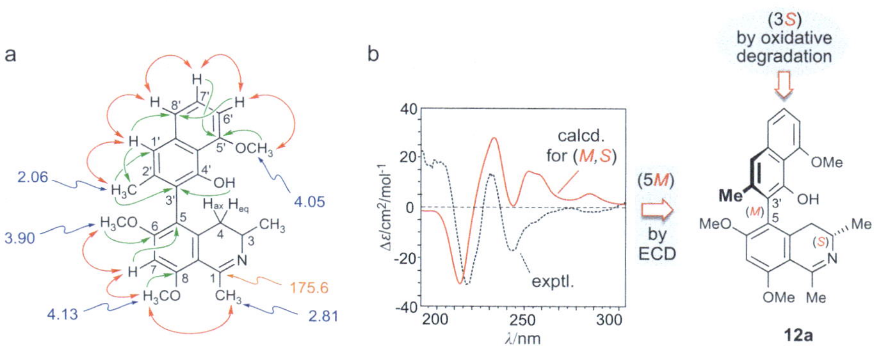

Fig. 100 Ancistrotanzanine A (**12a**), the first 5,3′-coupled naphthylisoquinoline alkaloid discovered in Nature, from *A. tanzaniensis*: **a** NOESY (red double arrows) and HMBC (green single arrows) interactions relevant for its constitution; **b** assignment of the absolute configuration of **12a** by AM1-Boltzmann-based ECD calculations using the CNDO/2S approach and comparison with the experimental ECD curve

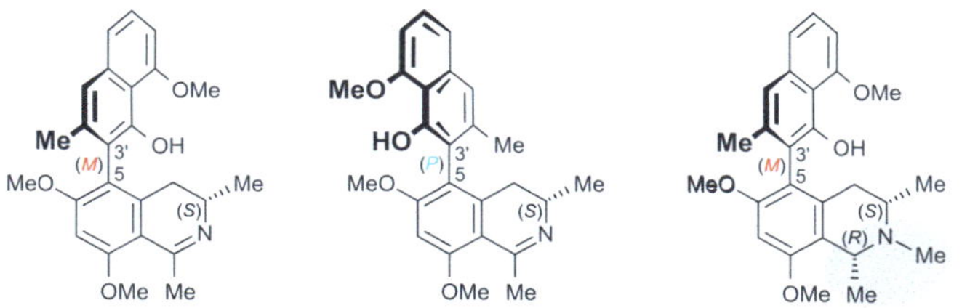

12a **12b** **205**

Fig. 101 Only three known [60, 155] representatives of naphthylisoquinoline alkaloids with a 5,3′-biaryl linkage between the naphthalene and isoquinoline portion: ancistrotanzanine A (**12a**), its atropisomer ancistrotectorine D (**12b**), and ancistrotectorine C (**205**)

(3*S*)-configured naphthylisoquinoline alkaloids have been isolated from *A.* aff. *tectorius*, but no oxidative degradation was performed for a firm assignment, nor were there specific NOE interactions reported establishing the relative configuration at the axis versus the center [60]. The ECD spectrum of ancistrotectorine D (**12b**) showed Cotton effects opposite to those of ancistrotanzanine A (**12a**), thus revealing **12a** and **12b** to be atropo-diastereomers, and establishing the absolute axial configuration of **12b** as (*P*) [60]. The chemical shifts of the H-1 (δ_H = 3.67 ppm) and H-3 (δ_H = 2.24 ppm) signals indicated a relative *cis*-array between the two methyl groups of the stereogenic centers at C-1 and C-3 in ancistrotectorine C (**205**), and thus led to the assignment of the absolute (1*R*,3*S*)-configuration in the isoquinoline portion of this alkaloid. No NOE measurements were done to confirm the relative configuration at C-1 versus C-3. The ECD spectrum of **205** was similar to that measured for ancistrotanzanine A (**12a**), thus establishing ancistrotectorine C (**205**) to be (*M*)-configured at the biaryl axis [60].

Ancistrotanzanine A (**12a**) showed good inhibitory activities against the pathogens of leishmaniasis and Chagas' disease, *L. donovani* and *T. cruzi*, with half-maximum inhibitory concentrations (IC_{50}) in the low micromolar range, while it displayed only moderate activities against *P. falciparum* parasites [155]. Furthermore, all of the three 5,3′-coupled naphthylisoquinoline alkaloids exhibited rather low antiproliferative activities against the leukemia cell lines HL-60 and K562, and the human macrophage cell line U937, with IC_{50} values ranging between 7.45 and 15.0 μ*M* [60].

8.5.4 Ancistrotectoriline A and Related 5,8′-Coupled Naphthylisoquinoline Alkaloids, the Only Five Representatives of this Coupling Type Isolated from an Asian *Ancistrocladus* Liana

Ancistrotectoriline A and Related Alkaloids

Only very few 5,8′-linked naphthylisoquinoline alkaloids (five compounds) have so far been isolated from Asian *Ancistrocladus* species. Chemotaxonomically remarkable, the alkaloids of this coupling type have as yet been identified exclusively as minor compounds in *A.* aff. *tectorius* from Hainan Island (Fig. 102) [56, 60, 62]. This is in a sharp contrast to the alkaloid patterns of African *Ancistrocladus* species, in particular those from Cameroon, Nigeria, or the Congo Basin, which are dominated by a high number of 5,8′-coupled naphthylisoquinoline alkaloids [17, 156, 159, 161, 163–181] (in total more than 100 compounds, among them 44 michellamine- and mbandakamine-type dimers [17]) (see Sect. 2). These alkaloids are often the main constituents of the plants, and a noticeable number are hybrid-type alkaloids, i.e., with (*R*)-configuration at C-3 and an oxygen function at

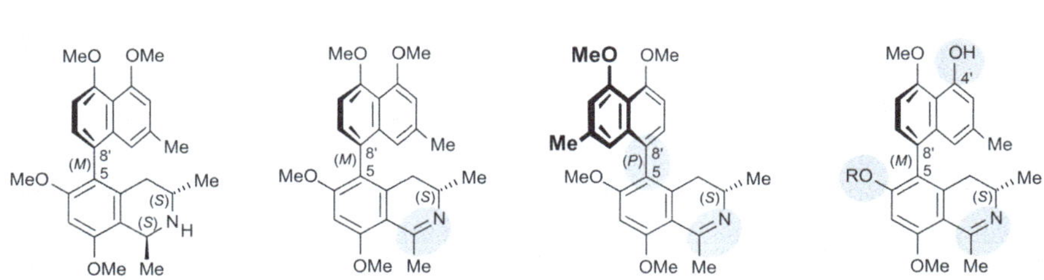

209 **218a** **218b** **124**: R = Me **219**: R = H

Fig. 102 Only five 5,8′-coupled naphthylisoquinoline alkaloids have been identified so far in Asian *Ancistrocladus* lianas, with all of them isolated from *A.* aff. *tectorius* from the Hainan Province in China: ancistrotectoriline A (**209**), ancistrotanzanine B (**218a**) and its atropo-diastereomer ancistroealaine A (**218b**), ancistectorine D (**124**) and its 6-*O*-demethyl analogue **219**. The structural modifications in **218a**, **218b**, **124**, and **219** in comparison to the structure of ancistrotectoriline A (**209**) are underlaid in light gray

C-6; such naphthylisoquinoline alkaloids have never been found so far in any *Ancistrocladus* plant from Asia.

The first 5,8′-linked naphthylisoquinoline alkaloid ever discovered in an Asian *Ancistrocladus* species was ancistrotectoriline A (**209**). It was isolated from the stems and leaves of *A.* aff. *tectorius* from the Chinese Province of Hainan by Ye et al. [56] and has later also been identified by the authors' group in the leaves of the East African species *A. tanzaniensis* [155]. The relative *trans*-configuration at C-1 versus C-3 of **209** was deduced only from the chemical shift of H-3 (δ_H = 3.30 ppm), which was in the typical range of *trans*-substituted 1,3-dimethyltetrahydroisoquinolines; but no NOE experiments, as usual for the structural elucidation, were described for the analysis of the relative stereoarray at C-1 versus C-3 [56]. The absolute configuration at C-1 and C-3 of **209** was tentatively assumed by Ye et al. [56] as (1*S*,3*S*) considering the fact that (nearly) all of the Asian Ancistrocladaceae plants strictly produce typical Ancistrocladaceae-type alkaloids with (*S*)-configuration at C-3. This assumption was later firmly proven in the authors' group by applying their ruthenium-mediated oxidative degradation to an authentic sample [155] kindly provided by the Ye group [56]. It was further in agreement with a NOESY interaction between Me-1 and H-3, indicative of a relative *trans*-configuration in the tetrahydroisoquinoline portion (Fig. 103a), likewise monitored in the authors' group [155] in the course of NMR investigations to prove the identity of the material obtained from *A. tanzaniensis* with ancistrotectoriline A (**209**) isolated from *A. tectorius* by the Ye group. Furthermore, 2D NMR measurements in the authors' group [155] revealed cross-peaks between H-7′ and H-4_{eq} (hence, both being above the isoquinoline plane) and between H-1′ and H-4_{ax} (both below the isoquinoline plane) (Fig. 103a) and, thus, permitted assignment of the configuration at the biaryl axis relative to the stereogenic centers. Given the (*S*)-configuration at C-1 and C-3, as established by the oxidative degradation procedure, the axial configuration of ancistrotectoriline A (**209**) was attributed as (*M*) [155]. This result was confirmed by ECD spectroscopy [155], as obvious from the virtually identical ECD spectrum of **209** compared to that of the known [192]

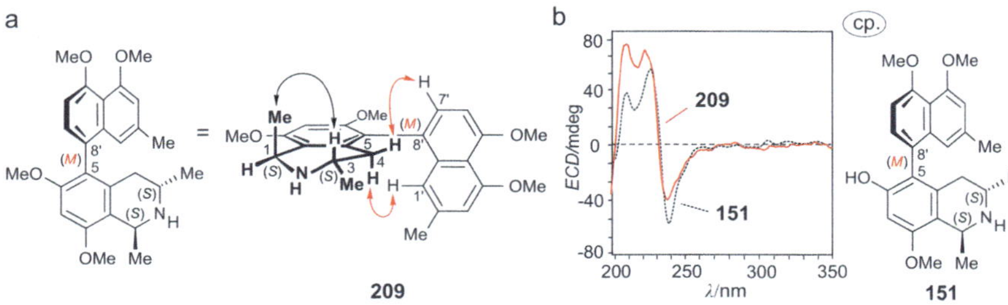

Fig. 103 Assignment of the complete stereostructure of ancistrotectoriline A (**209**) [56, 155] **a** by specific NOESY interactions defining the relative configuration at C-3 versus C-1 in the tetrahydroisoquinoline portion (black double arrow) and the relative configuration at the axis (red double arrows), and **b** the absolute axial configuration by comparison of its ECD spectrum with that of the known, structurally related alkaloid ancistrobrevine B (**151**) [192]

5,8′-coupled and (*M*)-configured alkaloid ancistrobrevine B (**151**) (Fig. 103b). The attribution of the absolute axial configuration was in accordance with the likewise ECD-based assignment by the Ye group [56], thus permitting an unambiguous attribution of the full stereostructure of ancistrotectoriline A (**209**), which occurs in both *A. tectorius* and *A. tanzaniensis*.

The second 5,8′-coupled alkaloid, the naphthyldihydroisoquinoline ancistrotanzanine B (**218a**) (Fig. 102), which was identified in the aerial parts of *A.* aff. *tectorius* by the Hua group [60], had already been known from the African liana *A. tanzaniensis*, and likewise isolated from leaf material [155]. Phytochemical studies on the twigs and stems of *A. tectorius* by the authors' group [62] furnished three further 5,8′-linked naphthyldihydroisoquinolines (Fig. 102); two of these, ancistectorine D (**124**) and its 6-*O*-demethyl analogue **219**, were new. The structure of ancistectorine D (**124**) closely resembles that of ancistrotanzanine B (**218a**), differing only by the presence of a free phenolic hydroxy group at C-4′ on the naphthalene ring. The third compound was the known atropo-diastereomer of **218a**, ancistroealaine A (**218b**). The latter alkaloid had previously been discovered as a main metabolite in *A. ealaensis* [174] and in a botanically so far undescribed *Ancistrocladus* liana [171] from the Congo Basin.

Ancistrotanzanine B (**218a**) and ancistroealaine A (**218b**) had previously been identified as promising antitrypanosomal and antileishmanial agents [146, 147, 149, 155, 174]. Similar to ancistrotanzanine B (**218a**), the new alkaloid ancistectorine D (**124**) displayed only moderate to weak activities against the NF54 strain (sensitive to all known drugs) of *P. falciparum* [62]. The antitrypanosomal activities of **124** against the pathogen of African sleeping sickness (*T. brucei rhodesiense*) were found to be moderate to weak, too, while ancistectorine D (**124**) exerted pronounced inhibitory effects against *T. cruzi*, the causative agent of Chagas' disease, and against *L. donovani*, responsible for visceral leishmaniasis (also known as kala-azar) [62].

The half-maximum inhibitory (IC_{50}) concentration value of **124** (4.44 μ*M*) against *T. cruzi* was nearly the same as those of ancistrotanzanine B (**218a**) (3.58 μ*M*) [155] and ancistroealaine A (**218b**) (5.60 μ*M*) [174]. The inhibitory activity of **124** (IC_{50}, 1.20 μ*M*) against *L. donovani* were distinctly better than those of **218a** (IC_{50}, 3.81 μ*M*), by a factor of 3.2, and even better than those of **218b** (IC_{50}, 9.82 μ*M*), by a factor of 8.0 [62]. Only weak antiparasitic effects, by contrast, with IC_{50} values of 25.9 and 16.3 μ*M* against *T. cruzi*, and IC_{50} values of 52.3 and 43.4 μ*M* against *L. donovani*, were exerted by the 6-*O*-demethyl and 6,5′-*O*,*O*-didemethyl analogues of ancistroealaine A (**218b**), discovered in a botanically as yet undescribed Congolese *Ancistrocladus* liana [171]. These results suggested that a high degree of O-methylation might be an important structural parameter for the antileishmanial and antitrypanosomal activities of 5,8′-coupled naphthyldihydroisoquinoline alkaloids.

Additionally, ancistectorine D (**124**) showed promising antileukemic activities. It inhibited the viability of drug-sensitive CCRF-CEM leukemic cells (IC_{50}, 4.5 μ*M*) and of their multidrug-resistant P-glycoprotein-overexpressing subline, CEM/ADR5000 (IC_{50}, 25.8 μ*M*), similar to the effect induced by the standard antileukemic drug doxorubicin [62]. Remarkably, **124** showed a very low cross-resistance (5.75-fold) toward CEM/ADR5000 cells compared to doxorubicin (ca. 1770-fold) [62, 461], thus indicating that ancistectorine D (**124**) might be a promising candidate to treat resistant and refractory tumors.

Total Synthesis of Ancistrotanzanine B and Ancistroealaine A by Atroposelective Biaryl Coupling Using Chiral Catalysts

In view of their promising antileishmanial properties, ancistrotanzanine B (**218a**) and its atropo-diastereomer ancistroealaine A (**218b**) were rewarding synthesis targets [337, 462]. The construction of the rotationally hindered biaryl axis of these two 5,8′-coupled naphthylisoquinoline alkaloids was achieved by Suzuki cross-coupling of the boronic acid **220** with the iodinated isoquinoline **221** (see Fig. 104). Particularly remarkable was the fact that the coupling step still worked in the presence of the free imino function and when chiral palladium catalysts were used. By application of $Pd(PPh_3)_4$ as an achiral catalyst, only weak, internal asymmetric inductions by the stereogenic center in the dihydroisoquinoline part were observed, leading to an atropisomeric ratio of 55:45 in favor of the (*P*)-isomer ancistroealaine A (**218b**). Quite good diastereoselectivities, however, now in favor of the (*M*)-isomer ancistrotanzanine B (**218a**) (*dr* up to 75:25), were obtained when using chiral catalysts prepared from $Pd_2(dba)_3$ and chiral ligands like, e.g., the ferrocene derivative (R_c,S_p)-**222** or (*P*)-BINAP ((*P*)-**223**) (Fig. 104) [462].

Using the (S_c,R_p)-enantiomer of ferrocene **222** as the catalyst, no asymmetric induction was observed (*dr* = 51:49), along with a lower chemical yield of only 34% (Table 2) [462]. The catalyst derived from (*M*)-BINAP ((*M*)-**223**), by contrast, gave a yield of 45%, with a *dr* of 61:39 in favor of ancistrotanzanine B (**218a**), suggesting that the use of (*P*)-BINAP ((*P*)-**223**) would permit an even higher diastereomeric ratio, this time hopefully in favor of the (*P*)-atropisomer

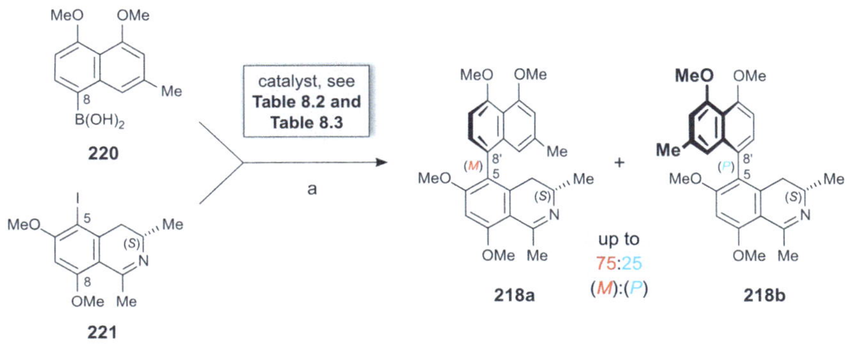

Fig. 104 Atroposelective biaryl coupling using chiral catalysts: total synthesis of ancistrotanzanine B (**218a**) and its (*P*)-configured atropo-diastereomer, ancistroealaine A (**218b**). Reaction conditions: (a) catalyst (see Tables 2 and 3), satd. aq. $NaHCO_3$, toluene, reflux, 10 h

Table 2 Suzuki coupling reactions of the naphthalene **220** with the dihydroisoquinoline **221** to give ancistrotanzanine B (**218a**) ((*M*)-isomer) and ancistroealaine A (**218b**) ((*P*)-isomer) using chiral ferrocene and BINAP ligands

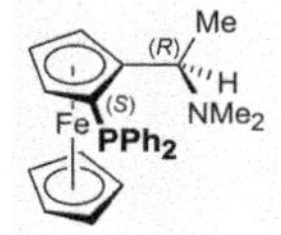

(R_c,S_p)-**222**

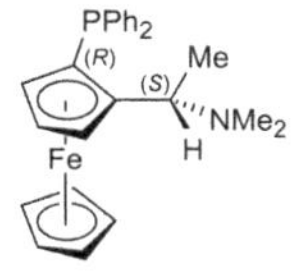

(S_c,R_p)-**222**

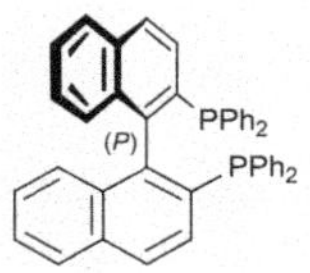

(*P*)-**223**

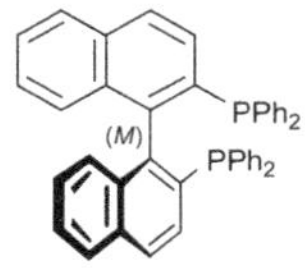

(*M*)-**223**

Catalytic system[a]	Yield/%	(*M*):(*P*)
0.1 equiv. of $Pd(PPh_3)_4$	50	45:55
0.05 equiv. of $Pd_2(dba)_3$, (R_c,S_p)-**222**	38	75:25
0.1 equiv. of $Pd_2(dba)_3$, (*P*)-**223**	50	75:25
0.05 equiv. of $Pd_2(dba)_3$, (S_c,R_p)-**222**	34	51:49
0.1 equiv. of $Pd_2(dba)_3$, (*M*)-**223**	45	61:39

[a]Further coupling conditions: **220** (1.5 equiv.), **221** (1.0 equiv.), catalyst, satd. aq. $NaHCO_3$, toluene/H_2O, reflux, 10 h

ancistroealaine A (**218b**). Indeed, an enhanced isomeric ratio of 75:25 was obtained for the construction of the biaryl axis (along with an improved chemical yield of 50%), but again in favor of the (*M*)-isomer ancistrotanzanine B (**218a**) (Table 2) [462]. Thus, all attempts to overcome the strong internal asymmetric induction exerted by the stereocenter at C-3 and to achieve a preferential formation of the (*P*)-atropisomer **218b**, by just using the other enantiomers of the respective catalysts, **222** or **223**, failed.

Similar coupling results were obtained in the presence of novel P,N-ligands of type (*M*)-**224** and (*M*)-**225** possessing both, stereogenic centers and axes (Table 3), as developed in the authors' group [337]. With the more electron-rich derivative (*M*,*R*)-**224** (R = OMe) of these phosphine amines, much better coupling yields (89%) were attained than with the commercially available chiral ligands ferrocene (R_c,S_p)-**222** and (*P*)-BINAP ((*P*)-**223**) (<50%) (Table 2). But again, the P,N-ligands delivered the (*M*)-configured ancistrotanzanine B (**218a**) as the main diastereomer in each case, albeit with only moderate stereocontrol (best *dr* = 67:33 for (*M*,*R*)-**225** with R = Me) (Table 2) [337].

Although being diastereomeric to each other, the target molecules ancistrotanzanine B (**218a**) and ancistroealaine A (**218b**) showed virtually identical chromatographic behaviors, both on normal and reversed-phase silica gel [462], as already seen in previous cases [32, 47, 184]. This hampered the detection of the isomeric ratios achieved by the use of various chiral catalysts as outlined in Fig. 105 in the Suzuki cross-coupling reaction. Chromatography on a chiral phase (Chiralcel® OD-H), thereby actually treating **218a** and **218b** like enantiomers, gave two peaks expected for the diastereomeric products (Fig. 105, center) [462].

Table 3 Suzuki coupling reactions of the naphthalene **220** with the dihydroisoquinoline **221** to give ancistrotanzanine B (**218a**) ((*M*)-isomer) and ancistroealaine A (**218b**) ((*P*)-isomer) using different axially chiral phosphineamine ligands

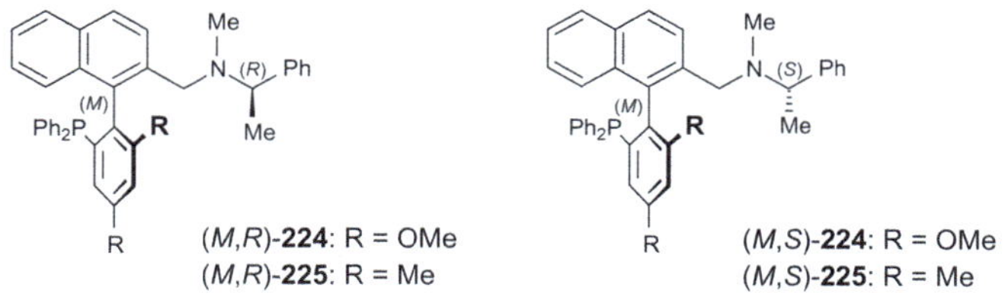

Catalytic system[a]	Yield/%	(*M*):(*P*)
0.1 equiv. of $Pd_2(dba)_3$, (*M,R*)-**224**	65	63:37
0.1 equiv. of $Pd_2(dba)_3$, (*M,S*)-**224**	89	57:43
0.1 equiv. of $Pd_2(dba)_3$, (*M,R*)-**225**	44	67:33
0.1 equiv. of $Pd_2(dba)_3$, (*M,S*)-**225**	58	62:38

[a]Further coupling conditions: **220** (1.5 equiv.), **221** (1.0 equiv.), catalyst, satd. aq. $NaHCO_3$, toluene/H_2O, reflux, 16 h

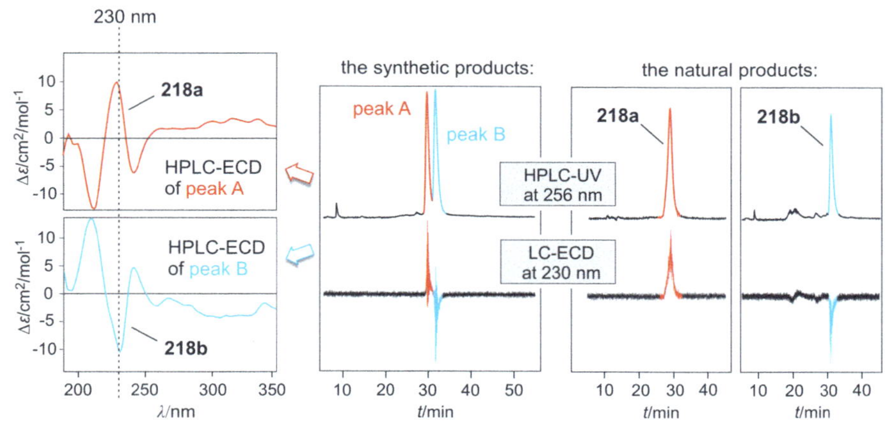

Fig. 105 Stereoanalysis of ancistrotanzanine B (**218a**) (= peak **A**, (*M*)-configured) and ancistroealaine A (**218b**) (= peak **B**, (*P*)-configured), by resolution on a chiral phase (Chiralcel® OD-H), and determination of their stereoisomeric ratios, online, by LC-UV and LC-ECD coupling to be 55:45 (**218a**:**218b**)

Their atropisomeric nature was easily recognized by their opposite ECD curves as monitored online, right from the peaks, by LC-ECD coupling (Fig. 105, left). Acquisition of the ECD effect at only one single wavelength (λ = 230 nm) corresponding to the ECD ellipticity maximum of ancistrotanzanine B (**218a**) and, simultaneously, the minimum for ancistroealaine A (**218b**) resulted in the expected positive peak for **218a** and a negative one for **218b** (see Fig. 105, center) [462].

In summary, the use of chiral catalysts in Suzuki cross-coupling led to good asymmetric inductions in favor of the (*M*)-atropisomer **218a** (up to 75:25), although the application of the other, oppositely configured ligand enantiomers did not permit the directed preparation of the (*P*)-isomer **218b** [337, 462]. This synthesis route provided an efficient and plant-independent access to ancistrotanzanine B (**218a**) and ancistroealaine A (**218b**), thus fulfilling a prerequisite for a further development of these two naphthylisoquinoline alkaloids as new antileishmanial leads. But it also showed that in many cases, intermolecular coupling procedures are not as efficient as, e.g., the lactone method, which permits the atroposelective synthesis of the one or, optionally, the other atropisomer.

Atroposelective Total Synthesis of 5,8′-Coupled Naphthylisoquinoline Alkaloids Without the Use of an External Chiral Source

A further approach aiming at the atroposelective total synthesis of ancistrotanzanine B (**218a**) and ancistroealaine A (**218b**) (Fig. 106) and, by chemical modification of **218a** and **218b**, also paving the way to other, structurally related 5,8′-coupled naphthylisoquinoline alkaloids, e.g., ancistrotectoriline A (**209**) (Fig. 107), has recently been reported by Cheon and coworkers [120, 121]. The strategy is based on a Suzuki–Miyaura coupling reaction between the known [108, 182] naphthyl pinacol boronate **226** and the new aryl iodide **227** as a still monocyclic isoquinoline precursor equipped with an (*S*)-2-(*N*-acetylamino)propyl group at the *ortho*-position, using the chiral aminopropyl substituent as the only atroposelectivity-controlling group, without the necessity of having an external chiral ligand and/or a chiral auxiliary. By application of $Pd(OAc)_2$ as the catalyst in combination with a certain Buchwald-type phosphine ligand and an appropriate base, the coupling step delivered the (*M*)-configured biaryl compound **228a** or, optionally, the (*P*)-configured coupling product **228b** in high asymmetric inductions and with excellent chemical yields, from the joint precursors **226** and **227** [120, 121]. Using *t*-BuXPhos as the ligand and K_3PO_4 as the base, the (*M*)-configured biaryl **228a** was obtained almost exclusively (*dr* >93:7) [120], while the cross-coupling reaction of **226** and **227** in the presence of the phosphine ligand SPhos and $Ba(OH)_2$ as the base led to a good atropisomeric ratio in favor of the (*P*)-isomer **228b** (*dr* = 83:17) [121] (Fig. 106). N-Deprotection of **228a** by Pd-catalyzed hydrogenolysis provided the acetamide **229a**, whose Bischler–Napieralski ring closure afforded the desired target molecule ancistrotanzanine B (**218a**) in an excellent yield of 93% [120]. In a similar way, **228b** was efficiently converted to the acetamide **229b**, and Bischler–Napieralski cyclization gave ancistroealaine A (**218b**) in 95% yield (Fig. 106) [121]. The spectroscopic data of the two atropo-diastereomers, **218a** and **218b**, thus obtained were in good agreement with those described previously in the literature [120, 121].

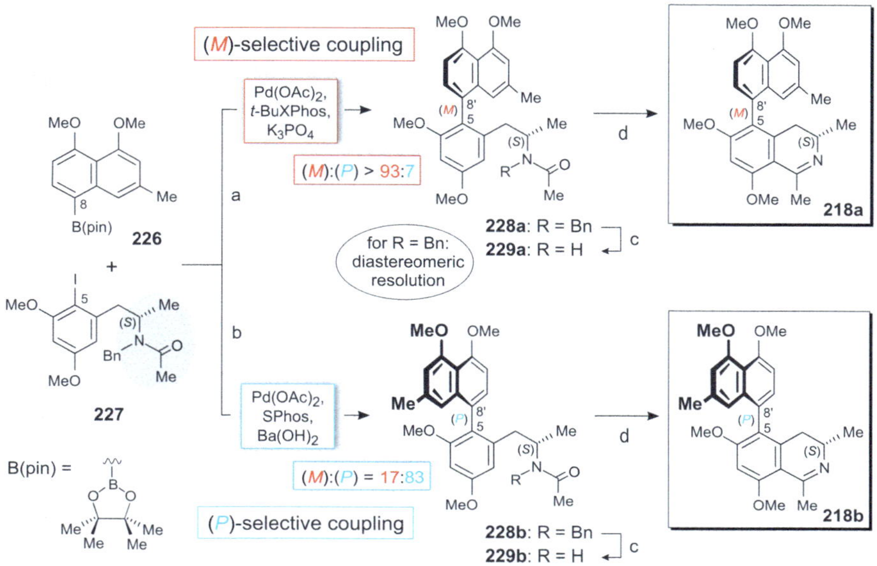

Fig. 106 Asymmetric total synthesis of ancistrotanzanine B (**218a**) and ancistroealaine A (**218b**) by Suzuki–Miyaura cross-coupling of the naphthyl pinacol boronate **226** and the aryl iodide **227**, using the chiral 2-(*N*-acetyl-*N*-benzylamino)propyl unit (underlaid in light gray) as an atroposelectivity-controlling group, switching the atroposelectivity by the choice of an appropriate catalytic system. Reaction conditions: (a) $Pd(OAc)_2$, *t*-BuXPhos, K_3PO_4, 2,6-di-*tert*-butyl-4-methylphenol ("butylated hydroxytoluene", BHT), DMF, 100°C, 92% ((*M*):(*P*) >13:1); (b) $Pd(OAc)_2$, SPhos, $Ba(OH)_2$, BHT, DMF, 100°C, 93% ((*P*):(*M*) = 5:1); (c) H_2, Pd/C, MeOH, r.t., 74% (for **229a**), 68% (for **229b**); (d) $POCl_3$, MeCN, 80°C, 93% (for **218a**), 95% (for **218b**)

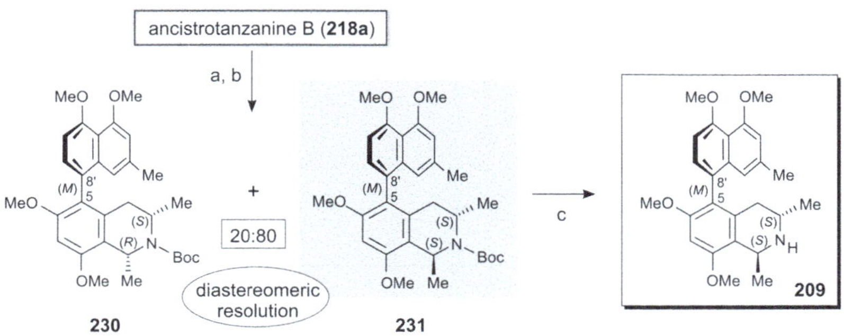

Fig. 107 Semi-synthetic conversion of ancistrotanzanine B (**218a**) into ancistrotectoriline A (**209**). Reaction conditions: (a) $LiAlH_4$, $AlEt_3$, THF, −78°C to r.t.; (b) Boc_2O, triethylamine, CH_2Cl_2, r.t. (*dr* = 4:1), 64% (for **231**), 16% (for **230**) (yields over two steps); (c) HCl, MeOH, r.t., 95%

As summarized in Table 4, the atroposelectivity of the Suzuki–Miyaura cross-coupling reaction between the naphthyl pinacol boronate **226** and the isoquinoline precursor **227** strongly depended on the optimal catalytic system, but the correct substituents on the nitrogen atom of the 2-aminopropyl group in **227** also were of major importance. No asymmetric induction was observed when using $Pd(PPh_3)_4$ as the catalyst and an aryl iodide bearing a 2-(*N*,*N*-dibenzylamino)propyl substituent, leading to the formation of the two respective atropo-diastereomers as an inseparable 1:1 mixture [120]. The presence of a free NH group in an aryl iodide substituted with a simple 2-(*N*-acetylamino)propyl group strongly hampered the cross-coupling reaction, providing the desired coupling product in only low chemical yields accompanied by unwanted by-products [120]. With a combination of Pd $(OAc)_2$ as the catalyst and P(*o*-tol)$_3$ as the ligand, the Suzuki–Miyaura reaction between **226** and **227** bearing a 2-(*N*-acetyl-*N*-benzylamino)propyl substituent gave the (*M*)-configured biaryl compound **228a** as the major product (*dr* = 78:22) in 92% yield. The atropo-controlled syntheses of ancistrotanzanine B (**218a**) and ancistroealaine A (**218b**) required carefully optimized, specifically adapted reaction parameters in the coupling reaction in view of the strong influence of the phosphine ligands [463, 464], which exerted a significant influence on the cross-coupling reaction between **226** and **227**, due to their stereoelectronic properties and the steric bulkiness around the phosphine atom [120, 121]. The application of ligands with a simple phenyl or *ortho*-diisopropyl substituted phenyl ring system in the bottom unit like, e.g., JohnPhos and XPhos gave moderate to good diastereoselectivities, usually in favor of the (*M*)-configured coupling product **228a** (see Table 4).

Table 4 Achiral phosphine ligands and reaction conditions used for Suzuki–Miyaura cross-coupling reactions of naphthyl pinacol boronate **226** and the aryl iodide **227** to give the biaryl coupling products (*M*)-**228a** and (*P*)-**228b**

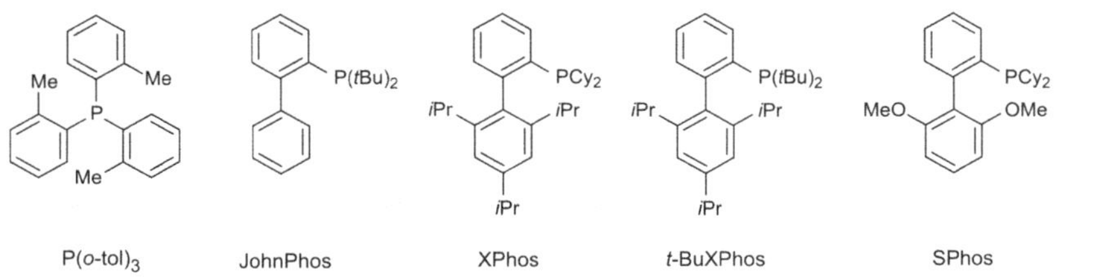

Ligand, base[a]	Yield/%	**228a**:**228b** ((*M*):(*P*))
P(*o*-tol)$_3$, K_3PO_4	92	78:22
XPhos, K_3PO_4	94	63:37
JohnPhos, K_3PO_4	91	78:22
t-BuXPhos, K_3PO_4	92	93:7
SPhos, K_3PO_4	95	25:75
SPhos, $Ba(OH)_2$	93	17:83

[a]Further coupling conditions: **226** (0.30 mmol), **227** (0.20 mmol), $Pd(OAc)_2$ (10 mol%) as the catalyst, ligand (20 mol%), base (1.0 mmol), BHT (0.14 mmol), and DMF at 100°C for 12 h

By using *t*-BuXPhos as the ligand and K_3PO_4 as the base, the asymmetric induction significantly increased with the high bulkiness of the alkyl groups on the phosphine atom, providing the (*M*)-configured coupling product **228a** (Fig. 106) as the major product (*dr* = 93:7) in 89% yield (after resolution by column chromatography) [120]. When using SPhos however, with its 2,6-dimethoxyphenyl group on the bottom ring of the ligand, the (*P*)-configured coupling product **228b** was obtained as the major product (*dr* = 75:25). Improvement of the atroposelectivity of the coupling reaction was then achieved by using $Ba(OH)_2$ as the base, providing the required the (*P*)-isomer **228b** (Fig. 106) in an enhanced ratio of 83:17. When running the coupling reaction on a gram scale, **228b** was obtained in 80% yield stereochemically pure after column chromatography on silica gel [121].

Furthermore, conversion of the dihydroisoquinoline unit of ancistrotanzanine B (**218a**) into a tetrahydroisoquinoline ring system provided synthesis access to other (*M*)-configured and likewise O-permethylated 5,8′-coupled naphthylisoquinolines, among them the above-described leaf and twig metabolite of *A. tectorius*, ancistrotectoriline A (**209**) (Sect. 8.5.4.1) and related, yet merely artificial compounds [120], so far not detected in Nature. Reduction of **218a** with $LiAlH_4/AlEt_3$ resulted in the formation of a mixture of the target molecule **209** as the major product, along with the corresponding *cis*-isomer, 1-*epi*-ancistrotectoriline A (80:20, *trans*:*cis*) (Fig. 107), but the authors did not succeed in resolving the two stereoisomers. For this reason, the mixture was reacted with di-*tert*-butyl dicarbonate to give **230** and **231**, of which the *trans*-carbamate **231** finally afforded pure ancistrotectoriline A (**209**) in high yields, after chromatographic separation and removal of the *N*-Boc group (Fig. 107) [120]. Starting from ancistroealaine A (**218a**), a series of O,N-permethylated naphthyltetrahydroisoquinolines with a (*P*)-configuration at the 5,8′-biaryl linkage was semi-synthetically prepared in a stereochemically homogeneous form, among them some compounds known to be produced by African *Ancistrocladus* species [121]. In summary, the strategy provides an efficient novel methodology for the construction of 5,8′-coupled naphthylisoquinoline alkaloids, yet so far restricted to target molecules of this coupling type that possess methoxy groups at C-6, C-4′, and C-5′.

8.5.5 Ancistrotectoriline B, Ancistectorines B_1 and B_2, and Related 7,1′-Coupled Naphthyltetrahydroisoquinoline Alkaloids

From the twigs and leaves of *A.* aff. *tectorius* from the Chinese Hainan Island, six 7,1′-coupled naphthylisoquinoline alkaloids were isolated [56, 60–62] (Fig. 108), among them three new representatives, ancistrotectoriline B (**232**) [56, 60], ancistectorine B_1 (**233**) [61], and ancistectorine B_2 (**210**) [62], along with three known naphthylisoquinolines, 7-*epi*-ancistrobrevine D (**171b**) [60], its 6-*O*-demethyl-8-*O*-methyl analogue **179**, and 6-*O*-methylancistectorine B_1 (**157**) [62]. Two of them, **171b** and **179**, had already been discovered in the Vietnamese liana *A. cochinchinensis* [45, 46] (see Sect. 7.2). The third compound, 6-*O*-methylancistectorine B_1 (**157**) [62], had so far been known only as a semi-synthetic

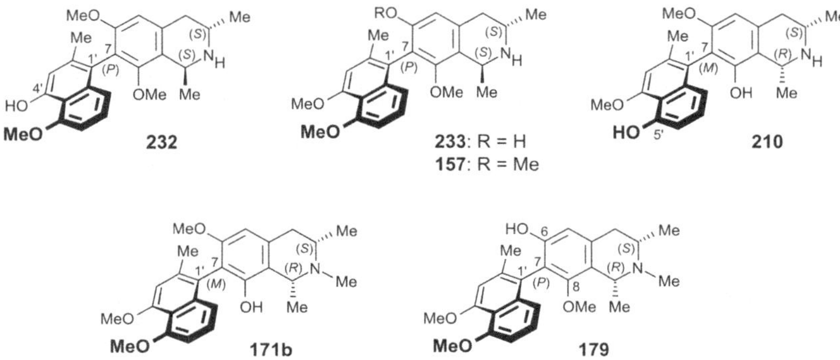

Fig. 108 A series of six 7,1′-coupled naphthylisoquinoline alkaloids from *A.* aff. *tectorius* from Hainan Province, China. Included are the three new alkaloids ancistrotectoriline B (**232**) as well as ancistectorines B_1 (**233**) and B_2 (**210**), isolated along with the known [60, 62] compounds 7-*epi*-ancistrobrevine D (**171b**), 6-*O*-demethyl-8-*O*-methyl-7-*epi*-ancistrobrevine D (**179**), and 6-*O*-methylancistectorine B_1 (**157**), which was previously known only by semi-synthesis [190]

compound obtained by *trans*-selective reduction of ancistrocladisine (**2a**) (isolated from the Indian liana *A. heyneanus*), by using $LiAlH_4/AlMe_3$ [190]. Prior to this phytochemical study, it had not been detected in any of the investigated *Ancistrocladus* plants, and was, thus, now identified as a true natural product for the first time [62]. Remarkably, all six alkaloids (see Fig. 108) have the same stereo-orientation at the biaryl axis, although—due to the formalism of the C.I.P. priority rules—displaying different (*M*)/(*P*)-stereodescriptors at the axis.

8.5.6 Ancistrotectoquinones A and B, Novel-Type Quinoid Natural Products and Related 7,3′-Coupled Naphthylisoquinoline Alkaloids

Ancistrotectorine and Ancistrotanzanine C

From the leaf, twig and stem bark extracts of the Chinese liana *A.* aff. *tectorius*, ancistrotectorine (**17a**) [58, 60] and its 6-*O*-demethyl analogue ancistrotanzanine C (**114**) [465] were isolated. The two alkaloids displayed a rare 7,3′-coupling position of the biaryl axis. The only other known 7,3′-linked naphthylisoquinoline alkaloids isolated from an Asian *Ancistrocladus* species were ancistrocladidine (**3a**) and ancistroheynine B (**172**). As reported in the previous Sect. 4.7, ancistroheynine B (**172**), which can likewise be addressed as the 6-*O*-demethyl analogue of **3a**, had been detected in the leaves of the Indian liana *A. heyneanus* as a major metabolite, along with its parent compounds **3a** and **114** [41]. Ancistrotectorine (**17a**) and ancistrocladidine (**3a**) had for the first time been isolated from *A.* aff. *tectorius* from Thailand [51, 52] (see Sect. 8.3). Ancistrotectorine (**17a**) had also been identified in *A. guineënsis* from Nigeria [159], and it had been found in *A. tanzaniensis*, along with its 6-*O*-demethyl analogue ancistrotanzanine C (**114**) [189].

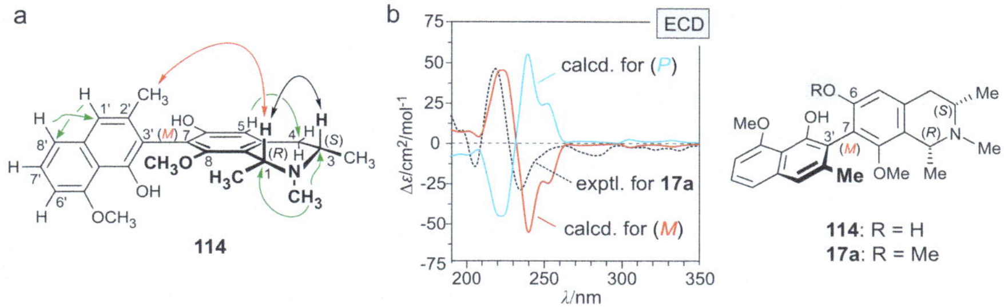

Fig. 109 Ancistrotectorine (**17a**) and ancistrotanzanine C (**114**) from the aerial parts of the Chinese liana *A.* aff. *tectorius*: Illustrated are **a** HMBC (green single arrows) interactions evidencing the constitution as well as specific NOESY interactions supporting the relative configuration at C-3 versus C-1 (black double arrow) and the configuration at the axis (red double arrow) of ancistrotanzanine C (**114**) relative to C-1; **b** experimental ECD spectrum of ancistrotectorine (**17a**) and the spectra calculated for the absolute (*P*)- and (*M*)-stereostructures by means of the semiempirical CNDO/2S and the AM1-Boltzmann approach [76, 159], displaying the (*M*)-configuration at the biaryl axis of **17a**

Ancistrotectorine (**17a**) and ancistrotanzanine C (**114**) display a relative *cis*-array of the methyl groups at C-1 and C-3, chemically stabilized against oxidation by an *N*-methyl function. The absolute configuration in their tetrahydroisoquinoline molecular half was attributed to be (1*R*,3*S*) [41, 51, 159, 189], due to specific NOESY correlations in combination with the results from the ruthenium-mediated oxidative degradation of **17a** and **114**. The absolute axial configuration of ancistrotectorine (**17a**) and ancistrotanzanine C (**114**), which display virtually identical ECD spectra, was deduced to be (*M*) [51, 159]. Comparison of the experimental ECD spectrum of ancistrotectorine (**17a**) with the curves obtained by quantum-chemical ECD calculations at the CNDO/2S-level, both, for the (*M*)-configured compound **17a** and for its (*P*)-diastereomer **17b**, clearly confirmed this assignment (Fig. 109). The ECD spectrum theoretically predicted for the (7*M*,1*R*,3*S*)-atropo-diastereomer was in good agreement with the ECD curve experimentally measured for ancistrotectorine (**17a**), whereas the ECD spectrum calculated for the (7*P*,1*R*,3*S*)-atropo-diastereomer (not shown) was virtually opposite to that of **17a** [159] (Fig. 109).

Ancistrotectoquinones A and B, the First Alkaloids with a 1,4-Naphthoquinone Portion Coupled to an Isoquinoline Moiety

Metabolic profiling of a leaf extract of *A. tectorius* led to the discovery of a new, orange-colored compound exhibiting a UV maximum around 400 nm, comparable to the UV spectrum of plumbagin (**37**). NMR measurements hinted at the presence of a 7,3′-coupled alkaloid with a structurally unusual naphthalene moiety [63]. In cparticular, the two carbonyl carbon signals (δ_C = 186.3 and 184.4 ppm) in the ^{13}C NMR spectrum and signals of only four aromatic protons indicating the lack of

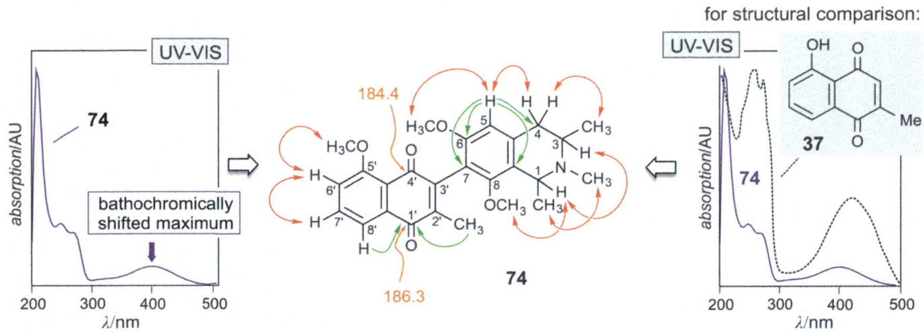

Fig. 110 Ancistrotectoquinone A (**74**), the first alkaloid with a 1,4-naphthoquinone portion coupled to an isoquinoline moiety. UV-vis comparison with the non-coupled naphthoquinone plumbagin (**37**) as well as selected ^{13}C NMR data (δ_C/ppm), and NOESY (red double arrows) and HMBC (green single arrows) interactions relevant for the constitution and the relative configuration at the stereogenic centers

one aromatic proton compared to all previously reported C,C-coupled naphthylisoquinoline alkaloids, suggested the new compound to possess a 2′-methyl-5′-methoxy-1′,4′-naphthoquinone portion [63] (see Fig. 110). The relative configuration of the methyl groups at C-1 versus C-3 was deduced as being *cis* from specific NOE correlations between H-1 and H-3, and the absolute configuration was assigned as (*R*) at C-1 and (*S*) at C-3 by ruthenium-mediated oxidative degradation. The new alkaloid was named ancistrotectoquinone A (**74**) [63].

Despite the *ortho*-substituents next to the biaryl axis of **74**, which should exert a sufficiently high degree of steric hindrance, the alkaloid **74** was configurationally only semi-stable at the axis (Fig. 111). The reduced atropisomerization barrier is mainly the result of mesomeric stabilization of the transition state of the axial rotation by a push–pull effect of the carbonyl group at C-1′ of the naphthoquinone half with the methoxy groups at C-6 and C-8 in the isoquinoline portion, due to the −*M* effect of the carbonyl group and the +*M* effect of the methoxy substituents. As a consequence, ancistrotectoquinone A (**74**) undergoes slow rotation about the biaryl axis at room temperature, thus occurring as a mixture of two configurationally semi-stable atropo-diastereomers, **74a** and **74b** (Fig. 111) [63].

After HPLC resolution of the two rotational isomers of ancistrotectoquinone A (**74a/74b**), online chiroptical assignment of the absolute axial configuration of **74a** and **74b** succeeded by comparison of their LC-ECD spectra with the ECD curve of the likewise 7,3′-coupled, yet configurationally stable alkaloid ancistrotectorine (**17a**) [63, 159], which is (*M*)-configured (see section above). Since the chromophores of ancistrotectoquinone A (**74a/74b**) were not fully comparable to those of **17a**, quantum-chemical ECD calculations by the ZINDO/S-CI approach were performed for the (*M*)- and (*P*)-configured atropo-diastereomers of **74** [63]. The ECD spectrum predicted for **74a** (*M*) matched well with the experimental one of the more rapidly eluting peak **A**, while the ECD curve measured for the slower peak **B** was in good agreement with the one calculated for **74b** (*P*) [63] (Fig. 111).

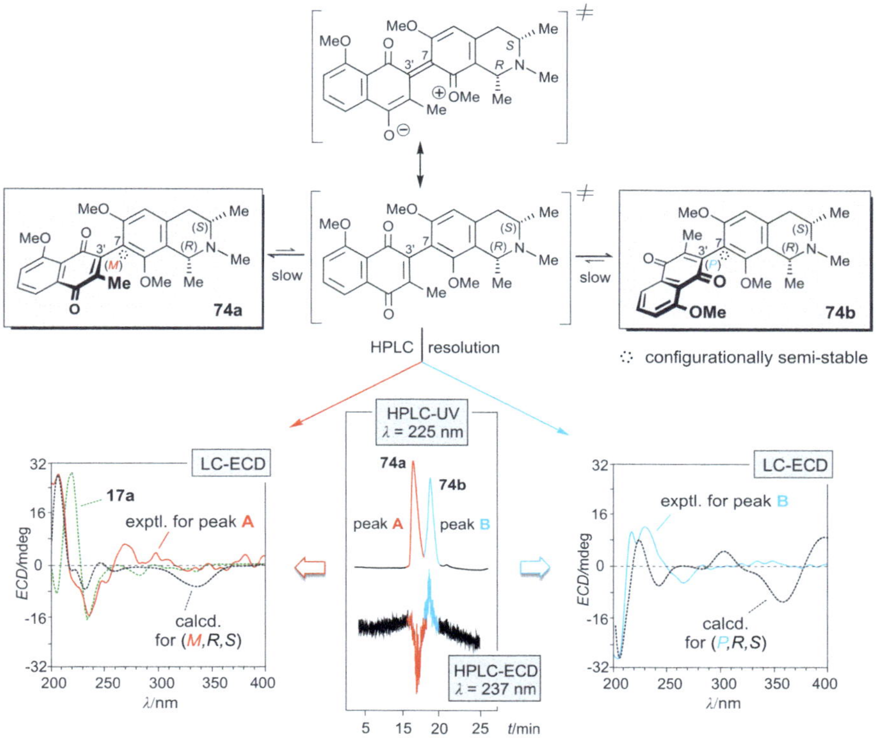

Fig. 111 Ancistrotectoquinone A (**74**), a new subtype of naphthylisoquinoline alkaloids isolated from the leaves of *A.* aff. *tectorius* (China), occurring as a pair of configurationally semi-stable and, thus, slowly interconverting atropo-diastereomers **74a** and **74b**. Also shown is the assignment of the absolute configuration of the two rotational isomers of **74** at the biaryl axes by comparison of their experimental ECD spectra with the computationally predicted ECD curves (ZNDO/S-CI approach) and the spectrum of the related ancistrotectorine (**17a**)

Kinetic experiments [63] on freshly separated samples enriched with the respective (*P*)- or (*M*)-atropisomer of ancistrotectoquinone A (**74**) were analyzed to determine the atropisomerization barrier of (*M*)-ancistrotectoquinone A (**74a**) to give the (*P*)-diastereomer **74b** (Fig. 112). The Gibbs free energies of activation were measured to be $\Delta G^{\#}_1 = 106.6$ kJ mol^{-1} (from (*M*) to (*P*)) and $\Delta G^{\#}_2 = 106.6$ kJ mol^{-1} (from (*P*) to (*M*)). These values were just above the defined energy barrier of well-separable stable isomers [63] and fitted well with the slow interconversion of the two atropo-diastereomers **74a** and **74b** at room temperature. They were also in a good agreement with the results from DFT calculations using B3LYP-D3/def2-TZVP in combination with COSMO solvent corrections, leading to calculated values of $\Delta G^{\#}_1 = 103.3$ kJ mol^{-1} (from (*M*) to (*P*)) and $\Delta G^{\#}_2 = 101.5$ kJ mol^{-1} (from (*P*) to (*M*)). These calculated values even reproduced

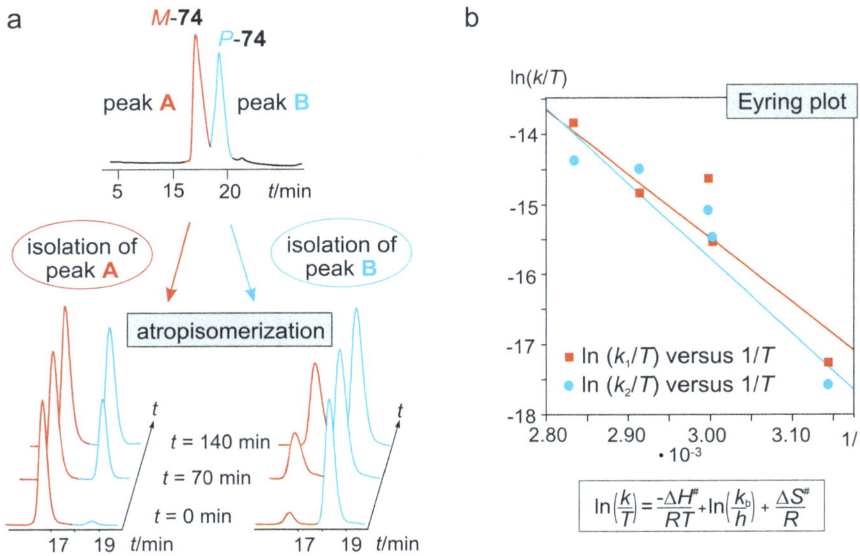

Fig. 112 **a** Analysis of the interconversion of peaks **A** (= (*M*)-**74**) and **B** (= (*P*)-**74**) of ancistrotectoquinone A (**74**), by analytical HPLC on a reversed-phase C_{18} column; **b** kinetic experiments to determine the rotational barrier of **74**: Eyring plots at four different temperatures with best-fit straight lines for the atropisomerization of (*M*)-**74** to (*M*)/(*P*)-**74** (■), and for the reverse reaction of (*P*)-**74** to (*M*)/(*P*)-**74** (●); the slopes are equal to ($\Delta H^{\#}/R$), the intercepts are equal to (23.76 + $\Delta S^{\#}/R$)

the slight preference for the (*M*)-atropo-diastereomer **74a** as obvious from a ratio of 57:43 ((*M*):(*P*)) as experimentally measured in the atropisomeric equilibrium at room temperature [63].

Similar results were obtained for the *O*-demethyl analogue of ancistrotecto-quinone A (**74a/74b**), which was isolated from stem material of *A.* aff. *tectorius*. This compound, named ancistrotectoquinone B (**234**), was found to rotate more rapidly about the biaryl axis due to the reduced steric hindrance displayed by the hydroxy group at C-6 compared to the slightly larger methoxy residue in **74**, and facilitated by hydrogen bonding of this OH group with the carbonyl oxygen at C-4′ [63]. The online LC-ECD spectra of the two nearly baseline-resolved atropo-diastereomers **234a** and **234b** were virtually identical to those of **74a** and **74b**, respectively (Fig. 113). The more rapidly eluting atropisomer **234a** (peak **A**) was attributed to be (*M*)-configured, while the more slowly eluting one **234b** (peak **B**) was assigned as having the (*P*)-configuration [63].

The question might arise whether such naphthylisoquinoline alkaloids like **74** and **234**, with a quinoid structural motif, are genuine natural products or whether they are just the result of an autoxidation process of their phenolic precursors like ancistrotectorine (**17a**) and ancistrotanzanine C (**114**). To further investigate this issue, **17a** was treated with Fremy's salt as an oxidant, smoothly affording the

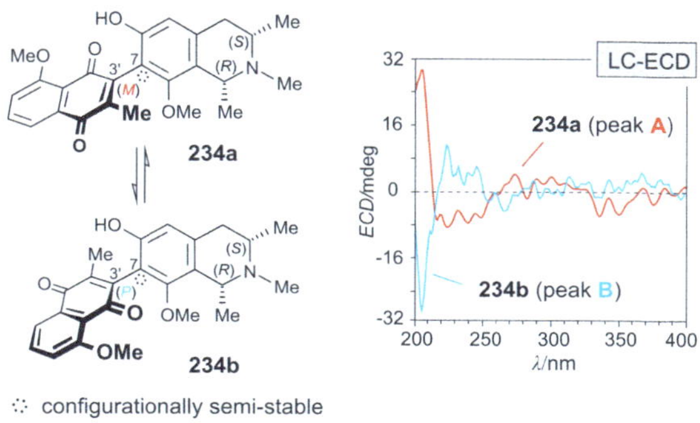

Fig. 113 Ancistrotectoquinone B (**234**) from the stems of *A.* aff. *tectorius*, occurring as a pair of configurationally semi-stable atropo-diastereomers: experimental LC-ECD spectra of the two atropo-diastereomers **234a** (more rapidly eluting peak **A**) and **234b** (more slowly eluting peak **B**)

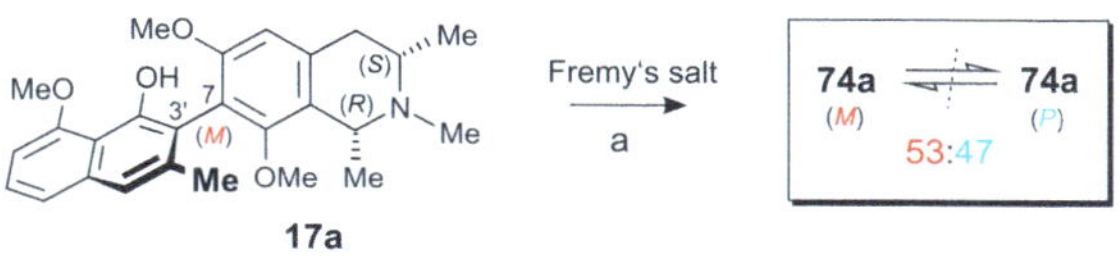

Fig. 114 Biomimetic conversion of the atropisomerically pure ancistrotectorine (**17a**) into a mixture of (*M*)- and (*P*)-ancistrotectoquinones A (**74a** and **74b**) using Fremy's salt as an oxidizing agent. Reaction conditions: (a) $(KSO_3)_2NO$ (= potassium nitrosodisulfonate, Fremy's salt), KH_2PO_4, acetone/water (1:2), room temperature, 30 min, 90%

likewise (*M*)-configured ancistrotectoquinone A (**74a**), which subsequently yielded the expected atropisomeric equilibrium mixture **74a** ⇆ **74b** (Fig. 114) [63]. Under these forced chemical conditions, the biomimetic conversion from **17** to **74a/74b** can thus be easily achieved, but it is unlikely that this reaction will happen spontaneously during work-up and purification procedures, because ancistrotectorine (**17a**) is stable at room temperature in the presence of air and light, so no formation of ancistrotectoquinone A (**74a/74b**) was observed, nor any decomposition [63]. This finding makes it most feasible that quinoid naphthylisoquinoline alkaloids are indeed genuine natural products, formed from their phenolic precursors by an enzymatic pathway in the living plants before the extraction process.

Summarizing, the new ancistrotectoquinones A (**74a/74b**) and B (**234a/234b**) [63] are the first representatives of a novel subclass of naphthylisoquinoline alkaloids possessing a *para*-quinone functionality in the naphthalene ring system. From *A. abbreviatus* from Coastal West Africa, two likewise oxygenated, yet 5,1′-coupled and *ortho*-quinoid naphthylisoquinoline alkaloids are known, ancistrobreviquinones A (**235**) and B (**236**) (Fig. 115) [193]. A similar naphthoquinone molecular half, but with the *ortho*-quinone functionality located in the methyl-free ring, has also been found as part of the naphthylnaphthoquinone triphyoquinone A

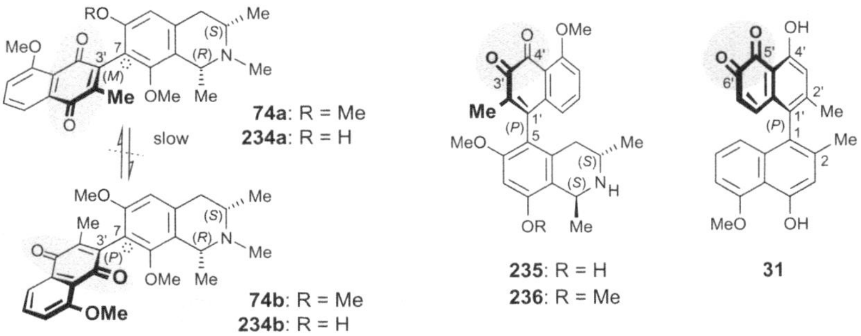

Fig. 115 Ancistrotectoquinones A (**74a/74b**) and B (**234a/234b**) with a *para*-quinone unit and further secondary metabolites possessing an *ortho*-quinone molecular entity discovered in Ancistrocladaceae and Dioncophyllaceae lianas from West Africa: ancistrobreviquinones A (**235**) and B (**236**) from the roots of *A. abbreviatus* and triphyoquinone A (**31**) from root cultures of *T. peltatum*

(**31**) (Fig. 115) [216]. This compound was isolated from the root cultures of the West African Dioncophyllaceae liana *T. peltatum*, but it was not found in the plant itself. A specific difference compared to the ancistrotectoquinones A (**74a/74b**) and B (**234a/234b**) from the Chinese liana *A.* aff. *tectorius* is the fact that all these compounds, **235**, **236**, and **31**, are rotationally stable at the biaryl axis and atropisomerically pure.

8.5.7 Ancistrotectoriline C and Ancistrotectorines A and B, the First 7,6′-Coupled Ancistrocladaceae-Type Naphthylisoquinoline Alkaloids

Phytochemical studies by Tang et al. [59] and by Jiang et al. [60] on the Chinese liana *A.* aff. *tectorius* led to the isolation of the very first (and so far only) 7,6′-linked naphthylisoquinoline alkaloids belonging to the subclass of Ancistrocladaceae-type compounds (i.e., possessing an oxygen function at C-6 and the (*S*)-configuration at C-3; the latter was, however, assigned only tentatively, simply based on the fact that all other known naphthylisoquinoline alkaloids from *A.* aff. *tectorius* possess this structural feature [13, 17, 18, 31, 33, 50–58, 61–64]). Three such compounds were isolated from the stems and leaves as minor metabolites and were named ancistrotectorines A (**237**) and B (**18**) [60] and ancistrotectoriline C (**206**) [59] (Fig. 116). As exemplified here for ancistrotectorine A (**237**) (Fig. 116a), the coupling site of the biaryl axis was deduced to be at C-7 in the isoquinoline portion due to the low-field-shifted signals of H-4 (δ_H = 2.60 and 2.63 ppm) indicating that they were not subject to a shielding effect from the naphthalene subunit. This was in accordance with the upfield-shifted signal of MeO-8 (δ_H = 3.18 ppm) [60]. With the presence of a spin system of three neighboring protons in the naphthalene portion, and a not high-field-shifted signal of

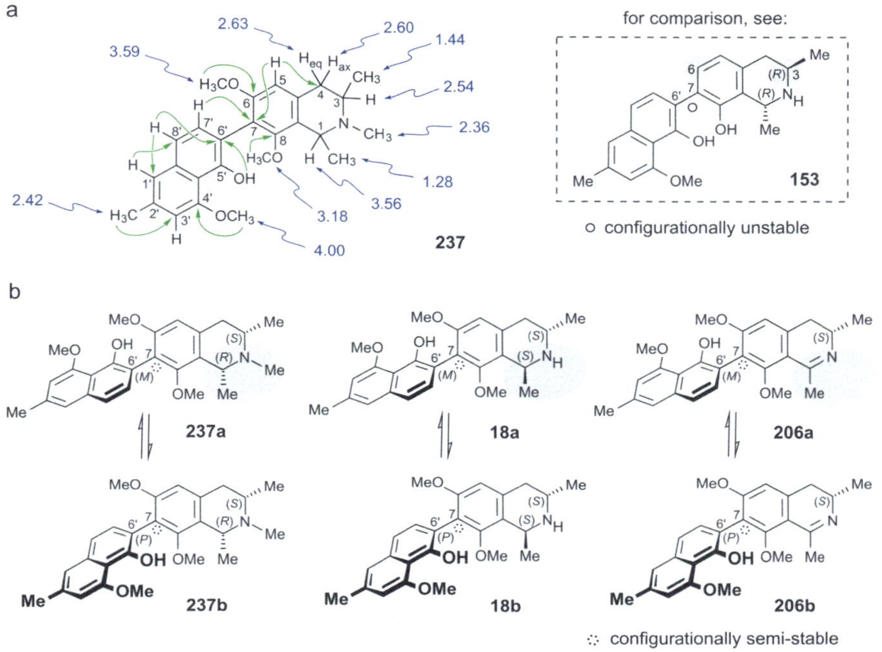

Fig. 116 **a** Selected ^{1}H NMR data and HMBC interactions (green single arrows) indicative of the constitution and relative configuration at C-1 versus C-3 of ancistrotectorine A (**237**); **b** the only three representatives of Ancistrocladaceae-type naphthylisoquinoline alkaloids known so far to exhibit a 7,6′-coupling type: ancistrotectorines A (**237a/237b**) and B (**18a/18b**) and ancistrotectoriline C (**206a/206b**). Their biaryl axes are configurationally semi-stable, thus occurring as pairs of slowly interconverting atropo-diastereomers. For comparison, see the structure of the Dioncophyllaceae-type alkaloid dioncophylline B (**153**) with its configurationally unstable biaryl axis, which can freely rotate

Me-2′ (δ_H = 2.42 ppm), the biaryl axis had to be located in the methyl-free naphthalene ring. HMBC interactions from H-8′ to C-1′, and from H-1′ to C-8′ excluded C-8′ to be the axis-bearing carbon atom. The resulting 7,6′-coupling type was further confirmed by HMBC cross-peaks from H-8′ to C-6′ and from both, H-7′ and H-5 to C-7 [59, 60].

Two of the new alkaloids, ancistrotectorines A (**237**) and B (**18**) [60], were naphthyltetrahydroisoquinolines, differing only by the relative configurations of the stereogenic centers at C-1 versus C-3 (see Fig. 116b). Ancistrotectorine A (**237**) displayed a relative *cis*-array of the two methyl groups at C-1 and C-3, stabilized by an *N*-methyl function, while the relative configuration of ancistrotectorine B (**18**) at C-1 versus C-3 was deduced to be *trans*. The third example within this series of 7,6′-coupled Ancistrocladaceae-type alkaloids, ancistrotectoriline C (**206**), was a naphthyldihydroisoquinoline [59]. All of the compounds were equipped with the same naphthalene molecular half.

Prior to these findings, only six other 7,6′-coupled naphthylisoquinoline alkaloids had been discovered in Nature like e.g., dioncophylline B (**153**) (Fig. 116) from the West African liana *T. peltatum* [153, 219], but they were all Dioncophyllaceae-type compounds [13, 14], i.e., lacking an oxygen function at C-6 and being (*R*)-configured at C-3. Because of the presence of only two *ortho*-substituents next to the biaryl linkage, **153** possesses a configurationally unstable axis that can freely rotate at room temperature, so that its atropo-diastereomers are not chromatographically or spectroscopically distinguishable or even isolable.

The Ancistrocladaceae-type alkaloids ancistrotectorines A (**237**) and B (**18**) and ancistrotectoriline C (**206**), by contrast, possessing three *ortho*-substituents next to the biaryl axis, exhibited two sets of proton signals with identical coupling patterns and coupling constants in their ^{1}H NMR spectra [59, 60], thus suggesting the presence of atropo-diastereomers. The additional *O*-methyl group at C-6 in the isoquinoline moieties of **237**, **18**, and **206**, as compared to dioncophylline B (**153**) [219], exerted some steric hindrance, and thus, **237**, **18**, and **206** were found to undergo a slow rotation about the biaryl axis at room temperature. The isomerization was slow enough to permit a resolution of the atropo-diastereomers of **237**, **18**, and **206** by HPLC on a reversed-phase C_{18} column, providing in each case two peaks in a ratio of 1:1. The two pure, yet configurationally semi-stable rotational isomers were observed to change again into a 1:1 mixture of slowly interconverting atropo-diastereomers (Fig. 116b) [59, 60].

Ancistrotectorines A (**237a/b**) and B (**18a/b**) displayed potent growth-retarding activities against human leukemia cells, here HL-60, K562, and U937 cell lines, with half-maximum inhibitory concentrations (IC_{50}) in the low micromolar range [60].

8.5.8 Ancistectorine C_1, the Only Known Example of a 7,8′-Coupled Naphthylisoquinoline Alkaloid Isolated from *A.* aff. *tectorius*

Naphthylisoquinoline alkaloids belonging to the subclass of 7,8′-coupled compounds have been discovered in Asian *Ancistrocladus* plants only rarely. So far, only six such alkaloids were identified, among them the new naphthyltetrahydroisoquinoline ancistectorine C_1 (**71a**), isolated as a minor metabolite from the twigs of *A.* aff. *tectorius* [61]. The compound displays a relative *cis*-array of the two methyl groups at C-1 and C-3, stabilized by an *N*-methyl substituent. By the oxidative degradation procedure, **71a** was determined to be (1*R*,3*S*)-configured, and in conjunction with this result, the absolute axial configuration was determined to be (*M*), from an NOE interaction between H-1′ and Me-1 (Fig. 117). The assignment was confirmed by comparison of the ECD spectrum of **71a** with that of the known [13, 43, 191] and structurally closely related, but (*P*)-configured alkaloid 6-*O*-demethylancistrobrevine A (**71b**) earlier isolated from *A. abbreviatus* (from Ivory Coast). The ECD curves of the two compounds were fully opposite to each other. Ancistectorine C_1 (**71**) is the as yet only example of a 7,8′-linked naphthylisoquinoline alkaloid found in *A.* aff. *tectorius* from the Chinese Hainan Province [61].

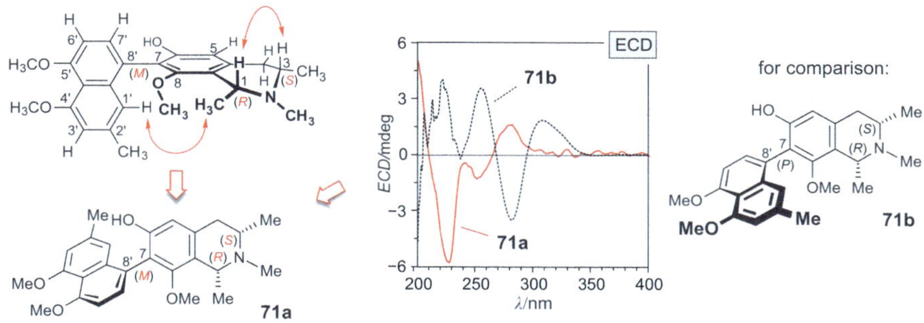

Fig. 117 New 7,8′-coupled naphthylisoquinoline alkaloid ancistectorine C_1 (**71a**) from the Chinese liana *A.* aff. *tectorius*. NOESY interactions indicative of the relative configurations at C-1 and C-3 and at the biaryl axis, and assignment of the absolute axial configuration of **71a** by comparison of its ECD spectrum with that of the known [13, 191] related, yet oppositely configured alkaloid 6-*O*-demethylancistrobrevine A (**71b**)

8.5.9 N,C-Coupled Naphthyldihydroisoquinolinium Alkaloids: Ancistrocladinium A and B and Related Compounds

Occurrence of Ancistrocladinium A Together with Its Mono- and Bisphenolic Analogues

Phytochemical analysis of the twig and leaf extracts of the Chinese liana *A.* aff. *tectorius* revealed the presence of the N,C-coupled cationic naphthyldihydroisoquinolinium alkaloid ancistrocladinium A (**6a/6b**) along with its mono- and bisphenolic analogues **123a/123b** and **174a/174b** (Fig. 118) [18, 33]. Remarkably, the isolation of all these N,8′-linked naphthylisoquinoline alkaloids from one single *Ancistrocladus* species had previously only been reported for the Vietnamese liana *A. cochinchinensis* [47] (see Sect. 7.3). As earlier in *A. cochinchinensis*, ancistrocladinium A (**6a**) was found to occur in twig extracts, along with its (*P*)-configured atropo-diastereomer N,8′-*epi*-ancistrocladinium A (**6b**) in a ratio of 4:1 ((*M*):(*P*)). These findings were comparable to those for the two atropo-diastereomers of the monophenolic analogue, **123a** and **123b**, which had been detectable in the leaf extracts in a ratio of 3:1 ((*M*):(*P*)) [33]. The bisphenolic alkaloid **174** was only present in the leaves, and only as a minor metabolite, with the (*M*)-configured atropo-diastereomer **174a** again being the predominant one, while its (*P*)-atropo-diastereomer **174b** was observable only in traces [33]. Unfortunately, as already described in Sect. 7.3, all attempts to resolve these rotational isomers have failed so far [18, 33, 47].

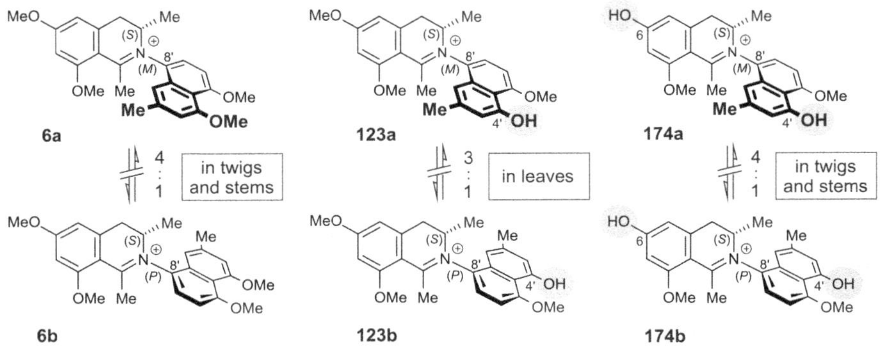

Fig. 118 Ancistrocladinium A (**6**) and its mono- and bisphenolic derivatives **123** and **174**, as detected in the twigs and leaves of the Chinese liana *A.* aff. *tectorius*, occurring as pairs of atropo-diastereomers, with a distinct preference for the respective (*M*)-configured isomer. The structural modifications in **123** and **174** in comparison to ancistrocladinium A (**6**) are underlaid in light gray

First Total Synthesis of Ancistrocladinium A and B and the Related, as Yet Unnatural N,3′-Coupled Analogue Ancistrocladinium C, Starting from the Chiral Pool

The N,8′-linked ancistrocladinium A (**6a/6b**) and the N,6′-coupled ancistrocladinium B (**14a/b**), both showing the rare phenomenon of N,C-axial chirality, were attractive synthesis target molecules, also with respect to their excellent antiparasitic properties against *Leishmania* parasites [18, 32, 47, 66–72]. These two naphthyldihydroisoquinolinium alkaloids display pronounced activities against *L. donovani* (amastigotes), the pathogen responsible for visceral leishmaniasis, with IC_{50} values of 1.71 μ*M* (for **6a/6b**) [32, 47] and 2.71 μ*M* (for **14a/14b**) [32]. Against *L. major* (promastigotes), a pathogen that causes cutaneous leishmaniasis, ancistrocladinium A (**6a/6b**) (IC_{50}, 7.32 μ*M*) and ancistrocladinium B (**14a/14b**) (IC_{50}, 6.62 μ*M*) exerted activities better than that of miltefosine (IC_{50} = 31.9 μ*M*), which was used as a reference compound and positive control [32, 47, 66]. The two N,C-coupled alkaloids were thus considered to be promising lead structures for the development of new leishmanicidal agents, also with respect to the fact that the inhibitory activities seemed to be highly specific. Thus, ancistrocladinium B (**14a/14b**) exhibited only moderate effects against the pathogens causing Chagas' disease (*T. cruzi*) and African sleeping sickness (*T. brucei rhodesiense*) [32], while ancistrocladinium A (**6a/6b**) displayed a fivefold higher activity against *T. rhodesiense*, but was inactive against *T. cruzi* [32, 47]. Against the malaria parasite *P. falciparum* (K1 strain), **6a/6b** and **14a/14b** revealed only moderate to weak activities [18, 32, 47].

For a facile availability of ancistrocladinium A (**6a/6b**) and B (**14a/14b**) in sufficient quantities, e.g., for further pharmacological studies, an efficient strategy for their flexible convergent synthesis was developed, involving a short sequence of

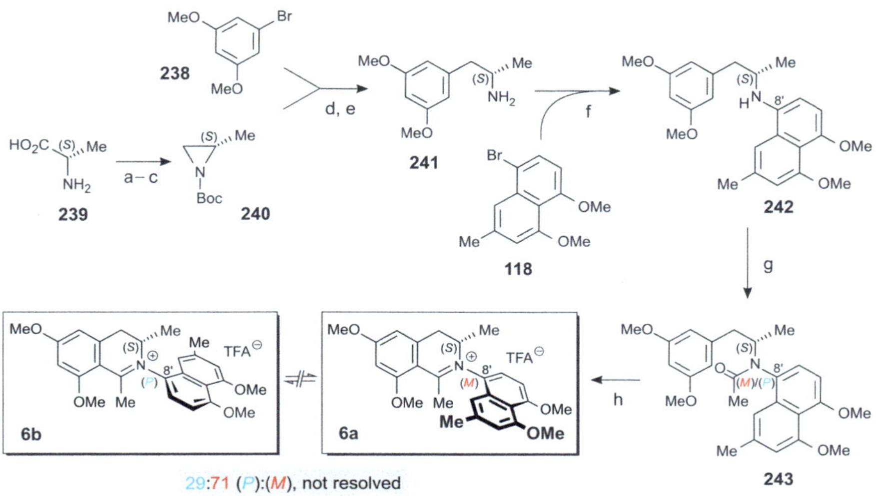

Fig. 119 Preparation of ancistrocladinium A (**6a/6b**) by the reaction sequence: Buchwald–Hartwig amination → N-acetylation → Bischler–Napieralski cyclization—preferentially leading to the naturally predominant (*M*)-atropo-diastereomer **6a**. Reaction conditions: (a) $LiAlH_4$, THF, 95%; (b) Boc_2O, MeCN, 97%; (c) TosCl, KOH, THF, 98%; (d) Mg, I_2, $(BrCH_2)_2$, THF, CuBr·SMe_2, 91%; (e) TFA, CH_2Cl_2, r.t., quant.; (f) $Pd_2(dba)_3$, *rac*-BINAP, Ko*t*-Bu, toluene, 110°C, 63%; (g) AcCl, DMAP, toluene, 94%; (h) $POCl_3$, MeCN, 88%

eight linear steps. The coupling of the two molecular portions, namely, the attachment of the (*S*)-configured amine **241** to the appropriately brominated naphthalene building block (**118** and **245** for the synthesis of **6a/6b** and **14a/14b**, respectively) at a late stage of the synthesis, by Buchwald–Hartwig amination and Bischler–Napieralski cyclization, provided ancistrocladinium A (**6a/6b**), with a significant atroposelectivity in favor of the naturally predominant (*M*)-isomer **6a** ((*M*):(*P*) = 79:21) (see Fig. 119), while ancistrocladinium B (**14a/14b**) was obtained, as expected, as a mixture of its two configurationally semi-stable atropo-diastereomers, **14a** and **14b** ((*M*):(*P*) = 54:46), which slowly interconvert at room temperature, by rotation about the N,C-axis (see Fig. 120) [101].

An essential prerequisite for a more general applicability of the developed method was the elaboration of a suitable strategy for the preparation of the known [242, 466] 1-aryl-2-propylamines such as **241** (Fig. 119) as the key precursors in the total synthesis of N,C-coupled naphthyldihydroisoquinolines. For the enantioselective preparation of the (*S*)-configured 1-aryl-2-propylamine **241**, an approach via chiral aziridines initially developed by Hoye et al. [108, 467] for the directed synthesis of (3*R*)-configured naphthylisoquinoline alkaloids was adapted to the preparation of (3*S*)-configured target molecules in the authors' group, now via the (*S*)-enantiomer **240** of the *N*-Boc-activated aziridine. This key precursor was obtained in 90% overall yield by reduction of L-alanine (**239**) with $LiAlH_4$ and *N*-functionalization with $(Boc)_2O$, followed by aziridine ring closure through an

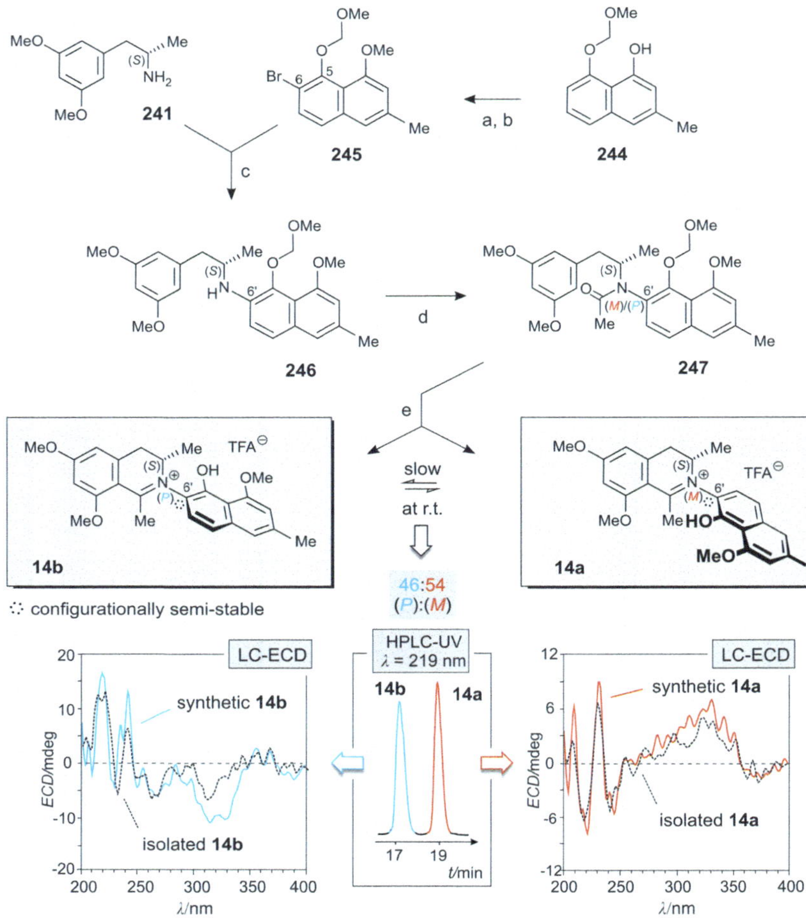

Fig. 120 First and as yet only total synthesis of ancistrocladinium B (**14a/14b**) and stereochemical assignment of the two diastereomers **14a** and **14b** of synthetic **14a/14b** by LC-ECD coupling and comparison of the obtained ECD spectra with those of the isolated alkaloid **14a/b**. Reaction conditions: (a) Me_2SO_4, K_2CO_3, NH_3, acetone, 88%; (b) *n*-BuLi, TMEDA, THF, $(CBrCl_2)_2$, −10°C → r.t., 74%; (c) $Pd_2(dba)_3$, *rac*-BINAP, KO*t*Bu, toluene, 110°C, 49%; (d) AcCl, DMAP, toluene, 86%; (e) $POCl_3$, MeCN, 95% → HPLC resolution on a reversed-phase C_{18} column

intramolecular S_N2 displacement of an in situ generated tosyloxy group. Regioselective ring opening of **240** by a phenyl cuprate generated in situ from the phenyl bromide **238**, followed by deprotection of the amino function yielded the targeted primary amine **241** in an enantiopure form in quantitative yield (Fig. 119). This highly economic strategy (five steps, 83% yield) [101] was an important improvement for the synthesis access to this important building block **241** compared to the conventional route used before (seven steps, 28% yield [108, 242,

467]), now permitting the preparation of enantiomerically pure **241** on a multigram scale from low-cost enantiopure starting materials from the chiral pool. The pathway has meanwhile enabled easy synthetic access to ancistrocladinium A (**6a/6b**) and B (**14a/14b**) and to related compounds, but also to a broad series of analogues with simplified structures.

A decisive step in the synthesis of ancistrocladinium A (**6a/6b**) (Fig. 119) was the Buchwald–Hartwig amination of the brominated naphthalene building block **118** with the isoquinoline precursor **241** to provide the secondary amine **242**. Despite the electron-rich character of the 8-bromonaphthalene **118**, N-arylation of the arylisopropylamine **241** afforded the desired key intermediate **242** in very good yields of 63% when using $Pd_2(dba)_3$ as the catalyst, BINAP as the chelating ligand, and KO*t*Bu as the base. Subsequent N-acetylation of **242** gave the amide **243**, for which its cyclization under Bischler–Napieralski conditions delivered the target molecule **6a/6b** as a 79:21 mixture in favor of the (*M*)-isomer **6a** in an excellent yield of 83% [101]. As already reported for the natural product **6a/6b** (see section above), all attempts to resolve the atropo-diastereomers, **6a** and **6b**, failed, even on chiral HPLC phases.

The synthetic product was identical in all physical, spectroscopic, and chiroptical data with the natural material of ancistrocladinium A (**6a/6b**) previously isolated from the twigs and stems of *A. tectorius* [33] and other *Ancistrocladus* plants from Africa and Asia [32, 47, 161, 176], including the prevalence of the naturally predominant (*M*)-atropo-diastereomer **6a** (see Fig. 118) in the atropo-diastereomeric mixture obtained by total synthesis.

Following the same strategy as described above, yet starting from the 6-brominated naphthalene **245** as the precursor, ancistrocladinium B (**14a/14b**) also was synthesized, again by a Buchwald–Hartwig amination → Bischler–Napieralski ring closure sequence (Fig. 120) [101]. The desired naphthalene building block **245** was obtained by analogy to a literature procedure [411, 412, 468], by introduction of the bromine at C-6 using a directed *ortho*-metalation (D*o*M) strategy [469, 470], assisted by the methoxymethyl (MOM) group at the oxygen function at C-5. Without the formation of any regioisomeric or overhalogenated by-products, the 6-bromonaphthalene **245** was obtained in very good yields (74%), by reaction of **244** with *n*-butyllithium/*N*,*N*,*N'*,*N'*-tetramethylethylenediamine (TMEDA) at −10°C, followed by treatment with 1,2-dibromotetrachloroethane [101]. Linkage of the isoquinoline building block **241** to the 6-bromonaphthalene **245** under the same Buchwald–Hartwig conditions as before (Fig. 119) afforded the secondary amine **246**, which, upon N-acetylation to give **247** and Bischler–Napieralski ring closure with concurrent cleavage of the MOM directing group in the naphthalene unit, delivered ancistrocladinium B (**14a/14b**) in 82% yield, as a 54:46-mixture of its two configurationally semi-stable rotational isomers, **14a** and **14b** (Fig. 120) [101].

As already described for **14a/14b** isolated from the Malaysian liana *A. benomensis* (see Sect. 6.3), HPLC-ECD investigations—now on the synthetic material of ancistrocladinium B (**14a/14b**)—expectedly gave rise to two almost mirror-imaged full LC-ECD spectra (Fig. 120). The ECD curve of the more rapidly eluting isomer perfectly matched the experimental spectrum of the (*P*)-configured atropo-diastereomer **14b** of natural ancistrocladinium B (**14a/14b**), while the ECD spectrum of the slower isomer showed an excellent agreement with the one monitored for the (*M*)-atropo-diastereomer **14a** of the authentic alkaloid [101].

By the highly convergent route presented in Fig. 120, ancistrocladinium B (**14a/14b**) was thus obtained in high chemical yields. The synthetic material was fully identical in all respects to the authentic natural product from previous isolation work [18, 32, 161], thus fully confirming the structure of this N,6′-coupled naphthyldihydroisoquinoline, including its absolute configuration [18, 32, 78, 83].

In addition, the merely artificial N,3′-linked ancistrocladinium C (**251**) (Fig. 121) [101] was prepared, exemplifying the extension of the synthesis pathway described above to representatives of N,C-coupled naphthyldihydroisoquinolines with further, biosynthetically imaginable coupling types as yet not found in Nature. The secondary amine **250**, as required for the synthesis of the sterically more demanding ancistrocladinium C (**251**), was prepared by Buchwald–Hartwig amination of the primary amine **241** with the naphthalene building block **249**, now equipped with the halogen atom at C-3. Starting from the known [101, 348] naphthol **97**, the 3-iodonaphthalene precursor **249** was obtained by conversion of **97** into the methoxymethyl derivative **248**, and subsequent introduction of the iodine substituent using a D*o*M reaction [469, 470], similar to the strategy already described above for the preparation of the 6-brominated naphthalene **245** (see Fig. 120). N-Acetylation of the secondary amine **249**, followed by Bischler–Napieralski cyclization of the resulting acetamide (not shown), finally provided the desired target molecule ancistrocladinium C (**251**) (Fig. 121), as a mixture of its configurationally stable atropo-diastereomers, **251a** and **251b** ((*M*):(*P*) = 40:60, in 53% chemical yield) [101].

Previous attempts to establish the nitrogen-aryl bond formation of the primary amine **241** with the corresponding 3-bromonaphthalene (not shown) had failed, despite extended variation of the reaction parameters, apparently for lack of reactivity due to an increased electron density at the targeted coupling site [101].

After their successful resolution by HPLC on a reversed-phase C_{18} column, an unambiguous assignment of the absolute configuration of the two atropo-diastereomers of ancistrocladinium C (**251**) was achieved by 2D NMR investigations in combination with quantum-chemical ECD calculations for **251a** and **251b** and comparison of the predicted ECD spectra with the experimental ones. In view of the stereogenic center at C-3 known to be (*S*)-configured, the relative and, thus, absolute configuration at the biaryl axis of the major isomer **251b** (= peak **A**) was attributed to be (*P*), due to NOESY interactions between H-3 and the protons of the methyl group at C-2′ (see Fig. 121, bottom right). In a similar manner, NOESY correlations between the methyl substituents at C-3 and C-2′, again in conjunction with the known (*S*)-configuration at C-3, permitted assignment of the absolute configuration at the iminium-aryl axis of the minor diastereomer **251a** (= peak **B**) as (*M*) [101].

The ECD spectra of the two atropo-diastereomers were monitored online, directly from the respective HPLC peaks **A** and **B** (see Fig. 121), by HPLC-ECD coupling in the stopped-flow mode. In accordance with the dominant ECD effect of

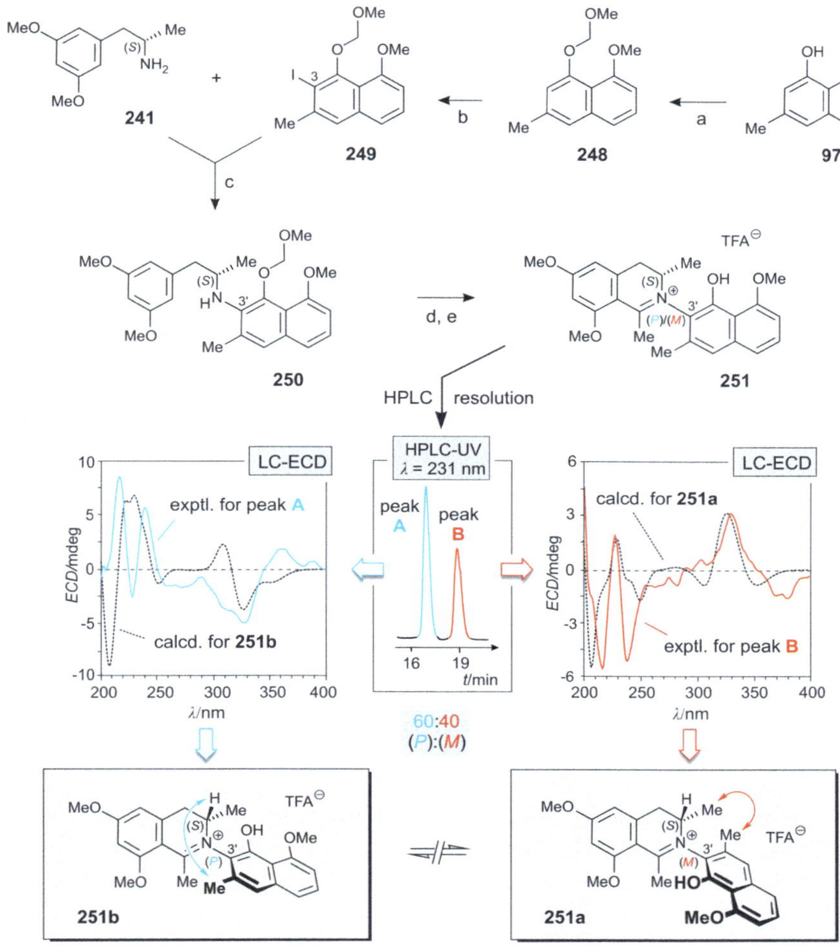

Fig. 121 Total synthesis of the N,3′-coupled naphthyldihydroisoquinoline ancistrocladinium C (**251**) (so far unknown in Nature) and assignment of the relative and absolute axial configurations of the two atropo-diastereomers **251a** and **251b** by NOE investigations and quantum-chemical ECD calculations using TDB3LYP/6-31G*. Reaction conditions: (a) MOMCl, NaH, DMF, diethylether, r.t., 87%; (b) *n*-BuLi, TMEDA, I_2, *n*-hexane, −10°C, 45%; (c) $Pd_2(dba)_3$, *rac*-BINAP, KO*t*Bu, toluene, 110°C, 42%; (d) AcCl, DMAP, toluene, 78%; (e) $POCl_3$, MeCN, 78%

the element of axial chirality in **251** and due to the presence of opposite configurations at the N,C-axis, the ECD curves were fully opposite to each other. A clear attribution of the complete stereostructures of the two atropo-diastereomers of ancistrocladinium C (**251**) was achieved by the analysis of the conformational space based on DFT (B3LYP/6-31G*) calculations in combination with single-point energy (RI-SCS-MP2/TZVP) simulations. As presented in Fig. 121, the ECD curve predicted for the (*P*)-configured atropo-diastereomer **251b** was in accordance with the experimental ECD spectrum of the more rapidly eluting isomer on HPLC analysis, while the ECD curve calculated for the (*M*)-configured atropo-diastereomer **251a** matched the ECD spectrum monitored for the slower isomer **251a**, thus confirming the above-described assignment by NOESY investigations [101].

Both atropo-diastereomers of ancistrocladinium C, **251a** and **251b**, exerted significant inhibitory effects against *L. major* promastigotes, with IC_{50} values of 41.8 µ*M* (for **251a**) and 37.3 µ*M* (for **251b**), nearly as potent as those displayed by the reference compound miltefosine (IC_{50}, 31.9 µ*M*). The activities were, however, drastically weaker than those of the natural products ancistrocladinium A (**6a/6b**) and B (**14a/14b**) (IC_{50}, 7.32 µ*M* and 6.62 µ*M*, respectively, see above). Ancistrocladinium C (**251**) exhibited virtually no activities against *P. falciparum* and *T. brucei rhodesiense*, or against *T. cruzi* (Chagas' disease), showing the antiparasitic activities of **251** to be structure-dependent and, thus, specific [101].

Again Starting from the Chiral Pool: Directed Synthesis of All Four Pure Stereoisomers of Ancistrocladinium A

In view of the excellent inhibitory activity of natural ancistrocladinium A (**6a/b**) as an atropo-diastereomeric mixture against *Leishmania* parasites [18, 32, 47, 66–72] (see previous section), the preparation and separate biotesting of the two rotational isomers **6a** and **6b** in a stereochemically pure form seemed highly rewarding. For this purpose, a directed synthesis access to **6a** and **6b** was inevitable due to the fact that a chromatographic resolution of the two atropo-diastereomers had failed [32, 47], even on a chiral phase. Since the naturally predominating (*M*)-atropo-diastereomer **6a** was also obtained preferentially by Bischler–Napieralski cyclization of the acetamide precursor **243** [101], the origin of the atroposelectivity during this ring closure reaction was investigated more closely. NMR and HPLC analysis revealed that the amide precursor **243** already consists of two configurationally stable diastereomers (Fig. 122), despite the presence of only one stereogenic center [69]. Unlike in the case of related amides [471–476], they were not the usual (*E*)/(*Z*) isomers, but already displayed the phenomenon of atropisomerism at the later N,C-axis due to rotational restriction about the *N*-carbonyl bond. Fortunately, baseline resolution of the two diastereomers of **243** was achieved by HPLC on an achiral reversed-phase C_{18} column [69], even permitting an efficient separation of the two compounds at a preparative scale. Online ECD analysis directly from the HPLC peaks gave rise to nearly opposite ECD spectra (Fig. 122), thus corroborating the presence of two rotational isomers, with the expected chiroptical predominance of the imine-naphthalene axis over the stereogenic center at C-3. Comparison of the experimental ECD curves with the ones predicted for the two amide diastereomers of **243** by quantum-chemical ECD calculations revealed the faster atropisomer (= peak **A**) to be (*P*)-configured, i.e., possessing structure **243b**, and the slower one (= peak **B**) to display the (*M*)-configuration, i.e., possessing

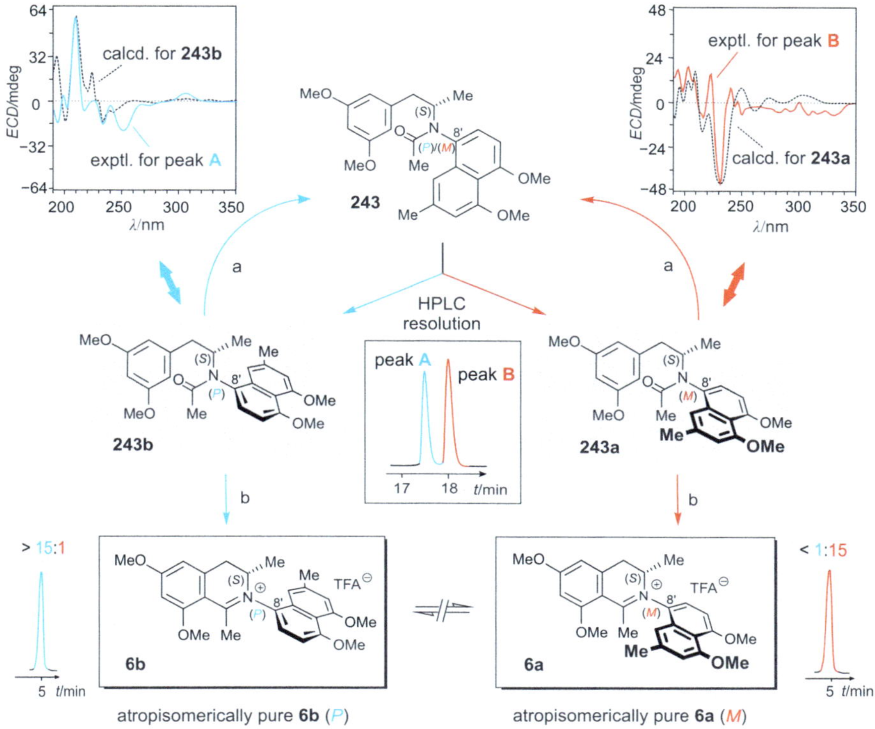

Fig. 122 Highly efficient, directed synthesis of the pure atropisomers of ancistrocladinium A, **6a** and **6b**, by resolution of the already rotationally hindered, and thus atropo-diastereomeric acetamide precursors **243a** and **243b**, followed by cyclization to the corresponding target molecules and re-use of the respective undesired atropisomers **243b** and **243a**. Reaction conditions: (a) recycling through thermal equilibration: MeCN, 95°C; (b) resolution and cyclization: $POCl_3$, MeCN, 80°C

structure **243a** (Fig. 122) [69]. The two atropisomerically pure amides were configurationally stable at room temperature, but were found to gradually interconvert at elevated temperature (above ca. 80°C).

As a further confirmation of the above-presented assignment, the stereochemically pure (*M*)-configured amide **243a** was specifically converted into the likewise (*M*)-configured target molecule ancistrocladinium A (**6a**), without any substantial isomerization at the N,C-axis during the ring closure reaction. Moreover, even the minor atropisomeric acetamide **243b** was not lost, but was recyclized by thermal equilibration to give an (*M*)/(*P*)-mixture by briefly heating it up to 95°C, followed by resolution of the two atropisomers and renewed Bischler–Napieralski cyclization of **243a** to give **6a** (Fig. 122) [69]. Optionally, the (*P*)-isomer of ancistrocladinium A, **6b**, can be prepared in an analogous manner, now by cyclization of the pure (*P*)-configured acetamide **243b**, again with the option to recycle the now undesired

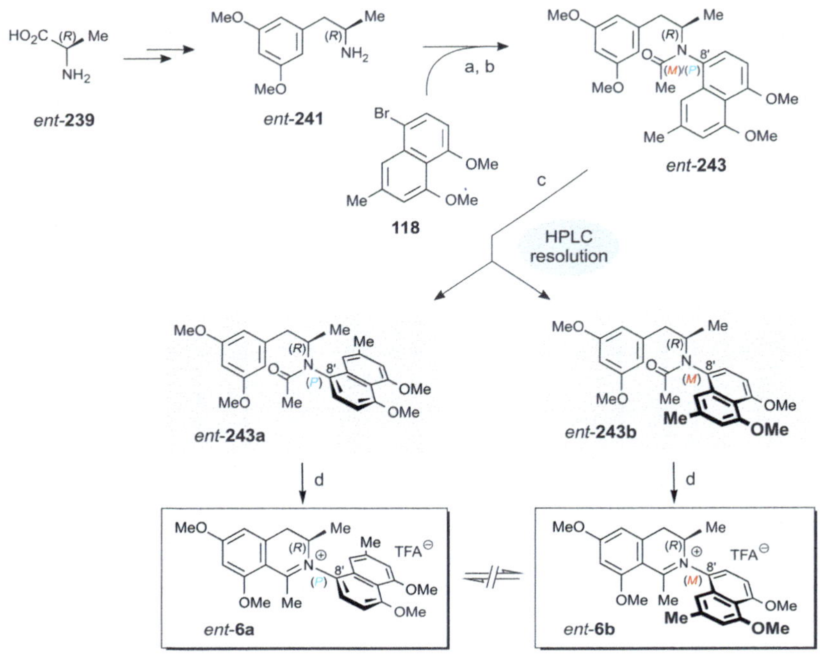

Fig. 123 Synthesis and resolution of the already rotationally hindered, and thus atropo-diastereomeric acetamide precursors *ent*-**243a** and *ent*-**243b**, and their stereospecific Bischler–Napieralski ring closure to give the (unnatural) (3*R*)-enantiomers of ancistrocladinium A, *ent*-**6a** (*P*) and *ent*-**6b** (*M*), each in an atropisomerically pure form. Reaction conditions: (a) $Pd_2(dba)_3$, *rac*-BINAP, Ko*t*-Bu, toluene, 110°C, 61%; (b) AcCl, DMAP, toluene, 95%; (c) chromatographic resolution by preparative HPLC on an achiral reversed-phase C_{18} column; (d) $POCl_3$, MeCN, 80°C, 81% (for *ent*-**6a**) and 88% (for *ent*-**6b**)

(*M*)-isomer, **243a**, thus delivering **6b**, which is the minor isomer in the plants, in a stereochemically homogeneous form (Fig. 122) [69].

In conclusion, the preparation of stereochemically pure atropo-diastereomeric amides **243a** and **243b** and their subsequent cyclization provided for the first time access to the desired target molecules, the (*M*)-configured **6a** or, optionally, the (*P*)-configured **6b**, in an atropisomerically homogeneous form, with excellent atropo-diastereomeric purities (Fig. 122). This is of particular value in view of the failed attempts to resolve the two naturally occurring atropo-diastereomers of ancistro-cladinium A, **6a** and **6b**, by chromatography [69].

The strategy also paved the way for the synthesis of the two (as yet unnatural) (3*R*)-analogues, *ent*-**6a** ((*R,P*)-configured) and *ent*-**6b** ((*R,M*)-configured) (Fig. 123), and, thus, for the availability of the complete series of all four possible stereoisomers of ancistrocladinium A, e.g., for a comparative evaluation of their anti-infective activities. Starting from D-alanine (*ent*-**239**), Pd-catalyzed N,C-coupling of the primary amine *ent*-**241** with the 8-bromonaphthalene **118**, followed by N-acetylation of the resulting secondary amine, yielded the acetamide *ent*-**243**, for which the chromatographic resolution gave its pure atropo-diastereomers, *ent*-**243a** and *ent*-**243b** [69, 101]. Using the same reaction conditions as previously applied in the case of the enantiomeric, natural series, i.e., with (*S*)-configuration at C-3, the desired target compounds, *ent*-**6a** (*P*)- and *ent*-**6b** (*M*)-configured, were obtained by Bischler–Napieralski ring closure of the individual acetamide isomer [69, 101].

Among the four stereoisomers of ancistrocladinium A (**6**), the naturally predominant atropo-diastereomer **6a** (*M*,*S*) displayed the highest inhibitory activity against *L. donovani* (IC_{50}, 0.081 μ*M*), distinctly more potent than natural **6a/6b** as a 3:1-mixture of **6a** and **6b** (IC_{50}, 1.71 μ*M*)—and better than the standard miltefosine (IC_{50}, 0.174 μ*M*). Moreover, **6a** showed a very good selectivity index (SI) of ca. 450, due to its low cytotoxicity against mammalian host cells (rat skeletal myoblast L6 cells) [69], so that this N,C-coupled alkaloid can be considered as a promising lead according to the TDR/WHO guidelines [238]. The antileishmanial effects exerted by the likewise natural, but minor compound **6b**, and by the unnatural isomers, *ent*-**6a** (*P*)- and *ent*-**6b** (*M*)-configured, with (*R*)-configuration at C-3, were much weaker, with IC_{50} values of 0.401 μ*M* (for **6b**), 0.233 μ*M* (for *ent*-**6a**), and 0.588 μ*M* (for *ent*-**6b**) [69]. Furthermore, strong inhibitory effects of the natural (*M*,*S*)-atropo-diastereomer **6a** and its unnatural (*P*,*R*)-enantiomer *ent*-**6a** were monitored against *Trypanosoma* parasites. Here again, natural **6a** was the most active compound, with IC_{50} values of 0.098 μ*M* (*T. brucei rhodesiense*) and 0.99 μ*M* (*T. cruzi*), similar to the standard compounds melarsoprol (IC_{50}, 0.010 μ*M*, for *T. brucei rhodesiense*) and benznidazole (IC_{50}, 3.21 μ*M*, for *T. cruzi*), or even better [69].

Directed Synthesis of the Pure Atropo-Diastereomers of the Unnatural N,1′-Coupled Naphthylisoquinoline Ancistrocladinium D, Starting from the Chiral Pool

The general applicability of the synthesis strategy presented above, even to the preparation of sterically more congested N,C-coupled naphthyldihydroisoquinolines, was demonstrated by the directed synthesis of the pure two atropo-diastereomers, **255a** and **255b**, of the as yet unnatural N,1′-linked ancistrocladinium D (**255**) (Fig. 124) [69, 101]. The structure of **255** differs from that of the N,8′-coupled ancistrocladinium A (**6**) only by the position of the methyl group on the naphthalene portion. In comparison to **6**, the position of Me-2′ in **255** is now next to the N,C-linkage, thus leading to a significantly enhanced steric hindrance at the axis.

The synthesis of ancistrocladinium D (**255**) required the preparation of the bromine-activated naphthalene **253**. The regioselective introduction of the bromine substituent at C-1 succeeded by an electrophilic bromination reaction using a slight excess of bromine at −40°C, directly on the known [348] 4,5-dimethoxy-substituted naphthalene **252**, in the presence of sodium acetate. The undesired dibrominated

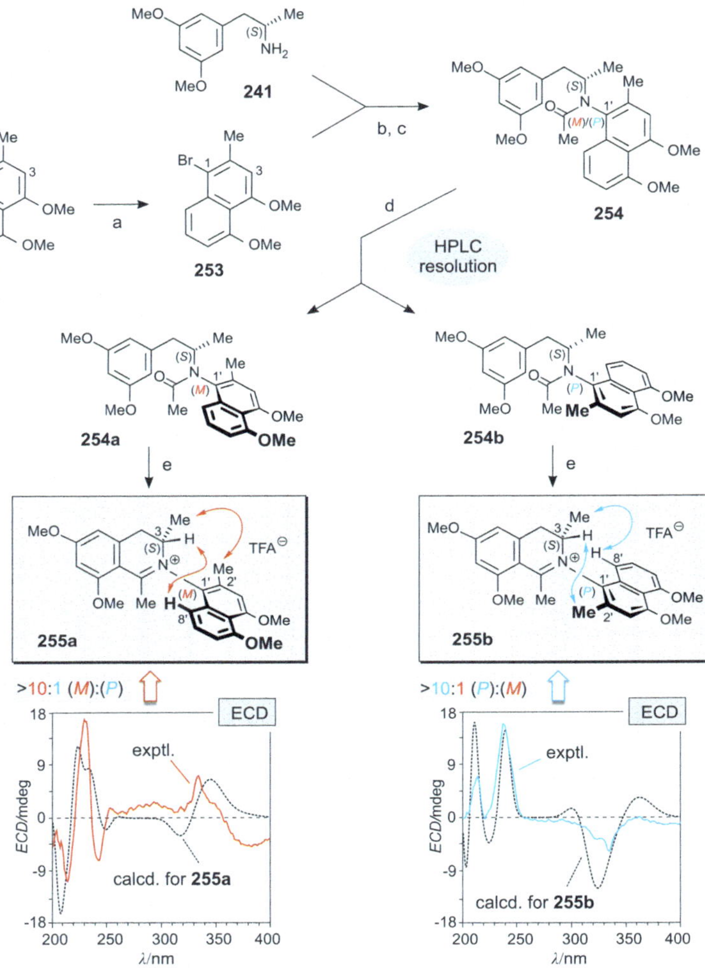

Fig. 124 Total synthesis of the unnatural N,1′-coupled naphthyldihydroisoquinoline ancistrocladinium D (**255**) in its two atropisomeric forms, **255a** (*M*) and **255b** (*P*), and assignment of their relative and absolute axial configurations by NOE investigations and quantum-chemical ECD calculations using MRCI-B3LYP/G/SV(P). Reaction conditions: (a) Br_2, NaOAc (1.2 equiv.), CH_2Cl_2, −40°C, 87%; (b) $Pd_2(dba)_3$, *rac*-BINAP, KO*t*Bu, toluene, 110°C, 56%; (c) AcCl, DMAP, toluene, 73%; (d) chromatographic resolution by preparative HPLC on an achiral C_{18} column; (e) $POCl_3$, MeCN, 80°C, 81% (for **255a**) and 82% (for **255b**)

naphthalene (with an additional bromine substituent at C-3), which was formed only as a by-product, was easily removed by column chromatography, affording the targeted pure building block **253** in 87% yield [101]. Arylation of the primary amine **241** with the 1-bromonaphthalene **253**, followed by N-acetylation of the resulting secondary amine (not shown) provided the acetamide **254** [101]. Its resolution by HPLC on an achiral C_{18} phase gave the pure, configurationally stable rotational isomers **254a** and **254b**. Specific Bischler–Napieralski cyclization of the individual acetamide precursors, **254a** and **254b**, eventually delivered the target compounds, **255a** and **255b**, respectively, each obtained as atropisomerically pure material (Fig. 124) [69].

The relative and thus, also the absolute axial configurations of the two atropo-diastereomers of ancistrocladinium D (**255**) were attributed by long-range NOESY interactions, and further confirmed by the comparison of the experimental ECD spectra of the pure isolated isomers with those obtained for **255a** and **255b** by quantum-chemical ECD calculations. In the case of **255a**, NOESY correlations between H-3 and H-8′, in conjunction with the known (*S*)-configuration at C-3, permitted assignment of the absolute configuration at the iminium-aryl axis as (*M*) (Fig. 124, left). The configuration at the biaryl axis of **255b** relative to the center at C-3 was established by NOESY cross-peaks between H-3 and the protons of the methyl group at C-2′, from which, due to the given (3*S*)-configured dihydroisoquinoline subunit, the absolute axial configuration was attributed as (*P*) (Fig. 124, right) [69, 101]. This stereochemical assignment was in accordance with quantum-chemical ECD calculations on **255a** and **255b** based on a combined approach of DFT (B3LYP/6-31G*) [239–241] and MRCI [477] methods. As illustrated in Fig. 124, a nearly perfect match was observed between the experimental ECD curves with the ones computed for the two atropo-diastereomers of ancistrocladinium D (**255**) [69, 101].

While the naturally occurring ancistrocladinium A (**6**) exerted excellent antiparasitic activities, in particular against *L. donovani*, with the main atropo-diastereomer **6a** showing the best inhibitory effect (IC_{50}, 0.081 μ*M*), its as yet unnatural regioisomer ancistrocladinium D (**255**) was substantially less active, in both atropisomeric forms, exhibiting only weak antileishmanial activities, with IC_{50} values in the micromolar range (e.g., for **255a**: IC_{50}, 7.43 μ*M*) [69]. The results underline the strong impact of the different coupling types of the two regioisomers **6** and **255** (N,8′ versus N,1′) and their molecular shapes, on the resultant antiparasitic activity.

Synthesis of the Stereoisomers of Ancistrocladinium A with Catalytic Enantioselective Construction of C-3

An alternative approach to ancistrocladinium A (**6**) based on the construction of the secondary amine **242** by asymmetric reductive amination of the arylpropanone **115** with the 8-aminonaphthalene **120** was subsequently published by Kim and Cheon [118, 119]. Starting from the known [256] *N*-Boc-protected naphthylamine **119**, the free primary amine **120** was prepared by treatment of **119** with TFA to remove the Boc group. Since **120** easily underwent decomposition during work-up and purification (see also Sect. 4.8), it was generated in situ and directly subjected to the reductive amination of ketone **115**, using the chiral phosphoric acid **256** (derived from (*M*)-BINOL) as a catalyst and the Hantzsch ester **257** as an organic hydride source. Running the asymmetric reduction of **115** with **120** under optimized

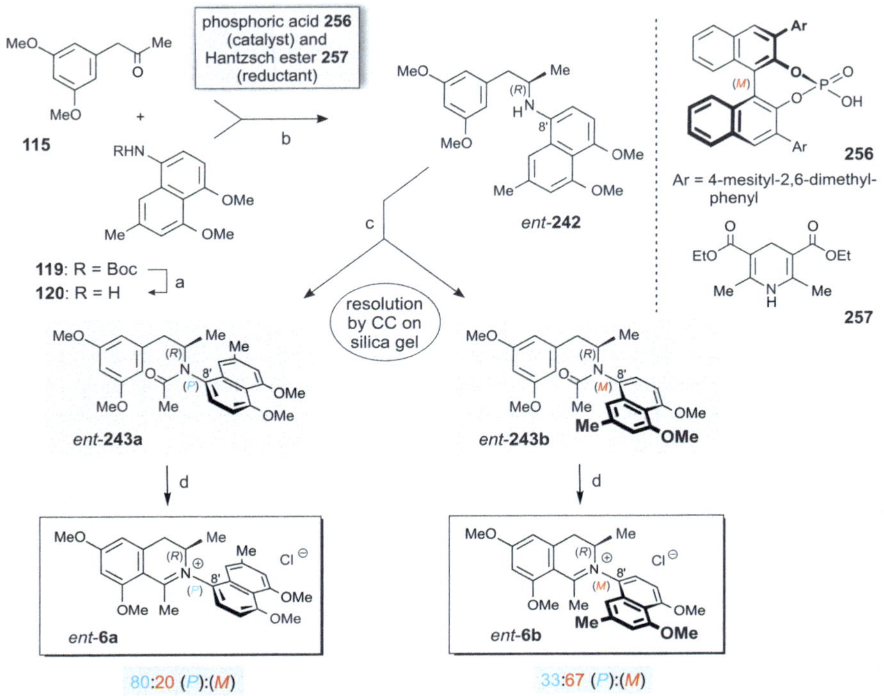

Fig. 125 Asymmetric total synthesis of *ent*-ancistrocladinium A (*ent*-**6**) by enantioselective reductive amination of the arylpropanone **115** with the 8-aminonaphthalene **120** using the chiral phosphoric acid **256** as a catalyst and the Hantzsch ester **257** as a hybrid source, followed by a Bischler–Napieralski ring closure of the pure resolved atropisomers of *ent*-**243**. Reaction conditions: (a) TFA, CH_2Cl_2, r.t. → 1 N NaOH (pH 11), 64%; (b) **257** (1.5 equiv.), **256** (10 mol %), 5 Å molecular sieve, TBME, 40°C, 40 h, 76%, 85% *ee*; (c) AcCl, DMAP, toluene, reflux, 3 h, 56% (for *ent*-**243a**) and 35% (for *ent*-**243b**) after resolution by chromatography on silica gel (*n*-hexane/ethyl acetate, 4:1, *v*/*v*); (d) $POCl_3$, MeCN, reflux, 3 h, 81% (for *ent*-**6a**) and 84% (for *ent*-**6b**)

conditions, namely, with the phosphoric acid **256** bearing 4-mesityl-2,6-dimethyl-phenyl substituents at the 3,3′-positions, with 5 Å molecular sieves as the dehydration agent, and with *tert*-butyl methyl ether (TBME) as the solvent, the secondary amine *ent*-**242** was obtained in 76% yield and 85% *ee* (Fig. 125) [118]. Thus, this reaction step happened to provide the 'wrong' enantiomer of amine **242**, i.e., the key building block *ent*-**242** leading to unnatural *ent*-ancistrocladinium A (*ent*-**6**), with (*R*)-configuration at C-3. Assignment of the configuration at C-3 was done by comparison of the optical properties of *ent*-**242** with those of the known, (3*S*)-configured compound **242** [71, 101].

N-Acetylation of *ent*-**242** afforded the acetamide *ent*-**243** as a mixture of atropo-diastereomers, *ent*-**243a** ((*P*)-configured at the N,C-axis) and *ent*-**243b**

((*M*)-configured), which were resolved by chromatography on normal-phase silica gel. Bischler–Napieralski ring closure of pure *ent*-**243a** ((*P*)-isomer) using $POCl_3$ in refluxing acetonitrile, yielded *ent*-ancistrocladinium A (*ent*-**6**), yet as a mixture of its atropo-diastereomers, *ent*-**6a** and *ent*-**6b**, with chloride as the counteranion, and an atroposelectivity in favor of the (*P*)-isomer, *ent*-**6a** (80:20 (*M*):(*P*)) [118]. Cyclization of pure *ent*-**6b** ((*M*)-isomer) expectedly gave the (*M*)-atropisomer of *ent*-ancistrocladinium A, *ent*-**6b**, as a major product, in a ratio of 75:25 (*M*):(*P*). As a result of an atropisomerization during purification of *ent*-**6b** by filtration over Sephadex gel, the atropisomeric purity was further diminished to 67:33 (*M*):(*P*), regardless of the atropisomeric composition of the starting materials [118].

Kim and Cheon [119] finally also achieved the total synthesis of the two naturally occurring stereoisomers of authentic ancistrocladinium A, **6a** and **6b**, each in an atropisomerically pure form, following the strategy described above, but appropriately adapted, as outlined in Fig. 126. In this improved second approach, asymmetric reductive amination of the arylpropanone **115** with the 8-amino-naphthalene **120** provided the enantiopure secondary amine **242**, this time installing the configuration at C-3 correctly as (*S*), by using the chiral phosphoric acid **258** (derived from (*P*)-BINOL) as the catalyst [119]. N-Acetylation of **242** afforded the expected mixture of the two diastereomeric amides **243a/b**, which were resolved by column chromatography on silica gel to give the individual rotational isomers **243a** and **243b** (both in 90% *ee*), in 51% and 36% yields, respectively.

Running the Bischler–Napieralski cyclization separately on the two pure amides **243a** and **243b**, using $POCl_3$, resulted in a significant loss of stereochemical purity at the N,C-axis, as already observed in their previous synthesis of *ent*-ancistrocladinium A (*ent*-**6**) [118] (see Fig. 125), again giving rise to mixtures of the (*M*)- and (*P*)-diastereomers of ancistrocladinium A, **6a** and **6b**. Thus, treatment of the amide **243a** ((*M*)-configured) with $POCl_3$ at 100°C delivered ancistrocladinium A (**6a**) as the major atropo-diastereomer, along with the (*P*)-isomer **6b**, in a ratio of 80:20. Under the same conditions, the Bischler–Napieralski cyclization of **243b** led to **6a** and **6b** as a 25:75 mixture, now in favor of the (*P*)-isomer **6b**. At temperatures below 80°C, however, the reaction of the amide oxygen in **243a/243b** with $POCl_3$ was found to proceed sluggishly [119]. A substantial improvement toward the stereospecific Bischler–Napieralski cyclization of each of the amides **243a** and **243b** was achieved by using triflic anhydride, instead of $POCl_3$, as the condensation reagent. The optimized protocol permitted running the ring closure reaction of the pure individual amides **243a** and **243b** even at −78°C, affording ancistrocladinium A (**6a**) and its atropo-diastereomer **6b**, after recrystallization, as their triflate salts, in a stereochemically pure form [119]. The authors did not report on the methods used to stereochemically characterize the amide precursors **243a** and **243b**, so one may assume that the absolute configurations at axes and centers were deduced from the specific conversion to the target compounds **6a** and **6b** (assigned by comparison of the optical rotation values known from those reported in the literature [71, 101]).

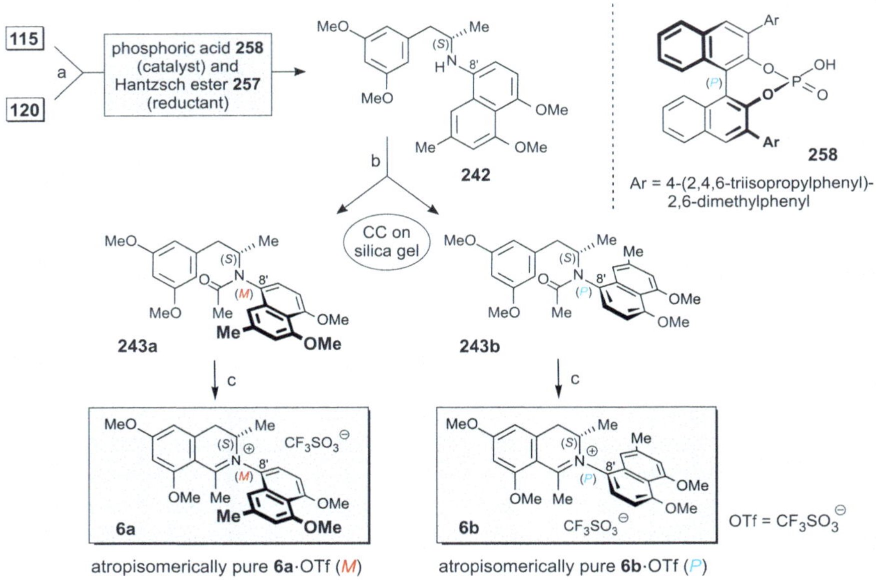

Fig. 126 An improved approach to the total synthesis of the two pure (now correctly C-3-configured) stereoisomers of natural ancistrocladinium A, **6a** and **6b**, by Kim and Cheon [119], with enantioselective construction of C-3, followed by stereospecific Bischler–Napieralski cyclization of the individual acetamides **243a** and **243b** with triflic anhydride to afford **6a** and **6b**, respectively. Reaction conditions: (a) **257** (1.5 equiv.), **258** (10 mmol%), 4 Å molecular sieve, *m*–xylene, 40°C, 40 h, 71%, 90% *ee*; (b) AcCl, DMAP, toluene, reflux, 3 h, 51%, 90% *ee* (for **243a**) and 36%, 90% *ee* (for **243b**) after resolution by chromatography on silica gel (*n*-hexane/ethyl acetate, 4:1, *v*/*v*); (c) 2-chloropyridine (1.1 equiv.), Tf_2O (1.1 equiv.), CH_2Cl_2, −78°C to r.t., 2 h, 88% (for **6a**) and 85% (for **6b**)

8.5.10 Ancistrocyclinones A and B, Unprecedented Pentacyclic N,C-Coupled Naphthylisoquinoline Alkaloids

From the twigs and stem extracts of *A. tectorius* collected in the Ledong region of the Chinese Island Hainan two further cyclized and, thus, even pentacyclic—N,C-coupled naphthylisoquinoline alkaloids were isolated, named ancistrocyclinones A (**7**) and B (**15**) (Fig. 127) [18, 33]. The UV spectra of these deep-red compounds were quite different from those of other naphthylisoquinoline alkaloids, because the third absorption maximum (usually at λ_{max} = 306 nm) was significantly bathochromically shifted to ca. 500 nm, indicating the presence of structurally unique N,C-coupled compounds with an extended conjugated π-system (Fig. 127). As deduced from HRESIMS and from the number of signals in the ^{13}C NMR spectrum, the molecular formula of **7** suggested a cationic structure, $C_{25}H_{24}NO_4^+$. Similar to other N,C-coupled alkaloids, **7** exhibited six protons in the aromatic region, of which one was significantly downfield-shifted, resonating at δ_H = 9.19 ppm. Further NMR data distinguishing **7** from usual naphthylisoquinoline alkaloids were the lack of the

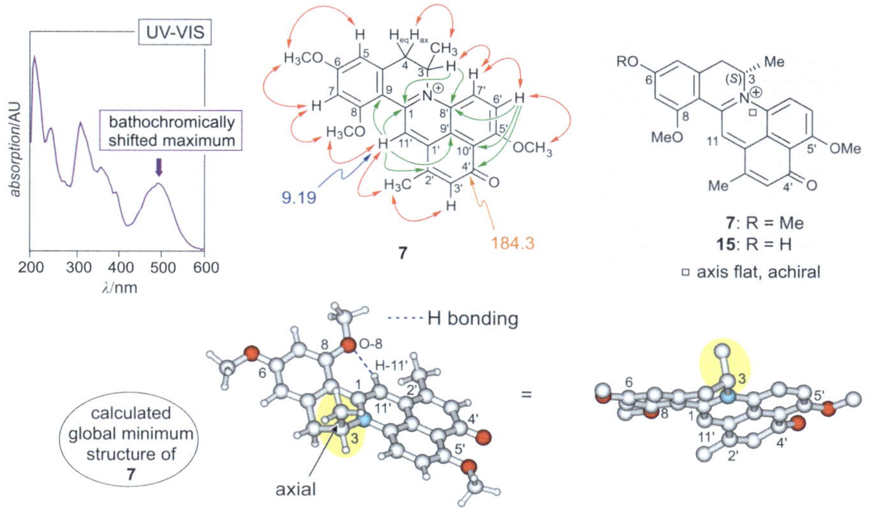

Fig. 127 Ancistrocyclinones A (**7**) and B (**15**), pentacylic N,C-coupled naphthylisoquinoline alkaloids from *A. tectorius*: UV spectra, selected ^{1}H and ^{13}C NMR data (δ/ppm), and NOESY (red double arrows) and HMBC (green single arrows) interactions indicative of the unprecedented molecular framework of **7**, and calculated global minimum structure of the (*S*)-enantiomer of the helically distorted molecule **7**, with the methyl group at C-3 adopting an *axial* position. For comparison, see the UV curves of the C,C-coupled naphthylisoquinoline dioncophylline A (**16**) and the N,C-coupled naphthyldihydroisoquinoline 4′-*O*-demethylancistrocladinium A (**123**), as depicted in Fig. 85

signal for the methyl group at C-1, but instead the presence of a carbonyl resonance at δ_C = 184.3 ppm. HMBC couplings of H-6′ to C-8′, C-10′, and C-4′, established the carbonyl carbon to be C-4′. From an HMBC interaction from H-3 to C-8′, and from an NOE correlation between H-3 and H-7′, the position of the coupling site of the two molecular portions was assigned as N-2 in the dihydroisoquinolinium ring, and C-8′ in the naphthalene portion. The HMBC cross-peaks from the downfield-shifted proton at δ_H = 9.19 ppm to C-1 and C-9 in the isoquinoline unit and to C-9′ and C-2′ in the naphthalene moiety assigned that proton signal to be H-11′, and revealed the dihydroisoquinolinium unit to be linked to the naphthalene moiety by a C–C bond from Me-1 to C-11′. NOESY correlations in the series of {OMe-8 ↔ H-11′ ↔ Me-2′ ↔ H-3′}, and from H-3 to H-7′ confirmed the unique pentacyclic structure of ancistrocyclinone A (**7**) (Fig. 127) [18, 33]. Such an additional six-membered ring, with C-11′ as the bridge between the two molecular units, was unprecedented and had never been found within the class of naphthylisoquinoline alkaloids before. The cyclization largely planarizes the N,C-axis in **7**, which is thus, despite the overall helical array (see below), no longer an element of stable or semi-stable axial chirality —in contrast to all other N,C-coupled naphthylisoquinoline alkaloids.

DFT calculations using the B3LYP/6-31G* method [239–241] clearly demonstrated that ancistrocyclinone A (**7**) is not flat, but helically distorted, with the

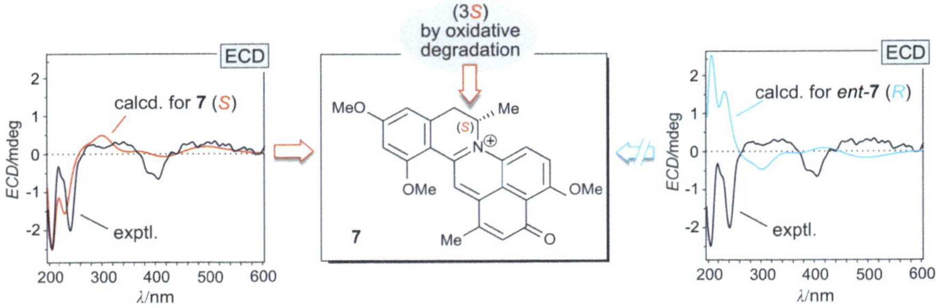

Fig. 128 Assignment of the absolute configuration of ancistrocyclinone A (**7**) by comparison of its experimental ECD curve with the spectra calculated for (*S*)-**7** and (*R*)-*ent*-**7**, using TDB2PLYP/def2-TZV(-f), and by ruthenium-mediated oxidative degradation

methyl group at C-3 adopting an *axial* position. The methoxy group at C-8 is fixed at one dihedral angle, as a result of steric interactions with the ring system of **7** and because of an intramolecular hydrogen bond between its oxygen and the proton at C-11′ (Fig. 127). Owing to repulsive interactions with the carbonyl function at C-4′, the methoxy substituent at C-5′ also adopts only one defined orientation [18, 33].

The absolute configuration of ancistrocyclinone A (**7**) at C-3 was determined by comparison of its experimental ECD spectrum with those quantum-chemically calculated for the (*S*)- and the (*R*)-enantiomers of **7**, using the TDB2PLYP/def2-TZVP(-f) method [33]. The curve predicted for (*S*)-**7** was fully in accordance with that of the experimental spectrum, while that obtained for *ent*-**7** ((*R*)-configured)) was mirror-image like, thus clearly evidencing ancistrocyclinone A to be (*S*)-configured at C-3 (see Fig. 128). This stereochemical assignment was confirmed by ruthenium-mediated oxidative degradation of **7**, affording 3-aminobutyric acid exclusively as its (*S*)-enantiomer, as assessed by GC-MSD analysis of the Mosher derivative of its methyl ester [33, 80].

The second pentacyclic N,C-coupled 3,4-naphthyldihydroisoquinolinium cation, likewise isolated from twig and stem extracts of *A. tectorius*, was chromatographically and spectroscopically similar to ancistrocyclinone A (**7**). NMR and MS analysis revealed this new metabolite to be the 6-*O*-demethyl analogue of **7** (for the structure, see Fig. 127). It was named ancistrocyclinone B (**15**) [33]. The natural counteranions of **7** and **15** are as yet unknown [18, 33].

Ancistrocyclinones A (**7**) and B (**15**) are the first (and so far only) representatives of a new subtype of alkaloids displaying a molecular framework reminiscent of that of berberine alkaloids, such as berberine (**259**) [478–481] itself (Fig. 129), but, in contrast to the essentially flat molecule of **259**, the ancistrocyclinones are characterized by an intriguing helical 3D structure, as presented in Fig. 127 [18, 33]. The ancistrocyclinones and berberines are illustrative examples evidencing the 'chemical creativity' of Nature, here in the convergence of biosynthesis pathways to give polycyclic isoquinoline alkaloids. Naphthylisoquinolines are formed from acetate/malonate units exclusively, via joint polyketide precursors [131–134], whereas berberines (like all the other isoquinoline alkaloids from the plant kingdom) are

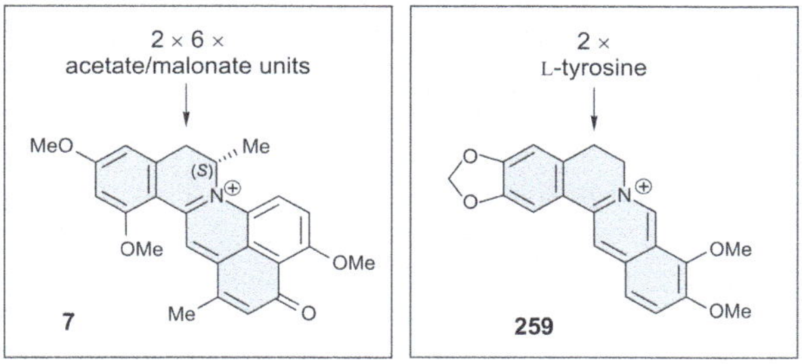

Fig. 129 Convergence in the biosynthesis of plant metabolites: polyketide- versus amino acid-derived origin of polycyclic isoquinoline alkaloids, as exemplified for berberine (**259**) and ancistrocyclinone A (**7**), which display similar molecular architectures

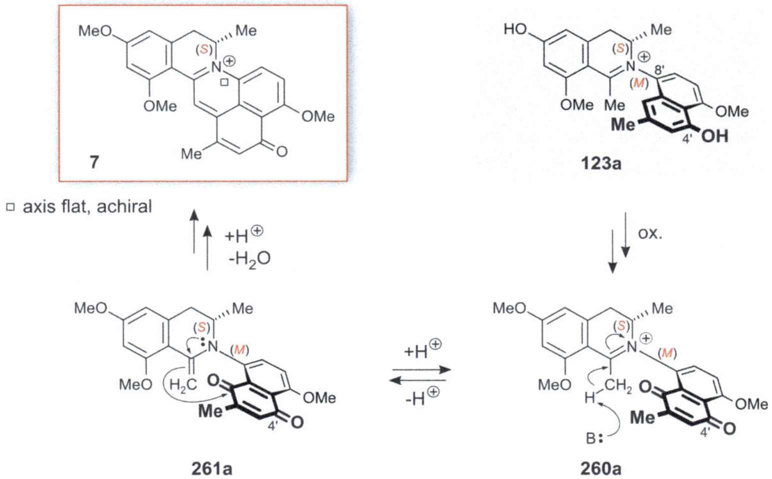

Fig. 130 Proposed biosynthesis pathway to ancistrocyclinone A (**7**) by an aldol-related cyclization of the quinoid intermediates **260a** and **261a** starting from 4′-*O*-demethylancistrocladinium A (formulated for the (*M*)-atropisomer, **123a**, as the precursor, as an example)

built up from aromatic amino acids such as tyrosine [135–140] (see also Sect. 2.3). Thus, although structurally quite similar, berberine (**259**) and ancistrocyclinones A (**7**) and B (**15**) originate from completely different biosynthesis routes.

The biosynthesis of ancistrocyclinone A (**7**) may be rationalized by an aldol-type cyclization of a (still hypothetical) N,C-coupled quinoid naphthylisoquinoline like **260a**, via the respective enamine **261a**. In agreement with this assumption is the co-occurrence of the related alkaloids 4′-*O*-demethylancistrocladinium A (**123a**) and its atropo-diastereomer **123b** in *A.* aff. *tectorius* (in a 4:1 ratio) [33], which are both, imaginable precursors to ancistrocyclinone A (**7**). As exemplified in Fig. 130 for the (*M*)-configured atropisomer **123a**, this plant metabolite, with its free hydroxy group at C-4′ in the naphthalene portion, might be oxidized to provide the corresponding naphthoquinone **260a**. Subsequent condensation of its tautomeric analogue, the enamine **261a**, would directly afford the pentacyclic ancistrocyclinone A (**7**).

Following the proposed biosynthesis pathway, the same product **7** would also arise when starting from the (*P*)-configured N,8′-*epi*-4′-*O*-demethylancistrocladinium A (**123b**), since the cyclization reaction leads to a configurationally unstable and, thus, achiral axis. Biomimetic conversion of **123a/123b** (atropisomeric mixture) to ancistrocyclinone A (**7**), with lead tetraacetate as the oxidant, supported the postulated route via the quinoid intermediate **260**, as evident from the isolation of substantial quantities of **260** (23% yield, obtained by preparative HPLC), along with the target compound **7** (36%) [18, 33].

Ancistrocyclinone A (**7**) displayed cytotoxic activities against drug-sensitive human lymphoblastic CCRF-CEM leukemia cells (IC_{50}, 16.2 μ*M*) and their multidrug-resistant subline, CEM/ADR5000 (IC_{50}, 25.3 μ*M*), with comparable efficacies. The antiproliferative effects of the assumed quinoid precursor **260a** toward CCRF-CEM cells were in the same range (IC_{50}, 18.8 μ*M*), but the inhibition of the viability of CEM/ADR5000 cells was distinctly lower (IC_{50}, 57.4 μ*M*) [33]. Ancistrocladinium A (**6a/6b**) and its monophenolic analogue **123a/123b**, in contrast, exerted strong growth-retarding activities against CCRF-CEM cells at a submicromolar range, with IC_{50} values of 1.40 μ*M* for **6a/6b** and 1.50 μ*M* for **123a/123b**, but they exhibited only very weak cytotoxic effects toward the CEM/ADR5000 leukemia subline, with an inhibition of growth of the tumor cells by <20% [33].

8.5.11 Shuangancistrotectorines A–E, Dimeric Naphthylisoquinoline Alkaloids with Three Consecutive Chiral Biaryl Axes

From air-dried twigs of *A.* aff. *tectorius* from the Chinese Hainan Island five trace alkaloids exhibiting novel dimeric structures were discovered [17, 31]. No dimers had so far been reported to occur in this *Ancistrocladus* species from any other location in Southeast Asia. All of these compounds (Fig. 131), named shuangancistrotectorines A–E (**262a**, **262b**, **5**, **207a**, and **207b**) [31], have in common that they consist of two identical naphthylisoquinoline portions, which are linked via the sterically constrained 3′,3″- or 1′,1″-positions of their naphthalene units, thus giving rise to a rotationally hindered central binaphthalene axis. Hence, they possess three consecutive stereogenic axes, and, including the up to four stereogenic centers, they have up to seven stereogenic units in total [17, 31]. The only other dimer known from an Asian *Ancistrocladus* species, ancistrogriffithine A (**148**) from the Thai liana *A. griffithii* (see Sect. 8.5), is likewise built up from two identical monomeric naphthylisoquinoline alkaloids, but is characterized by a 6′,6″-coupled central binaphthalene linkage, which can freely rotate and, thus, is not an additional element of chirality [17, 42, 43, 82]. All of the five shuangancistrotectorines are typical Ancistrocladaceae-type alkaloids with an oxygen function at C-6 and (*S*)-configuration at C-3 [17, 31].

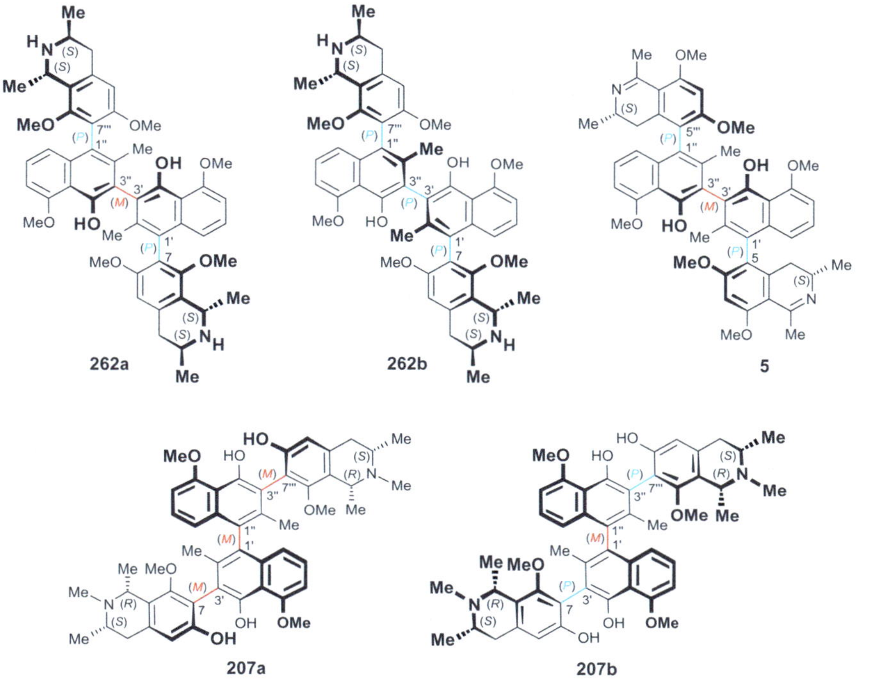

Fig. 131 Shuangancistrotectorines A–E (**262a**, **262b**, **5**, **207a**, and **207b**) from *A.* aff. *tectorius* from Hainan Province, China—an overview

One of the major peaks of these trace dimers was resolved into two very similar compounds, which had identical molecular formulas of $C_{50}H_{58}N_2O_8$, as deduced from HRESIMS due to the detection of a diprotonated molecular ion $[M + 2]^{2+}$ (*m*/*z* 407.2096), while MALDI-TOF-MS afforded a mono-ionized molecular peak $[M + H]^+$ (*m*/*z* 819.4109) (Fig. 132) [31]. NMR measurements showed only a half set of signals, which indicated that the two new dimers were built up from two equivalent 7,1′-coupled monomeric subunits, but left open the question of whether the two halves were homomorphous or enantiomorphous to each other. The latter possibility (i.e., the presence of achiral *meso*-compounds) was excluded by the optical activities of the compounds and their chiroptical properties, showing that they were C_2-symmetric. The extremely upfield-shifted signals at $\delta_H = 1.80$ ppm for Me-2′ (as compared to ca. $\delta_H = 2.16$ ppm if next to one axis, and ca. $\delta_H = 2.45$ ppm if not next to an axis) indicated the immediate proximity of two aryl substituents, thus suggesting a 3′,3″-coupling between the two identical molecular halves [31]. This was confirmed by a ROESY correlation of the proton signal of Me-2′ with that of OH-4′ (visible only in $CDCl_3$) (Fig. 132). 1D and 2D investigations showed that the two new dimers possessed the same symmetric constitution and the same relative configurations at all of the stereogenic centers and axes in their monomeric halves. Even the absolute configurations of the naphthylisoquinoline moieties were identical, as

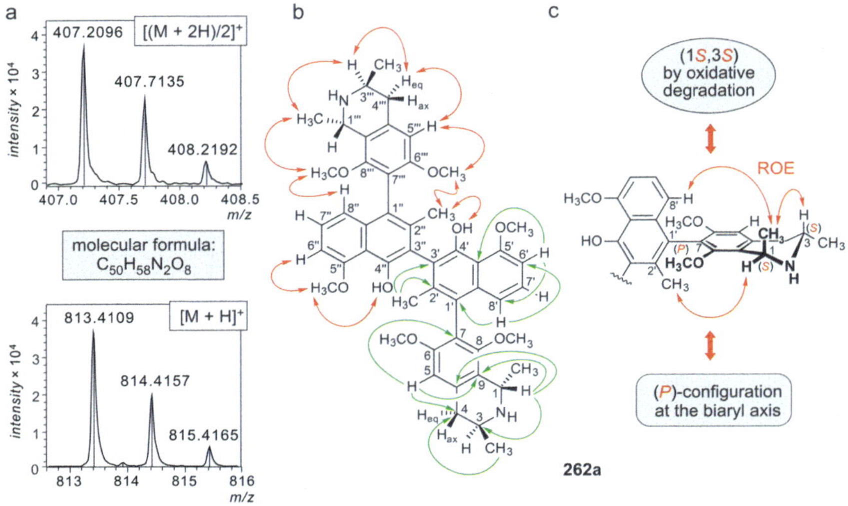

Fig. 132 Shuangancistrotectorine A (**262a**) from *A.* aff. *tectorius* from Hainan Province, China: **a** HRESIMS analysis yielding a doubly protonated ion $[M + 2H]^{2+}$ (*m/z* 407.2096) (top) and the monoprotonated molecular ion $[M + H]^+$ (*m/z* 813.4019) (bottom); **b** NMR ROESY (red double arrows) and HMBC (green single arrows) interactions providing the constitution of **262a** (including the position of the central axis and the outer ones); **c** absolute configuration of **262a** at the axes and centers in the two monomeric halves deduced from selective ROESY interactions in combination with the results of the oxidative degradation

deduced from the oxidative degradation procedure, in conjunction with ROESY measurements [31]. In the chiral tetrahydropyrido part, the two new dimers were characterized by a relative *trans*-array of the methyl groups at C-1 and C-3, with the absolute configuration being (1*S*,3*S*) in the two molecular portions according to the oxidative degradation procedure. The absolute configurations at the outer axes in the two monomeric halves were both assigned to be (*P*) due to ROESY correlations between H-8′ and Me-1′ and between Me-2′ and H-1 [31] (Fig. 132).

ECD measurements showed that the new dimers differ from each other only in the chirality of their central binaphthalene axis as obvious from their virtually mirror-imaged ECD spectra, reflecting the opposite configuration at the main chromophore, the chiral binaphthalene entity. The two new atropo-diastereomeric naphthylisoquinoline dimers were named shuangancistrotectorines A (**262a**) and B (**262b**) according to the name of the plant and the Chinese word *shuang* (= pair, couple) [31]. Based on the C_2-symmetric structure of **262a** and **262b**, the configuration at the central biaryl axis was, first of all, tentatively assigned by application of the Exciton Chirality method [311–314]. Accordingly, shuangancistrotectorine A (**262a**), which has a negative couplet, should be (*M*)-configured, while shuangancistrotectorine B (**262b**), with its positive couplet should have the (*P*)-configuration [31]. This assignment was unambiguously confirmed by quantum-chemical ECD calculations using the TD-B3LYP/6-31* (TD = time dependent) approach [240, 241, 482, 483]. The ECD spectrum calculated for the (*P*,*M*,*P*)-isomer matched well

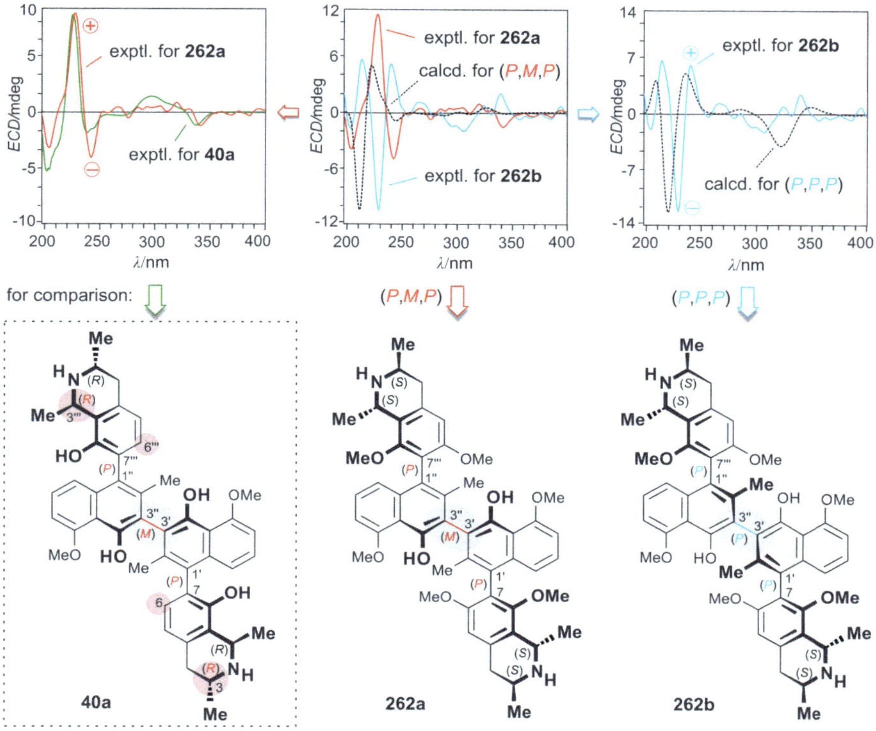

Fig. 133 Assignment of the absolute axial configuration of shuangancistrotectorine A (**262a**) and its atropo-diastereomer shuangancistrotectorine B (**262b**), by comparison of the experimental LC-ECD spectra (monitored in stopped-flow mode) with the ECD curves calculated for the (*P*,*M*,*P*)- and the (*P*,*P*,*P*)-configured atropo-diastereomers by using TD-B3LYP/6-31G*. For comparison, see the structurally related Dioncophyllaceae-type dimer jozimine A_2 (**40a**) and its experimental ECD spectrum

with the experimental ECD curve measured for shuangancistrotectorine A (**262a**), while the spectrum computed for (*P*,*P*,*P*) was in very good agreement with the one monitored for shuangancistrotectorine B (**262b**) (Fig. 133) [31].

The ECD investigations on **262a** and **262b** clearly demonstrated that the chiroptical properties of dimeric naphthylisoquinoline alkaloids are dominated by the orientation of the two naphthalene chromophores to each other, and thus by the configuration at the central biaryl linkage. The strong impact of the central axis, over all other six stereogenic elements—the four stereocenters and even the two outer axes—on the ECD behavior had also been observed for further representatives of 3′,3″-coupled dimeric naphthylisoquinoline alkaloids [122–124]. The central binaphthalene core of **262a** and **262b** closely resembled that of the structurally related and likewise C_2-symmetric, but Dioncophyllaceae-type dimer jozimine A_2 (**40a**) (i.e., (3*R*)-configured, no oxygen function at C-6), which was identified in two Congolese *Ancistrocladus* species [122, 123] and in the West

African liana *A. abbreviatus*, along with a series of three related compounds [124]. Thus, as expected, the ECD spectra of jozimine A_2 (**40a**), which is (*M*)-configured at the central biaryl axis, and shuangancistrotectorine A (**262a**) were found to be nearly identical, whereas **40a** and shuangancistrotectorine B (**262b**) exhibited virtually opposite ECD curves (Fig. 133) [122].

HRESIMS analysis of the third compound isolated from the twig extract of *A.* aff. *tectorius* gave a molecular formula of $C_{50}H_{52}N_2O_8$ (*m/z* 809.3 $[M + H]^+$ and *m/z* 405.2022 $[M + 2]^{2+}$) [31]. The ^1H and ^{13}C NMR data again showed only a half set of signals, indicating that this new dimer is symmetrical, like also the dimers **262a** and **262b** described above. Due to the chirality of the central axis (and because of the optical activity of **262a**, **262b**, and the new dimer), the presence of a *meso*-compound, with two enantiomorphous halves, was excluded and, thus, all these dimers were C_2-symmetric, consisting of two homomorphous halves. The ^1H NMR chemical shifts of the methyl groups at C-1 and C-3 in the new metabolite were typical of a dihydroisoquinoline alkaloid. Specific HMBC correlations revealed the monomeric halves of this new dimer to be 5,1′-coupled, and, again similar to **262a** and **262b**, established the two identical monomeric portions to be connected via a central 3′,3″-bond [17, 31]. This was also confirmed by the extreme high-field shift ($\delta_H = 1.74$ ppm) for the Me-2′ and Me-2″ signals next to two axes, hinting at the proximity of even two aromatic systems, as also found in **262a** and **262b** (see above).

Ruthenium-mediated oxidative degradation of the dimer afforded (*S*)-3-aminobutyric acid, thus establishing the absolute configurations at both, C-3 and C-3‴, as (*S*). The configuration at the axes in the monomeric portions relative to the stereogenic centers became evident from long-range ROESY correlations between H_{eq}-4 and H-8′, and between H_{ax}-4 and Me-2′ (Fig. 134). This, in combination with the absolute configuration at the two stereogenic centers in the isoquinoline subunits, showed that the two naphthalene-isoquinoline linkages of the 5,1′-coupled monomers had to be (*P*)-configured [17, 31].

ECD calculations permitted assignment of the absolute configuration of the new dimer at the central axis. Since the configurations at the outer axes and at the stereocenters at C-3 and C-3‴ were already known, only two stereoarrays (*P*,*M*,*P*) and (*P*,*P*,*P*), were possible; these structures were subjected to a PM3-based conformational analysis. The global minima thus obtained were further optimized with B3LYP/6-31G* and subsequent TDB3LYP/6-31G* calculations affording two ECD spectra, which, by comparison with the experimental one, clearly revealed the new naphthylisoquinoline dimer to be (*M*)-configured at the central biaryl axis (Fig. 134) [31]. In consequence, the compound was assigned as having the full absolute stereostructure **5**, with a (3*S*,5*P*,3′*M*,5‴*P*,3‴*S*)-configuration, as displayed in Fig. 134. The new naphthylisoquinoline dimer **5** was named shuangancistrotectorine C. Interestingly, and in contrast to the shuangancistrotectorines A (**262a**) and B (**262b**), no other stereoisomers of **5** were discovered in the extracts of *A.* aff. *tectorius* [31], suggesting the involvement of a highly specific coupling enzyme in the biosynthesis of this compound.

The attribution of the absolute configuration of shuangancistrotectorine C (**5**) at the central axis is a nice example that only quantum-chemical ECD calculations will

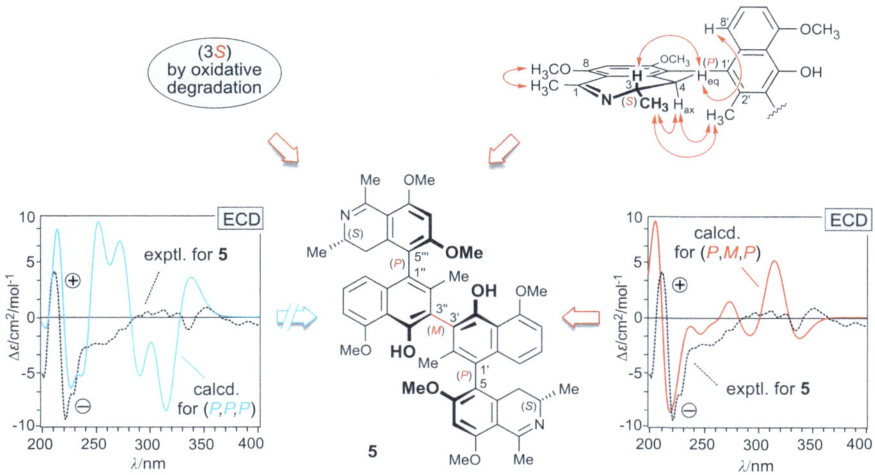

Fig. 134 Shuangancistrotectorine C (**5**), a C_2-symmetric dimer consisting of two 5,1′-coupled naphthylisoquinoline molecular portions. Absolute configuration at the axis and centers in the two monomeric halves deduced from oxidative degradation and selective ROESY interactions (top). Assignment of the absolute configuration at the central axis by comparison of the experimental ECD curve of **5** with the spectra calculated for the (*P*,*M*,*P*)- and (*P*,*P*,*P*)-atropo-diastereomers by using TD-B3LYP/6-31G* (bottom)

give reliable results, whereas the Exciton Chirality approach might lead to wrong assignments for this type of compounds [31, 484]. For the (*P*,*M*,*P*)-configured isolated natural product **5** itself, the assignment using the Exciton Chirality method indeed confirmed the central axis to be (*M*)-configured due to a negative couplet (that is, a first negative Cotton effect at higher wavelength followed by a positive second one). The ECD curve calculated for the (*P*,*P*,*P*)-configured atropo-diastereomer, however, likewise showed a negative exciton couplet, thus, according to the Exciton Chirality method, this would have led to a wrong assignment [31].

From a more polar fraction, yet another new naphthylisoquinoline dimer, compound **207a**, was identified displaying a molecular formula of $C_{50}H_{57}N_2O_8$, as evident from a MALDI-TOF-MS signal at *m*/*z* 813.4 $[M + H]^+$ and the HRESIMS peak at *m*/*z* 407.2091 $[M + 2]^{2+}$. The 1D and 2D NMR data suggested the presence of a further C_2-symmetric naphthylisoquinoline dimer [17, 31]. By detailed NMR analysis, the monomeric half was identified as an N-methylated 7,3′-coupled naphthyltetrahydroisoquinoline, leaving C-1′ and C-1″ as the only possible coupling positions of the central biaryl axis. The ROESY correlation between H-1 and H-3 established the relative configuration at the stereocenters C-1 and C-3 to be *cis* [31]. The absolute configurations were assigned as (1*R*) and (3*S*) by oxidative degradation, which yielded *N*-methyl-D-alanine and *N*-methyl-(3*S*)-aminobutyric acid, thus simultaneously confirming the relative *cis*-configurations assigned by NMR spectroscopy. The axial configuration relative to the stereogenic centers was determined by a ROESY interaction between Me-2′ and H-1, which, together with the absolute

(*R*)-configuration at C-1, indicated the outer axes to be (*M*)-configured [31] (Fig. 135a). In view of the symmetric structure of the alkaloid, both monomeric halves were assigned as (*R*,*S*,*M*)-configured, thus revealing the co-occurring monomer ancistrotanzanine C (**114**) (see Sect. 8.5.6.1), also well-known from *A. tanzaniensis* [189] and *A. heyneanus* [41], to be the molecular subunit of the isolated dimer.

In addition, a further naphthylisoquinoline dimer, compound **207b**, slightly less polar than the previous one, **207a**, was detected [17, 31]. According to HRESIMS, it had the same molecular formula of $C_{50}H_{58}N_2O_8$ and displayed ^{1}H and ^{13}C NMR signals and HMBC cross-peaks identical to those of **207a**, including the absolute configurations at the stereogenic centers, thus suggesting the presence of an atropo-diastereomer of **207a**. However, different from the atropisomeric pair consisting of the shuangancistrotectorines A (**262a**) and B (**262b**), which displayed mirror-imaged ECD spectra, the ECD spectra of the two new dimers **207a** and **207b** were almost identical (Fig. 135b) [17, 31]. Detailed analysis of the ROESY spectrum of **207b** showed a strong diastereomer-specific ROESY interaction between H-8″ and MeO-8, which was absent in **207a** (Fig. 135a). This interaction clearly confirmed that both spin systems of **207b** were on the same side of the naphthalene part (i.e., both above or both below). With respect to the C_2-symmetry of **207b**, its three consecutive biaryl axes could thus only be (*P*,*M*,*P*)- or (*M*,*P*,*M*)-configured, whereas **207a** could only be (*M*,*M*,*M*)- or (*M*,*P*,*M*)-configured [31].

Quantum-chemical ECD calculations finally led to a reliable attribution of the absolute stereostructures of **207a** and **207b** [31]. The global minima of the remaining potential stereostructures of **207a** and **207b** (that is for dimers with (*M*,*M*,*M*), (*M*,*P*,*M*), and (*P*,*M*,*P*)-configuration at the three biaryl axes, and with (1*R*,3*S*)-configuration at the stereogenic centers) were calculated with the PM3 Hamiltonian, followed by a further optimization of the lowest-energy conformers by B3LYP/6-31G*. The excited states of these structures were determined by using the TDB3LYP/6-31G* method [240, 241, 482, 483]. As outlined in Fig. 135b, the experimental ECD spectrum of **207a** matched the curve calculated for the (*M*,*M*,*M*)-isomer nearly perfectly, whereas only the ECD spectrum computed for the (*P*,*M*,*P*)-configuration agreed with the experimental ECD curve monitored for **207b** [17, 31].

In conclusion, the central biaryl axes of the two naphthylisoquinoline dimers **207a** and **207b** were deduced to be (*M*)-configured based on experimental and quantum-chemical ECD investigations. By ROESY measurements in combination with the oxidative degradation procedure the two homomorphous monomeric portions were attributed to be (1*R*,3*S*,7*M*)- (for **207a**) and (1*R*,3*S*,7*P*)-configured (for **207b**). In continuation of the series of naphthylisoquinoline dimers discovered in the twigs of *A.* aff. *tectorius*, these two novel dimers were named shuangancistrotectorines D and E, respectively [31].

The shuangancistrotectorines A–E (**262a**, **262b**, **5**, **207a**, and **207b**) displayed promising inhibitory effects against the chloroquine-resistant K1 strain of the malaria parasite *P. falciparum* [31, 151]. In particular, the shuangancistrotectorines A (**262a**), B (**262b**), and D (**207a**) exhibited very good antiplasmodial activities, which were even more potent than that of the standard drug, chloroquine. Interestingly, shuangancistrotectorine E (**207b**) was significantly less active than its atropo-diastereomer

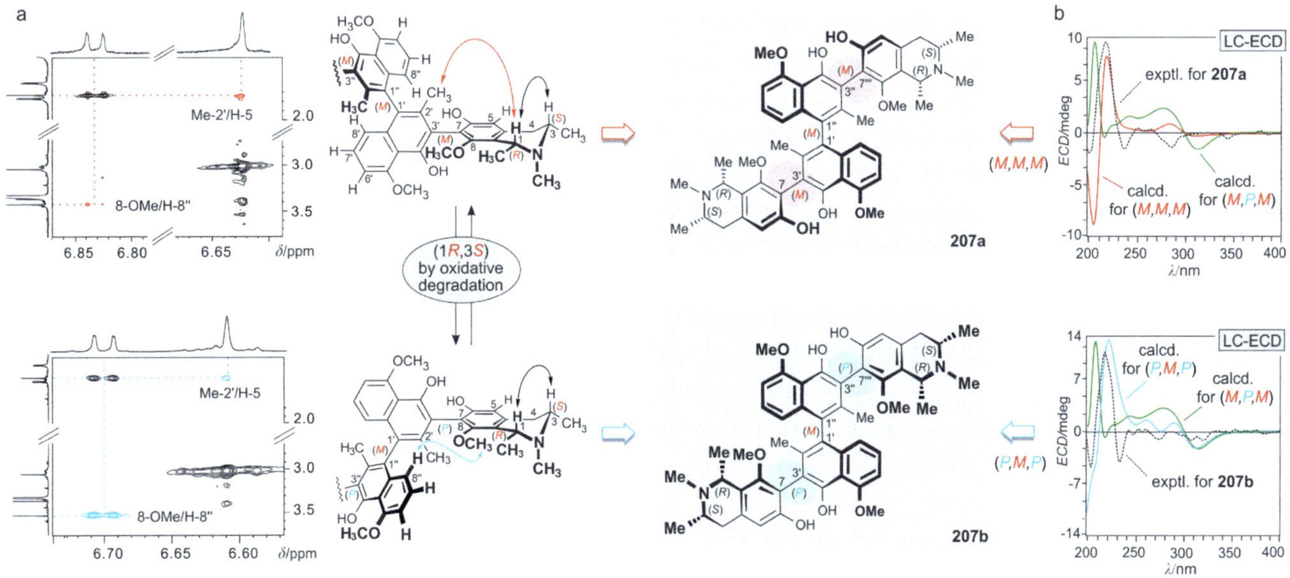

Fig. 135 Shuangancistrotectorines D (**207a**) and E (**207b**) from the Chinese liana *A.* aff. *tectorius*: **a** selected ROESY correlations in **207a** and **207b** and ruthenium-mediated oxidative degradation giving rise to the configurations at centers and axes; **b** assignment of the absolute configuration at the central axis of these two dimers by comparison of their experimental LC-ECD spectra (monitored in stopped-flow mode) with those obtained by quantum-chemical ECD calculations for (*M*,*M*,*M*), (*M*,*P*,*M*), and (*P*,*M*,*P*) by using TD-B3LYP/6-31G*

shuangancistrotectorine D (**207a**), thus again evidencing the importance of axial chirality for the biological properties of naphthylisoquinoline alkaloids.

In summary, the above-described discovery of five novel-type naphthylisoquinoline dimers from *A. tectorius* from Hainan Island strongly underlines the outstanding position of this taxon within the entire Ancistrocladaceae plant family as a most productive species and as a particularly rich source of numerous structurally unique alkaloids that have been found to occur in this Chinese liana, exclusively. Besides the more common C,C-coupled alkaloids (ca. 45) [49–64] and the novel-type dimers [17, 31], three further structurally unprecedented subtypes of naphthylisoquinoline alkaloids were identified, the first quinoid compounds, ancistrotectoquinones A (**74a/b**) and B (**234a/b**) [63], present in the plant as pairs of configurationally semi-stable atropo-diastereomers, N,C-coupled cationic representatives (four compounds and their diastereomers) [18, 33], among them ancistrocladinium A (**6a/6b**) and B (**14a/14b**), and even two pentacyclic N,C-linked alkaloids, the ancistrocyclinones A (**7**) and B (**15**) [18, 33].

The results presented in Sect. 8 clearly show the necessity to gain further reliable knowledge on the remarkable abundance of well-defined *Ancistrocladus* taxa in Southeast Asia and their detailed distribution. The high chemical diversity and the greatly differing patterns of naphthylisoquinoline alkaloids discovered in plants assigned as *A.* aff. *tectorius* (the so-called '*A. tectorius* complex' [36]) emphasize the demand for further research efforts on plants from different local sites, in particular with respect to the still insufficient taxonomic classification of the *Ancistrocladus* genus in Southeast Asia. This is also of great importance for the reproducibility of phytochemical investigations on Asian *Ancistrocladus* species in general. Thus, more-in-depth botanical, phylogenetical, and phytochemical studies on *Ancistrocladus* lianas from Asia will remain a rewarding task in the future.

9 Tables of the Naphthylisoquinoline Alkaloids and Related Compounds Isolated from Asian *Ancistrocladus* Species

So far, about 18 *Ancistrocladus* species (although nowadays reduced to a number of only six botanically acknowledged taxa, see Sect. 3) have been described for Asia [19, 21, 34]. Most of them are found in dry or wet lowland or submontane evergreen forests in Southeast Asia, from Malaysia, Singapore, Sumatra, and (in part) Borneo, through China (Hainan Islands), Vietnam, Laos, Cambodia, Thailand, Myanmar, and the Andaman and Nicobar Islands of India, with two outliers in India and Sri Lanka [19].

The Asian *Ancistrocladus* plants produce a plethora of structurally unique mono- and dimeric naphthylisoquinoline alkaloids (see Fig. 136). Phytochemical investigations on *A. heyneanus* (India) [1–5, 26–29, 37–41], *A. hamatus* (Sri Lanka) [5, 23, 24, 485] (Sect. 4), *A. griffithii* (Thailand) [42, 43, 82] (Sect. 5), *A. benomensis* (Malaysia) [30, 32, 44] (Sect. 6), *A. cochinchinensis* (Vietnam) [45–48, 418, 419] (Sect. 7), and *A. tectorius* (from China, Malaysia, Thailand, Laos, and Vietnam)

(Sect. 8) [31, 33, 49–64] furnished in total an impressive number of 108 naphthylisoquinoline alkaloids and biosynthetically related metabolites. Most of them were C,C-linked naphthylisoquinoline alkaloids (74 compounds), with representatives of all C,C-coupling types (5,1′, 5,3′, 5,8′, 7,1′, 7,3′, 7,6′, and 7,8′) that have so far been discovered in Nature. Twelve representatives with an unusual *N*-iminium-aryl axis such as ancisheynine (**78a/78b**) [18, 40, 256] have been identified, among them unprecedented further cyclized—and, thus, pentacyclic—N,C-coupled naphthylisoquinoline alkaloids like ancistrocyclinone A (**7**) [18, 33]. Furthermore, *A. tectorius* produces quinoid alkaloids (four compounds) that consist of a 1,4-naphthoquinone portion coupled to an isoquinoline moiety like ancistrotectoquinone A (**74a/74b**) [63] (see Fig. 136). Likewise discovered in Asian *Ancistrocladus* plants were constitutionally symmetric dimers (six compounds), with unique structures such as ancistrogriffithine A (**148**) [42, 43, 82], consisting of two 7,8′-linked monomeric naphthylisoquinolines with two stereogenic axes, or shuangancistrotectorine C (**5**) [17, 31], with even three consecutive chiral biaryl linkages, because in this case the central binaphthalene axis is an additional element of chirality (see Fig. 136). Besides the naphthylisoquinoline alkaloids themselves, also a series of related free, non-coupled isoquinolines (five examples) [48, 54] and naphthalenes (eleven compounds) [13, 46, 48, 249, 250, 315, 418, 419, 486] have been isolated. Depending on the individual structures, some of the naphthylisoquinoline alkaloids from Asian plants showed strong cytotoxic activities against human leukemia cells [62, 64], or they displayed pronounced inhibitory effects against the malaria parasite *P. falciparum* [31, 39, 41, 43, 61, 62, 189, 191] and *L. donovani*, the pathogen of visceral leishmaniasis [32, 47, 66–72].

The alkaloid patterns of the Asian *Ancistrocladus* species are characterized by the (nearly) exclusive production of naphthylisoquinolines with (*S*)-configuration at C-3 and an oxygen function at C-6, so-called Ancistrocladaceae-type compounds [13, 15, 17, 18]. There are, however, five non-hydrogenated alkaloids such as 5′-*O*-demethyl-*ent*-dioncophylleine A (**166**) [44] (see Fig. 136) discovered in the Malaysian liana *A. benomensis* (see Sect. 6) unexpectedly lacking an oxygen function at C-6, which thus, from their constitution, belong to the subclass of Dioncophyllaceae-type naphthylisoquinoline alkaloids. They might also be (previous) Ancistrocladaceae-Dioncophyllaceae-inverse hybrid-type compounds (i.e., (3*S*)-configured at C-3 with no oxygen substituent at C-6), but this cannot be concluded since there is no stereocenter at C-3. Apart from these five alkaloids, all the other naphthylisoquinolines identified from *A. heyneanus* [1–5, 26–29, 37–41], *A. hamatus* [5, 23, 24, 484], *A. griffithii* [42, 43, 82], *A. cochinchinensis* [45–48], *A. tectorius* [31, 33, 49–64], and also from *A. benomensis* [30, 32] are typical Ancistrocladaceae-type compounds. Among the 5,1′-coupled naphthylisoquinoline alkaloids, which are predominantly produced by the Asian *Ancistrocladus* lianas investigated so far, are the prominent and widely distributed atropo-diastereomers ancistrocladine (**1a**) and hamatine (**1b**) [1–5, 23, 24] (see Fig. 136), and a whole series of related analogues (34 compounds) like the non-hydrogenated ancistrobenomine A (**4**) [30], which has a hydroxymethylene group at C-3. Representatives of less frequently occurring coupling types such as the 7,1′-linked ancistrocladisine (**2a**) [5, 26–28], the 7,3′-coupled alkaloid

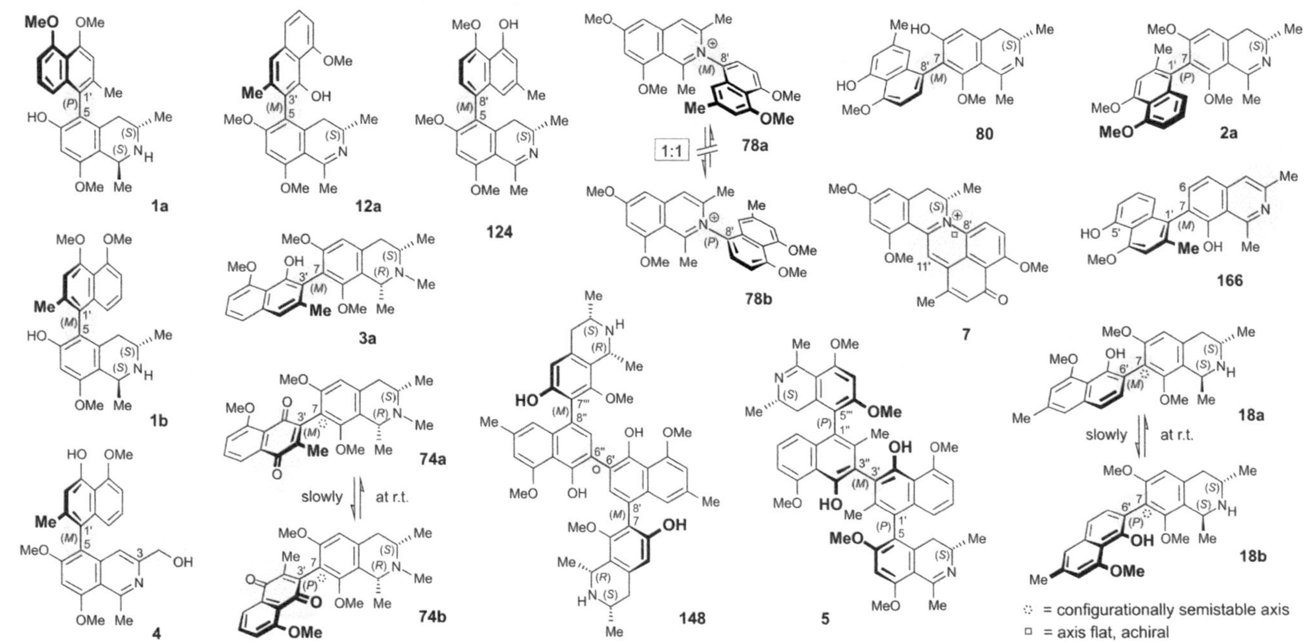

Fig. 136 Mono- and dimeric naphthylisoquinoline alkaloids from Asian *Ancistrocladus* species—a selection of C,C- and N,C-coupled representatives. The different coupling types and unprecedented structural modifications are marked in gray, such as the hydroxymethylene function at C-3 in ancistrobenomine A (**4**), the 1,4-naphthoquinone portion in ancistrotectoquinone A (**74a/74b**), which occurs as a pair of slowly interconverting atropo-diastereomers, or the pentacyclic ring system of ancistrocyclinone A (**7**)

ancistrotectorine (**17a**) [51, 159], and the 7,8′-coupled ancistroheynine A (**80**) [39] were also identified in Asian species (see Fig. 136). Chemotaxonomically remarkable, 5,8′-coupled naphthylisoquinolines such as ancistectorine D (**124**) [62] (see Fig. 136) have been found quite rarely in Asian *Ancistrocladus* species, where only five examples [56, 60, 62] have so far been detected, while alkaloids with this coupling type represent the by far largest group of constituents in most of the Central African taxa [17, 156, 159, 161, 163–181].

Due to recent comprehensive phytochemical investigations mainly by Ye et al. [56, 59], Hua et al. [60], and by the authors' group [31, 33, 61–64], more than 60 naphthylisoquinoline alkaloids have so far been isolated from *Ancistrocladus* aff. *tectorius* collected from the Chinese Hainan Island. A whole series of new, structurally even unique alkaloids solely known to occur in this Chinese liana have been identified, among them the first 7,6′-linked Ancistrocladaceae-type naphthylisoquinoline alkaloids such as ancistrotectorine B (**18a/18b**) [60], which undergoes a slow rotation about its biaryl axis, thus occurring as a mixture of two configurationally semi-stable and, thus, slowly interconverting atropo-diastereomers. Likewise remarkable was the discovery of ancistrotanzanine A (**12a**), along with two further analogues showing the extremely rare 5,3′-coupling type [60]. Prior to the isolation of these three alkaloids from aerial parts of *A. tectorius*, **12a** had been the only 5,3′-linked naphthylisoquinoline known from Nature, identified for the first time in the East African species *A. tanzaniensis* [155].

Tables 5, 6, 7, 8, 9, 10, 11, 12, 13 and 14 comprise a complete list of all of the 74 monomeric C,C-coupled naphthylisoquinoline alkaloids identified so far in the roots, stems, twigs, and leaves of *Ancistrocladus* species from India, Sri Lanka, and Southeast Asia (for reviews, see: [12, 13, 15, 16, 18]), compiled according to their molecular architectures and coupling sites. Tables 15 and 16 show the series of N,C-coupled naphthylisoquinoline alkaloids (twelve representatives) detected in Asian *Ancistrocladus* lianas [18, 32, 33, 47], while Table 17 refers to the six constitutionally symmetric naphthylisoquinoline dimers discovered in *A. griffithii* (one example) [17, 42, 43, 82] and *A. tectorius* (five examples) [17, 31]. Tables 18 and 19 summarize the bicyclic naphthalenes (naphthoquinones and tetralones) (eleven compounds) [13, 46, 48, 249, 250, 315, 418, 419, 486] and free, non-coupled isoquinolines (five representatives) [48, 54] isolated from Asian *Ancistrocladus* plants. The tables provide a compilation of the structures of the presented secondary metabolites discovered in Asian *Ancistrocladus* species, their physical data (melting points, optical rotations), their natural occurrence, and their bioactivities; also included are the spectroscopic, chemical, and computational methods used for their structural elucidation, with special emphasis on analytical procedures applied for the assignment of their complete stereostructures. Where appropriate, the key steps of diverse synthesis approaches are briefly described, as applied in the total synthesis of selected C,C- and N,C-coupled naphthylisoquinoline alkaloids. A synopsis on the basic structures of all of the secondary metabolites isolated from Asian *Ancistrocladus* lianas is given in Fig. 137, with a reference to the respective Tables 5, 6, 7, 8, 9, 10, 11, 12, 13, 14, 15, 16, 17, 18 and 19, in which the individual compounds of the different subgroups are presented in more detail.

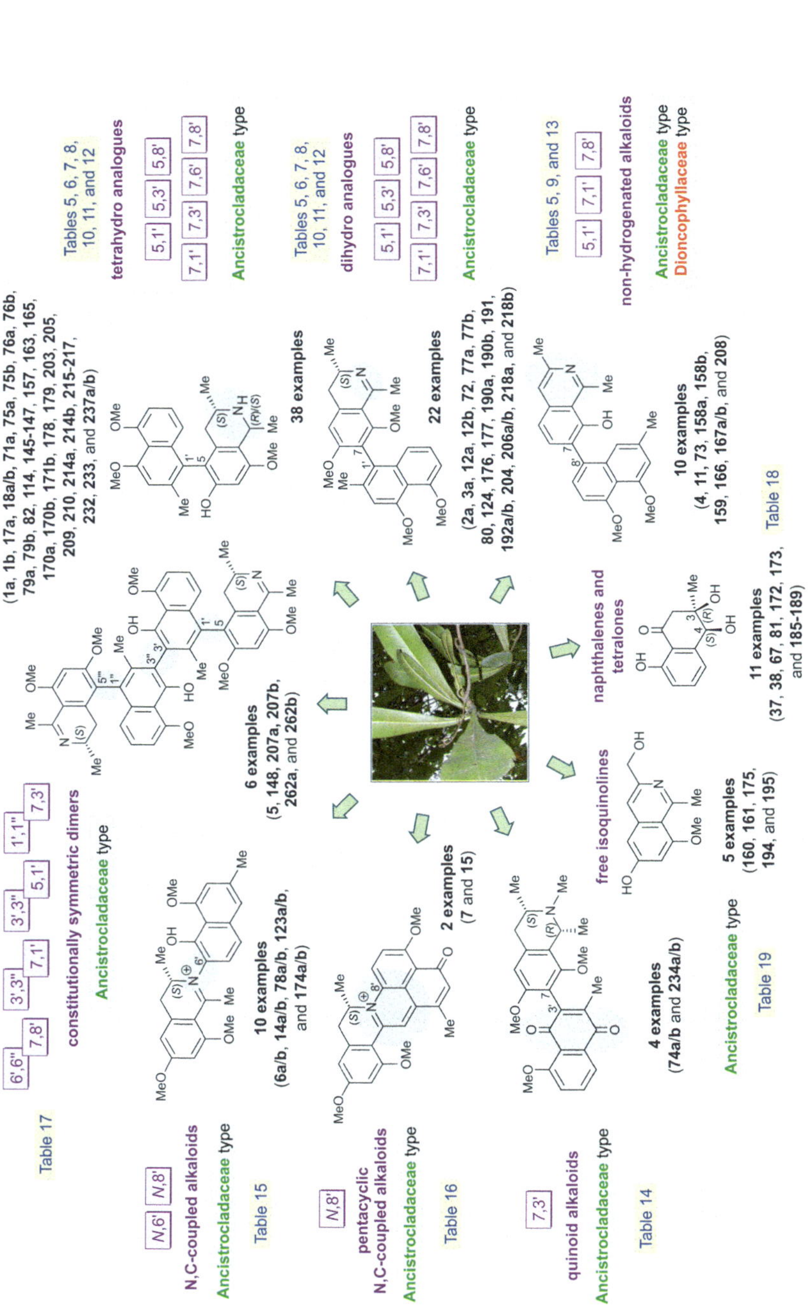

Fig. 137 Structural diversity of naphthylisoquinoline alkaloids and related compounds identified in Asian *Ancistrocladus* species; for reasons of clarity, chiral axes and some (few) of the stereogenic centers in the presented basic structures are drawn "flat". The photograph shows an adult liana of *Ancistrocladus* aff. *tectorius* in its natural habitat in South-Central Vietnam (Photograph: M. Dreyer)

Table 5 5,1′-coupled Ancistrocladaceae-type naphthylisoquinoline alkaloids

Name, formula (mol. wt.), and structure Physical data [*M.p.*/°C, $[\alpha]_D^T$/°cm^2 g^{-1} 10^{-1} (solvent)]	Producing plants and their organs	Isolation, structure elucidation, and further investigations Partial and/or total synthesis	Biological activities
Ancistrocladine (**1a**) $C_{25}H_{29}NO_4$ (407.5) 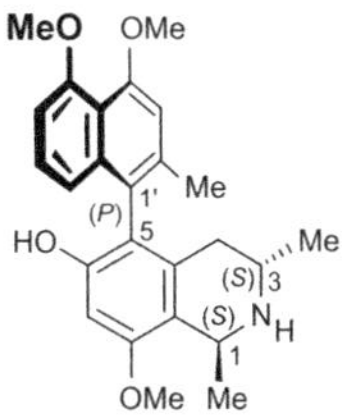 Amorphous solid [1], white heavy cubes (from MeOH): mp 265–267 (dec) [2]; $[\alpha]_D^{25}$ −32.4 (pyridine) [5] As its hydrochloride salt, white needles (from MeOH/acetone): mp 220–224 (dec); $[\alpha]_D^{25}$ −25.5 (*c* 2.29, MeOH) [1, 2] As its hydrobromide salt, white needles (from MeOH): mp 229–231 (dec); $[\alpha]_D^{25}$ −25.51 (*c* 2.29, MeOH) [1] Amorphous solid: mp 265 [487] Amorphous solid: mp 262–263; $[\alpha]_D^{23}$ −20.5 (*c* 014, $CHCl_3$) [50] Colorless amorphous solid (from MeOH): mp 260–263; $[\alpha]_D^{20}$ −15.4 (*c* 0.08, MeOH) [61] Crystalline powder (from MeOH/H_2O): mp 261–263; $[\alpha]_D^{25}$ −21.0 (*c* 0.07, $CHCl_3$) [25] White powder (from EtOH): mp 263–265; $[\alpha]_D^{25}$ −20.0 (*c* 0.15, $CHCl_3$) [192] Amorphous solid (from MeOH): mp 260–263;	*A. heyneanus* (India): roots and aerial parts [1–5, 487] *A. hamatus* (Sri Lanka): roots [5, 316, 485] *A. griffithii* (Thailand): twigs [43] *A. tectorius* (Laos): roots [49] *A. tectorius* (PR China): stems, twigs, and branches [50, 57, 61, 62, 489] *A. robertsoniorum* (Kenya): stem bark [25] *A. abbreviatus* (Ivory Coast): stem bark [13, 192] *A. barteri* (Ivory Coast): roots [13, 194] *A. ileboensis* (DR Congo): root bark [182] Two botanically yet undescribed Congolese *Ancistrocladus* species: root bark [122] and roots [171]	Isolated for the first time from the Indian liana *A. heyneanus* [1–5] Occurrence in *Ancistrocladus* plants along with its atropo-diastereomer hamatine (**1b**) [5, 13, 23, 25, 43, 50, 60, 61, 122, 171, 192, 194, 485], except for *A. heyneanus* [1, 2, 5] and *A. ileboensis* [182] From *A. griffithii*: identification online by LC-MS, and by LC-UV-ECD, by comparison of the structural data of the detected compound with those of reference material [43] *Structural Elucidation*: Constitution and complete stereostructure by MS, NMR, ECD, and X-ray analysis—structure and substitution pattern of ancistrocladine (**1a**) mainly deduced from the spectral data, from degradation experiments, and from spectroscopic studies on related derivatives [1–5] Assignment of the relative *trans*-configuration at the stereocenters at C-1 and C-3 by NMR investigations of **1a** and related derivatives [4, 5] Assignment of the absolute configuration at C-3 as (*S*) by exhaustive ozonolysis to give the stereochemically known compound (3*S*)-aminobutyric acid [4, 5] Determination of the axial configuration by ECD measurements and by application of the Exciton Chirality method on the corresponding fully dehydrogenated, merely axially chiral naphthylisoquinoline ancistrocladeine (**162**), prepared from 6-*O*-methylancistrocladine (**76a**) by dehydrogenation [4, 5] X-ray diffraction analysis of ancistrocladine hydrobromide, evidencing the constitution and the relative configuration [4] Determination of the relative axial configuration by NOE and ROE experiments [73]	Moderate to weak inhibitory activities in vitro against asexual erythrocytic forms of *Plasmodium falciparum* (strains: NF54/64 and K1) [198, 200] No inhibitory activity against asexual erythrocytic forms of the rodent malaria parasite *P. berghei* (Anka strain) [199] No inhibitory activity against exoerythrocytic stages of *P. berghei* (Anka) in vitro in human hepatoma cells (Hep G2) [201] Antitumoral activity of the crude extract of the Chinese *Ancistrocladus* species *A. tectorius* in vivo [50], but no activity of ancistrocladine (**1a**), the main component of the alkaloid fraction, against the human tumor cell line INA-6 (derived from a patient with multiple myeloma) [61] and only very weak activity against drug-sensitive lymphoblastic CCRF-CEM leukemia cells [62]

(continued)

Table 5 (continued)

Name, formula (mol. wt.), and structure Physical data [$M.p.$/°C, $[\alpha]_D^T$/°cm^2 g^{-1} 10^{-1} (solvent)]	Producing plants and their organs	Isolation, structure elucidation, and further investigations Partial and/or total synthesis	Biological activities
$[\alpha]_D^{20}$ −16.0 (c 0.09, MeOH) [122] Colorless amorphous solid (from CH_2Cl_2): mp 260; $[\alpha]_D^{20}$ −19.0 (c 0.10, MeOH) [171] cf.: Compd. prepared by stereoselective total synthesis: mp 261; $[\alpha]_D^{20}$ −23.0 (c 0.16, MeOH) [92, 93] 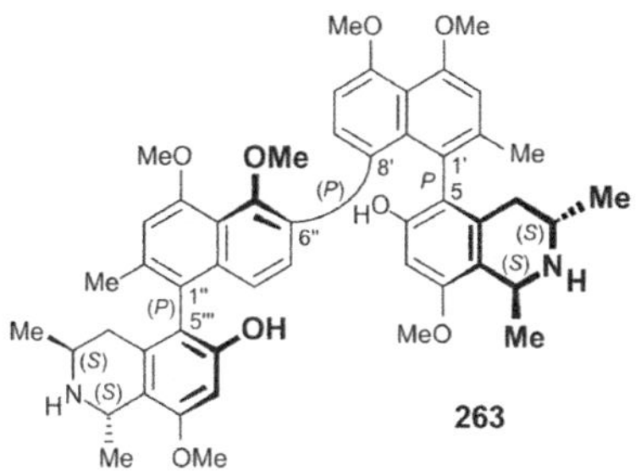 **263** Jozimine B (**263**): Amorphous solid (from MeOH): mp ≥ 229 (dec); $[\alpha]_D^{25}$ −14.7 (c 0.107, $CHCl_3$) [488]		Determination of the axial configuration as (P) by semiempirical ECD calculations using the CNDO/2S method [414, 490] Ab initio calculations of the optical rotatory power [ORP] by means of the Hartree–Fock [HF] and the Density Functional Theory (DFT) methods [491] *Further Investigations*: Localization and identification of ancistrocladine (**1a**) in the outer cortex of the roots of *A. heyneanus* by non-invasive FT-Raman microspectroscopy [349] *Total Synthesis and Related Work*: Stereocontrolled total synthesis of ancistrocladine (**1a**) by applying the lactone method [84, 85, 87–90]: by intramolecular mixed aryl coupling of the ester-type prefixed molecular moieties and reductive lactone cleavage giving rise to **1a** and its likewise natural atropo-diastereomer hamatine (**1b**) [84, 90, 91]; even more stereo-economical: the directed, atropo-divergent synthesis of **1a** and **1b** via the respective cyclic biaryl ethers [92] Oxidative dimerization of **1a** to give the unnatural unsymmetric quateraryl jozimine B (**263**, for its structure, see left column) [84, 86, 125, 488]	

(continued)

Table 5 (continued)

Name, formula (mol. wt.), and structure Physical data [*M.p.*/°C, $[\alpha]_D^T$/°cm^2 g^{-1} 10^{-1} (solvent)]	Producing plants and their organs	Isolation, structure elucidation, and further investigations Partial and/or total synthesis	Biological activities
Hamatine (1b) $C_{25}H_{29}NO_4$ (407.5) 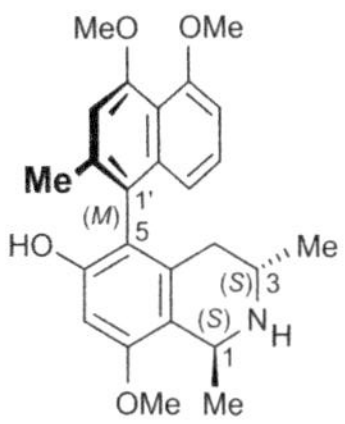 White cubes (from MeOH): mp 250–252; $[\alpha]_D^{20}$ +77.4 (*c* 0.86, $CHCl_3$) [5, 23] Amorphous solid: mp 250–252; $[\alpha]_D^{20}$ +66.1 (*c* 1.07, $CHCl_3$) [50] Colorless amorphous powder (from MeOH): mp 235–237; $[\alpha]_D^{20}$ +47 (*c* 0.09, MeOH) [61] Microcrystalline powder (from MeOH/H_2O): mp 240–242; $[\alpha]_D^{25}$ +16.0 (*c* 0.32, $CHCl_3$) [25] Amorphous solid (from MeOH/acetone): mp 247–248; $[\alpha]_D^{25}$ +64.0 (*c* 0.24, $CHCl_3$) [192] Amorphous solid (from MeOH): mp 235–237; $[\alpha]_D^{20}$ +46.0 (*c* 0,24, $CHCl_3$) [122] Colorless amorphous solid (from CH_2Cl_2): $[\alpha]_D^{20}$ +10.51 (*c* 0.10, MeOH) [171] cf.: Compd. prepared by stereoselective total synthesis: mp 250; $[\alpha]_D^{20}$ +68.0 (*c* 0.04, $CHCl_3$) [92, 93]	*A. hamatus* (Sri Lanka): roots [5, 23, 24, 485] *A. tectorius* (PR China): stem bark and twigs [50, 60–62] *A. griffithii* (Thailand): twigs [43] *A. robertsoniorum* (Kenya): stem bark [25] *A. abbreviatus* (Ivory Coast): stem bark [191, 192] *A. barteri* (Ivory Coast): roots [13, 194] *A. congolensis* (DR Congo): root bark [173] Two botanically yet undescribed Congolese *Ancistrocladus* species: root bark [122] and roots [171]	For the first time discovered in *A. hamatus* from Sri Lanka [5, 23, 24] Occurrence in *Ancistrocladus* plants, along with its atropo-diastereomer ancistrocladine (**1a**) [5, 13, 23, 25, 43, 50, 60, 61, 122, 171, 192, 194, 485] From *A. griffithii*: identification online, by LC-MS, and by LC-UV-ECD, by comparison of the spectroscopic data with those of authentic reference material [43] *Structural Elucidation*: Constitution and complete stereostructure of hamatine (**1b**) by MS, NMR, and ECD [5, 23, 24]—comparison of the spectral data with those obtained from ancistrocladine (**1a**) [5], stereochemically well known from partial and total synthesis [91, 92] Absolute configuration at C-3 established as (*S*) by chemical degradation [23, 79, 80] Attribution of the absolute configuration at C-1 as (*S*) by NMR spectroscopy [23] in combination with oxidative degradation and chemical modification [5, 24] Determination of the relative axial configuration by long-range NOESY experiments [73] Assignment of the absolute axial configuration by ECD measurements and by application of the Exciton Chirality method to the corresponding non-hydrogenated, merely axially chiral naphthylisoquinoline 6-*O*-methylhamateine (**73**), as obtained by 6-*O*-methylation of hamatine (**1b**) and subsequent dehydrogenation [23] Determination of the axial configuration as (*M*) by semiempirical ECD calculations using the CNDO/2S method [490]	Moderate inhibitory activity against asexual erythrocytic forms of *Plasmodium falciparum* (NF54 strain) [200] No inhibitory activity of a ca. 1:10 mixture of the atropo-diastereomers ancistrocladine (**1a**) and hamatine (**1b**) against the development of exoerythrocytic stages of *P. berghei* (Anka) in vitro in human hepatoma cells (Hep G2) [201] Moderate inhibitory activity against the human tumor cell line INA-6 (derived from a patient with multiple myeloma) [61] Weak inhibitory activity against drug-sensitive acute lymphoblastic CCRF-CEM leukemia cells [62] Moderate preferential cytotoxicity against PANC-1 human pancreatic cancer cells under 'austerity' (i.e., nutrient-deprived) conditions [191] Moderate to weak growth-inhibitory effects against human leukemia HL-60 and myelogenous leukemia K562 cells, and against human lung lymphoblast U937 cell lines [60]

(continued)

Table 5 (continued)

Name, formula (mol. wt.), and structure Physical data [*M.p.*/°C, $[\alpha]_D^T$/°cm^2 g^{-1} 10^{-1} (solvent)]	Producing plants and their organs	Isolation, structure elucidation, and further investigations Partial and/or total synthesis	Biological activities
		Ab initio calculations of the optical rotatory power (ORP) by means of the Hartree–Fock (HF) and the Density Functional Theory (DFT) methods [491] Additional confirmation of the structural assignment by conversion of hamatine (**1b**) into its 6-*O*-methyl and *N*-methyl analogues: comparison of the data (MS, NMR, ECD) with those obtained from the respective derivatives of ancistrocladine (**1a**) [23] ——— *Total Synthesis and Related Work*: Stereocontrolled total synthesis of hamatine (**1b**) by applying the lactone method [84, 85, 87–90]: by intramolecular mixed aryl coupling of the ester-type prefixed molecular moieties with subsequent reductive lactone cleavage giving rise to **1b** and its likewise natural atropo-diastereomer ancistrocladine (**1a**) [90, 91]; even more stereo-economical: the directed, atropo-divergent synthesis of **1a** and **1b** via the respective cyclic biaryl ethers [92]	

(continued)

Table 5 5,1′-coupled Ancistrocladaceae-type naphthylisoquinoline alkaloids

Name, formula (mol. wt.), and structure Physical data [*M.p.*/°C, $[\alpha]_D^T$/°cm^2 g^{-1} 10^{-1} (solvent)]	Producing plants and their organs	Isolation, structure elucidation, and further investigations Partial and/or total synthesis	Biological activities
6-*O*-Methylancistrocladine (76a) $C_{26}H_{31}NO_4$ (421.5) 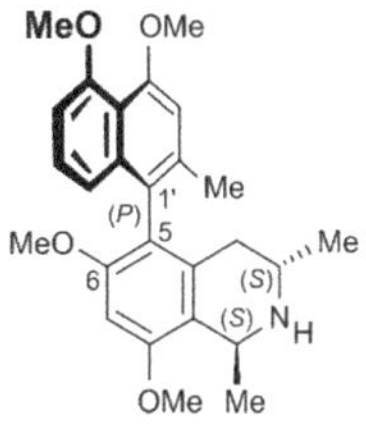 Amorphous solid (from MeOH); $[\alpha]_D^{20}$ +3.3 (*c* 0.12, CH_2Cl_2) [182, 458] cf.: Compd. prepared synthetically: White fluffy needles (from MeOH): mp 200–202 [2] As its hydrochloride salt: white needles (from acetone): mp 315–317 (dec); $[\alpha]_D^{25}$ −56.1 (*c* 1.9, $CHCl_3$) [2] As its hydrochloride salt: colorless prisms (from acetone/*n*-hexane): mp 270–274 (dec); $[\alpha]_D^{25}$ −38.5 (*c* 0.76, $CHCl_3$) [103, 319]	*A. heyneanus* (India): roots [5, 316] *A. hamatus* (Sri Lanka): stem bark [5, 485] *A. tectorius* (PR China): twigs [64] *A. abbreviatus* (Ivory Coast): roots and stem bark [13] *A. ileboensis* (DR Congo): root bark [182, 458]	Isolated for the first time from the Indian liana *A. heyneanus* [5, 316] Metabolic screening of root bark extracts of the newly described [162] Central African plant *A. ileboensis*: detection of **76a** by capillary electrophoresis coupled to ion trap mass spectrometry (CE-MS/MS) [458] *Structural Elucidation*: Constitution and complete stereostructure by comparison of the MS, NMR, and ECD data of **76a** with those of the structurally closely related, likewise (*P*)-configured ancistrocladine (**1a**) [1, 2, 5, 182, 458] *Total Synthesis and Related Work*: Semi-synthesis of **76a** from **1a** via *N*-formyl-6-*O*-methylancistrocladine (not shown) [2] Total synthesis of **76a** by asymmetric construction of the biaryl linkage, using Meyers' oxazoline method, by coupling a naphthalene precursor activated by a chiral oxazoline moiety to a Grignard reagent derived from a protected isoquinoline precursor [103, 319] Formal total synthesis of **76a** by preparation of a suitably substituted chiral naphthyl-3,5-dimethoxybenzaldehyde precursor by asymmetric construction of the biaryl axis between a naphthalene activated by a chiral oxazoline and an aryl Grignard reagent [321]	Biological activities not determined for lack of isolated material

(continued)

Table 5 (continued)

Name, formula (mol. wt.), and structure Physical data [$M.p.$/°C, $[\alpha]_D^T$/°cm^2 g^{-1} 10^{-1} (solvent)]	Producing plants and their organs	Isolation, structure elucidation, and further investigations Partial and/or total synthesis	Biological activities
6-*O*-Methylhamatine (76b) $C_{26}H_{31}NO_4$ (421.5) 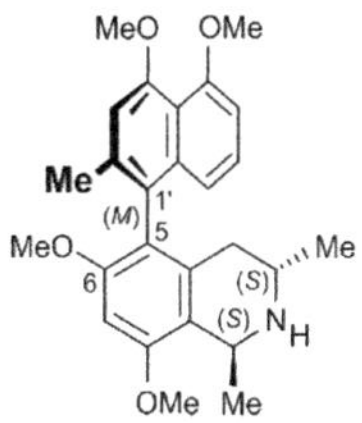 Amorphous solid (from MeOH): $[\alpha]_D^{20}$ +26.2 (*c* 0.15, CH_2Cl_2) [182, 458] cf.: Compd. prepared synthetically: As its hydrochloride salt: white needles (from MeOH/acetone): mp 318–322 (dec); $[\alpha]_D^{20}$ −20.53 [*] (*c* 0.76, $CHCl_3$) [23] As its hydrochloride salt: colorless prims (from CH_2Cl_2/*n*-hexane): mp 271–273 (dec); $[\alpha]_D^{25}$ +28.0 (*c* 0.40, MeOH) [103] [*] The published sign of this optical rotation may be a misprint	*A. hamatus* (Sri Lanka): stem bark [5, 485] *A. tectorius* (PR China): twigs [64] *A. abbreviatus* (Ivory Coast): stem bark [13] and roots [191] *A. congolensis* (DR Congo): root bark [173] *A. ileboensis* (DR Congo): root bark [182, 458] Two botanically yet undescribed Congolese *Ancistrocladus* species: roots [171] and root bark [161]	Isolated for the first time from the Sri Lankan liana *A. hamatus* [5] Metabolic screening of root bark extracts of the newly described [162] Central African plant *A. ileboensis*: detection of **76b** by capillary electrophoresis coupled to ion trap mass spectrometry (CE-MS/MS) [458] *Structural Elucidation*: Constitution and complete stereostructure by comparison of the MS, NMR, and ECD data of **76b** with those of the structurally closely related, likewise (*M*)-configured alkaloid hamatine (**1b**) [5, 161, 171] *Total Synthesis and Related Work*: Semi-synthesis of **76b** from hamatine (**1b**) by 6-*O*-methylation using diazomethane [23] Total synthesis of **76b** by asymmetric construction of the biaryl linkage by applying Meyers' asymmetric oxazoline method [103, 319]	Moderate preferential cytotoxicity against PANC-1 human pancreatic cancer cells under 'austerity' (i.e., nutrient-deprived) conditions [161, 191] Pronounced—and selective—inhibitory activity in vitro against *Plasmodium falciparum* (strain: NF54) [191] Inactive against *Trypanosoma cruzi*, *T. brucei rhodesiense*, and *Leishmania donovani* [191]

(continued)

Table 5 (continued)

Name, formula (mol. wt.), and structure Physical data [*M.p.*/°C, $[\alpha]_D^T$/°cm^2 g^{-1} 10^{-1} (solvent)]	Producing plants and their organs	Isolation, structure elucidation, and further investigations Partial and/or total synthesis	Biological activities
Ancistectorine A$_2$ (75a) $C_{25}H_{29}NO_4$ (407.5) 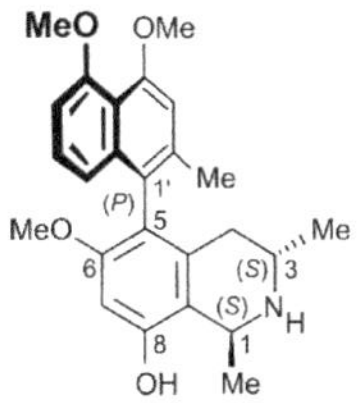 Compd. can likewise be addressed as 6-*O*-methyl-8-*O*-demethyl-ancistrocladine Colorless amorphous powder (from MeOH): mp 235; $[\alpha]_D^{20}$ −17.0 (*c* 0.09, MeOH) [61]	*A. tectorius* (PR China): twigs [61]	Occurrence in the Chinese liana *A. tectorius*, along with its atropo-diastereomer 5-*epi*-ancistectorine A$_2$ (**75b**) [61] *Structural Elucidation*: Constitution, coupling position, and relative *trans*-configuration of the two methyl groups at C-1 versus C-3 evidenced by HRESIMS and by ^{1}H and ^{13}C NMR, including specific HMBC and NOE correlations [61] Absolute configurations at C-1 and C-3 established as (*S*) by oxidative degradation [61] Assignment of the absolute axial configuration as (*P*) due to specific NOE interactions in combination with the absolute (1*S*,3*S*)-stereoarray in the tetrahydroisoquinoline portion, and by the nearly identical ECD spectrum of **75a** compared to that of the known, likewise 5,1′-coupled and (*P*)-configured alkaloid ancistrocladine (**1a**) [1, 2, 5, 61, 191]	Strong—and selective—inhibitory activity against the K1 strain [61] of *Plasmodium falciparum*, combined with a very weak cytotoxicity, leading to a high selectivity index, thus to be considered as a lead compound according to the TDR-WHO guidelines [238] Inactive against *Trypanosoma cruzi*, *T. brucei rhodesiense*, *T. brucei brucei*, and *Leishmania donovani* [61]
5-*epi*-Ancistectorine A$_2$ (75b) $C_{25}H_{29}NO_4$ (407.5) 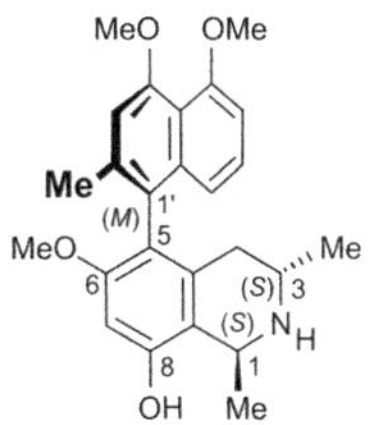 Compd. can likewise be addressed as 6-*O*-methyl-8-*O*-demethylhamatine Colorless amorphous powder (from MeOH): mp 230; $[\alpha]_D^{20}$ +24.4 (*c* 0.10, MeOH) [61]	*A. tectorius* (PR China): twigs [61] *A. abbreviatus* (Ivory Coast): roots [191]	Isolated for the first time from *A. tectorius* from the Chinese Hainan Island [61] Occurrence in the Chinese liana *A. tectorius*, along with its atropo-diastereomer, ancistectorine A$_2$ (**75a**) [61] *Structural Elucidation*: Constitution, coupling position, and relative *trans*-configuration of the two methyl groups at C-1 versus C-3 evidenced by HRESIMS and by ^{1}H and ^{13}C NMR, including specific HMBC and NOE correlations [61, 191] Absolute configurations at C-1 and C-3 established as (*S*) by oxidative degradation [61, 191] Assignment of the absolute axial configuration as (*M*) due to specific NOE interactions in combination with the absolute (1*S*,3*S*)-stereoarray in the tetrahydroisoquinoline portion, and by the nearly identical ECD spectrum of **75b** compared to that of the structurally closely related known, likewise	Strong—and selective—inhibitory activity against the K1 [61] and the NF54 strains [191] of *Plasmodium falciparum*, combined with a very weak cytotoxicity, leading to a high selectivity index, thus considered as a lead compound according to the TDR-WHO guidelines [238] Inactive against *Trypanosoma cruzi*, *T. brucei rhodesiense*, and *Leishmania donovani* [61] Weak inhibitory activity in vitro against the clinically relevant canine hematozoan parasite *Babesia canis* [459] Strong preferential cytotoxicity against PANC-1 human pancreatic cancer cells

(continued)

Table 5 (continued)

Name, formula (mol. wt.), and structure Physical data [*M.p.*/°C, $[\alpha]_D^T$/°cm^2 g^{-1} 10^{-1} (solvent)]	Producing plants and their organs	Isolation, structure elucidation, and further investigations Partial and/or total synthesis	Biological activities
		5,1′-coupled and (*M*)-configured alkaloid hamatine (**1b**) [23, 24, 43, 61, 191]	under 'austerity' (i.e., nutrient-deprived) conditions [191]
4′-*O*-Demethylancistrocladine (**170a**) $C_{24}H_{27}NO_4$ (393.5) 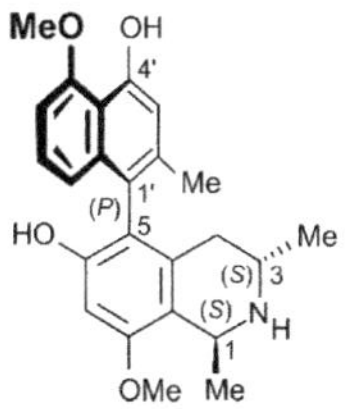 Gum (from MeOH/CH_2Cl_2): $[\alpha]_D^{20}$ +2.50 (*c* 1.00, MeOH) [54]	*A. tectorius* (Malaysia): stem bark [54] *A. tectorius* (PR China): twigs [64] *A. cochinchinensis* (Vietnam): root bark and leaves [47] An as yet botanically unknown Congolese *Ancistrocladus* species: roots [161]	Isolated for the first time from *A. tectorius* from Malaysia collected in Mersing Johor [54] Occurrence in the Chinese liana *A. tectorius* along with its atropo-diastereomer 4′-*O*-demethylhamatine (**170b**) [59] *Structural Elucidation*: Metabolic screening of root bark and leaf extracts of the Vietnamese species *A. cochinchinensis* using the analytical triad LC-MS/MS, LC-NMR (^{1}H, WET-ROESY), and LC-ECD: constitution and complete stereostructure of **170a** elucidated online, directly from crude extracts [47] Absolute configurations at C-1 and C-3 established as (*S*) by oxidative degradation [47, 64, 161] Constitution and absolute stereostructure by comparison of the MS, NMR, and ECD data of **170a** with those of the structurally closely related, likewise (*P*)-configured alkaloid ancistrocladine (**1a**), revealing **170a** to differ from its parent compound **1a** by the absence of an *O*-methyl group in the 4′-position [54, 64, 161]	Strong preferential cytotoxicity against PANC-1 human pancreatic cancer cells under 'austerity' (i.e., nutrient-deprived) conditions [161]

(continued)

Table 5 (continued)

Name, formula (mol. wt.), and structure Physical data [$M.p.$/°C, $[\alpha]_D^T$/°cm^2 g^{-1} 10^{-1} (solvent)]	Producing plants and their organs	Isolation, structure elucidation, and further investigations Partial and/or total synthesis	Biological activities
4′-*O*-Demethylhamatine (170b) $C_{24}H_{27}NO_4$ (393.5) 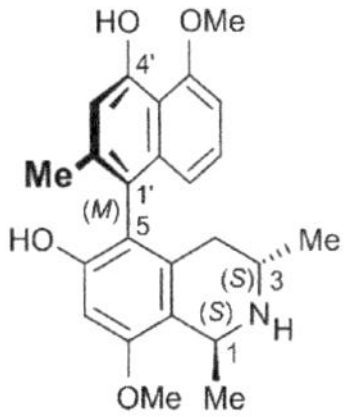 Pale-yellow amorphous powder (from MeOH): $[\alpha]_D^{20}$ +53.6 (*c* 0.14, MeOH) [59]	*A. tectorius* (PR China): stems and leaves [59]	Occurrence in the Chinese liana *A. tectorius*, along with its atropo-diastereomer 4′-*O*-demethyl-ancistrocladine (**1a**) [59] *Structural Elucidation*: Constitution and complete stereostructure by comparison of the MS, NMR, and ECD data of **170b** with those of the structurally closely related, likewise (*M*)-configured alkaloid hamatine (**1b**), revealing **170b** to differ from its parent compound **1b** by the absence of an *O*-methyl group in the 4′-position [59]	Biological activities not determined for lack of isolated material
4′-*O*-Demethyl-6-*O*-methylancistrocladine (214a) $C_{25}H_{29}NO_4$ (407.5) 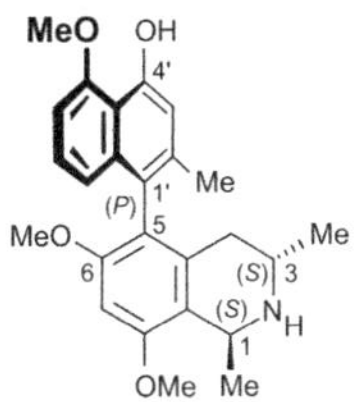 Light-yellow powder (from $CHCl_3$): $[\alpha]_D^{25}$ −3.14 (*c* 0.38, MeOH) [56] Yellow amorphous solid (from MeOH): $[\alpha]_D^{23}$ −16.7 (*c* 0.07 MeOH) [459]	*A. tectorius* (PR China): stem bark, twigs, and leaves [56] *A. abbreviatus* (Ivory Coast): roots [459]	Isolated for the first time from *A. tectorius* from the Chinese Hainan Island [56] *Structural Elucidation*: Constitution, coupling position, and relative *trans*-configuration at C-1 versus C-3 by HRESIMS and by ^{1}H and ^{13}C NMR, including HMBC and NOESY correlations, indicative of the presence of a 5,1′-linked naphthyltetrahydroisoquinoline with a free hydroxy group at C-4′ [56, 459] Absolute configurations at C-3 established as (*S*) by oxidative degradation [459] Assignment of the axial configuration as (*P*) due to specific NOESY correlations in combination with the absolute (1*S*,3*S*)-stereoarray in the tetrahydroisoquinoline portion, and by the nearly identical ECD spectrum of **214a** compared to that of the likewise (*P*)-configured alkaloid ancistrocladine (**1a**) [56, 459]	Moderate inhibitory activity in vitro against the clinically relevant canine hematozoan parasite *Babesia canis* [459]

(continued)

Table 5 (continued)

Name, formula (mol. wt.), and structure Physical data [*M.p.*/°C, $[\alpha]_D^T$/°cm^2 g^{-1} 10^{-1} (solvent)]	Producing plants and their organs	Isolation, structure elucidation, and further investigations Partial and/or total synthesis	Biological activities
4′-*O*-Demethyl-6-*O*-methylhamatine (214b) $C_{25}H_{29}NO_4$ (407.5) 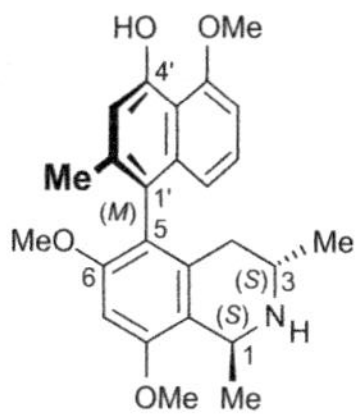 Light-yellow powder (from $CHCl_3$): $[\alpha]_D^{25}$ +12.4 (*c* 0.36, MeOH) [56]	*A. tectorius* (PR China): stem bark, twigs, and leaves [56] *A. abbreviatus* (Ivory Coast): roots [191]	Isolated for the first time from *A. tectorius* from the Chinese Hainan Island [56] *Structural Elucidation*: Constitution, coupling position, and relative *trans*-configuration at C-1 versus C-3 by HRESIMS and by ^{1}H and ^{13}C NMR, including HMBC and NOESY correlations indicative of the presence of a 5,1′-linked naphthyltetrahydroisoquinoline with a free hydroxy group at C-4′ [56, 191] Absolute configuration at C-3 established as (*S*) by oxidative degradation [191] Assignment of the axial configuration as (*M*) due to specific NOESY correlations in combination with the absolute (1*S*,3*S*)-stereoarray in the tetrahydroisoquinoline portion, and by the nearly identical ECD spectrum of **214b** compared to that of the likewise (*M*)-configured alkaloid hamatine (**1b**) [23, 24, 56, 191]	Strong preferential cytotoxicity against PANC-1 human pancreatic cancer cells under 'austerity' (i.e., nutrient-deprived) conditions [191] Weak inhibitory activity in vitro against the clinically relevant canine hematozoan parasite *Babesia canis* [459]

(continued)

Table 5 (continued)

Name, formula (mol. wt.), and structure Physical data [*M.p.*/°C, $[\alpha]_D^T$/°cm^2 g^{-1} 10^{-1} (solvent)]	Producing plants and their organs	Isolation, structure elucidation, and further investigations Partial and/or total synthesis	Biological activities
4′-*O*-Demethylancistectorine A$_2$ (82) $C_{24}H_{27}NO_4$ (393.5) 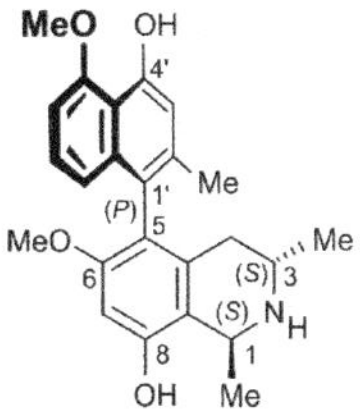 Compd. can likewise be addressed as 4′,8-*O*,*O*-didemethyl-6-*O*-methylancistrocladine White amorphous solid (from MeOH): mp 181; $[\alpha]_D^{20}$ −15.1 (*c* 0.10, MeOH) [64]	*A. tectorius* (PR China): twigs [64] *A. heyneanus* (India): roots [70]	Isolated for the first time from *A. tectorius* from the Chinese Hainan Island [64] *Structural Elucidation*: Constitution, coupling position, and relative *trans*-configuration at C-1 versus C-3 by HRESIMS and by ^{1}H and ^{13}C NMR, including HMBC and NOESY correlations, revealing **82** to differ from its parent compound, ancistectorine A$_2$ (**75a**) [61], by the absence of an *O*-methyl group in the 4′-position [64] Absolute configurations at C-1 and C-3 established as (*S*) by oxidative degradation [64] Assignment of the absolute axial configuration as (*P*) due to specific NOE interactions in combination with the relative *trans*-configuration at C-1 versus C-3, and by the very similar ECD spectrum of **82** compared to that of the likewise (*P*)-configured parent compound ancistectorine A$_2$ (**75a**) [61], and, as expected, virtually opposite to that of its (*M*)-configured atropo-diastereomer, 5-*epi*-ancistectorine A$_2$ (**75b**) [61, 64]	Biological activities not determined for lack of isolated material

(continued)

Table 5 (continued)

Name, formula (mol. wt.), and structure Physical data [*M.p.*/°C, $[\alpha]_D^T$/°cm^2 g^{-1} 10^{-1} (solvent)]	Producing plants and their organs	Isolation, structure elucidation, and further investigations Partial and/or total synthesis	Biological activities
5′-*O*-Demethylhamatine (217) $C_{24}H_{27}NO_4$ (393.5) 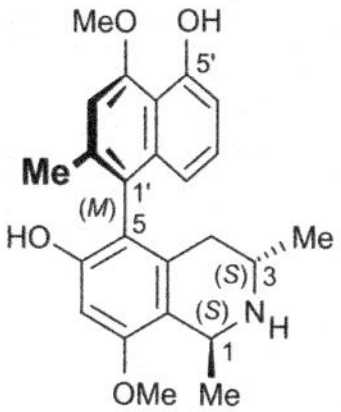 Colorless amorphous solid (from acetonitrile): mp 174–180; $[\alpha]_D^{25}$ +26.0 (*c* 0.10, MeOH) [171]	*A. tectorius* (PR China): twigs [64] *A. congolensis* (DR Congo): root bark [173] An as yet botanically unknown Central Congolese *Ancistrocladus* species: roots [171]	Isolated for the first time from a botanically yet undescribed *Ancistrocladus* taxon collected in the rainforest Yeteto of the Ikela region in the Democratic Republic of the Congo [171] *Structural Elucidation*: Constitution, coupling position, and relative *trans*-configuration at C-1 versus C-3 by HRESIMS and by ^{1}H and ^{13}C NMR, including HMBC and NOESY correlations, revealing **217** to differ from its parent compound hamatine (**1b**) by the absence of an *O*-methyl group in the 5′-position [64, 171, 173] Absolute configurations at C-1 and C-3 established as (*S*) by oxidative degradation [171] Assignment of the axial configuration as (*M*) due to specific NOESY correlations in combination with the absolute (1*S*,3*S*)-stereoarray in the tetrahydroisoquinoline portion, and by the nearly identical ECD spectrum of **217** compared to that of the likewise (*M*)-configured alkaloid hamatine (**1b**) [64, 171, 173]	Weak to moderate inhibitory activities in vitro against *Plasmodium falciparum* (K1 strain), *Leishmania donovani*, *Trypanosoma cruzi*, and *T. brucei rhodesiense* [171] Weak inhibitory activity against drug-sensitive lymphoblastic CCRF-CEM leukemia cells [64]

(continued)

Table 5 (continued)

Name, formula (mol. wt.), and structure Physical data [*M.p.*/°C, $[\alpha]_D^T$/°cm^2 g^{-1} 10^{-1} (solvent)]	Producing plants and their organs	Isolation, structure elucidation, and further investigations Partial and/or total synthesis	Biological activities
Isoancistrocladine (**79a**) $C_{25}H_{29}NO_4$ (407.5) 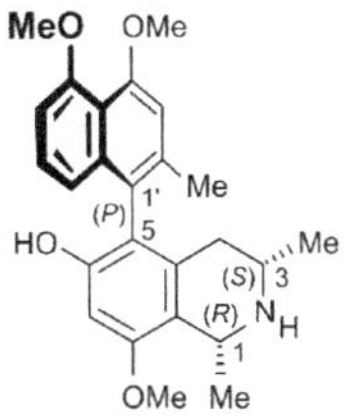 Colorless crystals (from MeOH): mp 230–232; $[\alpha]_D^{20}$ +59.5 (*c* 0.60, $CHCl_3$) [38] cf.: Compd. prepared by semi-synthesis: Pale-yellow needles (from MeOH): mp 235–236; $[\alpha]_D^{25}$ +61.8 (*c* 1.76, $CHCl_3$) [4, 37]	*A. heyneanus* (India): roots [38]	*Structural Elucidation*: Constitution and complete stereostructure by comparison of the MS, NMR, and ECD data of **79a** with those of the structurally closely related, likewise 5,1′-linked and (*P*)-configured alkaloid ancistrocline (**163**), which is stereochemically well known from semi- and total synthesis [53, 84, 85], revealing **79a** to differ from its parent compound **163** by the absence of an *N*-methyl group [4, 38] Additional proof of the absolute axial configuration by quantum-chemical ECD calculations and comparison of the predicted spectrum with the experimental ECD curve of **79a** [492] *Partial Synthesis*: Semi-synthesis of **79a** by *cis*-selective reduction of the corresponding dihydroisoquinoline alkaloid ancistrocladinine (**77a**) using $NaBH_4$ [4, 5, 37]	Biological activities not determined for lack of isolated material

(continued)

Table 5 (continued)

Name, formula (mol. wt.), and structure Physical data [$M.p.$/°C, $[\alpha]_D^T$/°cm^2 g^{-1} 10^{-1} (solvent)]	Producing plants and their organs	Isolation, structure elucidation, and further investigations Partial and/or total synthesis	Biological activities
Isohamatine (**79b**) $C_{25}H_{29}NO_4$ (407.5) 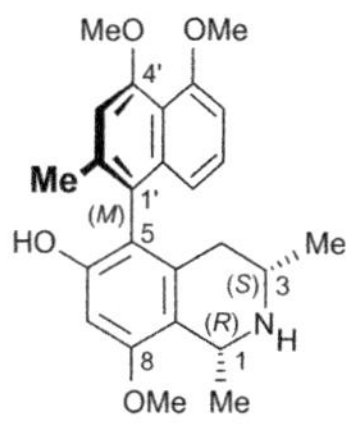 Compd. can likewise be addressed as 4-*O*-methyl-5-*epi*-ancis-tectorine A_1 or as 8-*O*-methyl-5-*epi*-ancistectorine A_2 ——— Pale-yellow amorphous powder (from MeOH): $[\alpha]_D^{25}$ −52.7 (*c* 0.20, MeOH) [48]	*A. cochinchinensis* (Vietnam): leaves [48]	*Structural Elucidation*: Constitution, coupling position, and relative *cis*-configuration of the two methyl groups at C-1 versus C-3 evidenced by HRESIMS and by ^{1}H and ^{13}C NMR, including specific HMBC and NOE correlations [48] Assignment of the absolute axial configuration as (*M*) due to the nearly identical ECD spectrum of **79b** compared to that of the structurally closely related, likewise 5,1′-coupled, and also (*M*)-configured, alkaloid 5-*epi*-ancistectorine A_2 (**75b**) [48, 61] Further structural proof by comparison of the spectroscopic data (MS, NMR, ECD) of **79b** with those obtained from the related known [61] 5,1′-linked alkaloids ancistectorines A_1 (**203**) and A_2 (**75a**) [48] Additional proof of the absolute axial configuration by quantum-chemical ECD calculations and comparison of the predicted spectrum with the experimental ECD curve of **79b** [492] ——— *Partial synthesis*: Semi-synthesis of **79b** by *cis*-selective reduction of the corresponding dihydroisoquinoline alkaloid hamatinine (**77b**) using $NaBH_4$ [492]	Strong antiproliferative activity against several human cancer cell lines, among them HL-60 (promyelocytic leukemia), LU-1 (lung carcinoma), and SK-MEL-2 (melanoma) cells [48]

(continued)

Table 5 (continued)

Name, formula (mol. wt.), and structure / Physical data [*M.p.*/°C, $[\alpha]_D^T$/°cm^2 g^{-1} 10^{-1} (solvent)]	Producing plants and their organs	Isolation, structure elucidation, and further investigations / Partial and/or total synthesis	Biological activities
Ancistrocline (163) $C_{26}H_{31}NO_4$ (421.5) 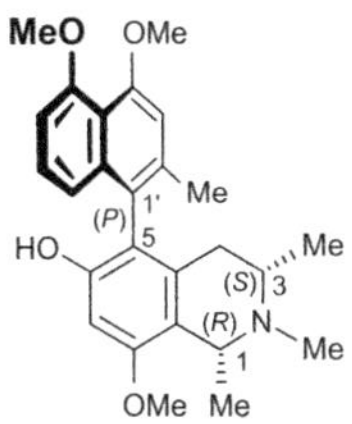 Amorphous solid (from MeOH): mp 227–228; $[\alpha]_D^{25}$ +61.7 (*c* 2.11, $CHCl_3$) [50] Colorless amorphous powder (from MeOH): mp 220–222; $[\alpha]_D^{20}$ +58.6 (*c* 1.30, $CHCl_3$) [61] Amorphous solid (from MeOH): mp 223–224; $[\alpha]_D^{25}$ +62.0 (*c* 0.15, $CHCl_3$) [465, 492] Amorphous solid (from MeOH): 229–233 [58] cf.: Compd. prepared by semi-synthesis [53] and by total synthesis [83, 84, 493]: mp 223–224; $[\alpha]_D^{25}$ +59.1 (*c* 0.23, $CHCl_3$)	*A. tectorius* (PR China): stems, stem bark, twigs, and branches [50, 53, 58, 61, 62, 64, 465, 489]	*Structural Elucidation*: Constitution, coupling position, and relative *cis*-configuration of the two methyl groups at C-1 versus C-3 evidenced by HRESIMS and by ^{1}H and ^{13}C NMR, including specific HMBC and NOE correlations [50, 53, 58, 61, 465, 493] Absolute configuration at C-3 established as (*S*) by oxidative degradation [53, 61, 62, 64, 465] Assignment of the absolute axial configuration as (*P*) due to specific NOE interactions in conjunction with the absolute (1*R*,3*S*)-stereoarray in the tetrahydroisoquinoline portion [53, 61, 62, 64, 465], and by the nearly identical ECD spectrum of **163** compared to that of the known, 5,1′-coupled—and likewise (*P*)-configured—alkaloid ancistrocladine (**1a**) [5, 61, 465, 493] *Total Synthesis and Related Work*: Semi-synthesis of **163** by stereoselective reduction of ancistrocladinine (**77a**) using $NaBH_4$, followed by reductive *N*-methylation [53] Convergent total synthesis of the (unnatural) 6-*O*-methyl-derivative of ancistrocline (**163**) by applying Meyers' oxazoline method—confirmation of the absolute stereostructure of the synthesis target molecule at the stereocenters by ruthenium-mediated oxidative degradation and, at the axis, by ECD calculations [322] Total synthesis of ancistrocline (**163**) by applying the lactone methodology [83, 84, 493]	Moderate inhibitory activity against the human tumor cell line INA-6 (derived from a patient with multiple myeloma) [61] Weak inhibitory activity against drug-sensitive lymphoblastic CCRF-CEM leukemia cells [62] Pronounced inhibitory effect against *Staphylococcus aureus* [489]

(continued)

Table 5 (continued)

Name, formula (mol. wt.), and structure Physical data [*M.p.*/°C, $[\alpha]_D^T$/°cm^2 g^{-1} 10^{-1} (solvent)]	Producing plants and their organs	Isolation, structure elucidation, and further investigations Partial and/or total synthesis	Biological activities
Ancistrobertsonine B (**216**) $C_{27}H_{33}NO_4$ (435.6) 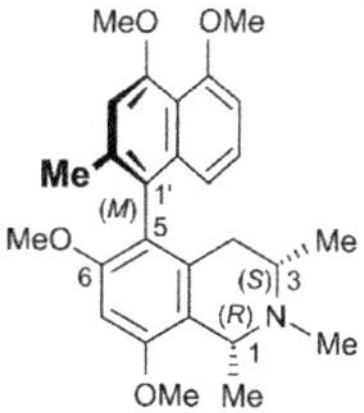 Compd. can likewise be addressed as 6-*O*-methyl-5-*epi*-ancistrocline Amorphous light-yellow solid (from *n*-hexane/*i*-propanol): mp 192–193; $[\alpha]_D^{25}$ +4.0 (*c* 0.09, $CHCl_3$) [190] Amorphous solid: mp 190–191 [58]	*A. tectorius* (PR China): stems, branches, and leaves [58, 60] *A. robertsoniorum* (Kenya): stem bark [190]	For the first time isolated from the Kenyan liana *A. robertsoniorum* by high-speed countercurrent chromatography (HSCCC) [190] *Structural Elucidation*: Constitution and coupling position deduced from HRESIMS and from ^{1}H and ^{13}C NMR, including specific HMBC and long-range ROESY interactions, indicative of a 5,1′-coupled *O*,*N*-permethylated naphthyltetrahydroisoquinoline alkaloid [58, 60, 190] Absolute configuration at C-3 established as (*S*) by oxidative degradation [190] Absolute configuration at C-1 determined as (*R*) due to the relative *cis*-configuration at C-1 versus C-3 as deduced from NOE interactions in combination with the given absolute (*S*)-configuration at C-3 [190] Assignment of the absolute axial configuration as (*M*) due to specific ROESY correlations in conjunction with the known absolute (1*R*,3*S*)-stereoarray in the tetrahydroisoquinoline part, confirmed by the nearly mirror-image like ECD spectrum of **216** compared to that of the related, likewise 5,1′-coupled alkaloid ancistrocline (**163**), which is (*P*)-configured [53]	Moderate to weak activity in vitro against *Plasmodium falciparum* (strains: K1, NF54) [190] Moderate to weak growth-inhibitory effects against human leukemia HL-60, K562, and human macrophage U937 cell lines [60]

(continued)

Table 5 (continued)

Name, formula (mol. wt.), and structure Physical data [$M.p.$/°C, $[\alpha]_D^T$/°cm^2 g^{-1} 10^{-1} (solvent)]	Producing plants and their organs	Isolation, structure elucidation, and further investigations Partial and/or total synthesis	Biological activities
Ancistectorine A$_1$ (203) $C_{24}H_{27}NO_4$ (393.5) 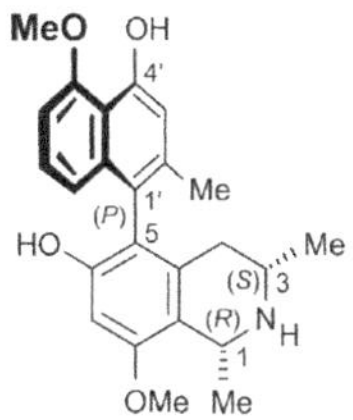 Compd. can also be considered as 4′-*O*-demethylisoancistrocladine ——— Colorless amorphous powder (from MeOH): mp 190; $[\alpha]_D^{25}$ +19.1 (*c* 0.10, MeOH) [61]	*A. tectorius* (PR China): twigs [61]	*Structural Elucidation*: Constitution, coupling position, and relative *cis*-configuration of the two methyl groups at C-1 versus C-3 evidenced by HRESIMS and by ^{1}H and ^{13}C NMR, including specific HMBC and NOE correlations [61] Absolute configuration at C-3 established as (*S*) by oxidative degradation [61] Assignment of the absolute axial configuration as (*P*) due to specific NOE interactions in combination with the absolute (1*R*,3*S*)-stereoarray in the tetrahydroisoquinoline portion, and by the virtually identical ECD spectrum of **203** compared to that of the known [50, 53], structurally related and likewise 5,1′-coupled and (1*R*,3*S*,*P*)-configured ancistrocline (**163**) [61]	Weak inhibitory activity in vitro against *Plasmodium falciparum* (strain: K1) [61] No inhibitory activities against *Trypanosoma cruzi*, *T. brucei rhodesiense*, and *Leishmania donovani* [61]
***N*-Methylancistectorine A$_1$ (215)** $C_{25}H_{29}NO_4$ (407.5) 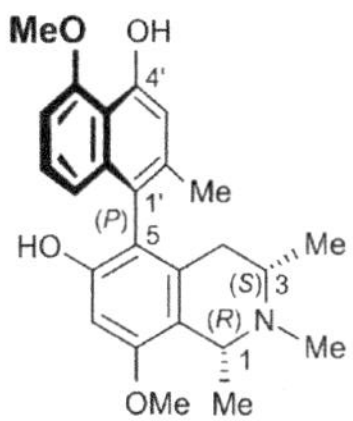 Compd. can likewise be considered as 4′-*O*-demethylancistrocline ——— Colorless amorphous powder (from MeOH): mp 131; $[\alpha]_D^{25}$ +81.3 (*c* 0.10, MeOH) [61]	*A. tectorius* (PR China): twigs [61]	*Structural Elucidation*: Constitution and complete stereostructure deduced from HRESIMS and from ^{1}H and ^{13}C NMR, including specific HMBC and long-range NOESY interactions, revealing **215** to differ from its parent compound ancistectorine A$_1$ (**203**) by the presence of an *N*-methyl function [61] Assignment of the absolute axial configuration as (*P*) due to the relative configuration of centers versus axis known from NMR in conjunction with the absolute (1*R*,3*S*)-stereoarray in the tetrahydroisoquinoline part, corroborated by the nearly identical ECD spectrum of **215** compared to that of its parent compound ancistectorine A$_1$ (**203**) [61]	Strong—and selective—inhibitory activity against the K1 strain [61] of *Plasmodium falciparum*, combined with a very weak cytotoxicity, leading to a high selectivity index, thus to be considered as a lead compound according to the TDR-WHO guidelines [238] Inactive against *Trypanosoma cruzi*, *T. brucei rhodesiense*, *T. brucei brucei*, and *Leishmania donovani* [61] Inactive against bacterial organisms like *Staphylococcus aureus*, *S. epidermidis*, *Enterococcus faecalis*, *E. faecium*, *Escherichia coli*, *Pseudomonas aeruginosa*, *Yersinia pseudotuberculosis*, and *Y. pestis* [61]

(continued)

Table 5 (continued)

Name, formula (mol. wt.), and structure Physical data [*M.p.*/°C, $[\alpha]_D^T$/°cm^2 g^{-1} 10^{-1} (solvent)]	Producing plants and their organs	Isolation, structure elucidation, and further investigations Partial and/or total synthesis	Biological activities
5′-*O*-Demethylancistrocline (165) $C_{25}H_{29}NO_4$ (407.5) 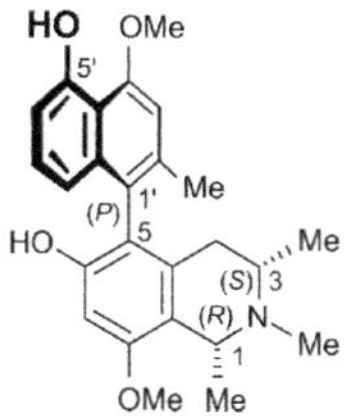 Brownish amorphous solid (from MeOH): mp 213; $[\alpha]_D^{25}$ +61.7 (*c* 0.01, MeOH) [30]	*A. benomensis* (Malaysia): stem bark [30]	*Structural Elucidation*: Constitution and coupling position deduced from HRESIMS and from ^{1}H and ^{13}C NMR, including specific HMBC and long-range NOESY interactions, revealing **165** to differ from its parent compound ancistrocline (**163**) by the absence of an *O*-methyl group in the 5′-position [30] Absolute configuration at C-3 established as (*S*) by oxidative degradation [30] Assignment of the absolute configuration at C-1 as (*R*) due to the relative *cis*-configuration at C-1 versus C-3 as deduced from NOESY interactions and the given absolute (*S*)-configuration at C-3 [30] Assignment of the absolute axial configuration as (*P*) due to NOESY correlations in conjunction with the known absolute (1*R*,3*S*)-stereoarray in the tetrahydroisoquinoline part, and by the nearly identical ECD spectrum of **165** compared to that of the related 6-*O*-methylancistrocline (**164**), stereochemically well known from total synthesis [30, 322]	Moderate inhibitory activity in vitro against *Plasmodium falciparum* (K1 strain) [30] Weak inhibitory activities in vitro against *Trypanosoma brucei rhodesiense*, *T. cruzi*, and *Leishmania donovani* [30]

(continued)

Table 5 (continued)

Name, formula (mol. wt.), and structure Physical data [*M.p.*/°C, $[\alpha]_D^T$/°cm^2 g^{-1} 10^{-1} (solvent)]	Producing plants and their organs	Isolation, structure elucidation, and further investigations Partial and/or total synthesis	Biological activities
Ancistrocladinine (77a) $C_{25}H_{27}NO_4$ (405.5) 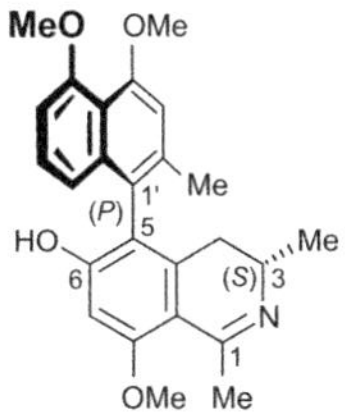 Amorphous solid (from MeOH): mp 235–238; $[\alpha]_D^{25}$ −321.8 (*c* 1.06, pyridine) [37] Amorphous solid (from MeOH): mp 254–257 [58] cf.: Compd. prepared by total synthesis: White needles (from acetone): mp 255–258; $[\alpha]_D^{20}$ −148.9 (*c* 0.48, pyridine) [104, 320]	*A. heyneanus* (India): roots [5, 37] and aerial parts [487] *A. cochinchinensis* (Vietnam): leaves [45, 46] *A. tectorius* (PR China): stems, branches, and twigs [58, 60]	Isolated for the first time from the Indian liana *A. heyneanus* [5, 37] Occurrence in the Vietnamese species *A. cochinchinensis* and in the Chinese liana *A. tectorius*, along with its atropo-diastereomer hamatinine (**77b**) [45, 46, 58, 60] Isolation from *A. cochinchinensis* only as a mixture of the two atropo-diastereomers, ancistrocladinine (**77a**) and hamatinine (**77b**), in a ca. 1:1 ratio [45, 46] *Structural Elucidation*: Constitution and coupling position by HRESIMS and UV and by ^{1}H and ^{13}C NMR [5, 37, 45, 58, 60], including HMBC and NOESY interactions [45], indicative of a 5,1′-coupled naphthyldihydroisoquinoline alkaloid with a free hydroxy function at C-6 Absolute configuration at C-3 established as (*S*) by oxidative degradation [45, 79, 80] Confirmation of the structural assignment by conversion of ancistrocladinine (**77a**) into structurally and stereochemically well-known naphthylisoquinoline alkaloids: (a) by treatment with $NaBH_4$ into isoancistrocladine (**79a**) [37, 38, 53, 322], and (b) by reduction with zinc and aqueous sulfuric acid into ancistrocladine (**1a**) [4, 5] *Total Synthesis*: Preparation of ancistrocladinine (**77a**) by asymmetric construction of the biaryl linkage by applying Meyers' oxazoline method [104, 320]	Moderate to weak growth-inhibitory effects against human leukemia HL-60, K562, and human macrophage U937 cell lines [60]

(continued)

Table 5 (continued)

Name, formula (mol. wt.), and structure Physical data [*M.p.*/°C, $[\alpha]_D^T$/°cm^2 g^{-1} 10^{-1} (solvent)]	Producing plants and their organs	Isolation, structure elucidation, and further investigations Partial and/or total synthesis	Biological activities
Hamatinine (77b) $C_{25}H_{27}NO_4$ (405.5) 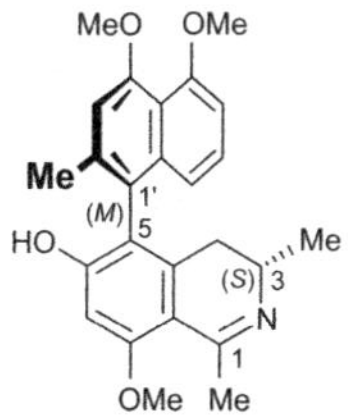 Amorphous powder (from MeOH): mp 254–257 [58] Yellow amorphous powder (from MeOH): $[\alpha]_D^{20}$ +69.7 (*c* 0.10, MeOH) [171]	*A. cochinchinensis* (Vietnam): leaves [45, 46] *A. tectorius* (PR China): stems, branches, and twigs [58, 60] An as yet botanically unknown Congolese *Ancistrocladus* species: roots [171]	Isolated for the first time from the Vietnamese liana *A. cochinchinensis* by column chromatography (CC) on silica gel [45, 46] Occurrence in the Vietnamese species *A. cochinchinensis* and in the Chinese liana *A. tectorius*, along with its atropo-diastereomer ancistrocladinine (**77a**) [45, 46, 58, 60] Isolation from *A. cochinchinensis* only as a mixture of the two atropo-diastereomers, ancistrocladinine (**77a**) and hamatinine (**77b**), in a ca. 1:1 ratio [45, 46] Isolation of **77b** in a stereochemically pure form from a botanically yet unknown Congolese *Ancistrocladus* species by fast centrifugal partition chromatography (FCPC) and by preparative HPLC [171] *Structural Elucidation*: Constitution and coupling position by HRESIMS and by ^{1}H and ^{13}C NMR, including HMBC, HMQC, H,H-COSY, and NOE interactions, indicative of the presence of a 5,1′-coupled naphthyldihydroisoquinoline alkaloid [45, 171] Absolute configuration at C-3 established as (*S*) by oxidative degradation [45, 171] Assignment of the absolute axial configuration as (*M*) due to specific long-range NOE interactions, in combination with the absolute (*S*)-configuration at C-3 [45, 171]	Weak to moderate inhibitory activities in vitro against *Plasmodium falciparum* (K1 strain), *Leishmania donovani*, *Trypanosoma cruzi*, and *T. brucei rhodesiense* [171] Moderate to weak growth-inhibitory effects against human leukemia HL-60, K562, and human macrophage U937 cell lines [60]

(continued)

Table 5 (continued)

Name, formula (mol. wt.), and structure Physical data [*M.p.*/°C, $[\alpha]_D^T$/°cm^2 g^{-1} 10^{-1} (solvent)]	Producing plants and their organs	Isolation, structure elucidation, and further investigations Partial and/or total synthesis	Biological activities
6-*O*-Methylhamatinine (176) $C_{26}H_{29}NO_4$ (419.5) 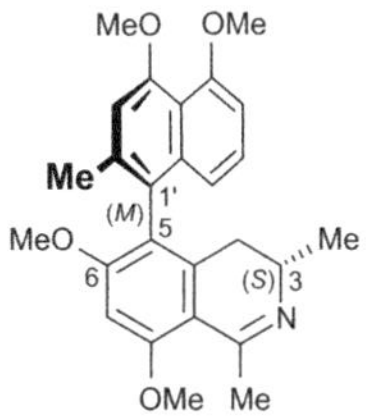 Amorphous solid (from acetone/MeOH): $[\alpha]_D^{23}$ +34.1 (*c* 1.00, $CHCl_3$) [45] Colorless amorphous powder (from MeOH): $[\alpha]_D^{20}$ +15.6 (*c* 0.10, MeOH) [64]	*A. cochinchinensis* (Vietnam): leaves [45, 46] *A. tectorius* (PR China): twigs [64] *A. abbreviatus* (Ivory Coast): roots [191] An as yet botanically unknown Central Congolese *Ancistrocladus* species: roots [171]	Isolated for the first time from the Vietnamese liana *A. cochinchinensis*, by column chromatography (CC) on silica gel [45, 46] Isolation from a botanically so far undescribed Congolese *Ancistrocladus* species (Yeteto rainforests near the town of Ikela) by fast centrifugal partition chromatography (FCPC) in combination with preparative HPLC [171] *Structural Elucidation*: Constitution and coupling position by HRESIMS and by ^{1}H and ^{13}C NMR, including HMBC, HMQC, H,H-COSY, and NOE interactions, indicative of the presence of an *O*-permethylated 5,1′-linked naphthyldihydroisoquinoline [45, 46, 171, 191] Absolute configuration at C-3 determined as (*S*) by oxidative degradation [45, 46, 171, 191] Assignment of the absolute axial configuration as (*M*) due to specific long-range NOE interactions, in combination with the absolute (*S*)-configuration at C-3 [45, 46, 171] Additional proof of the absolute axial configuration by comparison of the ECD spectrum of **176** with that of the (*P*)-configured naphthyldihydroisoquinoline ancistrocladinine (**77a**), which was found to be opposite, as expected [45, 46]	Weak to moderate inhibitory activities in vitro against *Plasmodium falciparum* (K1 strain), *Leishmania donovani*, *Trypanosoma cruzi*, and *T. brucei rhodesiense* [171] Pronounced inhibitory activity in vitro against the clinically relevant canine hematozoan parasite *Babesia canis* [459] Pronounced preferential cytotoxicity against PANC-1 human pancreatic cancer cells under 'austerity' (i.e., nutrient-deprived) conditions [191]

(continued)

Table 5 (continued)

Name, formula (mol. wt.), and structure Physical data [$M.p.$/°C, $[\alpha]_D^T$/°cm^2 g^{-1} 10^{-1} (solvent)]	Producing plants and their organs	Isolation, structure elucidation, and further investigations Partial and/or total synthesis	Biological activities
8-*O*-Demethyl-6-*O*-methylhamatinine (191) $C_{25}H_{27}NO_4$ (405.5) 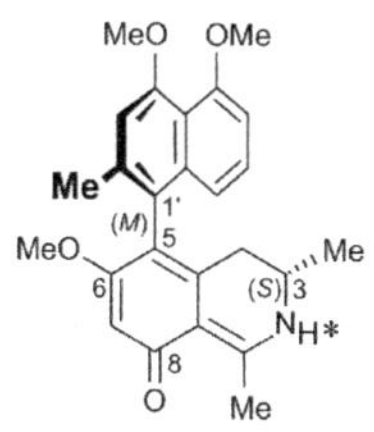 * H atom not drawn in the report [55], but plausible from the MS data Amorphous solid (from $MeOH/CHCl_3$): $[\alpha]_D^{20}$ −5.8 (c 0.24, MeOH) [55]	*A. tectorius* (Thailand): twigs and leaves [55]	*Structural Elucidation*: Isolation by gel permeation chromatography on Sephadex LH-20 and by flash chromatography on amino-bonded silica gel [55] Constitution and coupling position by HREIMS and by ^{1}H and ^{13}C NMR, including HMBC, HMQC, H,H-COSY, and NOE interactions, indicative of the presence of a 5,1′-linked naphthyldihydroisoquinoline alkaloid possessing a free hydroxy function at C-8 [55] Reported occurrence of **191** predominantly in a tautomeric form with a keto function at C-8 in the isoquinoline subunit, as deduced from an intense downfield shift in ^{13}C NMR for C-8 (δ_C = 180.89 ppm [55]), compared to the chemical shift values of around δ_C = 155–165 ppm as displayed by related 5,1′- or 5,8′-linked naphthylisoquinoline alkaloids with a hydroxy group at C-8 [61, 156, 164, 167, 168, 182, 191]—in accordance with the appearance of a major absorption band at λ = 384 nm [55], never reported for related alkaloids before Absolute configuration at C-3 determined as (*S*) by oxidative degradation [55, 79, 80] Assignment of the absolute axial configuration as (*M*) by the nearly identical ECD spectrum of **191** compared to that of the structurally closely related known 5,1′-coupled, and likewise (*M*)-configured alkaloid hamatine (**1b**) [23, 24, 55]	Inactive against HIV in vitro [55] Potent cytotoxic activity and a paclitaxel-resistance reversing effect of an 80% ethanol extract of *A. tectorius* in PR-HepG2 cells [494]

(continued)

Table 5 (continued)

Name, formula (mol. wt.), and structure Physical data [*M.p.*/°C, $[\alpha]_D^T$/°cm^2 g^{-1} 10^{-1} (solvent)]	Producing plants and their organs	Isolation, structure elucidation, and further investigations Partial and/or total synthesis	Biological activities
Ancistectorine A$_3$ (204) $C_{24}H_{25}NO_4$ (391.5) 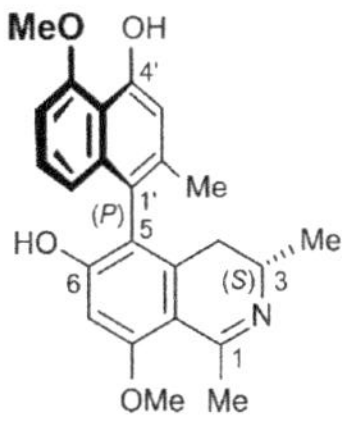 Compd. can likewise be considered as 4′-*O*-demethyl-ancistrocladinine ——— Colorless amorphous powder (from MeOH): mp 140; $[\alpha]_D^{20}$ −146.0 (*c* 0.10, MeOH) [61]	*A. tectorius* (PR China): twigs [61]	*Structural Elucidation*: Constitution and coupling position by HRESIMS, and by ^{1}H and ^{13}C NMR, including specific HMBC and NOE correlations, indicative of the presence of a 5,1′-linked naphthyldihydro-isoquinoline alkaloid equipped with hydroxy groups at both, C-6 and C-4′ [61] Absolute configuration at C-3 established as (*S*) by oxidative degradation [61] Assignment of the absolute axial configuration as (*P*) due to specific long-range NOE interactions, in combination with the absolute configuration at C-3 assigned as (*S*) [61] Confirmation of the axial configuration by ECD spectroscopy [61] as obvious from the mirror-image-like ECD spectrum of **204** in comparison to that of the known, likewise 5,1′-coupled, but (*M*)-configured 5′-*O*-demethylhamatinine (**177**) [171]	Weak inhibitory activity in vitro against *Plasmodium falciparum* (K1 strain) [61] No inhibitory activity against *Trypanosoma cruzi*, *T. brucei rhodesiense*, and *Leishmania donovani* [61]
6-*O*-Methylancistectorine A$_3$ (190a) $C_{25}H_{27}NO_4$ (405.5) 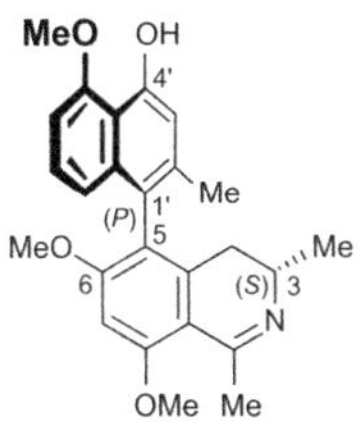 Compd. can likewise be considered as 4′-*O*-demethyl-6-*O*-methylancistrocladinine ——— Colorless amorphous powder (from MeOH): $[\alpha]_D^{20}$ −40.5 (*c* 0.10, MeOH) [64]	*A. tectorius* (PR China): twigs [64]	Isolated for the first time from *A. tectorius* from the Chinese Hainan Island [64] Identification of the atropo-diastereomer of **190a**, 4′-*O*-demethyl-6-*O*-methylhamatinine (**190b**), so far only in *A. tectorius* from Thailand [55] *Structural Elucidation*: Constitution and coupling position by HRESIMS and by ^{1}H and ^{13}C NMR, including specific HMBC and NOE correlations, indicative of the presence of a 5,1′-linked naphthyldihydro-isoquinoline with a free hydroxy group at C-4′ [64] Absolute configuration at C-3 established as (*S*), by oxidative degradation [64] Assignment of the absolute axial configuration as (*P*) due to specific NOESY interactions and the given absolute configuration at C-3 [64]	Biological activities not determined for lack of isolated material

(continued)

Table 5 (continued)

Name, formula (mol. wt.), and structure Physical data [$M.p.$/°C, $[\alpha]_D^T$/°cm^2 g^{-1} 10^{-1} (solvent)]	Producing plants and their organs	Isolation, structure elucidation, and further investigations Partial and/or total synthesis	Biological activities
		Additional proof of the absolute axial configuration by comparison of the ECD spectrum of **190a** with that of the likewise (*P*)-configured parent compound ancistectorine A_3 [61, 64]	
4′-*O*-Demethyl-6-*O*-methylhamatinine (190b) $C_{25}H_{27}NO_4$ (405.5) 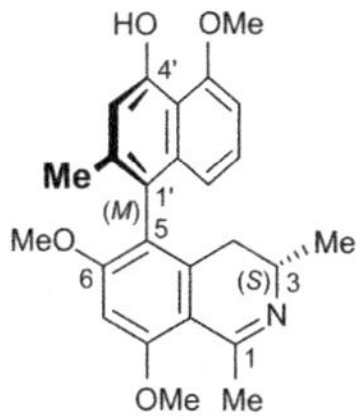 Compd. can likewise be considered as 5-*epi*-6-*O*-methylancistectorine A_3	*A. tectorius* (Thailand): twigs and leaves [55]	Identification of the atropo-diastereomer of **190b**, 6-*O*-methylancistectorine A_3 (**190a**) [64], so far only in *A. tectorius* from the Chinese Hainan Island Isolation of **190b** by gel permeation chromatography on Sephadex LH-20 followed by flash chromatography on amino-bonded silica [55] *Structural Elucidation*: Constitution and relative configuration by HRESIMS and by ^{1}H and ^{13}C NMR, and by HMBC and NOESY interactions indicative of the configuration at the biaryl axis relative to the stereogenic center [55] Absolute configuration at C-3 established as (*S*), by oxidative degradation [55] Assignment of the absolute axial configuration as (*M*) by comparison of the ECD spectrum of **190b** [55] with that of the likewise (*M*)-configured hamatine (**1b**) [23, 24]	Inactive against HIV in vitro [55] Potent cytotoxic activity and a paclitaxel-resistance reversing effect of an 80% ethanol extract of *A. tectorius* in PR-HepG2 cells [494]

(continued)

Table 5 (continued)

Name, formula (mol. wt.), and structure Physical data [$M.p.$/°C, $[\alpha]_D^T$/°cm^2 g^{-1} 10^{-1} (solvent)]	Producing plants and their organs	Isolation, structure elucidation, and further investigations Partial and/or total synthesis	Biological activities
Amorphous solid (from MeOH/$CHCl_3$): $[\alpha]_D^{20}$ +41.7 (*c* 0.29, MeOH) [55]			
8,4′-*O,O*-Didemethyl-6-*O*-methyl-ancistrocladinine (192a) 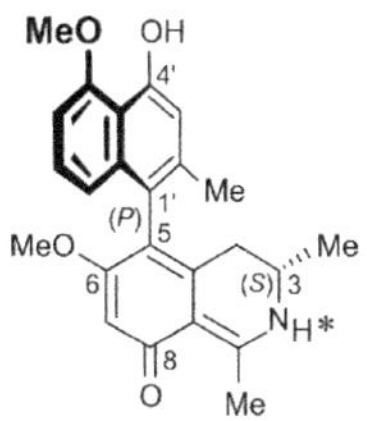 and **8,4′-*O,O*-Didemethyl-6-*O*-methyl-hamatinine (192b)** $C_{24}H_{25}NO_4$ (391.5) 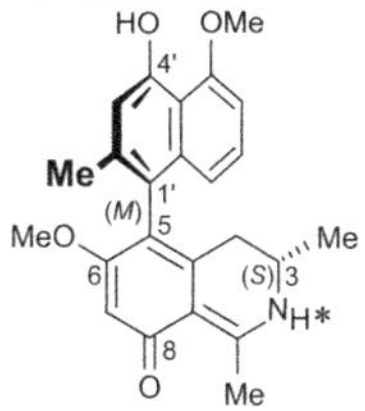 * H atom not drawn in the report [55], but plausible from the MS data	*A. tectorius* (Thailand): twigs and leaves [55]	Occurrence in *A. tectorius* from the Khao Sok National Park in Thailand as an inseparable mixture of two atropo-diastereomers, **192a** and **192b** [55] Isolation by gel permeation chromatography on Sephadex LH-20 followed by flash chromatography on amino-bonded silica [55] *Structural Elucidation*: Constitution by HRFABMS and by ^{1}H and ^{13}C NMR, and by difference NOESY experiments, suggesting the presence of two 5,1′-linked diastereomeric naphthyldihydroisoquinolines (doubling of almost all peaks in the ^{1}H NMR spectrum) with free hydroxy groups at C-8 and C-4′ [55] Reported occurrence of **192a/b** predominantly in the keto form at C-8 in the isoquinoline subunit as deduced from an intense downfield shift in ^{13}C NMR for C-8 (δ_C = 180.15 ppm)—an unprecedented occurrence of alkaloids in such a tautomeric form, as also reported for the co-occurring naphthyldihydroisoquinoline alkaloid **191** (see entry above) [55] Absolute configuration at C-3 established as (*S*), by oxidative degradation [55, 79, 80] For an additional structural proof, the inseparable mixture of the two atropo-diastereomers **192a** and **192b** was reduced using $NaBH_4$ to give the corresponding	Inactive against HIV in vitro [55] Potent cytotoxic activity and a paclitaxel-resistance reversing effect of an 80% ethanol extract of *A. tectorius* in PR-HepG2 cells [494]

(continued)

Table 5 (continued)

Name, formula (mol. wt.), and structure Physical data [*M.p.*/°C, $[\alpha]_D^T$/°cm² g⁻¹ 10⁻¹ (solvent)]	Producing plants and their organs	Isolation, structure elucidation, and further investigations Partial and/or total synthesis	Biological activities
Atropo-diastereomeric mixture (ratio: *ca.* 1:1), occurring as an amorphous solid (from MeOH/$CHCl_3$): $[\alpha]_D^{20}$ −90.3 (*c* 0.39, MeOH) [55]		*cis*-configured naphthyltetra-hydroisoquinolines, again as an atropo-diastereomeric mixture [55]	
5′-*O*-Demethylhamatinine (177) $C_{24}H_{25}NO_4$ (391.5) 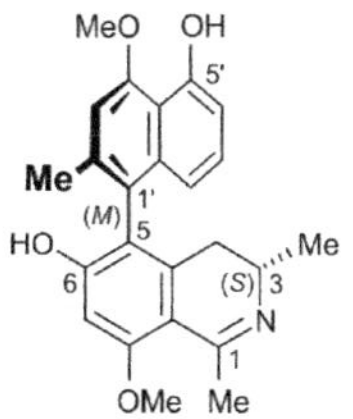 Yellow amorphous solid (from acetonitrile): $[\alpha]_D^{25}$ +25.0 (*c* 0.10, MeOH) [171]	*A. cochinchinensis* (Vietnam): root bark and leaves [47] *A. tectorius* (PR China): twigs [64] An as yet botanically unknown Congolese *Ancistrocladus* species: roots [171]	Isolated for the first time from a botanically so far undescribed *Ancistrocladus* liana collected in the rainforest Yeteto of the Ikela region in the Democratic Republic of the Congo [171] *Structural Elucidation*: Metabolic screening of root bark and leaf extracts of the Vietnamese species *A. cochinchinensis* using the analytical triad LC-MS/MS, LC-NMR (^{1}H, WET-ROESY), and LC-ECD: constitution and complete stereostructure of **177** elucidated online, directly from crude extracts [47] Constitution by HRESIMS, and by ^{1}H and ^{13}C NMR, including specific HMBC and NOESY interactions, revealing the presence of a 5,1′-linked naphthyldihydroisoquinoline alkaloid with free hydroxy functions at C-6 and C-5′ [47, 171] Absolute configuration at C-3 established as (*S*) by oxidative degradation [171] Assignment of the absolute axial configuration as (*M*) due to specific long-range NOESY correlations, in conjunction with the given (*S*)-configuration at C-3, and due to the nearly	Weak to moderate inhibitory activities in vitro against *Plasmodium falciparum* (K1 strain), *Leishmania donovani*, *Trypanosoma cruzi*, and *T. brucei rhodesiense* [171]

(continued)

Table 5 (continued)

Name, formula (mol. wt.), and structure Physical data [*M.p.*/°C, $[\alpha]_D^T$/°cm^2 g^{-1} 10^{-1} (solvent)]	Producing plants and their organs	Isolation, structure elucidation, and further investigations Partial and/or total synthesis	Biological activities
		identical ECD spectrum of **177** compared to that of the likewise 5,1′-coupled and (*M*)-configured hamatinine (**77b**) [45, 47, 171]	
6-*O*-Methylhamateine (73) $C_{26}H_{27}NO_4$ (417.5) 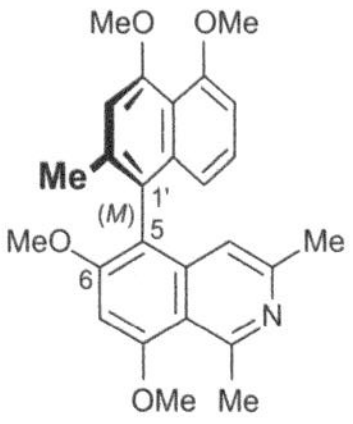 Crystalline solid (from $CHCl_3$/*n*-hexane): mp 233–236; $[\alpha]_D^{23}$ −41.4 (*c* 0.46, $CHCl_3$) [45] cf.: Compd. prepared synthetically: Needles (from diethyl ether): mp 240–241; $[\alpha]_D^{23}$ −50.46 (*c* 1.6, $CHCl_3$) [23] (*P*)-Isomer: unnatural 6-*O*-methylancistrocladeine, prepared synthetically [2, 103, 319] Amorphous solid (from diethyl ether): mp 240–242; $[\alpha]_D^{25}$ +58.87 (*c* 1.6, $CHCl_3$) [2] Prisms (from diethyl ether/*n*-hexane): mp 238–240; $[\alpha]_D^{23}$ +51.0 (*c* 0.30, $CHCl_3$) [103]	*A. cochinchinensis* (Vietnam): leaves [45, 46] *A. abbreviatus* (Ivory Coast): roots [186]	Discovered for the first time in the Vietnamese liana *A. cochinchinensis*, but no investigations on the enantiomeric purity of **73** isolated from this plant [45, 46] HPLC-ECD analysis on a chiral phase (Lux-Cellulose 1), revealing **73** isolated from the roots of *A. abbreviatus* to be enantiomerically pure [186] *Structural Elucidation*: Constitution and coupling position established by HRESIMS and by 1D and 2D NMR, revealing the presence of a 5,1′-linked non-hydrogenated and O-permethylated naphthylisoquinoline alkaloid [45, 46, 186] Absolute axial configuration assigned as (*M*) by ECD investigations using the Exciton Chirality method [23, 45] Additional proof of the axial configuration of **73** by comparison of its ECD spectrum with the spectra of the likewise 5,1′-linked, but (*P*)-configured fully dehydrogenated naphthyl-isoquinolines 6-*O*-methylancistrocladeine [2, 4] and ancistrocladeine (**162**) [186, 414]—showing, as expected, a virtually opposite ECD curve of **73** as compared to those of **162** and its 6-*O*-methyl analogue (not shown) *Synthesis and Related Work*: Semi-synthesis of **73** from 6-*O*-methylhamatine (**76b**) by dehydrogenation [23]	Strong cytotoxic activity against parental drug-sensitive CCRF-CEM leukemia cells and their multi-drug resistant subline CEM/ADR5000 in the low micromolar range [186]

(continued)

Table 5 (continued)

Name, formula (mol. wt.), and structure / Physical data [$M.p.$/°C, $[\alpha]_D^T$/°cm^2 g^{-1} 10^{-1} (solvent)]	Producing plants and their organs	Isolation, structure elucidation, and further investigations / Partial and/or total synthesis	Biological activities
		Total synthesis of the (*P*)-isomer of **73** (i.e., 6-*O*-methylancistrocladeine), by asymmetric construction of the biaryl linkage by using the Meyers' oxazoline method [103, 319]	
Ancistrobenomine A (**4**) $C_{25}H_{25}NO_5$ (419.5) 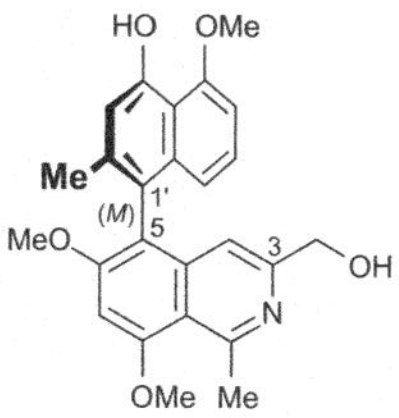 Pale-yellow powder (from MeOH): mp 270 (dec); $[\alpha]_D^{25}$ −21.0 (*c* 0.10, MeOH) [30]	*A. benomensis* (Malaysia): root bark [30] *A. cochinchinensis* (Vietnam): root bark and leaves [47] *A. tectorius* (PR China): twigs [64]	Isolated for the first time from *A. benomensis* from Peninsular Malaysia [30] *Structural Elucidation*: Metabolic screening of root bark and leaf extracts of the Vietnamese species *A. cochinchinensis* using the analytical triad LC-MS/MS, LC-NMR (^{1}H, WET-ROESY), and LC-ECD: constitution and complete stereostructure of **4** elucidated online, directly from crude extracts [47] Constitution and coupling position established by HRESIMS and by 1D and 2D NMR, revealing **4** to be a 5,1′-linked non-hydrogenated naphthylisoquinoline alkaloid possessing a hydroxymethylene residue at C-3 (instead of a methyl group) [30] Assignment of the absolute axial configuration as (*M*) due to a virtually opposite ECD spectrum of **4** compared to that of ancistrocladeine (**162**), a known [2], likewise fully dehydrogenated and 5,1′-coupled, but (*P*)-configured (and	Moderate to weak inhibitory activities in vitro against *Plasmodium falciparum* (K1 strain), *Trypanosoma cruzi*, and *T. brucei rhodesiense* [30] Nearly no inhibitory activities against *Leishmania donovani* [30]

(continued)

Table 5 (continued)

Name, formula (mol. wt.), and structure Physical data [*M.p.*/°C, $[\alpha]_D^T$/°cm^2 g^{-1} 10^{-1} (solvent)]	Producing plants and their organs	Isolation, structure elucidation, and further investigations Partial and/or total synthesis	Biological activities
		6-*O*-demethylated) naphthylisoquinoline obtained by semi-synthesis from natural ancistrocladine (**1a**) [30, 414] HPLC-ECD analysis on a chiral phase (Chiralcel® OD-H), revealing ancistrobenomine A (**4**) to occur enantiomerically pure (i.e., 100:0, (*M*):(*P*)) in its natural producers [30, 64]	
6-*O*-Demethylancistrobenomine A (159) $C_{24}H_{23}NO_5$ (405.5) 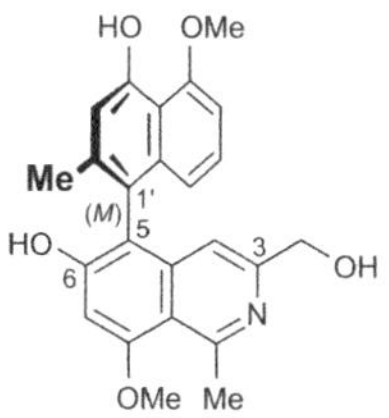 Pale-yellow amorphous solid (from MeOH): mp 169 (dec); $[\alpha]_D^{25}$ −34.4 (*c* 0.05, MeOH) [30]	*A. benomensis* (Malaysia): root bark [30] *A. cochinchinensis* (Vietnam): root bark and leaves [47]	Isolated for the first time from *A. benomensis* from Peninsular Malaysia [30] *Structural Elucidation*: Metabolic screening of root bark and leaf extracts of the Vietnamese species *A. cochinchinensis* using the analytical triad LC-MS/MS, LC-NMR (^{1}H, WET-ROESY), and LC-ECD: constitution and complete stereostructure of **159** elucidated online, directly from crude extracts [47] Constitution and complete stereostructure by HRESIMS, 1D and 2D NMR, oxidative degradation, and ECD—comparison of the data of **159** with those obtained from its likewise (*M*)-configured parent compound ancistrobenomine A (**4**), revealing **159** to differ from **4** by the presence of a free hydroxy function at C-6 [30] Similar to ancistrobenomine A (**4**), occurrence of **159** enantiomerically pure (i.e., 100:0, (*M*):(*P*)) in the plants, as determined by LC-ECD on a chiral phase (Chiralcel® OD-H) [30]	Moderate to weak inhibitory activities in vitro against *Plasmodium falciparum* (K1 strain), *Trypanosoma cruzi*, and *T. brucei rhodesiense* [30] No inhibitory activities against *Leishmania donovani* [30]

(continued)

Table 5 (continued)

Name, formula (mol. wt.), and structure Physical data [*M.p.*/°C, $[\alpha]_D^T$/°cm^2 g^{-1} 10^{-1} (solvent)]	Producing plants and their organs	Isolation, structure elucidation, and further investigations Partial and/or total synthesis	Biological activities
Ancistrobenomine B (11) $C_{26}H_{27}NO_5$ (433.5) 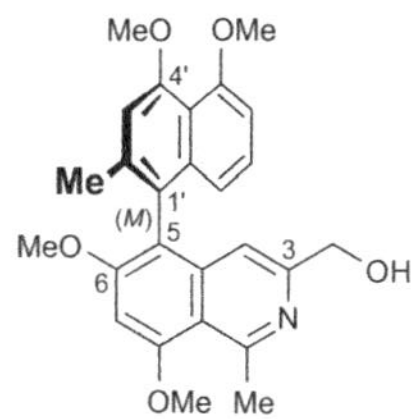 Compd. can likewise be considered as 4′-*O*-methylancistrobenomine A Pale-yellow amorphous powder (from MeOH): $[\alpha]_D^{20}$ +70.0 (*c* 0.44, MeOH) [64]	*A. tectorius* (PR China): twigs [64]	*Structural Elucidation*: Constitution and complete stereostructure by HRESIMS, 1D and 2D NMR, oxidative degradation, and ECD—comparison of the data of **11** with those obtained from its parent compound ancistrobenomine A (**4**), evidencing the presence of a non-hydrogenated and *O*-permethylated naphthylisoquinoline alkaloid with a hydroxymethylene group at C-3 [64] Assignment of the absolute axial configuration as (*M*) due to the nearly identical ECD spectrum of **11** compared to that of the likewise (*M*)-configured ancistrobenomine A (**4**) [64] Similar to ancistrobenomine A (**4**) [30], occurrence of **11** in the plants in an enantiomerically pure form (i.e., 100:0, (*M*):(*P*)), as determined by LC-ECD on a chiral phase (Chiralcel® OD-H) [64]	Moderate to weak inhibitory activities in vitro against *Plasmodium falciparum* (K1 and NF54 strains), *Trypanosoma cruzi*, and *T. brucei rhodesiense* [64] Nearly no inhibitory activities against *Leishmania donovani* [64] Strong growth-retarding activities against drug-sensitive CCRF-CEM leukemia cells and their multi-drug resistant subline CEM/ADR5000 [64]
Ancistrobenomine C (208) $C_{25}H_{25}NO_5$ (419.5) 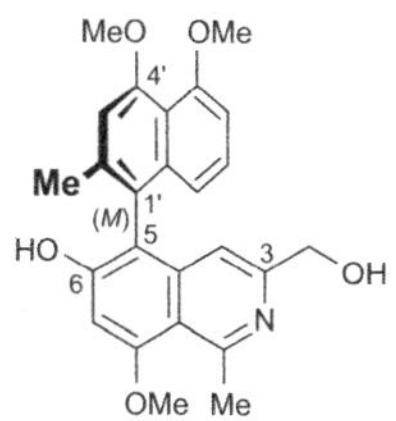 Compd. can likewise be considered as 6-*O*-demethylancistrobenomine B Pale-yellow amorphous powder (from MeOH): mp 168; $[\alpha]_D^{20}$ +33.2 (*c* 0.10, MeOH) [64]	*A. tectorius* (PR China): twigs [64]	*Structural Elucidation*: Constitution and complete stereostructure by HRESIMS, 1D and 2D NMR, oxidative degradation, and ECD—comparison of the data of **208** with those obtained from the likewise (*M*)-configured ancistrobenomine B (**11**), revealing **208** to differ from **11** by the presence of a free hydroxy function at C-6 [64] Similar to ancistrobenomine A (**4**) [30], occurrence of **208** in the plants in an enantiomerically pure form (i.e., 100:0, (*M*):(*P*)), as determined by LC-ECD on a chiral phase (Chiralcel® OD-H) [64]	Moderate to weak inhibitory activities in vitro against *Plasmodium falciparum* (K1 and NF54 strains), *Trypanosoma cruzi*, and *T. brucei rhodesiense* [64] Nearly no inhibitory activities against *Leishmania donovani* [64] Weak cytotoxic activity against drug-sensitive acute lymphoblastic CCRF-CEM leukemia cells [64]

Table 6 5,3′-coupled Ancistrocladaceae-type naphthylisoquinoline alkaloids

Name, formula (mol. wt.), and structure Physical data [*M.p.*/°C, $[\alpha]_D^T$/°cm^2 g^{-1} 10^{-1} (solvent)]	Producing plants and their organs	Isolation, structure elucidation, and further investigations Partial and/or total synthesis	Biological activities
Ancistrotanzanine A (12a) $C_{25}H_{27}NO_4$ (405.5) 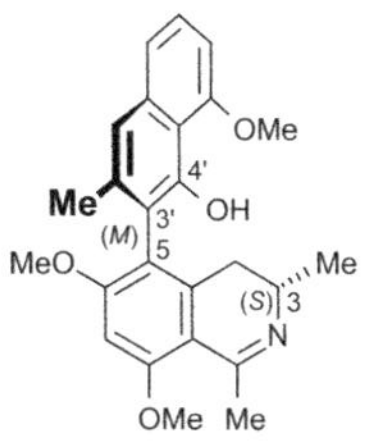 Pale-yellow powder (from MeOH): mp 243; $[\alpha]_D^{25}$ +69.5 (*c* 0.10, EtOH) [155]	*A. tectorius* (PR China): stems and leaves [60] *A. tanzaniensis* (Tanzania): leaves [155]	Isolated for the first time from *A. tanzaniensis* from the Uzungwa Mountains National Park in Central Tanzania [155] *Structural Elucidation*: Constitution and coupling position established by HRESIMS and by ^{1}H and ^{13}C NMR, including HMBC and NOESY correlations, indicating the presence of a 5,3′-linked naphthyldihydroisoquinoline alkaloid with a free hydroxy group at C-4′ [155] Absolute configuration at C-3 by ruthenium-mediated oxidative degradation [155] Assignment of the absolute axial configuration of ancistrotanzanine A (**12a**) by quantum-chemical ECD calculations using the semiempirical CNDO/2S Hamiltonian method, based on the AM1-Boltzmann approach, or applying an MD simulation using the MM3 force field—good agreement of the ECD spectrum calculated for (3*S*,5*M*)-**12** with the experimental ECD curve [155]	Excellent—and highly specific—inhibitory activity in vitro against *Leishmania donovani* [155] Pronounced inhibitory activity in vitro against *Trypanosoma cruzi*, but only weak activity against *T. brucei rhodesiense* [155] Moderate inhibitory activity in vitro against the chloroquine-resistant K1 strain of *Plasmodium falciparum*, but virtually no antiplasmodial activity against the chloroquine-sensitive 3D7 strain [155] Moderate growth-inhibitory effects against human leukemia HL-60, K562, and human macrophage U937 cell lines [60]
Ancistrotectorine D (12b) $C_{25}H_{27}NO_4$ (405.5) 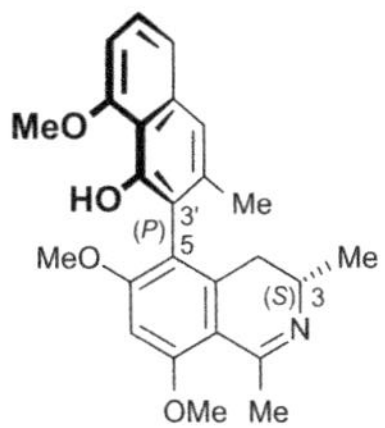 Compd. can likewise be considered as 5-*epi*-ancistrotanzanine A Light-yellow solid (from MeOH): $[\alpha]_D^{25}$ −45.1 (*c* 0.25, MeOH) [60]	*A. tectorius* (PR China): stems and leaves [60]	*Structural Elucidation*: Constitution and coupling position assigned by HRESIMS, by ^{1}H NMR and ^{13}C NMR chemical shifts, and by HMBC interactions revealing the presence of a naphthyldihydroisoquinoline alkaloid with a constitution identical to that of ancistrotanzanine A (**12a**) [60] Absolute configuration at C-3 assumed to be (*S*) [60], merely based on the observation that so far all naphthylisoquinoline alkaloids from *A. tectorius* were (*S*)-configured [13, 17, 18] Attribution of the absolute axial configuration as (*P*) by comparison of its ECD spectrum with that of the likewise 5,3′-coupled alkaloid ancistrotanzanine A (**12a**)—with the two alkaloids displaying nearly mirror-imaged ECD spectra, as expected [60]	Moderate growth-inhibitory effects against human leukemia HL-60, K562, and human macrophage U937 cell lines [60]

(continued)

Table 6 (continued)

Name, formula (mol. wt.), and structure Physical data [*M.p.*/°C, $[\alpha]_D^T$/°cm^2 g^{-1} 10^{-1} (solvent)]	Producing plants and their organs	Isolation, structure elucidation, and further investigations Partial and/or total synthesis	Biological activities
Ancistrotectorine C (205) $C_{26}H_{31}NO_4$ (421.5) 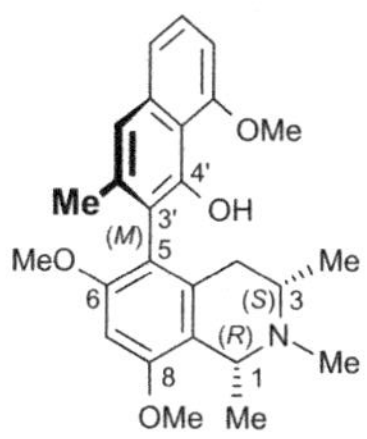 Light-yellow solid (from MeOH): $[\alpha]_D^{25}$ +45.7 (*c* 0.20, MeOH) [60]	*A. tectorius* (PR China): stems and leaves [60]	*Structural Elucidation*: Constitution and coupling position established by HRESIMS, and by 1D and 2D NMR, supporting the presence of a 5,3′-linked *cis*-configured and N-methylated naphthyltetrahydroisoquinoline with methoxy groups at C-6, C-8, and C-5′, and a free hydroxy function at C-4′ [60] Constitution and complete stereostructure established by HRESIMS, and by 1D and 2D NMR—comparison of the data of **205** with those of its parent compound ancistrotanzanine A (**12a**) [60, 155], revealing **205** to be an N-methylated and *cis*-configured 5,3′-coupled naphthyltetra-hydroisoquinoline alkaloid [60] Absolute configuration at C-3 assumed to be (*S*) [60], merely based on the observation that so far only (3*S*)-configured naphthylisoquinoline alkaloids have been isolated from *A. tectorius* [13, 17, 18] Assignment of the absolute configuration at C-1 as (*R*) due to the relative *cis*-configuration at C-1 versus C-3 as deduced from HMBC interactions and relying on the assumed absolute configuration at C-3 [60]	Moderate growth-inhibitory effects against human leukemia HL-60, K562, and human macrophage U937 cell lines [60]

Table 7 5,8′-coupled Ancistrocladaceae-type naphthylisoquinoline alkaloids

Name, formula (mol. wt.), and structure Physical data [*M.p.*/°C, $[\alpha]_D^T$/°cm² g⁻¹ 10⁻¹ (solvent)]	Producing plants and their organs	Isolation, structure elucidation, and further investigations Partial and/or total synthesis	Biological activities
Ancistrotectoriline A (209) $C_{26}H_{31}NO_4$ (421.5) 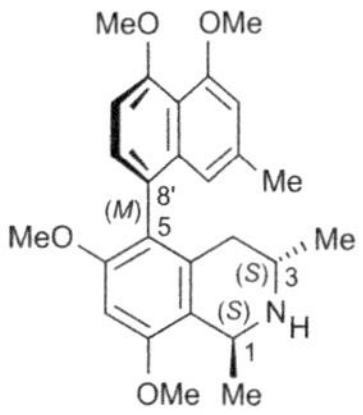 White amorphous solid (from $CHCl_3$): mp 103–105; $[\alpha]_D^{25}$ +1.34 (*c* 0.75, $CHCl_3$) [56] Yellow oil (from MeOH): $[\alpha]_D^{25}$ +5.67 (*c* 0.10, MeOH) [155] cf.: Ancistrotectoriline A (**209**) and its 1-epimer, prepared by asymmetric total synthesis [120]: 1-*epi*-Ancistrotectoriline A: White solid (from CH_2Cl_2/MeOH): mp 94–95; $[\alpha]_D^{25}$ −47.7 (*c* 1.00, MeOH) [120]	*A. tectorius* (PR China): leaves, stem bark, and twigs [56, 62] *A. tanzaniensis* (Tanzania): leaves [155] *A. ealaensis* (DR Congo): twigs [176] An as yet botanically unknown Congolese *Ancistrocladus* species: roots [161]	Isolated for the first time from *A. tectorius* from the Chinese Hainan Island [56] *Structural Elucidation*: Constitution and relative configurations at the three stereogenic elements (two centers and one axis) deduced from HRESIMS and from ^1H and ^{13}C NMR, including HMBC and specific NOE interactions, revealing the presence of an 8,5′-coupled *O*-permethylated and *trans*-configured naphthyltetrahydroisoquinoline [56, 155] Absolute configuration at C-1 and C-3 tentatively assumed by Ye et al. [56] as (1*S*,3*S*), and later experimentally confirmed applying the oxidative degradation procedure [80] by the authors' group [56, 155, 161] Assignment of the absolute axial configuration as (*M*) due the fact that **209** and the likewise 5,8′-coupled and (*M*)-configured alkaloids ancistrobrevine B (**151**) [192] and korupensamine B (**13b**) [156] displayed nearly identical ECD spectra [56, 155] Assignment of the absolute axial configuration as (*M*) due to specific long-range NOESY interactions, in conjunction with the given absolute configuration at C-3 (as known from the degradation) [56, 155, 161]	Moderate inhibitory activity in vitro against the chloroquine-resistant K1 strain of *Plasmodium falciparum*, but virtually no antiplasmodial activity against the chloroquine-sensitive 3D7 strain [155] Weak inhibitory activity against *Trypanosoma* parasites [56, 155] Virtually no inhibitory activities against *Leishmania donovani* [56, 155] Weak inhibitory activity against drug-sensitive acute lymphoblastic CCRF-CEM leukemia cells [62] Moderate to weak preferential cytotoxicity against PANC-1 human pancreatic cancer cells under 'austerity' (i.e., nutrient-deprived) conditions [161]

(continued)

Table 7 (continued)

Name, formula (mol. wt.), and structure Physical data [*M.p.*/°C, $[\alpha]_D^T$/°cm^2 g^{-1} 10^{-1} (solvent)]	Producing plants and their organs	Isolation, structure elucidation, and further investigations Partial and/or total synthesis	Biological activities
Ancistrotectoriline A (**209**): White solid (from CH_2Cl_2/MeOH): mp 100–101; $[\alpha]_D^{25}$ +7.8 (*c* 1.00, MeOH) [120]		*Total Synthesis and Related Work*: Asymmetric total synthesis of ancistrotectoriline A (**209**) by Suzuki–Miyaura cross-coupling between the respective naphthyl pinacol boronate and an aryl iodide, as a still monocyclic isoquinoline precursor equipped with an (*S*)-2-(*N*-acetylamino)propyl group at the *ortho*-position, using the chiral 2-aminopropyl substituent as an atroposelectivity-controlling group [120, 121]—using Pd $(OAc)_2$ as the catalyst in the presence of *t*-BuXPhos as the ligand and K_3PO_4 as the base, affording the desired (*M*)-configured biaryl coupling product as the major atropo-diastereomer (13:1, (*M*):(*P*)); total synthesis of **209** accomplished by Pd-catalyzed hydrogenolysis and Bischler–Napieralski ring closure; conversion of ancistrotanzanine B (**218b**) into the target molecule **209** by *trans*-selective reduction with $LiAlH_4$ in the presence of $AlEt_3$, giving **209** as the major product, along with its 1-epimer in a ratio of 4:1 (*trans*:*cis*); column chromatography after *N*-Boc protection and eventual removal of the Boc group providing pure ancistrotectoriline A (**209**) [120]	

(continued)

Table 7 (continued)

Name, formula (mol. wt.), and structure Physical data [*M.p.*/°C, $[\alpha]_D^T$/°cm^2 g^{-1} 10^{-1} (solvent)]	Producing plants and their organs	Isolation, structure elucidation, and further investigations Partial and/or total synthesis	Biological activities
Ancistrotanzanine B (218a) $C_{26}H_{29}NO_4$ (419.5) 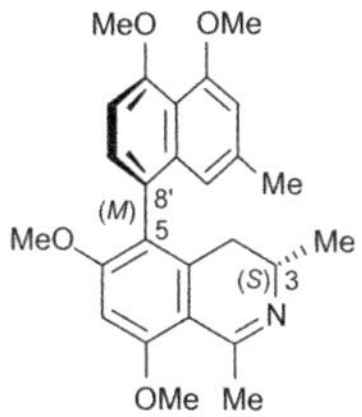 Compd. can likewise be considered as 5-*epi*-ancistroealaine A Yellow oil (from MeOH): $[\alpha]_D^{25}$ +44.0 (*c* 0.01, MeOH) [155] cf.: Ancistrotanzanine B (**218a**) and its atropo-diastereomer, ancistroealaine A (**218b**), prepared	*A. tectorius* (PR China): stems and leaves [60] *A. tanzaniensis* (Tanzania): leaves [155]	Isolated for the first time from *A. tanzaniensis* from the Uzungwa Mountains National Park of Central Tanzania [155] *Structural Elucidation*: Constitution and coupling position by HRESIMS and by ^{1}H and ^{13}C NMR, including specific HMBC and NOE correlations, indicative of the presence of an *O*-permethylated 5,8′-linked naphthyldihydroisoquinoline [155] Absolute configuration at C-3 established as (*S*) by oxidative degradation [60, 155] Assignment of the absolute axial configuration as (*M*) due to specific NOESY interactions, in conjunction with the absolute (*S*)-configuration at C-3 [155] Confirmation of the assigned axial configuration as (*M*) by comparison of the ECD spectrum of **218a** with that of ancistroealaine A (**218b**) [174], the two atropo-diastereomeric alkaloids displaying virtually opposite ECD curves, as expected [155]	Excellent—and highly specific—inhibitory activity in vitro against *Leishmania donovani* [155] Pronounced inhibitory activity in vitro against *Trypanosoma cruzi*, but only weak activity against *T. brucei rhodesiense* [155] Moderate inhibitory activity in vitro against the chloroquine-resistant K1 strain of *Plasmodium falciparum*, but virtually no antiplasmodial activity against the chloroquine-sensitive 3D7 strain [155] Moderate growth-inhibitory effects against human leukemia HL-60, K562, and human macrophage U937 cell lines [60]

(continued)

Table 7 (continued)

Name, formula (mol. wt.), and structure Physical data [*M.p.*/°C, $[\alpha]_D^T$/°cm^2 g^{-1} 10^{-1} (solvent)]	Producing plants and their organs	Isolation, structure elucidation, and further investigations Partial and/or total synthesis	Biological activities
by stereoselective total synthesis [120, 121, 337, 462]: Light-yellow powder (from EtOH): mp 92; $[\alpha]_D^{25}$ +42.8 (*c* 0.016, MeOH) [462] Yellow solid (from CH_2Cl_2/MeOH): mp 115–116; $[\alpha]_D^{25}$ +57.3 (*c* 1.00, MeOH) [120]		*Total Synthesis*: Total synthesis of ancistrotanzanine B (**218a**) and its atropo-diastereomer, ancistroealaine A (**218b**), by Suzuki coupling of the naphthalene and isoquinoline building blocks, with the coupling positions activated by an iodine substituent (for the isoquinoline part) and by a boronic acid entity (for the naphthalene moiety), leading to atropisomeric ratios of up to 75:25 in favor of ancistrotanzanine B (**218a**) when using chiral palladium catalysts [337, 462] Asymmetric total synthesis of ancistrotanzanine B (**218a**) by Suzuki–Miyaura cross-coupling between the respective naphthyl pinacol boronate and an aryl iodide as the still monocyclic isoquinoline precursor equipped with an (*S*)-2-(*N*-acetylamino)propyl group in the *ortho*-position, using the chiral aminopropyl substituent as an atroposelectivity-controlling group [120, 121]—coupling step delivering the targeted (*M*)-configured biaryl product (major atropo-diastereomer, 13:1, (*M*):(*P*)) by the use of $Pd(OAc)_2$ as the catalyst in the presence of *t*-BuXPhos as the ligand and K_3PO_4 as the base; total synthesis of **218a** accomplished by Pd-catalyzed hydrogenolysis (leading to the respective chiral acetamide) and Bischler–Napieralski ring closure [120]	

(continued)

Table 7 (continued)

Name, formula (mol. wt.), and structure Physical data [*M.p.*/°C, $[\alpha]_D^T$/°cm^2 g^{-1} 10^{-1} (solvent)]	Producing plants and their organs	Isolation, structure elucidation, and further investigations Partial and/or total synthesis	Biological activities
Ancistroealaine A (218b) $C_{26}H_{29}NO_4$ (419.5) 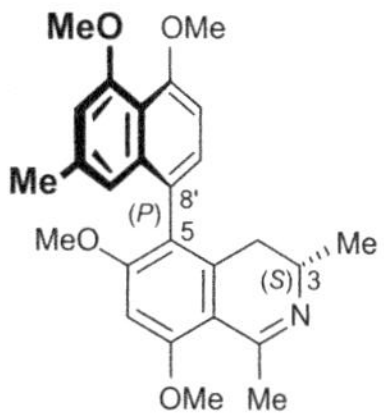 Light-yellow powder (from MeOH): mp 94–96; $[\alpha]_D^{25}$ −34.3 (*c* 0.55, EtOH) [174] Amorphous solid (from MeOH/CH_2Cl_2): mp 92; $[\alpha]_D^{25}$ −38.3 (*c* 0.45, EtOH) [167] Yellow amorphous powder (from acetonitrile): $[\alpha]_D^{20}$ −52.6 (*c* 0.10, MeOH) [171]	*A. tectorius* (PR China): twigs and stems [62] *A. ealaensis* (DR Congo): stem bark [174] *A. likoko* (DR Congo): roots [167, 495] An as yet botanically unknown Central Congolese *Ancistrocladus* species: roots [171]	Isolated for the first time from *A. ealaensis* collected in the Eala Botanical Garden, near the town of Mbandaka, Democratic Republic of the Congo [174] *Structural Elucidation*: Screening studies on root bark metabolites of the Congolese species *A. likoko* by LC-MS/MS and LC-NMR (applying H,H-TOCSY and ROESY experiments) in stopped-flow mode: elucidation of the constitution and relative configuration of **218b** online, directly from crude extracts [495] Constitution and relative stereostructure deduced from HREIMS, and from ^{1}H and ^{13}C NMR, including ^{1}H,^{1}H-COSY, and specific ROESY and HMBC interactions [174] Absolute configuration at C-3 established as (*S*) by oxidative degradation [171, 174] Absolute axial configuration assigned as (*P*) due to specific ROESY interactions in conjunction with the absolute (*S*)-configuration at C-3 [174]	Excellent inhibitory activity in vitro against *Leishmania donovani* [174], but virtually no inhibitory effects against *L. major* [66] Pronounced inhibitory activities in vitro against asexual erythrocytic forms of *Plasmodium falciparum* (strains: NF54/64 and K1) [174] Pronounced inhibitory activity in vitro against *Trypanosoma cruzi*, but only moderate activity against *T. brucei rhodesiense* [174]

(continued)

Table 7 (continued)

Name, formula (mol. wt.), and structure Physical data [*M.p.*/°C, $[\alpha]_D^T$/°cm^2 g^{-1} 10^{-1} (solvent)]	Producing plants and their organs	Isolation, structure elucidation, and further investigations Partial and/or total synthesis	Biological activities
cf.: Ancistroealaine A (**218b**) and its atropo-diastereomer, ancistrotanzanine B (**218a**), prepared by stereoselective total synthesis [120, 121, 337, 462]: Amorphous solid (from MeOH): mp 93; $[\alpha]_D^{25}$ −22.9 (*c* 0.025, EtOH) [462] Yellow solid (from CH_2Cl_2/MeOH): mp 179–180; $[\alpha]_D^{25}$ −73.2 (*c* 1.00, MeOH) [121]		Additional proof of the structural assignment by semi-synthetic transformation (see Scheme) of ancistroealaine A (**218b**) into the related naphthyltetrahydroisoquinoline alkaloid ancistrobertsonine C (**264**) [190], by *cis*-selective reduction (*dr* >95:5) with $NaBH_4$, followed by reductive N-methylation with formaldehyde-$NaBH_4$ to give **264** [121, 174, 190], identical with an authentic sample of natural ancistrobertsonine C (**264**), by HPLC co-elution and by ECD and NMR [174, 190] 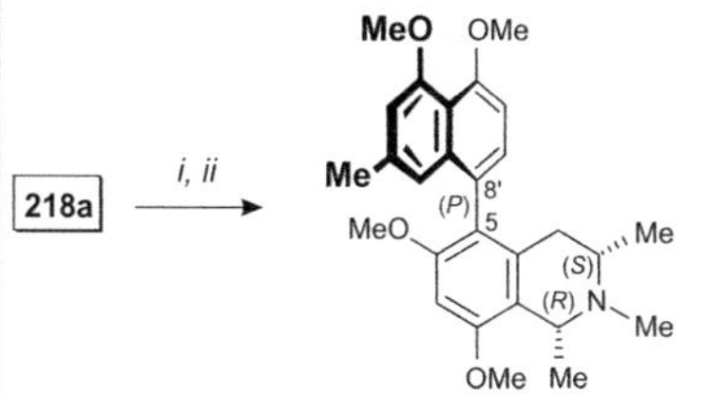 **264** Reaction conditions: (i) $NaBH_4$, MeOH, r.t.; (ii) HCHO, MeOH, r.t. → $NaBH_4$, MeOH, r.t.	

(continued)

Table 7 (continued)

Name, formula (mol. wt.), and structure Physical data [*M.p.*/°C, $[\alpha]_D^T$/°cm^2 g^{-1} 10^{-1} (solvent)]	Producing plants and their organs	Isolation, structure elucidation, and further investigations Partial and/or total synthesis	Biological activities
		Total Synthesis: Total synthesis of ancistroealaine A (**218b**) and its atropo-diastereomer, ancistrotanzanine B (**218a**), by Suzuki coupling of the naphthalene and isoquinoline portions, with the coupling positions activated by a boronic acid group (for the naphthalene moiety) and by an iodine substituent (for the isoquinoline part), leading to atropisomeric ratios of up to 75:25 in favor of ancistrotanzanine B (**218a**) when using chiral palladium catalysts [337, 462] Asymmetric total synthesis of ancistroealaine A (**218b**) by Suzuki–Miyaura cross-coupling between the respective naphthyl pinacol boronate and an aryl iodide as the still monocyclic isoquinoline precursor equipped with an (*S*)-2-(*N*-acetylamino)propyl group in the *ortho*-position, using the chiral aminopropyl substituent as an atroposelectivity-controlling group [120, 121]—coupling step achieved with $Pd(OAc)_2$ as the catalyst in the presence of SPhos as the ligand and $Ba(OH)_2$ as the base, affording the desired (*P*)-configured biaryl product as the major atropo-diastereomer (>5:1, (*M*):(*P*)); total synthesis of **218b** accomplished by Pd-catalyzed hydrogenolysis (leading to the respective acetamide) and Bischler–Napieralski ring closure [121]	

(continued)

Table 7 (continued)

Name, formula (mol. wt.), and structure Physical data [*M.p.*/°C, $[\alpha]_D^T$/°cm^2 g^{-1} 10^{-1} (solvent)]	Producing plants and their organs	Isolation, structure elucidation, and further investigations Partial and/or total synthesis	Biological activities
Ancistectorine D (124) $C_{25}H_{27}NO_4$ (405.2) 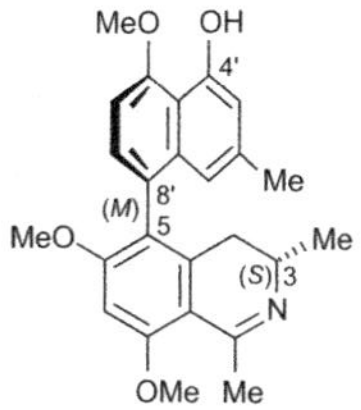 Compd. can likewise be addressed as 4′-*O*-demethyl-ancistrotanzanine B ——— Colorless amorphous powder (from MeOH/CH_2Cl_2): $[\alpha]_D^{20}$ +70.0 (*c* 0.44, MeOH) [62]	*A. tectorius* (PR China): twigs and stems [62]	*Structural Elucidation*: Constitution established by HRESIMS, and by ^{1}H NMR and ^{13}C NMR, including HMBC and NOESY interactions, indicative of an 5,8′-coupled naphthyldihydroisoquinoline with a free hydroxy function at C-4′ [62] Absolute configuration at C-3 established as (*S*) by oxidative degradation [62] Assignment of the absolute axial configuration as (*M*) due to specific long-range NOESY correlations and the given (*S*)-configuration at C-3, and by the nearly identical ECD spectrum of **124** compared to that of the likewise 5,8′-coupled and (*M*)-configured ancistrotanzanine B (**218b**) [62, 155]	Pronounced inhibitory activities in vitro against *Trypanosoma cruzi* and *Leishmania donovani*, but only moderate activities against the NF54 strain of *Plasmodium falciparum* [62] Strong cytotoxic activities against human leukemia (CCRF-CEM) and multidrug-resistant tumor cells (CEM/ADR5000) [62]
6-*O*-Demethylancistectorine D (219) $C_{24}H_{25}NO_4$ (391.2) 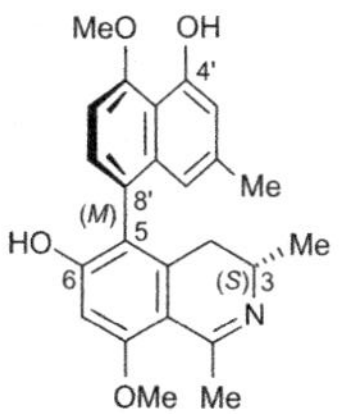 ——— Colorless amorphous powder (from MeOH): mp 168; $[\alpha]_D^{20}$ +33.2 (*c* 0.10, MeOH) [62]	*A. tectorius* (PR China): twigs and stems [62]	*Structural Elucidation*: Constitution and complete stereostructure by MS, NMR, and ECD, and by oxidative degradation—comparison of the data of **219** with those obtained from its likewise (3*S*,5*M*)-configured parent compound ancistectorine D (**124**), revealing **219** to differ from **124** by the presence of a free hydroxy function at C-6 [62]	Biological activities not determined for lack of isolated material

Table 8 7,1′-coupled Ancistrocladaceae-type naphthylisoquinoline alkaloids

Name, formula (mol. wt.), and structure Physical data [*M.p.*/°C, $[\alpha]_D^T$/°cm^2 g^{-1} 10^{-1} (solvent)]	Producing plants and their organs	Isolation, structure elucidation, and further investigations Partial and/or total synthesis	Biological activities
Ancistrogriffine B (**146**) $C_{24}H_{27}NO_4$ (393.5) 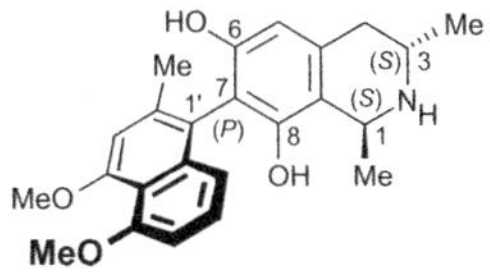 —	*A. griffithii* (Thailand): twigs [43]	*Structural Elucidation*: Elucidation of the constitution and complete stereostructure achieved online, by hyphenation of HPLC-MS, HPLC-NMR (^{1}H, WET-ROESY), and HPLC-ECD, directly from crude twig extracts, without prior isolation of the compound [43] HPLC-NMR experiments, including LC-ROESY measurements in stopped-flow mode, suggesting the presence of a 7,1′-coupled alkaloid with two hydroxy functions at C-6 and C-8 in the tetrahydroisoquinoline portions and with two methoxy groups in the naphthalene subunit [43] Relative configuration at C-1 versus C-3 established as *trans* from ROESY interactions, monitored online, between Me-1 and H-3 [43] Absolute configuration at C-3 established as (*S*) by ruthenium-mediated oxidative degradation [79, 80] of the crude plant extract, due to an exclusive formation (>99:1) of (3*S*)-aminobutyric acid, thus revealing the presence of di- or tetrahydroisoquinolines with an (*S*)-configuration at C-3 in the extract mixture, exclusively [43] Attribution of the absolute axial configuration as (*P*) [43] by comparison of the online LC-ECD spectrum of **146** with the ECD curve of the likewise 7,1′-coupled and (*P*)-configured (but 6-O-methylated) 6-*O*-methylancistectorine B_1 (**157**), which is well known from isolation [62] and synthesis work [190]	Biological activities not determined for lack of isolated material

(continued)

Table 8 (continued)

Name, formula (mol. wt.), and structure Physical data [*M.p.*/°C, $[\alpha]_D^T$/°cm^2 g^{-1} 10^{-1} (solvent)]	Producing plants and their organs	Isolation, structure elucidation, and further investigations Partial and/or total synthesis	Biological activities
Ancistectorine B₁ (233) $C_{25}H_{29}NO_4$ (407.5) 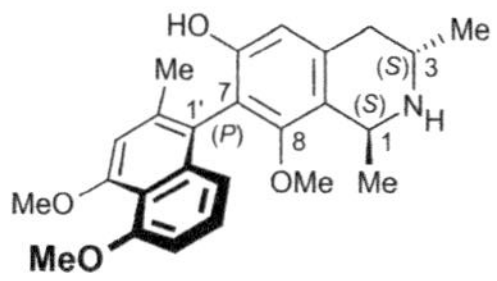 Colorless amorphous powder (from MeOH): mp 211; $[\alpha]_D^{20}$ +32.9 (*c* 0.09, MeOH) [61]	*A. tectorius* (PR China): stems and twigs [61]	*Structural Elucidation*: Constitution, coupling position, and relative *trans*-configuration of the two methyl groups at C-1 versus C-3 evidenced by HRESIMS and by ^{1}H and ^{13}C NMR, including specific long-range HMBC and NOE correlations [61] Absolute configuration at C-3 established as (*S*) by oxidative degradation [61] Assignment of the absolute axial configuration as (*P*) due to specific NOE interactions in combination with the absolute (1*S*,3*S*)-stereoarray in the tetrahydroisoquinoline portion, and by the nearly identical ECD spectrum of **233** compared to that of the structurally closely related known [56] 7,1′-coupled, and likewise (*P*)-configured, alkaloid ancistrotectoriline B (**232**) [61]	Weak inhibitory activity in vitro against *Plasmodium falciparum* (K1 strain) [61] No inhibitory activities against *Trypanosoma* and *Leishmania* parasites [61] No inhibitory activities against *Staphylococcus aureus*, *S. epidermidis*, *Enterococcus faecalis*, *E. faecium*, *Escherichia coli*, *Pseudomonas aeruginosa*, *Yersinia pseudotuberculosis*, and *Y. pestis* [61]
6-*O*-Methylancistectorine B₁ (157) $C_{25}H_{29}NO_4$ (407.5) 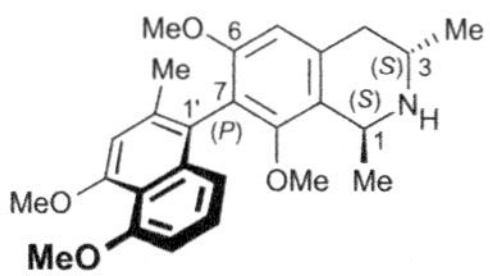 Colorless amorphous powder (from MeOH): mp 260–262; $[\alpha]_D^{20}$ +34.2 (*c* 0.10, $CHCl_3$) [62]	*A. tectorius* (PR China): stems and twigs [62]	*Structural Elucidation*: Constitution and complete stereostructure by MS, NMR, and ECD, and by oxidative degradation—comparison of the data of **157** with those obtained from its likewise (3*S*,7*P*)-configured parent compound ancistectorine B₁ (**233**) [61], revealing **157** to be an O-permethylated naphthylisoquinoline alkaloid, thus differing from **233** by the presence of an additional methoxy function at C-6 [62] Confirmation of the absolute axial configuration assigned as (*P*) by comparison of the ECD spectrum of **157** with that of the structurally closely related co-occurring alkaloid ancistectorine B₂ (**210**) (opposite descriptors of **157** and **210**	Moderate inhibitory activity in vitro against *Plasmodium falciparum* (K1 and NF54 strains) [62, 190] Weak inhibitory activities against *Trypanosoma* and *Leishmania* parasites [62]

(continued)

Table 8 (continued)

Name, formula (mol. wt.), and structure / Physical data [*M.p.*/°C, $[\alpha]_D^T$/°cm^2 g^{-1} 10^{-1} (solvent)]	Producing plants and their organs	Isolation, structure elucidation, and further investigations / Partial and/or total synthesis	Biological activities
cf.: Title compd. prepared semi-synthetically: By *trans*-selective reduction of ancistrocladisine (**2a**) using $LiAlH_4/AlMe_3$, giving a colorless solid (from $CHCl_3$/MeOH): mp 260–262; $[\alpha]_D^{20}$ +43.0 (*c* 0.13, $CHCl_3$) [190]		due to the priority rules of the Cahn–Ingold–Prelog denotation) [62] ——— *Partial Synthesis*: Semi-synthesis of 6-*O*-methylancistectorine B_1 (**157**) by *trans*-selective reduction of ancistrocladisine (**2a**) using $LiAlH_4/AlMe_3$ [190]	
Ancistrotectoriline B (**232**) $C_{25}H_{29}NO_4$ (407.5) 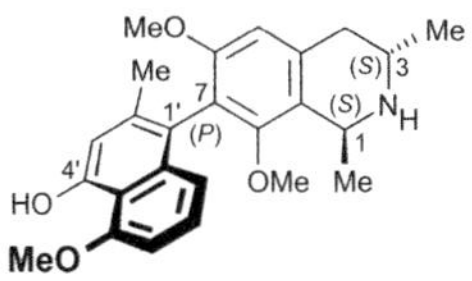 ——— Light-yellow powder (from $CHCl_3$): $[\alpha]_D^{20}$ +79.1 (*c* 0.46, $CHCl_3$) [56]	*A. tectorius* (PR China): stem bark, twigs, and leaves [56, 60]	*Structural Elucidation*: Constitution and relative configurations at the three stereogenic elements (two centers and one axis) deduced from HRESIMS and from ^{1}H and ^{13}C NMR, including HMBC and specific NOE interactions, revealing the presence of a 7,1′-coupled naphthyltetrahydroisoquinoline with a relative *trans*-configuration at C-1 versus C-3 [56, 60] Absolute configuration at C-1 and C-3 tentatively assumed by Ye et al. [56] as (1*S*,3*S*), but not evidenced experimentally Assignment of the absolute axial configuration as (*P*) due to a nearly mirror-image like ECD spectrum of **232** compared to that of the known, structurally related 7,1′-coupled alkaloid ancistrobrevine C (**266**) [185], thus revealing **232** to exhibit an opposite axial stereoarray; for formal reasons, however, the two alkaloids have the same (*P*)-descriptor, according to the Cahn-Ingold-Prelog denotation [56]	Moderate to weak growth-inhibitory effects against human leukemia HL-60, K562, and human macrophage U937 cell lines [60]

(continued)

Table 8 (continued)

Name, formula (mol. wt.), and structure Physical data [$M.p.$/°C, $[\alpha]_D^T$/°cm^2 g^{-1} 10^{-1} (solvent)]	Producing plants and their organs	Isolation, structure elucidation, and further investigations Partial and/or total synthesis	Biological activities
6-*O*-Demethyl-7-*epi*-ancistrobrevine D (178) $C_{25}H_{29}NO_4$ (407.5) 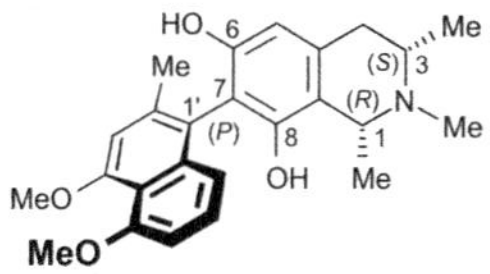 Colorless crystals (from $CHCl_3$/*n*-hexane): mp 229–232; $[\alpha]_D^{23}$ +51.8 (*c* 0.49, $CHCl_3$) [45]	*A. cochinchinensis* (Vietnam): leaves [45, 46]	*Structural Elucidation*: Constitution, coupling position, and relative *cis*-configuration of the two methyl groups at C-1 versus C-3 evidenced by HRESIMS and by ^{1}H and ^{13}C NMR, including specific HMBC and NOE correlations [45] Absolute configuration at C-3 established as (*S*) by oxidative degradation [45] Absolute configuration at C-1 assigned as (*R*) due to the relative *cis*-configuration at C-1 versus C-3 as deduced from NOE interactions and the given absolute (*S*)-configuration at C-3 [45] Absolute axial configuration assigned as (*P*) due to the absolute (1*R*,3*S*)-configuration at the stereocenters in the isoquinoline part (determined by degradation) and the relative configuration of centers versus axis (known from NMR) [45, 46] Additional proof of the absolute axial configuration established as (*P*) by the virtually opposite ECD spectrum of **178** compared to that of the known, structurally closely related ancistrobrevine D (**171a**) [13, 191, 420] and the nearly identical ECD spectrum of **178** as that of 7-*epi*-ancistrobrevine D (**171b**) [45, 161, 420], yet, for formal reasons, with opposite descriptors for **178** and **171b**, according to the Cahn-Ingold-Prelog priority rules [45, 46]	Biological activities not determined for lack of isolated material

(continued)

Table 8 (continued)

Name, formula (mol. wt.), and structure Physical data [*M.p.*/°C, $[\alpha]_D^T$/°cm^2 g^{-1} 10^{-1} (solvent)]	Producing plants and their organs	Isolation, structure elucidation, and further investigations Partial and/or total synthesis	Biological activities
6-*O*-Demethyl-8-*O*-methyl-7-*epi*-ancistrobrevine D (179) $C_{26}H_{31}NO_4$ (421.5) 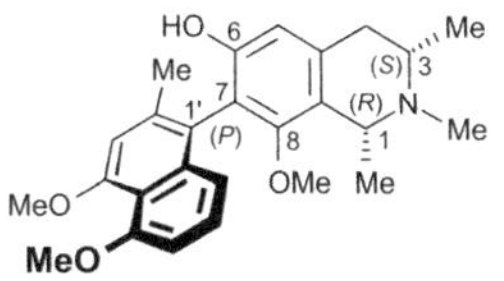 Amorphous solid (from $CHCl_3$/MeOH): $[\alpha]_D^{23}$ +51.8 (*c* 0.71, $CHCl_3$) [45]	*A. cochinchinensis* (Vietnam): leaves [45, 46] *A. tectorius* (PR China): stems and twigs [62]	Isolated for the first time from the Vietnamese liana *A. cochinchinensis* [45, 46] *Structural Elucidation*: Constitution and complete stereostructure by MS, NMR, and ECD, and by oxidative degradation—comparison of the data of **179** with those obtained from its likewise (3*S*,7*P*)-configured parent compound 6-*O*-demethyl-7-*epi*-ancistrobrevine D (**178**) [45, 46], revealing **179** to be an *N*-methylated and *cis*-configured 7,1′-coupled naphthyltetra-hydroisoquinoline alkaloid, differing from **178** by a methoxy function at C-8 (instead of a free hydroxy group) [45, 46, 62] Additional proof of the absolute axial configuration established as (*P*) by the nearly identical ECD spectrum of **179** compared to that of 7-*epi*-ancistrobrevine D (**171b**), yet, for formal reasons, with opposite descriptors for **179** and **171b**, according to the Cahn-Ingold-Prelog priority rules [45, 46, 62]	Biological activities not determined for lack of isolated material
7-*epi*-Ancistrobrevine D (171b) $C_{26}H_{31}NO_4$ (421.5) 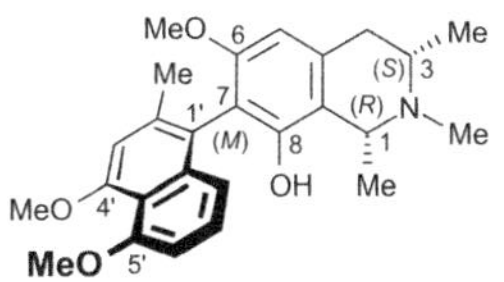 Colorless crystals (from $CHCl_3$/*n*-hexane): mp 177–178; $[\alpha]_D^{23}$ +46.1 (*c* 1.00, $CHCl_3$) [45]	*A. cochinchinensis* (Vietnam): leaves [45, 46] *A. tectorius* (PR China): stems, leaves, and twigs [60] An as yet botanically unknown Congolese *Ancistrocladus* species: root bark [161]	Isolated for the first time from the Vietnamese liana *A. cochinchinensis* [45, 46] *Structural Elucidation*: Constitution and coupling position deduced from HRESIMS and from ^{1}H and ^{13}C NMR, including HMBC and NOESY correlations, indicating the presence of a 7,1′-linked N-methylated naphthyltetrahydroisoquinoline with methoxy groups at C-4′, C-5′, and C-6 and a free hydroxy function at C-8 [45] Relative *cis*-configuration at C-1 versus C-3 deduced from specific NOE interactions [45]	Moderate preferential cytotoxicity against PANC-1 human pancreatic cancer cells under 'austerity' (i.e., nutrient-deprived) conditions [161] Moderate to weak growth-inhibitory effects against human leukemia HL-60, K562, and human macrophage U937 cell lines [60] Pronounced cytotoxicity of alkaloidal fractions of three Vietnamese *Ancistrocladus* species against

(continued)

Table 8 (continued)

Name, formula (mol. wt.), and structure Physical data [$M.p.$/°C, $[\alpha]_D^T$/°cm^2 g^{-1} 10^{-1} (solvent)]	Producing plants and their organs	Isolation, structure elucidation, and further investigations Partial and/or total synthesis	Biological activities
cf.: Title compd. prepared by stereoselective total synthesis giving a white solid (from MeOH): mp 166; $[\alpha]_D^{20}$ +46.4 (*c* 1.00, $CHCl_3$) [420]		Absolute configuration at C-3 established as (*S*) by ruthenium-mediated oxidative degradation [45] Assignment of the absolute axial configuration as (*M*) due to specific NOESY correlations in conjunction with the absolute (1*R*,3*S*)-stereoarray in the tetrahydroisoquinoline part [45] ——— *Total Synthesis*: Unambiguous structural proof by stereoselective total synthesis (by applying the lactone method [84, 85, 87–90]) of all the four possible *cis*-configured stereoisomers and comparison of their physical constants, ^{1}H NMR data, and their chromatographic behaviors on a chiral phase (Chiralcel® OD) with those of the natural product **171b** [420]	hepatocellular carcinoma (Hep-G2) and rhabdosarcoma (RD) cell lines [272]
Ancistectorine B$_2$ (210) $C_{24}H_{27}NO_4$ (393.2) 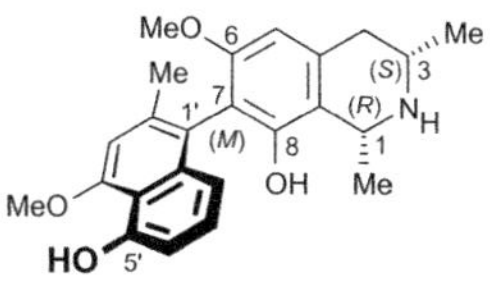 ——— Colorless amorphous powder (from $CHCl_3$): $[\alpha]_D^{20}$ +62.3 (*c* 0.77, MeOH) [62]	*A. tectorius* (PR China): stems and twigs [62]	*Structural Elucidation*: Constitution and relative configurations at the three stereogenic elements (two centers and one axis) deduced from HRESIMS and from ^{1}H and ^{13}C NMR, including HMBC and specific NOE interactions, revealing the presence of a 7,1′-coupled naphthyltetrahydroisoquinoline equipped with free hydroxy functions at C-8 and C-5′ [62] Absolute configuration at C-3 established as (*S*) by oxidative degradation and at C-1 as (*R*) due to the relative *cis*-configuration at C-1 versus C-3 (as known from NOE) [62] Assignment of the absolute axial configuration by the close resemblance of the ECD spectrum of **210** with that of the	Biological activities not determined for lack of isolated material

(continued)

Table 8 (continued)

Name, formula (mol. wt.), and structure Physical data [*M.p.*/°C, $[\alpha]_D^T$/°cm² g⁻¹ 10⁻¹ (solvent)]	Producing plants and their organs	Isolation, structure elucidation, and further investigations Partial and/or total synthesis	Biological activities
		related alkaloid 6-*O*-methylancistectorine B_1 (**157**), which occurs in the Chinese liana *A. tectorius*, too, yet, with opposite descriptors for **210** and **157**, according to the Cahn-Ingold-Prelog priority rules [62]	
Ancistrocladisine (**2a**) $C_{26}H_{29}NO_4$ (419.5) 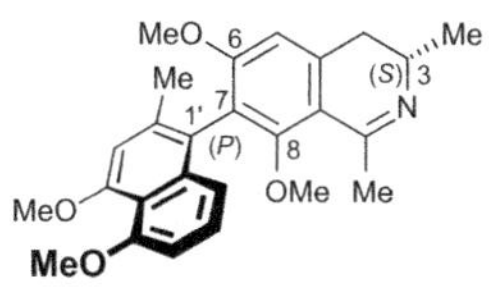 Pale-brown needles (from diethyl ether): mp 178–180 [5, 26, 487] As its hydrochloride salt, obtained as white needles (from MeOH/acetone): mp 220–222; $[\alpha]_D^{25}$ −16.1 (*c* 1.29, $CHCl_3$) [5, 94] Title compd. prepared by semi-synthesis giving an amorphous solid (from CH_2Cl_2): mp 176; $[\alpha]_D^{25}$ +6.0 (*c* 0.04, $CHCl_3$) [190] cf.: Ancistrocladisine (**2a**) and its atropo-diastereomer, 7-*epi*-ancistrocladisine (**2b**), prepared by atropo-divergent stereoselective total synthesis [93, 94]: Ancistrocladisine (**2a**): mp 178; $[\alpha]_D^{25}$ +7.8 (*c* 0.53, $CHCl_3$) [93]	*A. heyneanus* (India): roots [5, 26] and aerial parts [487] *A. hamatus* (Sri Lanka): roots [316]	Isolated for the first time from the Indian liana *A. heyneanus* [5, 26] *Structural Elucidation*: Constitution and complete stereostructure by MS, NMR, and ECD—evidence for the structure and substitution pattern of ancistrocladisine (**2a**) from its spectral data, from its Hofmann degradation products, and from chemical investigations [5, 26, 27] Assignment of the absolute configuration at C-3 as (*S*) by exhaustive ozonolysis to give L-β-amino-*n*-butyric acid [5, 27] Elucidation of the absolute axial configuration by application of the Exciton Chirality method after conversion of ancistrocladisine (**2a**) into its fully dehydrogenated analogue, didehydroancistrocladisine [5, 27] X-ray diffraction analysis (rigorous structural proof of the constitution and relative configuration) [28] *Total Synthesis and Related Work*: Semi-synthesis of ancistrocladisine (**2a**) by treatment of 1,2-didehydroancistrobertsonine D (**265**), which differs from **2a** only by the lack of an *O*-methyl group at C-6, with diazomethane [190]	Moderate inhibitory activity in vitro against *Plasmodium falciparum* (K1 and NF54 strains) [190] Moderate to low antifeedant and growth-retarding activity against the herbivorous insect *Spodoptera littoralis* [496]

(continued)

Table 8 (continued)

Name, formula (mol. wt.), and structure Physical data [*M.p.*/°C, $[\alpha]_D^T$/°cm^2 g^{-1} 10^{-1} (solvent)]	Producing plants and their organs	Isolation, structure elucidation, and further investigations Partial and/or total synthesis	Biological activities
Ancistrocladisine (**2a**) as its hydrochloride salt: $[\alpha]_D^{25}$ −20.5 (*c* 0.50, $CHCl_3$) [93] 7-*epi*-Ancistrocladisine (**2b**): mp 221–224; $[\alpha]_D^{25}$ −36.8 (*c* 0.36, $CHCl_3$) [93]		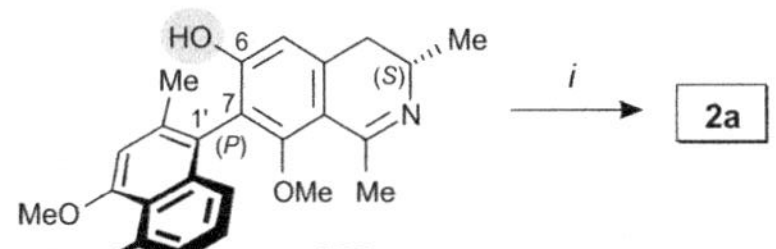 **265** (i) CH_2N_2 in Et_2O, MeOH-H_2O (9:1), r.t., 10 h Analogous semi-synthesis of *ent*-ancistrocladisine (*ent*-**2a**) by diazomethane methylation of ancistrobrevine C (**266**), which differs from *ent*-**2a** by the lack of an *O*-methyl group at C-8 [185] 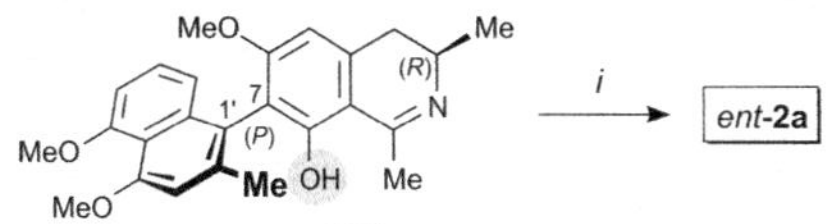 **266** Reaction conditions: (i) CH_2N_2 in, Et_2O, MeOH, r.t.; 2 d Synthesis of an inseparable mixture of all four possible stereoisomers of ancistrocladisine (**2a**) based on Meyers' oxazoline approach for the construction of the biaryl linkage; 'homogenization' of this mixture by dehydrogenation afforded uniformous, yet unnatural (and racemic!) dehydroancistrocladisine [317, 318] Stereocontrolled total synthesis of ancistrocladisine (**2a**) by applying the lactone method [84, 85, 87–90], by Pd-catalyzed intramolecular aryl coupling of the two ester-type prefixed molecular moieties and atroposelective lactone cleavage, giving rise to **2b** or, optionally, to its unnatural atropo-diastereomer, 7-*epi*-ancistrocladisine (**2b**) [84, 85, 90, 93, 94]	

Table 9 7,1′-coupled Dioncophyllaceae-type naphthylisoquinoline alkaloids

Name, formula (mol. wt.), and structure Physical data [*M.p.*/°C, $[\alpha]_D^T$/°cm^2 g^{-1} 10^{-1} (solvent)]	Producing plants and their organs	Isolation, structure elucidation, and further investigations Partial and/or total synthesis	Biological activities
Dioncophylleine A (158a) and ***ent*-Dioncophylleine A (158b)** $C_{24}H_{23}NO_3$ (373.5) 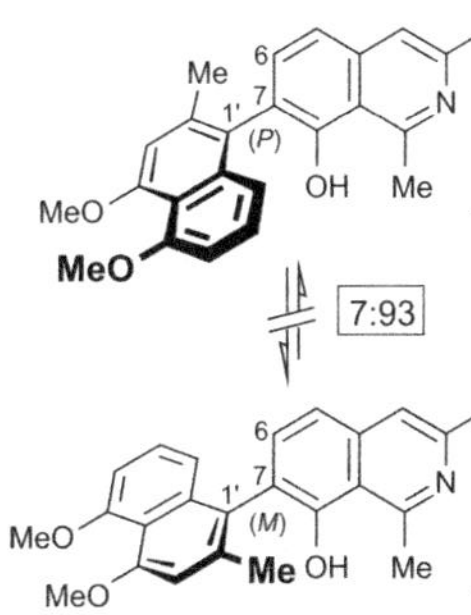 158a 158b Occurring as a scalemic mixture in *A. benomensis* [44] ——— White solid (from MeOH): mp 265; $[\alpha]_D^{25}$ −29.7 (*c* 0.08, MeOH) [44] cf.: (*P*)-Isomer dioncophylleine A (**158a**) prepared by semi-synthesis, obtained as a white solid (from EtOH): mp 272 [414, 496, 497]	*A. benomensis* (Malaysia): leaves [44] *A. abbreviatus* (Ivory Coast): roots [186], only **158b** identified in the plant	Isolated for the first time from the Malaysian liana *A. benomensis* [44] Categorization of **158a/b** as a Dioncophyllaceae-type naphthylisoquinoline alkaloid, since lacking an oxygen function at C-6, and thus, substantially differing from (nearly) all the other alkaloids found in Asian *Ancistrocladus* species with respect to the exclusive production of Ancistrocladaceae-type compounds [44] *Structural Elucidation*: Constitution established by HREIMS and by ^{1}H and ^{13}C NMR, indicating the presence of a non-hydrogenated naphthyliso-quinoline alkaloid—assignment of the coupling position and the location of a proton at C-6 typical of a Dioncophyllaceae-type compound by NOESY and HMBC interactions [44, 186] Attribution of the absolute axial configuration as (*M*) due to the virtually opposite ECD spectrum of **158b** compared to that of its (*P*)-configured enantiomer dioncophylleine A (**158a**), as prepared semi-synthetically, by dehydrogenation of dioncophylline A (**16**) [13, 44, 414] HPLC-ECD analysis on a chiral phase (Chiralcel® OD), revealing that *A. benomensis* produces *ent*-dioncophylleine A (**158b**) in a scalemic form, as a 93:7-mixture of **158b** and its (*P*)-configured enantiomer, dioncophylleine A (**158a**) [44]	Weak inhibitory activities against *Leishmania major* [66], *Plasmodium falciparum* (K1 strain), and *Trypanosoma brucei rhodesiense* [44] No inhibitory activity against *T. cruzi* [44] Moderate antiproliferative effects against lymphoblastic CCRF-CEM leukemia cells and their multi-drug resistant subline CEM/ADR5000 [186]

(continued)

Table 9 (continued)

Name, formula (mol. wt.), and structure Physical data [*M.p.*/°C, $[\alpha]_D^T$/°cm^2 g^{-1} 10^{-1} (solvent)]	Producing plants and their organs	Isolation, structure elucidation, and further investigations Partial and/or total synthesis	Biological activities
		Occurrence of **158b** in *A. abbreviatus* (Ivory Coast) in an enantiomerically pure form, as deduced from HPLC-ECD analysis on a chiral phase (Lux-Cellulose-1) [186]	
5′-*O*-Demethyl-*ent*-dioncophylleine A (166) $C_{23}H_{21}NO_3$ (359.4) 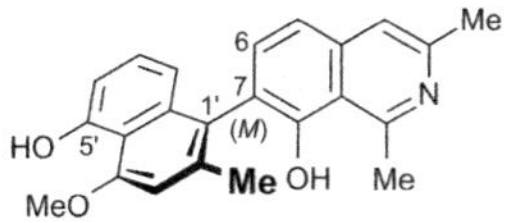 White solid (from MeOH): mp 195; $[\alpha]_D^{25}$ −27.3 (*c* 0.05, MeOH) [44]	*A. benomensis* (Malaysia): leaves [44]	*Structural Elucidation*: Constitution established by HREIMS and by ^{1}H and ^{13}C NMR, including NOESY and HMBC interactions—comparison of the data of **166** with those obtained from the related alkaloid *ent*-dioncophylleine A (**158b**), revealing the presence of a likewise non-hydrogenated Dioncophyllaceae-type naphthylisoquinoline, differing from **158b** by a phenolic hydroxy function at C-5′ (instead of a methoxy group) [44] Assignment of the absolute axial configuration as (*M*) due to a nearly identical ECD spectrum of **166** compared to that of the likewise (*M*)-configured *ent*-dioncophylleine A (**158b**) [44] In contrast to its parent compound *ent*-dioncophylleine A (**158b**), occurrence of **166** in *A. benomensis* in an enantiomerically pure form [264], as detected by LC-ECD on a chiral phase	Weak inhibitory activity in vitro against *Leishmania major* [66], *Plasmodium falciparum* (K1 strain) and *Trypanosoma brucei rhodesiense* [44] No inhibitory activities against *T. cruzi* [44]

Table 10 7,3′-coupled Ancistrocladaceae-type naphthylisoquinoline alkaloids

Name, formula (mol. wt.), and structure Physical data [*M.p.*/°C, $[\alpha]_D^T$/°cm^2 g^{-1} 10^{-1} (solvent)]	Producing plants and their organs	Isolation, structure elucidation, and further investigations Partial and/or total synthesis	Biological activities
Ancistrotectorine (17a) $C_{26}H_{31}NO_4$ (421.5) 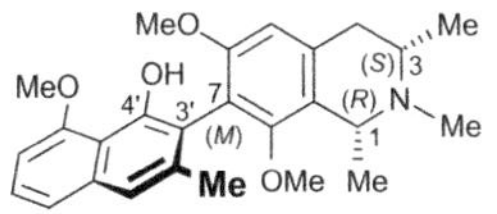 Pale-yellow needles (from acetone): mp 134–140; $[\alpha]_D^{25}$ 0.0 (CHCl$_3$) [51] Amorphous solid (from MeOH): mp 198–201 [58] Colorless needles (from EtOH): mp 150–151; $[\alpha]_D^{25}$ −3.6 (*c* 0.20, EtOH) [159] Amorphous solid (from MeOH): mp 149; $[\alpha]_D^{25}$ −4.1 (*c* 0.20, MeOH) [189] cf.: Title compd. prepared by regio- and atroposelective total synthesis [99]: Light-yellow solid (from MeOH): mp 149–151; $[\alpha]_D^{20}$ −1.6 (*c* 0.04, MeOH)	*A. tectorius* (Thailand): leaves [51] *A. tectorius* (PR China): stems, branches, and leaves [58, 60, 489] *A. tanzaniensis* (Tanzania): leaves [189] *A. guineënsis* (Nigeria): leaves [75, 159]	Isolated for the first time from *A. tectorius* collected at Sakae-Raj Environmental Research Station, Rachasima Province, Thailand [51] *Structural Elucidation*: Metabolic screening of a leaf extract of the Nigerian species *A. guineënsis* in stopped-flow mode by LC-NMR, applying H,H-TOCSY and ROESY experiments: constitution and relative configuration of **17a** elucidated online, directly from a crude extract [75] Constitution and relative configurations at the three stereogenic elements (two centers and one axis) deduced from HRESIMS and from ^{1}H and ^{13}C NMR, including HMBC and specific ROE interactions, revealing the presence of a 7,3′-coupled N-methylated and *cis*-configured naphthyltetrahydroisoquinoline equipped with one hydroxy function at C-4′ [51, 58, 159, 189] Single-crystal X-ray diffraction of ancistrotectorine (**17a**) (constitution and all relative configurations) [189] Attribution of the absolute axial configuration by comparison of the ECD spectrum of ancistrotectorine (**17a**) with that of the structurally similar, likewise 7,3′-coupled ancistrocladidine (**3a**) [27–29]; based on this finding, in combination with the X-ray structure analysis [51], assignment of the absolute stereostructure of **17a** [51, 159]	Moderate inhibitory activities against *Plasmodium falciparum* (K1 and 3D7 strains) [189] Weak inhibitory activities against *Trypanosoma cruzi* and *T. brucei rhodesiense* [189] No inhibitory activity against *Leishmania donovani* [189] Moderate growth-inhibitory effects against human leukemia HL-60, K562, and human macrophage U937 cell lines [60] Potent cytotoxic activity and a paclitaxel-resistance reversing effect of an 80% ethanol extract of *A. tectorius* (from Thailand) in PR-HepG2 cells [492]

(continued)

Table 10 (continued)

Name, formula (mol. wt.), and structure Physical data [*M.p.*/°C, $[\alpha]_D^T$/°cm^2 g^{-1} 10^{-1} (solvent)]	Producing plants and their organs	Isolation, structure elucidation, and further investigations Partial and/or total synthesis	Biological activities
		Absolute configuration at C-3 established by ruthenium-mediated oxidative degradation [80, 159, 189] Additional confirmation of the assignment of the absolute axial configuration of ancistrotectorine (**17a**) as (*M*) by quantum-chemical ECD calculations by means of the semiempirical CNDO/2S Hamiltonian method, based on the AM1-Boltzmann approach [76, 159], evidencing the (*M*)-configuration at the biaryl axis of **17a** ——— *Total Synthesis*: Atroposelective joint total synthesis of ancistrotectorine (**17a**) and the likewise 7,3′-coupled ancistrocladidine (**3a**) by applying the lactone method [84, 85, 87–90], here following the concept of 'inverse halogenation' (i.e., introduction of the halogen function for the activation of the coupling position into the phenolic isoquinoline and not into the naphthoate building block as normally): formation of the key biaryl lactone by intramolecular coupling of the ester-type prefixed molecular moieties by using the palladium-phosphinous acid catalyst (POPd)—atroposelective cleavage of the lactone auxiliary bridge using the CBS catalyst (*R*)-(–)-2-methyl-oxazaborolidine, delivering, with a high external asymmetric induction (*dr* = 94:6), the final joint intermediate **112a**, divergently leading to ancistrocladidine (**3a**) and ancistrotectorine (**17a**) [99]	

(continued)

Table 10 (Continued)

Name, formula (mol. wt.), and structure Physical data [*M.p.*/°C, $[\alpha]_D^T$/°cm^2 g^{-1} 10^{-1} (solvent)]	Producing plants and their organs	Isolation, structure elucidation, and further investigations Partial and/or total synthesis	Biological activities
Ancistrotanzanine C (114) $C_{25}H_{29}NO_4$ (407.5) 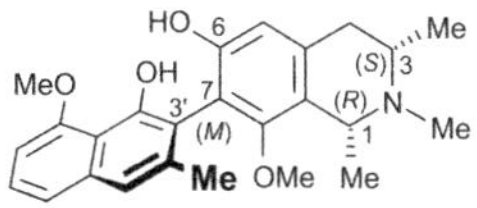 Compd. can also be addressed as 6-*O*-demethylancistrotectorine ——— Yellow oil: $[\alpha]_D^{25}$ −75.5 (*c* 0.01, $CHCl_3$) [41] Yellow oil: $[\alpha]_D^{25}$ −76.0 (*c* 0.01, $CHCl_3$) [189]	*A. heyneanus* (India): leaves [41] *A. tectorius* (PR China): twigs and leaves [466] *A. tanzaniensis* (Tanzania): leaves [189]	Isolated for the first time from *A. tanzaniensis* from the Uzungwa Mountains National Park of Central Tanzania [189] *Structural Elucidation*: Constitution and coupling position deduced from HRESIMS and from ^{1}H and ^{13}C NMR, including specific HMBC and long-range NOESY interactions, revealing **114** to differ from its parent compound ancistrotectorine (**17a**) by the presence of a hydroxy group (instead of a methoxy function) at C-6 [41, 189] Absolute configuration at C-3 established as (*S*) by oxidative degradation [41, 189] Absolute configuration at C-1 determined as (*R*) due to the relative *cis*-configuration at C-1 versus C-3 as deduced from NOE interactions, in combination with the given absolute (*S*)-configuration at C-3 [41, 189] Assignment of the absolute axial configuration as (*M*) due to specific NOESY correlations, in conjunction with the known absolute (1*R*,3*S*)-stereoarray in the tetrahydroisoquinoline part, confirmed by the nearly identical ECD spectrum of **114** compared to that of ancistrotectorine (**17a**), which is (*M*)-configured [41, 189]	Pronounced inhibitory activity in vitro against *Plasmodium falciparum* (K1 and 3D7 strains) [189] Weak to moderate inhibitory activities in vitro against *Trypanosoma brucei rhodesiense* and *T. cruzi* [41, 189] No inhibitory activities against *Leishmania donovani* [189] Significant free-radical scavenging activity of a methanolic leaf extract of *A. heyneanus* by inhibiting lipid peroxidation [498]
Ancistrocladidine (3a) $C_{25}H_{27}NO_4$ (405.5) 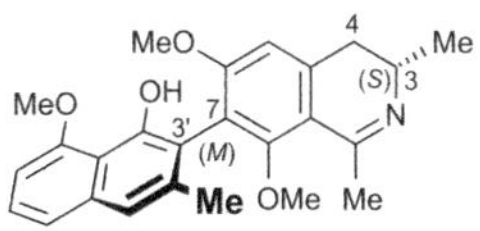 ——— Amorphous solid: mp 245–247 (dec); $[\alpha]_D^{25}$ −149.7 (*c* 0.025, MeOH) [5, 29]	*A. heyneanus* (India): leaves [41, 487] *A. tectorius* (Thailand): twigs and leaves [52] *A. tanzaniensis* (Tanzania): leaves [189]	Isolated for the first time from the Indian liana *A. heyneanus* [41] *Structural Elucidation*: Constitution and complete stereostructure by HRESIMS, 1D and 2D NMR, in particular by NOESY and INEPT measurements, and ECD—evidence for the structure and substitution pattern of ancistrocladidine (**3a**) from its spectral data, and from numerous chemical investigations [5, 27, 29, 41, 52, 189] Assignment of the absolute configuration at C-3 as (*S*) *by* exhaustive ozonolysis to give L-β-amino-*n*-butyric acid [5, 27]	Pronounced inhibitory activity in vitro against *Plasmodium falciparum* (K1 and 3D7strains) [189] Pronounced inhibitory activities in vitro against *Leishmania donovani* [189], but no inhibitory effects against *L. major* [66]

(continued)

Table 10 (continued)

Name, formula (mol. wt.), and structure Physical data [*M.p.*/°C, $[\alpha]_D^T$/°cm^2 g^{-1} 10^{-1} (solvent)]	Producing plants and their organs	Isolation, structure elucidation, and further investigations Partial and/or total synthesis	Biological activities
Amorphous solid: 248 [487] Amorphous solid (from MeOH): $[\alpha]_D^{25}$ −122.3 (*c* 0.05, MeOH) [41, 189] Pale-yellow needles (from acetone): mp 255–258; $[\alpha]_D^{25}$ −129.7 (*c* 0.064, $CHCl_3$) [52] cf.: Ancistrocladidine (**3a**) prepared by total synthesis: mp 253–254; $[\alpha]_D^{20}$ −136.0 (*c* 1,84, $CHCl_3$) [115, 116] Total synthesis of the unnatural 7-*epi*-ancistrocladidine (90% diastereomerically pure): $[\alpha]_D^{25}$ +40.4 (*c* 0.292, $CHCl_3$) [115, 116] cf.: Ancistrocladidine (**3a**) prepared by regio- and atroposelective total synthesis [99]: White solid (from MeOH): mp 232–234; $[\alpha]_D^{20}$ −118.6 (*c* 0.046, MeOH)		Elucidation of the absolute axial configuration by application of the Exciton Chirality method after conversion of ancistrocladidine (**3a**) into its fully dehydrogenated analogue, 3,4-didehydro-ancistrocladidine [5, 27] X-ray diffraction analysis, corroborating the constitution and relative configuration of **3a** [28] *Total Synthesis*: Total synthesis of ancistrocladidine (**3a**) and its atropo-diastereomer as a 1:1 mixture based on the formation of the sterically hindered biaryl linkage by Pinhey–Barton *ortho*-arylation of a respective naphthol building block equipped with an aryl lead triacetate unit; delivery of stereochemically pure ancistrocladidine (**3a**) by recrystallization out of the atropisomeric mixture [115, 116] Regio- and stereoselective total synthesis of ancistrocladidine (**3a**) and ancistrotectorine (**17a**) by applying the lactone method [84, 85, 87–90]: activation of the two molecular halves in an 'inverse-halogenated' form (i.e., with the halogen located in the phenolic part of the isoquinoline moiety), prior to the prefixation via an ester bridge; construction of the biaryl axis by intramolecular coupling using the POPd catalyst—atroposelective cleavage of the lactone auxiliary bridge using the CBS catalyst (*R*)-(−)-2-methyl-oxazaborolidine giving the final joint intermediate to ancistrocladidine (**3a**) and ancistrotectorine (**17a**) with a high external asymmetric induction (*dr* = 94:6), eventually, after oxidation with $KMnO_4$, leading to ancistrocladidine (**3a**) [99]	Moderate to weak inhibitory activities in vitro against *Trypanosoma* parasites [189] Inactive against HIV in vitro [55] Potent cytotoxic activity and a paclitaxel-resistance reversing effect of an 80% ethanol extract of *A. tectorius* in PR-HepG2 cells [493]

(continued)

Table 10 (continued)

Name, formula (mol. wt.), and structure Physical data [*M.p.*/°C, $[\alpha]_D^T$/°cm^2 g^{-1} 10^{-1} (solvent)]	Producing plants and their organs	Isolation, structure elucidation, and further investigations Partial and/or total synthesis	Biological activities
Ancistroheynine B (72) $C_{24}H_{25}NO_4$ (391.5) 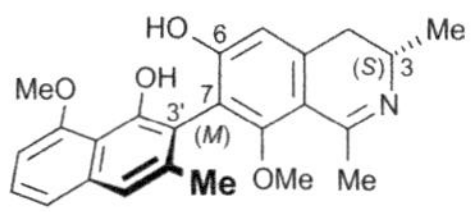 Compd. can also be addressed as 6-*O*-demethylancistrocladidine Yellow oil: $[\alpha]_D^{25}$ −194.9 (*c* 0.025, MeOH) [41]	*A. heyneanus* (India): leaves [41]	*Structural Elucidation*: Constitution and complete stereostructure established by HRESIMS, 1D and 2D NMR, oxidative degradation, and ECD—comparison of the data of **72** with those obtained from its parent compound ancistrocladidine (**3a**), revealing **72** to differ from **3a** by the presence of a free hydroxy function at C-6 [41]	Moderate to weak inhibitory activities in vitro against *Plasmodium falciparum* (K1 strain) and *Trypanosoma* parasites [91] Weak inhibitory activities in vitro against *Leishmania major* [66] and *L. donovani* [41] Significant free-radical scavenging activity of a methanolic leaf extract of *A. heyneanus* by inhibiting lipid peroxidation [498]

Table 11 7,6′-coupled Ancistrocladaceae-type naphthylisoquinoline alkaloids

Name, formula (mol. wt.), and structure Physical data [*M.p.*/°C, $[\alpha]_D^T$/° cm^2 g^{-1} 10^{-1} (solvent)]	Producing plants and their organs	Isolation, structure elucidation, and further investigations Partial and/or total synthesis	Biological activities
Ancistrotectorine A (237) $C_{26}H_{31}NO_4$ (421.5) 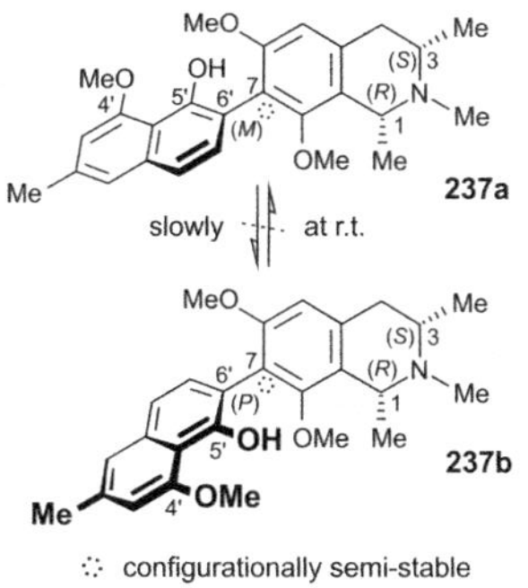 Light-yellow amorphous solid: $[\alpha]_D^{25}$ 0 (*c* 0.20, MeOH) [60]	*A. tectorius* (PR China): leaves and stems [60]	*Structural Elucidation*: Detection on TLC due to the positive reaction of ancistrotectorine A (**237**) to Dragendorff's reagent and isolation by preparative HPLC [60] Constitution deduced from HRESIMS, ^{1}H and ^{13}C NMR, and from HMBC interactions, revealing the presence of a 7,6′-coupled N-methylated naphthyltetrahydroisoquinoline alkaloid with a free hydroxy function at C-$_{5'}$ [60] Relative configuration at C-1 and C-3 assigned from the chemical shifts of H-1 (δ = 3.57 ppm) and H-3 (δ = 2.51 ppm) being in the range typical of a *cis*-configured 1,3-dimethyltetrahydroisoquinoline portion [60] Absolute configuration at C-1 and C-3 supposed by Jiang et al. [60] to be (1*R*,3*S*), based on the mere assumption that the configuration at C-3 should be (*S*) as for all other naphthylisoquinoline alkaloids so far isolated from *A. tectorius*, but not proven experimentally Medium steric hindrance at the biaryl axis due to the presence of an additional oxygen function at C-6 (Ancistrocladaceae-type alkaloid) as compared to the structurally similar and likewise 7,6′-coupled dioncophylline B (**153**) [219], leading to the occurrence of ancistrotectorine A (**237**) as a pair of two configurationally semi-stable, thus slowly interconverting atropo-diastereomers, **237a** and **237b** (ratio: ca. 1:1) [60]	Moderate to good growth-inhibitory effects against human leukemia HL-60, K562, and human macrophage U937 cell lines [60]

(continued)

Table 11 (continued)

Name, formula (mol. wt.), and structure Physical data [*M.p.*/°C, $[\alpha]_D^T$/° $cm^2\ g^{-1}\ 10^{-1}$ (solvent)]	Producing plants and their organs	Isolation, structure elucidation, and further investigations Partial and/or total synthesis	Biological activities
Ancistrotectorine B (18) $C_{25}H_{29}NO_4$ (407.5) 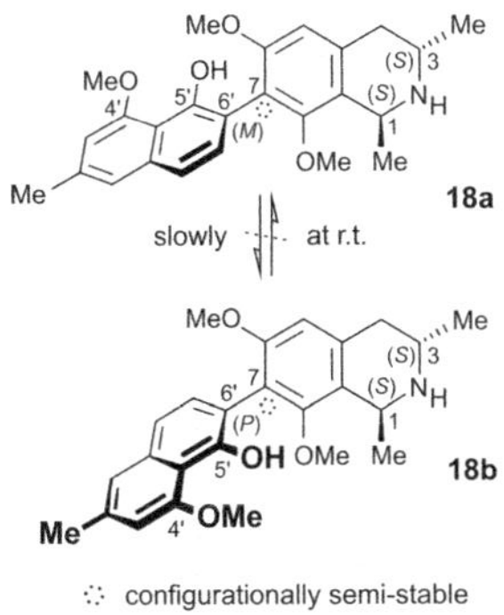 Light-yellow amorphous solid: $[\alpha]_D^{25}$ 0 (*c* 0.20, MeOH) [60]	*A. tectorius* (PR China): leaves and stems [60]	*Structural Elucidation*: Detection on TLC due to the positive reaction of ancistrotectorine B (**18**) to Dragendorff's reagent and isolation by preparative HPLC [60] Constitution established by HRESIMS, 1D and 2D NMR—comparison of the data of **18** with those obtained from ancistrotectorine A (**237**), indicating the presence of a 7,6′-coupled naphthyltetrahydroisoquinoline alkaloid with a free hydroxy function at C-5′, yet differing from **237** by the absence of an *N*-methyl function [60] Relative configuration at C-1 and C-3 assigned from the chemical shifts of H-1 (δ = 4.37 ppm) and H-3 (δ = 3.36 ppm) being in the range typical of a *trans*-configured 1,3-dimethyltetrahydroisoquinoline subunit [60] Absolute configuration at C-1 and C-3 supposed by Jiang et al. [60] to be (1*R*,3*S*), based on the finding that all alkaloids so far isolated from *A. tectorius* were exclusively (3*S*)-configured, but not proven experimentally Similar to ancistrotectorine A (**237**), only medium degree of steric hindrance at the biaryl axis of ancistrotectorine B (**18**), thus occurrence of **18** in the plant as a mixture of two configurationally semi-stable atropo-diastereomers, **18a** and **18b** (ratio: ca. 1:1) [60]	Moderate to good growth-inhibitory effects against human leukemia HL-60, K562, and human macrophage U937 cell lines [60]

(continued)

Table 11 (continued)

Name, formula (mol. wt.), and structure Physical data [$M.p./°C$, $[\alpha]_D^T/°$ $cm^2\ g^{-1}\ 10^{-1}$ (solvent)]	Producing plants and their organs	Isolation, structure elucidation, and further investigations Partial and/or total synthesis	Biological activities
Ancistrotectoriline C (206) $C_{25}H_{27}NO_4$ (405.5) 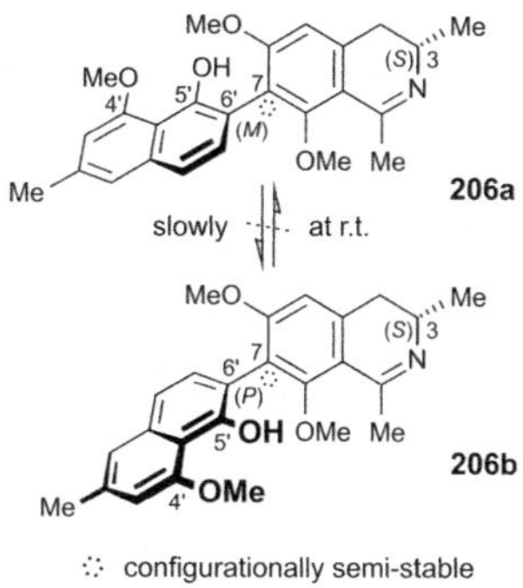 Pale-yellow amorphous powder: $[\alpha]_D^{20}$ 0 (c 0.086, $CHCl_3$) [59]	*A. tectorius* (PR China): leaves and stems [59]	*Structural Elucidation*: Detection on TLC due to the positive reaction of ancistrotectoriline C (**206**) to Dragendorff's reagent and isolation by CC and preparative TLC [59] Constitution deduced from HREIMS, ^1H and ^{13}C NMR, and from HMBC interactions, revealing the presence of a 7,6′-coupled naphthyldihydroisoquinoline alkaloid with a free phenolic hydroxy group at C-5′ [59] Absolute configuration at C-3 tentatively assigned by Ye et al. [59] as (*S*), based on the finding that all other alkaloids isolated so far from *A. tectorius* were exclusively (3*S*)-configured, but not proven experimentally Due to the medium degree of steric hindrance at the biaryl axis, there is the occurrence of ancistrotectoriline C (**206**) as a pair of slowly interconverting atropo-diastereomers, **206a** and **206b**, exhibiting two sets of very similar, but not superimposable NMR resonances—HPLC analysis revealing two close peaks in a ratio of ca. 1:1 [59]	Weak activity against the intracellular protozoan *Eimeria tenella*, the vector of avian coccidiosis in chicken [59] Moderate cytotoxicity to the coccidian parasites *Eimeria tenella* and *Neospora caninum* (leading to spontaneous abortion in dogs, cats, sheeps, and cattle) [59]

Table 12 7,8′-coupled Ancistrocladaceae-type naphthylisoquinoline alkaloids

Name, formula (mol. wt.), and structure Physical data [$M.p.$/°C, $[\alpha]_D^T$/° cm^2 g^{-1} 10^{-1} (solvent)]	Producing plants and their organs	Isolation, structure elucidation, and further investigations Partial and/or total synthesis	Biological activities
Ancistectorine C$_1$ (71a) $C_{26}H_{31}NO_4$ (421.5) 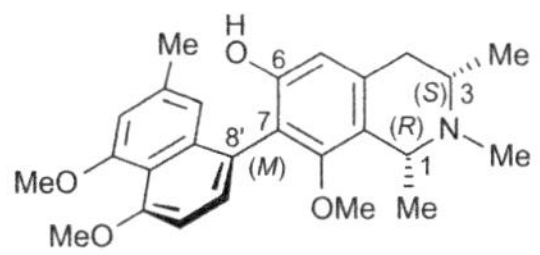 Colorless amorphous powder (from MeOH): mp 161; $[\alpha]_D^{20}$ +5.9 (c 0.09, MeOH) [61]	*A. tectorius* (PR China): twigs [61]	*Structural Elucidation*: Constitution, coupling position, and relative *cis*-configuration of the two methyl groups at C-1 versus C-3 evidenced by HRESIMS and by ^{1}H and ^{13}C NMR, including specific HMBC and NOE correlations [61] Absolute configuration at C-3 established as (*S*) by oxidative degradation [61] Assignment of the absolute axial configuration as (*M*) due to long-range NOESY correlations in combination with the known absolute (1*R*,3*S*)-stereoarray in the tetrahydroisoquinoline portion, and by the virtually opposite ECD spectrum of **71a** compared to that of its atropo-diastereomer, 6-*O*-demethylancistrobrevine A (**71b**), previously isolated from *A. abbreviatus* from Coastal West Africa [13, 191]	Weak inhibitory activity in vitro against the K1 strain of *Plasmodium falciparum* [61] No inhibitory activities against *Trypanosoma* and *Leishmania* parasites [61]
Ancistrogriffine A (145) $C_{25}H_{29}NO_4$ (407.5) 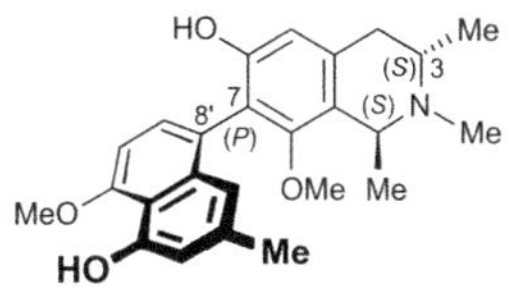 Colorless solid (from MeOH): mp 148; $[\alpha]_D^{25}$ +26.0 (c 0.40, MeOH) [43]	*A. griffithii* (Thailand): twigs [43]	*Structural Elucidation*: Constitution and complete stereostructure established online, directly from crude leaf extracts, using the analytical triad LC-MS/MS, LC-NMR (^{1}H, WET-ROESY, and TOCSY experiments), and LC-ECD [43] Absolute configuration at C-3 established as (*S*) by ruthenium-mediated oxidative degradation of the crude plant extract, due to an exclusive formation (>99:1) of (3*S*)-aminobutyric acid, thus revealing that only di- or tetrahydroisoquinoline substructures with an (*S*)-configuration at C-3 were present in the extract mixture [43, 80] Constitution and the relative *trans*-configuration previously deduced online, now confirmed by HRESIMS, and by ^{1}H and ^{13}C NMR (including HMQC, HMBC, and ROESY experiments) after isolation of ancistrogriffine A (**145**) in a pure form [43]	Pronounced inhibitory activities in vitro against *Plasmodium falciparum* (K1 strain) and *Leishmania donovani* [43] Moderate inhibitory activities in vitro against *Trypanosoma cruzi* and *T. brucei rhodesiense* [43]

(continued)

Table 12 (continued)

Name, formula (mol. wt.), and structure Physical data [*M.p.*/°C, $[\alpha]_D^T$/° $cm^2\ g^{-1}\ 10^{-1}$ (solvent)]	Producing plants and their organs	Isolation, structure elucidation, and further investigations Partial and/or total synthesis	Biological activities
		Attribution of the absolute axial configuration as (*P*) by stopped-flow online HPLC-ECD measurements and comparison of the ECD spectrum of ancistrogriffine A (**145**) [43] with that of the structurally related, and likewise 7,8′-coupled, 6-*O*-demethylancistrobrevine A (**71b**) [13, 171, 191]	
Ancistrogriffine C (**147**) $C_{24}H_{27}NO_4$ (393.5) 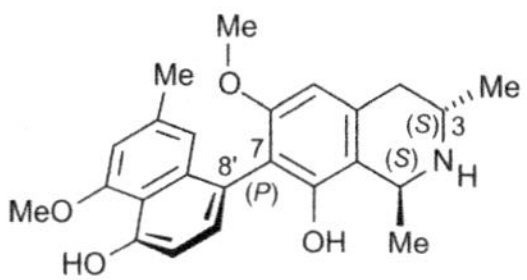 Light-yellow powder (from MeOH): mp 165; $[\alpha]_D^{25}$ +13.8 (*c* 0.25, MeOH) [43]	*A. griffithii* (Thailand): twigs [43]	*Structural Elucidation*: Constitution and stereostructure established online, directly from crude leaf extracts, by hyphenation of HPLC with MS/MS and with NMR (^{1}H, WET-ROESY experiments) [43] Absolute configuration at C-3 established as (*S*) by ruthenium-mediated oxidative degradation of the crude extract (see entry above) and of pure material of ancistrogriffine C (**147**) [43, 80] Preparative isolation of ancistrogriffine C (**147**) for a reliable and detailed analysis of further C–H correlations, leading to the confirmation of the constitution and the absolute configurations at the three stereogenic centers previously deduced online, now by HRESIMS and by ^{1}H and ^{13}C NMR (including HMQC, HMBC, and ROESY experiments) [43]	Moderate to weak inhibitory activities in vitro against *Plasmodium falciparum* (K1 strain), *Leishmania donovani*, *Trypanosoma cruzi*, and *T. brucei rhodesiense* [43]

(continued)

Table 12 (continued)

Name, formula (mol. wt.), and structure Physical data [*M.p.*/°C, $[\alpha]_D^T$/° $cm^2\ g^{-1}\ 10^{-1}$ (solvent)]	Producing plants and their organs	Isolation, structure elucidation, and further investigations Partial and/or total synthesis	Biological activities
Ancistroheynine A (80) $C_{24}H_{25}NO_4$ (391.5) 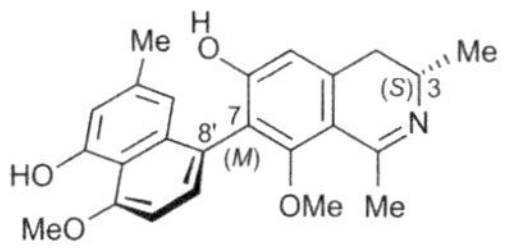 Yellow amorphous powder (from CH_2Cl_2/MeOH): $[\alpha]_D^{25}$ +68.0 (*c* 0.30, $CHCl_3$) [39]	*A. heyneanus* (India): fresh pith material prepared from shoots [39]	*Structural Elucidation*: Constitution and relative configuration at the two stereogenic elements (a center and an axis) deduced from HREIMS, ^1H and ^{13}C NMR, and from HMBC, HMQC, and H,H-COSY interactions [39] Absolute configuration at C-3 established as (*S*) by oxidative degradation [39] Absolute axial configuration determined as (*M*) by long-range NOE measurements in combination with the absolute (*S*)-configuration at C-3 in the dihydroisoquinoline portion [39] *Further Investigations*: Discovery of **80** by light and fluorescence microscopy as yellow-colored solutes in single parenchyma cells of fresh stem material of *A. heyneanus*, showing a bright-yellow fluorescence upon irradiation with UV light (365 nm) [39, 349] Enhanced accumulation of **80** in spherical inclusions [39], in particular in older cuttings of *A. heyneanus*, giving rise to the assumption that **80** might be formed as a stress metabolite, similar to the naphthoquinones plumbagin (**37**) and droserone (**38**) [132–134, 217, 249–255] In situ localization of ancistroheynine A (**80**) in fresh shoots and leaves of *A. heyneanus*, by non-invasive and, thus, non-destructive FT-Raman microspectroscopy [349] Non-invasive and non-destructive studies on the anatomy of the root and the shoot of *A. heyneanus* by NMR microscopy recording quantitative NMR relaxation time images to distinguish between different types of tissues [499] Non-invasive studies on the spatial distribution of metabolites in *A. heyneanus* by ^1H chemical shift imaging (CSI) NMR microscopy [500]: detection of ancistroheynine A (**80**) in the shoot [499]	Pronounced inhibitory activity in vitro against asexual erythrocytic stages of *Plasmodium falciparum* (strain NF54) [39]

Table 13 7,8′-coupled Dioncophyllaceae-type naphthylisoquinoline alkaloids

Name, formula (mol. wt.), and structure Physical data [*M.p.*/°C, $[\alpha]_D^T$/° cm^2 g^{-1} 10^{-1} (solvent)]	Producing plants and their organs	Isolation, structure elucidation, and further investigations Partial and/or total synthesis	Biological activities
Dioncophylleine D (167) $C_{24}H_{23}NO_3$ (373.5) 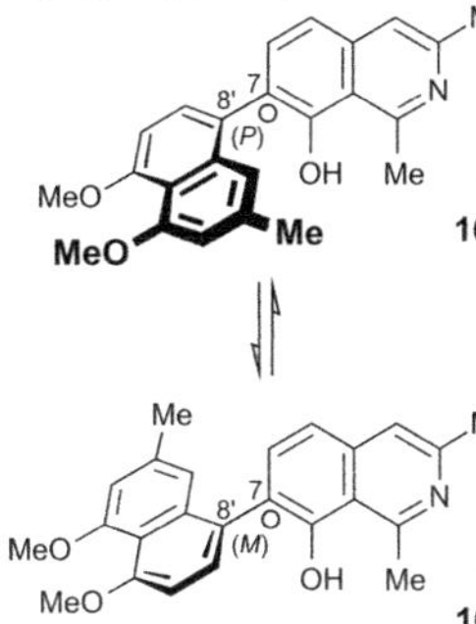 **167a** ⇌ **167b** ○ = configurationally unstable White amorphous solid (from MeOH): mp 212; $[\alpha]_D^{25}$ 0.0 (*c* 0.10, MeOH) [44]	*A. benomensis* (Malaysia): leaves [44]	*Structural Elucidation*: Constitution and coupling position established by HRESIMS and by 1D and 2D NMR, revealing **167** to be a 7,8′-linked non-hydrogenated naphthylisoquinoline alkaloid categorized as a Dioncophyllaceae-type naphthylisoquinoline alkaloid, since lacking an oxygen function at C-6—thus, substantially differing from (nearly) all the other alkaloids from Asian *Ancistrocladus* species, which produce Ancistrocladaceae-type alkaloids exclusively [44] Occurrence of dioncophylleine D (**167**) as a racemic mixture of its two interconverting atropo-enantiomers, **167a** and **167b**, which are configuratively unstable due to the lack of an oxygen function at C-6, and, thus, show a low rotational barrier at the biaryl axis [44] Dioncophylleine D (**167**), with its lack of any stereogenic element, as one of the extremely rare macroscopically achiral naphthylisoquinoline alkaloids [44]	Weak inhibitory activity in vitro against *Plasmodium falciparum* (K1 strain) [44] Moderate to weak inhibitory activity in vitro against *Trypanosoma brucei rhodesiense*, but inactive against *T. cruzi* [44] No inhibitory activity against *Leishmania major* [66]

Table 14 Naphthylisoquinoline alkaloids with a 1,4-naphthoquinone portion

Name, formula (mol. wt.), and structure Physical data [$M.p./°C$, $[\alpha]_D^T/°cm^2\ g^{-1}\ 10^{-1}$ (solvent)]	Producing plants and their organs	Isolation, structure elucidation, and further investigations Partial and/or total synthesis	Biological activities
Ancistrotectoquinone A (74) $C_{26}H_{29}NO_5$ (435.2) 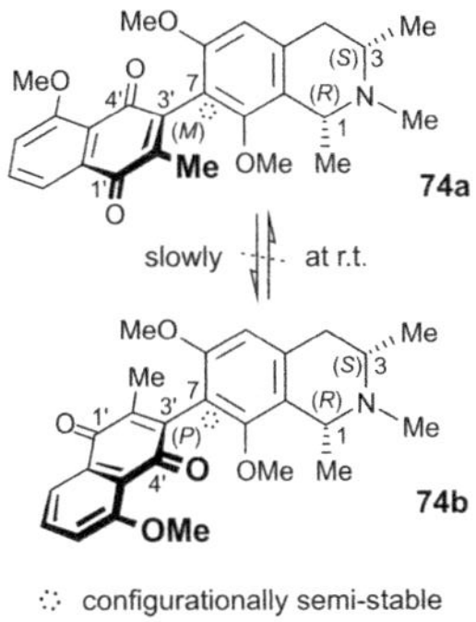 Orange-yellow amorphous powder [63] Resolution of the two configurationally semi-stable atropo-diastereomers (ca. 1:1 mixture) by HPLC on a reversed-phase C_{18} column: (*M*)-ancistrotectoquinone A (**74a**): $[\alpha]_D^{20}$ +7.3 (*c* 0.13, MeOH) (*P*)-ancistrotectoquinone A (**74b**): $[\alpha]_D^{20}$ +15.1 (*c* 0.13, MeOH)	*A. tectorius* (PR China): leaves [63]	*Structural Elucidation*: Constitution and coupling position established by HRESIMS and by ^1H and ^{13}C NMR, including HMBC and NOESY correlations, indicating the presence of a 7,3′-linked naphthylisoquinoline alkaloid consisting of a 1,4-naphthoquinone portion coupled to an N-methylated naphthyltetrahydroisoquinoline moiety [63] Absolute configuration at C-3 established as (*S*), by oxidative degradation [63] Absolute configuration at C-1 determined as (*R*) due to the relative *cis*-configuration at C-1 versus C-3 as assigned by specific NOESY interactions [63] Slow rotation of ancistrotectoquinone A (**74**) about the central axis at room temperature, thus occurring as a pair of two configurationally semi-stable and, thus, slowly interconverting atropo-diastereomers, **74a** and **74b** [63] After HPLC resolution of the two rotational isomers **74a** and **74b**, successful online chiroptical assignment of the absolute configurations at the biaryl axes of **74a** and **74b**, by comparison of their LC-ECD spectra with the ECD curve of the likewise 7,3′-coupled, yet configurationally stable, Ancistrocladaceae-type alkaloid ancistrotectorine (**17a**), which is (*M*)-configured at the biaryl axis [63] Determination of the atropisomerization rates of (*P*)-ancistrotectoquinone A (**74b**) to give the (*M*)-epimer **74a** and from **74a** to **74b**, by analytical HPLC on an achiral C_{18} phase: calculation of the Gibbs free energies of activation to be $\Delta G^{\#}_1 = 106.5$ kJ mol^{-1} and $\Delta G^{\#}_2 = 106.2$ kJ mol^{-1}; experimental values in agreement with the results from DFT calculations using B3LYP-D3/def2-TZVP and COSMO solvent corrections [63]	Biological activities not determined for lack of isolated material

(continued)

Table 14 (continued)

Name, formula (mol. wt.), and structure Physical data [*M.p.*/°C, $[\alpha]_D^T$/°cm^2 g^{-1} 10^{-1} (solvent)]	Producing plants and their organs	Isolation, structure elucidation, and further investigations Partial and/or total synthesis	Biological activities
		Further Investigations: Biomimetic oxidative conversion of ancistrotectorine (**17a**) [51, 159, 189] into the likewise (*M*)-configured ancistrotectoquinone A (**74a**) with Fremy's salt as the oxidant—giving rise to the expected atropisomeric equilibrium mixture **74a** ⇆ **74b** [63]	
Ancistrotectoquinone B (**234**) $C_{25}H_{27}NO_5$ (421.2) 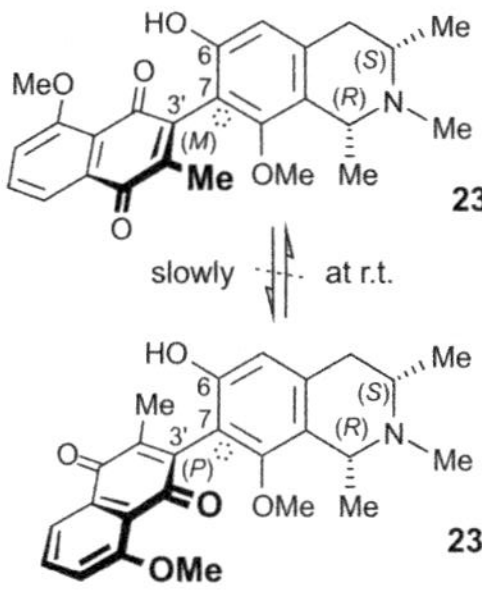 **234a** **234b** ∴ configurationally semi-stable Orange-yellow-colored oil [63]	*A. tectorius* (PR China): stems [63]	Ancistrotectoquinone B (**234**) as the only other known example of a naphthylisoquinoline alkaloid consisting of an isoquinoline subunit coupled to a 1,4-naphthoquinone portion (besides the co-occurring ancistrotectoquinone A, see entry above) [63] *Structural Elucidation*: Constitution and complete stereostructure assigned by HRESIMS, 1D and 2D NMR, oxidative degradation, and ECD—by comparison of the data of **234** with those obtained from its parent compound ancistrotectoquinone A (**74**), revealing **234** to differ from **74** by a free hydroxy function at C-6 (instead of a methoxy group as in **74**) [63] Ancistrotectoquinone B (**234**) occurs as a nearly 1:1 mixture of two slowly interconverting, and, thus, has configurationally semi-stable atropo-diastereomers, **234a** and **234b**, rotating even more rapidly, as a consequence of the further reduced steric hindrance due to the smaller hydroxy group at C-6 compared to the slightly larger methoxy residue in ancistrotectoquinone A (**74**) [63]	Biological activities not determined for lack of isolated material

Table 15 Naphthylisoquinoline alkaloids with a stereogenic *N*-iminium-aryl axis

Name, formula (mol. wt.), and structure Physical data [*M.p.*/°C, $[\alpha]_D^T$/°cm^2 g^{-1} 10^{-1} (solvent)]	Producing plants and their organs	Isolation, structure elucidation, and further investigations Partial and/or total synthesis	Biological activities
Ancistrocladinium A (6) $C_{26}H_{30}NO_4^+$ (420.5) 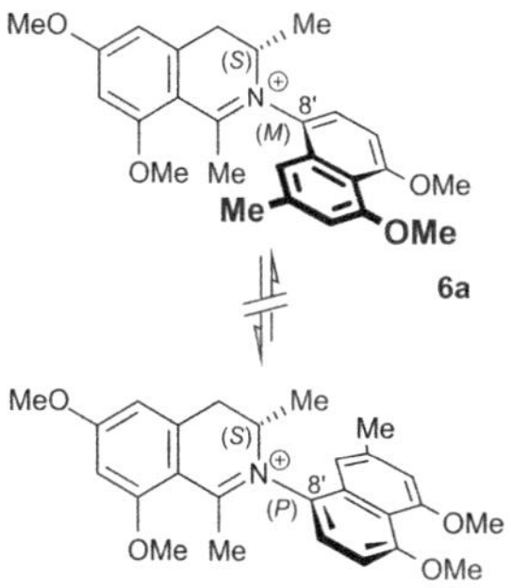 **6a** / **6b** Pale-yellow crystals (from MeOH): mp ≥ 230 (dec); $[\alpha]_D^{20}$ −6.0 (*c* 0.05, MeOH) [32] Total synthesis of ancistrocladinium A (**6a/b**), yielding a 2.5:1-ratio (*M*):(*P*) of the respective atropo-diastereomers [101]: Yellow solid (from MeCN/H_2O): mp ≥ 230 (dec); $[\alpha]_D^{20}$ −14.0 (*c* 0.17, MeOH) [101] Total synthesis of the unnatural (3*R*)-enantiomers of ancistrocladinium A, providing *ent*-ancistrocladinium A (*ent*-**6a/b**) in a 2.5:1-ratio (*M*):(*P*) of the respective atropo-diastereomers [101]: Yellow oil: $[\alpha]_D^{20}$ +10.0 (*c* 0.14, MeOH) [101]	*A. cochinchinensis* (Vietnam): stem bark [47] *A. tectorius* (Vietnam): leaves [32] *A. tectorius* (PR China): leaves, twigs, and stems [33] *A. ealaensis* (DR Congo): twigs [176] Two botanically yet undescribed Congolese *Ancistrocladus* species: leaves [32] and root bark [161]	Discovered for the first time in the leaves of an as yet botanically undescribed *Ancistrocladus* liana endemic to the Central Congo Basin [32, 161] Preparative isolation by liquid–liquid separation, fast centrifugal partition chromatography (FCPC), and preparative HPLC [32] Constitution and complete stereostructure of **6** established online from crude bark extracts of *A. cochinchinensis* [47] and *A. tectorius* [32], using the analytical triad LC-MS/MS, LC-NMR (^{1}H, WET-ROESY), and LC-ECD *Structural Elucidation*: Constitution and coupling position established by HRESIMS, ^{1}H NMR and ^{13}C NMR, including ROESY, NOESY, and HMBC interactions, supporting the presence of a cationic naphthyldihydroisoquinoline alkaloid with an unprecedented N,8′-linkage; natural counteranion as yet unknown [18, 32, 33, 176] Absolute configuration at C-3 established as (*S*) by oxidative degradation [32, 33, 47, 161, 176] Assignment of the absolute configuration at the biaryl axis as (*M*) due to specific long-range NOESY interactions across the N,C-axis, in conjunction with the given absolute configuration at C-3 (as known from the degradation) [18, 32, 47] Additional proof of the assignment of the absolute axial configuration of ancistrocladinium A (**6**) by quantum-chemical ECD calculations at the TDDFT (B3LYP/TZVP) level and comparison of the ECD spectra calculated for the (*M*)- and (*P*)-isomer **6a** and **6b** with the experimental one [32] Likewise detectable in the NMR spectra of ancistrocladinium A (**6a**): also the other, (*P*)-configured atropo-diastereomer **6b**, in the bark of *A. cochinchinensis* occurring in a ratio of 1:10 (*P*:*M*), in the leaf material of 1:3 (*P*):(*M*) [47], and in the twigs and stems of *A. tectorius* of 1:4 (*P*):(*M*) [18, 33]; failure to separate the atropo-diastereomers **6a** and **6b**, even on chiral HPLC phases [18, 47]	Moderate inhibitory activity in vitro against *Plasmodium falciparum* (K1 strain) [32, 69] Moderate to good inhibitory activity in vitro against *Trypanosoma brucei rhodesiense*, but inactive against *T. cruzi* [32, 69] Excellent inhibitory activity in vitro against *Leishmania donovani* and *L. major* [32, 69], making ancistrocladinium A (**6**) a promising candidate to be considered as a lead compound for new leishmanicidal drugs [18, 32, 66, 67, 69] Investigations on structure–activity relationships, on the molecular mechanism, and on the metabolism of leishmanicidal N,C-coupled arylisoquinolines structurally related to ancistrocladinium A (**6**) [18, 67, 68, 70–72] Considerable antiproliferative activity against drug-sensitive CCRF-CEM leukemia tumor cells [33]

(continued)

Table 15 (continued)

Name, formula (mol. wt.), and structure Physical data [*M.p.*/°C, $[\alpha]_D^T$/°cm^2 g^{-1} 10^{-1} (solvent)]	Producing plants and their organs	Isolation, structure elucidation, and further investigations Partial and/or total synthesis	Biological activities
Directed total synthesis of all four possible pure stereoisomers [69, 118, 119]: Ancistrocladinium A (**6a**): yellow oil: $[\alpha]_D^{20}$ −5.8 (*c* 0.08, MeOH) [69]; yellow solid (from *n*-hexane): $[\alpha]_D^{20}$ −107 (*c* 0.10, MeOH) [119] *N-epi*-Ancistrocladinium A (**6b**): yellow oil: $[\alpha]_D^{20}$ −3.6 (*c* 0.06, MeOH) [69]; yellow solid (from *n*-hexane): $[\alpha]_D^{20}$ −46 (*c* 0.17, MeOH) [119] *ent*-Ancistrocladinium A (*ent*-**6a**): yellow oil: $[\alpha]_D^{25}$ +38.9 (*c* 0.14, MeOH) [69] *ent-N-epi*-Ancistrocladinium A (*ent*-**6b**): yellow oil [69] 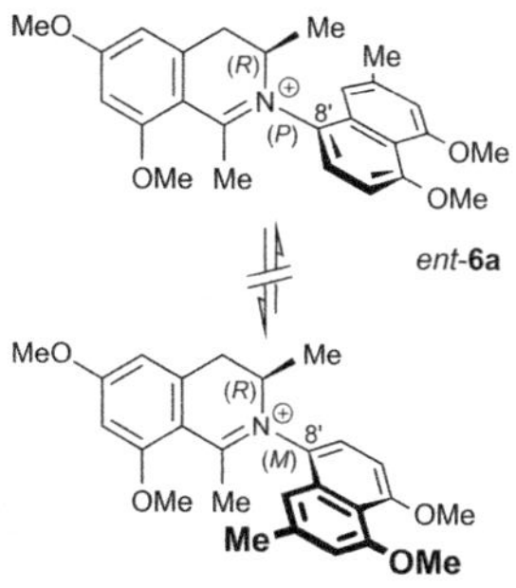 *ent*-**6a** *ent*-**6b** The two (as yet unnatural) (3*R*)-analogues, *ent*-**6a** and -*ent*-**6b** of ancistrocladinium A		*Total Synthesis*: Convergent total synthesis of ancistrocladinium A (**6**) in only eight linear steps without the need of protective groups: coupling of the two molecular portions at a late stage by a Buchwald–Hartwig amination, followed by a Bischler–Napieralski ring closure sequence, atropisomer-selectively leading to the naturally predominant (*M*)-atropo-diastereomer **6a**, preferentially [18, 101] Directed synthesis of all four possible stereoisomers of ancistrocladinium A (**6**) in a stereochemically pure form (center and axis), by resolution of the already rotationally hindered, and thus atropo-diastereomeric acetamide precursors and further cyclization to the respective target molecules **6a**, **6b**, *ent*-**6a**, and *ent*-**6b** without major loss of stereochemical purity [18, 69] Total synthesis of ancistrocladinium A (**6a**) and its atropo-diastereomer **6b** by catalytic asymmetric reductive amination of 1-arylpropanone with appropriate 8-aminonaphthalene using a Hantzsch ester as an organic hydride source and chiral phosphoric acid (derived from (*P*)-BINOL) as the catalyst to give the respective *N*-naphthylamine (90%, *ee*), followed by N-acetylation to yield the respective (*M*)- and (*P*)-configured tertiary amides as an atropisomeric mixture; separation by column chromatography and Bischler–Napieralski cyclization of each of these amides in the presence of triflic anhydride at −78°C, providing **6a** and **6b** in an atropisomerically pure form [119]—a previous attempt following this approach by using chiral phosphoric acid (derived from (*M*)-BINOL) as a catalyst having led to the 'wrong', i.e., (*R*)-configured *N*-naphthylamine: Bischler–Napieralski cyclization of the (*P*)-configured amide using $POCl_3$ giving *ent*-**6a**, with chloride as the counteranion, and with the (*P*)-isomer as the major product (4:1, *ent*-**6a**:*ent*-**6b**); analogously, cyclization of the (*M*)-acetamide leading to an atroposelectivity in favor of the (*M*)-isomer (3:1, *ent*-**6b**:*ent*-**6a**) [18, 118]	

(continued)

Table 15 (continued)

Name, formula (mol. wt.), and structure Physical data [*M.p.*/°C, $[\alpha]_D^T$/°cm² g⁻¹ 10⁻¹ (solvent)]	Producing plants and their organs	Isolation, structure elucidation, and further investigations Partial and/or total synthesis	Biological activities
4′-*O*-Demethylancistrocladinium A (123) $C_{25}H_{28}NO_4^+$ (406.5) 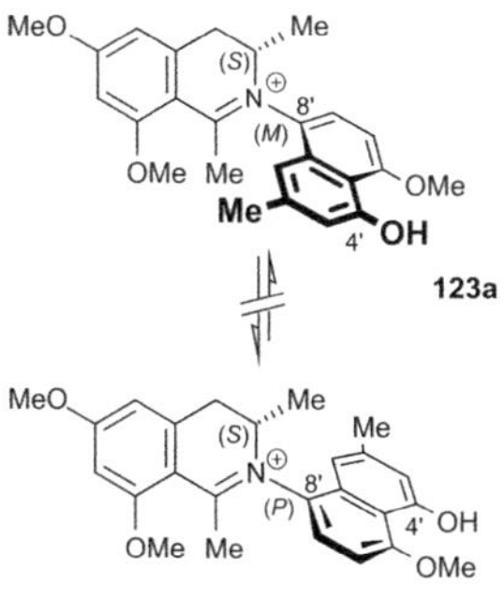 **123a** ⇌ **123b** Pale-yellow needles (from MeOH): mp ≥ 230 (dec); $[\alpha]_D^{20}$ −5.5 (*c* 0.05, MeOH) [47]	*A. cochinchinensis* (Vietnam): leaves and bark [47] *A. tectorius* (PR China): leaves, twigs, and stems [33]	Isolated for the first time from the Vietnamese liana *A. cochinchinensis* [47] *Structural Elucidation*: Constitution and complete stereostructure assigned by HRESIMS, 1D and 2D NMR, oxidative degradation, and ECD—by comparison of the data of **123** with those obtained from its parent compound ancistrocladinium A (**6**), revealing **123** to possess a free phenolic hydroxy group at C-4′ (instead of a methoxy group as in **6**); natural counteranion as yet unknown [18, 33, 47] Absolute configuration at C-3 established as (*S*), by oxidative degradation [33, 47] Absolute configuration at the biaryl axis assigned as (*M*) due to the nearly identical ECD spectrum of **123** compared to that of the likewise (*M*)-configured parent compound ancistrocladinium A (**6**) [18, 47] Isolation by preparative reversed-phase HPLC along with its (*P*)-configured atropo-diastereomer, from the bark of *A. cochinchinensis* in a ratio of 1:10 (*P*):(*M*), from leaf material in a ratio of 1:3 (*P*):(*M*) [47], and from twigs and stems of *A. tectorius* in a ratio of 1:4 (*P*):(*M*) [33]—unsuccessful attempts to separate the atropo-diastereomers, even on chiral HPLC phases [47]	Moderate inhibitory activity in vitro against *Plasmodium falciparum* (K1 strain) [47] Excellent inhibitory activity in vitro against *Trypanosoma cruzi*, about 70-fold higher than that of the standard drug benznidazole [18, 47] Excellent inhibitory activity in vitro against *Leishmania donovani*, much superior to the standard drug miltefosine [18, 47] Pronounced activity against the insect form of *Leishmania major*, suggesting a non-stage specific mode of action [47]

(continued)

Table 15 (continued)

Name, formula (mol. wt.), and structure Physical data [$M.p./°C$, $[\alpha]_D^T/°cm^2\ g^{-1}\ 10^{-1}$ (solvent)]	Producing plants and their organs	Isolation, structure elucidation, and further investigations Partial and/or total synthesis	Biological activities
4′,6-*O*,*O*-Didemethylancistrocladinium A (174) $C_{24}H_{26}NO_4^+$ (392.5) 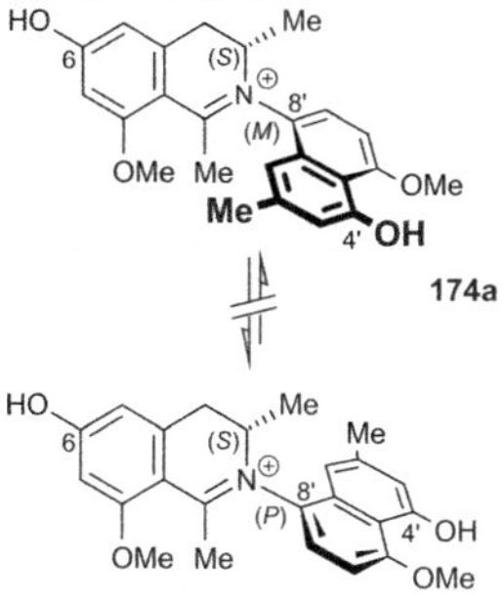 **174a** / **174b** Pale-yellow oil: $[\alpha]_D^{20}$ −4.4 (*c* 0.024, MeOH) [47]	*A. cochinchinensis* (Vietnam): leaves, bark [47] *A. tectorius* (PR China): leaves, twigs, and stems [33] *A. ealaensis* (DR Congo): twigs [176]	Isolated for the first time from the Vietnamese liana *A. cochinchinensis* [47] *Structural Elucidation*: Constitution and complete stereostructure assigned by HRESIMS, 1D and 2D NMR, oxidative degradation, and ECD—by comparison of the data of **174** with those obtained from its parent compound ancistrocladinium A (**6**), revealing **174** to differ from **6** by the presence of two free phenolic hydroxy groups at C-4′ and C-6 (instead of methoxy functions as in **6**), the natural counteranion being as yet unknown [18, 33, 47, 176] As for ancistrocladinium A (**6**) and its monophenolic analogue **123**, detection of the respective (*P*)-atropo-diastereomer only as a minor component (see entry above), and again without the possibility to resolve the two rotational isomers [18, 47]	Moderate to weak inhibitory activity in vitro against *P. falciparum* (K1 strain), *Trypanosoma cruzi*, *T. brucei rhodesiense*, *T. brucei brucei*, and *Leishmania donovani* amastigotes and *L. major* promastigotes [18, 47] No antiproliferative activity against drug-sensitive CCRF-CEM leukemia cells [33]

(continued)

Table 15 (continued)

Name, formula (mol. wt.), and structure Physical data [*M.p.*/°C, $[\alpha]_D^T$/°cm^2 g^{-1} 10^{-1} (solvent)]	Producing plants and their organs	Isolation, structure elucidation, and further investigations Partial and/or total synthesis	Biological activities
Ancistrocladinium B (**14**) $C_{25}H_{28}NO_4^+$ (406.5) 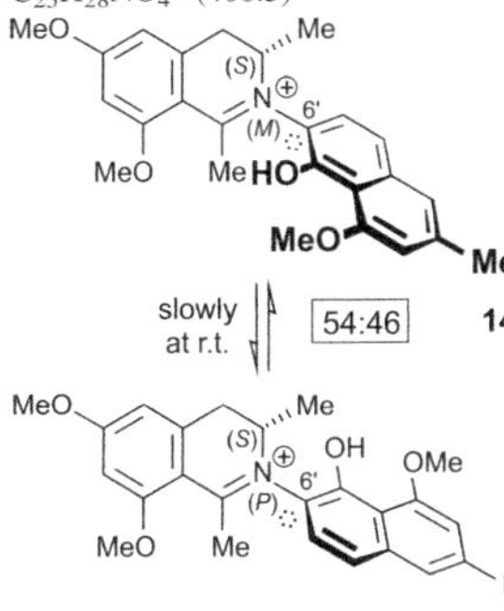 **14a**, **14b** ⁚⁚ configurationally semi-stable Pale-yellow crystals (from MeOH), atropisomeric mixture (ratio ca. 54:46, (*P*):(*M*)), chemical purity >99%: mp ≥ 230 (dec); $[\alpha]_D^{20}$ −8.0 (*c* 0.06, MeOH) [32] (*M*)-Ancistrocladinium B (**14a**), isolated by HPLC: chemical purity >99%; $[\alpha]_D^{20}$ −4.5 (*c* 0.04, MeOH) [32] N,6′-*epi*-Ancistrocladinium B (**14b**), isolated by HPLC: chem. purity >99%; $[\alpha]_D^{20}$ −11.3 (*c* 0.04, MeOH) [32] cf.: Title compd. prepared by total synthesis as a 46:54-mixture (*P*):(*M*) of its atropo-diastereomers: Yellow oil: $[\alpha]_D^{20}$ −19.0 (*c* 0.11, MeOH) [101]	*A. tectorius* (PR China): leaves, twigs, and stems [33] *A. benomensis* (Malaysia): leaves [32] *A. heyneanus* (India): leaves [32, 70] Two botanically yet undescribed Congolese *Ancistrocladus* species: leaves [32] and root bark [161]	Discovered for the first time from the leaves and root bark of an as yet botanically undescribed *Ancistrocladus* liana endemic to the Central Congo Basin [32, 161] Preparative isolation by liquid–liquid separation, fast centrifugal partition chromatography (FCPC), and preparative HPLC, giving rise to two configurationally semi-stable atropo-diastereomers, **14a** and **14b** (54:46), showing a slow interconversion at room temperature [18, 32] Constitution and complete stereostructure of **14** established online, directly from crude leaf extracts of *A. benomensis*, *A. heyneanus*, and *A. tectorius* [32], by using the analytical triad LC-MS/MS, LC-NMR (^{1}H, WET-ROESY), and LC-ECD *Structural Elucidation*: Constitution and coupling position established by HRESIMS and by ^{1}H and ^{13}C NMR, including HMBC and ROE interactions indicative of two interconverting, since configurationally semi-stable cationic naphthyldihydroisoquinoline alkaloids displaying an unprecedented N,6′-linkage—NMR investigations performed online, by LC-NMR (^{1}H, WET-ROESY) individually on each of the two atropo-diastereomers; natural counteranion as yet unknown [18, 32] Absolute configuration at C-3 established as (*S*) by oxidative degradation of the atropo-diastereomeric mixture [32] Absolute axial configuration deduced as (*P*) for the more rapidly eluting isomer, **14b**, and as *M* for the slower one, **14a** (by NOESY investigations on the pure atropo-diastereomers), due to specific NOE correlations across the N,C-axis in combination with the known (*S*)-configuration at C-3 [18, 32] Additional proof of the assignment of the absolute axial configuration of ancistrocladinium B (**14**) by quantum-chemical ECD calculations at the TDDFT (B3LYP/TZVP) level and comparison of the ECD spectra calculated for the (*M*)- and	Moderate inhibitory activities in vitro against *P. falciparum* (K1 strain), *Trypanosoma cruzi*, and *T. brucei rhodesiense* [32] Excellent inhibitory activity in vitro against *Leishmania donovani* and *L. major* [32]—ancistrocladinium B (**14**) as a promising candidate to be considered as a lead compound for new leishmanicidal drugs [18, 32, 66, 70–72]

(continued)

Table 15 (continued)

Name, formula (mol. wt.), and structure Physical data [*M.p.*/°C, $[\alpha]_D^T$/°cm^2 g^{-1} 10^{-1} (solvent)]	Producing plants and their organs	Isolation, structure elucidation, and further investigations Partial and/or total synthesis	Biological activities
		(*P*)-isomers, **14a** and **14b**, with the experimental ones obtained online by LC-ECD [18, 32, 76, 78] Determination of the atropisomerization rates of ancistrocladinium B (**14b**) and its (*M*)-epimer **14a**, by analytical HPLC on an achiral C_{18} phase (Gibbs free energy of activation: $\Delta G^{\#}_1$ = 105.9 kJ mol^{-1} and $\Delta G^{\#}_2$ = 105.7 kJ mol^{-1}) and by DFT calculations [18, 32] ——— *Total Synthesis*: Convergent total synthesis of ancistrocladinium B (**14**) in only eight linear steps without the need of protective groups: coupling of the two molecular portions at a late stage by a Buchwald–Hartwig amination, followed by a Bischler–Napieralski ring closure sequence, yielding ancistrocladinium B as a 54:46 (*P*):(*M*) mixture of the atropo-diastereomers **14a** and **14b**—stereochemical assignment of the two diastereomers by LC-ECD coupling and comparison of the obtained ECD spectra with those of the two genuine natural products [101]	

(continued)

Table 15 (continued)

Name, formula (mol. wt.), and structure Physical data [*M.p.*/°C, $[\alpha]_D^T$/°cm^2 g^{-1} 10^{-1} (solvent)]	Producing plants and their organs	Isolation, structure elucidation, and further investigations Partial and/or total synthesis	Biological activities
Ancisheynine (78) $C_{26}H_{28}NO_4^+$ (418.5) 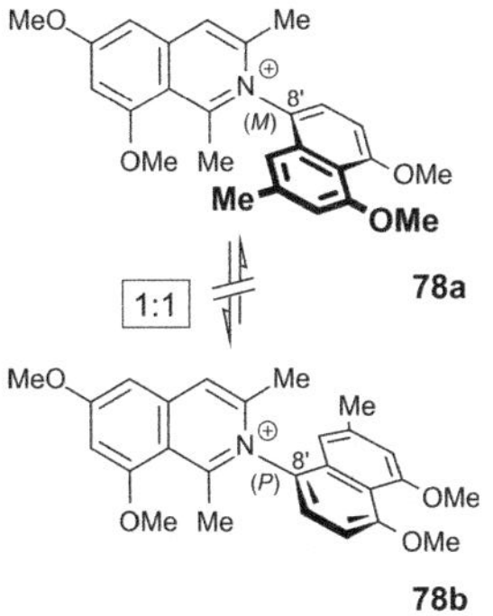 Yellow oil: $[\alpha]_D^{20}$ 0.0 (*c* 0.42, $CHCl_3$) (optically inactive) [40] cf.: Title compd. prepared by total synthesis (yielded as a racemate) [256]: Yellow powder (from $MeCN/H_2O$): mp ≥ 250 (dec)	*A. heyneanus* (India): aerial parts and leaves [40]	*Structural Elucidation*: Preparative isolation of ancisheynine (**78**) as a minor (or even trace) metabolite from aerial parts of the plant by ion exchange chromatography, followed by purification of alkaloid-enriched fractions using centrifugal thin layer chromatography and gel permeation chromatography on Sephadex LH-20—occurrence in the plant only in very small quantities, **78** (depending on the season) even not always detectable in the extracts [40, 256] Constitution and coupling position established by HRESIMS and by ^{1}H, ^{15}N, and ^{13}C NMR chemical shifts, including H,H-COSY, HMBC, ^{15}N HMBC and ROESY interactions, indicative of a cationic non-hydrogenated naphthylisoquinoline alkaloid with an unprecedented N,8′-biaryl linkage, the natural counteranion being as yet unknown [18, 40] *Total Synthesis and Stereoanalysis*: Convergent total synthesis of ancisheynine (**78**) based on the cyclocondensation of a monocyclic diketone (or the respective benzopyrylium salt) with an appropriate aminonaphthalene, leading to the fully dehydrogenated isoquinolinium moiety of **78**; use of trifluoroacetate as an electron-poor counteranion to enhance the stability of the target molecule [18, 256] Resolution and stereochemical assignment of the two atropo-enantiomers of ancisheynine, **78a** and **78b**, on a chiral phase online, by LC-ECD coupling in combination with quantum-chemical ECD calculations [18, 40, 83] Online LC-ECD analysis of fresh leaf extracts of *A. heyneanus*, revealing ancisheynine (**78**) to be present in the plant as a racemic mixture of its two configurationally stable atropo-enantiomers, **78a** and **78b** [40, 83] Rotational stability of ancisheynine (**78**) evidenced by reinjecting enantio-enriched samples, showing their constant enantiomeric ratios over time at r.t. [18, 83]	Weak to moderate inhibitory activity in vitro against *Leishmania donovani*, *Trypanosoma cruzi*, and *T. brucei rhodesiense* [18, 70] Biological testing of a whole series of simplified synthetic arylisoquinolinium analogues structurally related to ancisheynine (**78**), some of them exhibiting excellent antiprotozoal (in particular antileishmanial), antibiotic, and biofilm-inhibiting activities, and low in vitro toxicities (therapeutic indices >1000) [66–68, 70] Investigations on structure–activity relationships, on the molecular mechanism, and on the metabolism of synthetic leishmanicidal N,C-coupled arylisoquinolines structurally related to ancisheynine (**78**) [66–68, 70] Detailed investigations on the bactericidal and cytotoxic effects of a dimeric isoquinoline related to ancisheynine (**78**) in *Staphylococcus aureus* and *S. epidermidis* in combination with complementary analyses of gene expression and metabolite pathways in the host and in the pathogen [65, 70]

Table 16 Pentacyclic N,C-coupled naphthylisoquinoline alkaloids

Name, formula (mol. wt.), and structure Physical data [*M.p.*/°C, $[\alpha]_D^T$/°cm^2 g^{-1} 10^{-1} (solvent)]	Producing plants and their organs	Isolation, structure elucidation, and further investigations Partial and/or total synthesis	Biological activities
Ancistrocyclinone A (7) $C_{25}H_{24}NO_4^+$ (402.2) 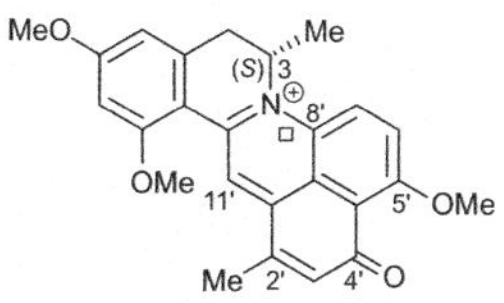 □ axis flat, achiral Deep-red amorphous solid: $[\alpha]_D^{20}$ +6.6 (*c* 0.05, MeOH) [33]	*A. tectorius* (PR China): leaves, twigs, and stems [33]	*Structural Elucidation*: Deep-red substance with a UV spectrum substantially different from that of all other naphthylisoquinoline alkaloids, with the third maximum of absorption (usually at ca. 306 nm) [13, 17, 18, 41, 47, 155, 189] significantly shifted bathochromically to a wavelength of ca. 500 nm, thus hinting at a unique structure with a large conjugated π-system [33] Constitution deduced from HRESIMS, and by ^{1}H and ^{13}C NMR, and by HMBC and NOE interactions, supporting the coupling position of the central axis (not a biaryl axis in a strict sense!) being at N-2 in the isoquinoline part, and at C-8′ in the naphthalene-derived moiety, thus indicating the presence of a 3,4-dihydroisoquinolinium cation, the natural counteranion being as yet unknown [33] HMBC correlations establishing the presence of an additional six-membered ring, with C-11′ as the bridge between the two usual bicyclic units, and thus, overall, revealing ancistrocyclinone A (**7**) to possess a pentacyclic ring system similar to that of the isoquinoline alkaloid berberine (**258**)—yet, in contrast to the essentially flat berberine [478], possessing an intriguing helical 3D structure [33] Absolute configuration at C-3 assigned as (*S*), by quantum-chemical ECD calculations using the B3LYP/6-31G* approach; confirmation of this assignment by ruthenium-mediated oxidative degradation of **7**, leading to the (*S*)-enantiomer of 3-aminobutyric acid exclusively [33] Convergence of different biosynthesis pathways to similar polycyclic isoquinoline alkaloids: formation of the berberine skeleton from aromatic amino acids [135–140], in contrast to the polyketide origin of ancistrocyclinone A (**7**), apparently via the co-occurring 4′-*O*-demethylancistrocladinium A (**123**) [33]	Significant inhibition of the viability of drug-sensitive human leukemia (CCRF-CEM) and multi-drug-resistant tumor cells (CEM/ADR5000) with comparable efficacies [33]

(continued)

Table 16 (continued)

Name, formula (mol. wt.), and structure Physical data [*M.p.*/°C, $[\alpha]_D^T$/°cm^2 g^{-1} 10^{-1} (solvent)]	Producing plants and their organs	Isolation, structure elucidation, and further investigations Partial and/or total synthesis	Biological activities
		Plausibility of the postulated biosynthesis pathway to ancistrocyclinone A (**7**) studied by biomimetic conversion of 4′-*O*-demethyl-ancistrocladinium A (**123**) to **7** via an N,C-coupled quinoid intermediate [33]	
Ancistrocyclinone B (15) $C_{24}H_{22}NO_4^+$ (388.2) 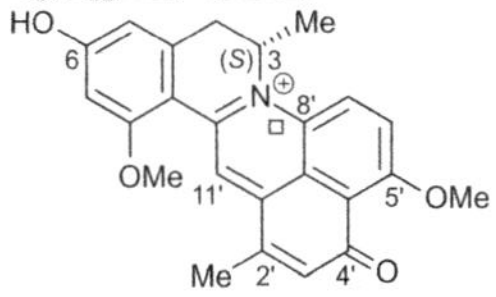 □ axis flat, achiral Deep-red amorphous solid [33]	*A. tectorius* (PR China): leaves, twigs, and stems [33]	*Structural Elucidation*: Constitution and complete stereostructure assigned by HRESIMS, 1D and 2D NMR, oxidative degradation, and ECD—comparison of the data of **15** with those from its parent compound ancistrocyclinone A (**7**), revealing **15** to possess a free phenolic hydroxy group at C-6 (instead of a methoxy group as in **7**), the natural counteranion being as yet unknown [33] Absolute configuration at C-3 established as (*S*), by oxidative degradation and from the similarity of the ECD spectra of **15** and **7** [33]	Biological activities not determined for lack of isolated material

Table 17 Constitutionally symmetric Ancistrocladaceae-type naphthylisoquinoline dimers

Name, formula (mol. wt.), and structure Physical data [*M.p.*/°C, $[\alpha]_D^T$/°cm² g⁻¹ 10⁻¹ (solvent)]	Producing plants and their organs	Isolation, structure elucidation, and further investigations Partial and/or total synthesis	Biological activities
6′,6″-dimerized naphthylisoquinoline alkaloids			
Ancistrogriffithine A (148) $C_{48}H_{52}N_2O_8$ (784.9) 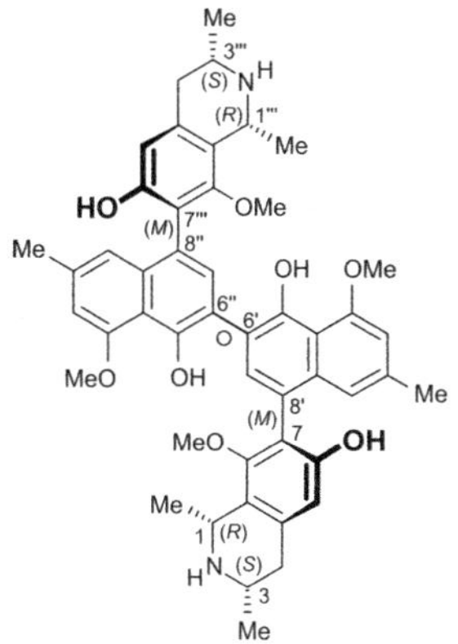 O = configurationally unstable Light-yellow powder (from MeOH): mp 230 (dec); $[\alpha]_D^{25}$ +73.5 (*c* 0.30, MeOH) [43]	*A. griffithii* (Thailand): leaves and twigs [42, 43, 82]	Hints at the presence of naphthylisoquinoline dimers in *A. griffithii* from a photometric chemical screening method based on the oxidation of their central binaphthalene core with lead tetraacetate giving a characteristic long-wavelength absorption in UV, with the maximum at 525 nm, due to the formation of intensely violet-blue-colored dinaphthoquinones [42] *Structural Elucidation*: Constitution and coupling positions established online, directly from crude leaf and twig extracts, by LC combined with MS/MS and NMR (including ¹H, 2D-WET-ROESY, and TOCSY experiments), indicative of the presence of a C_2-symmetric dimeric naphthylisoquinoline alkaloid consisting of two 7,8′-coupled monomers linked through the least hindered C-6′ and C-6″ positions in their naphthalene moieties, leading to a free rotation about the central axis [17, 42, 82] Relative configuration of the stereocenters C-1 versus C-3 and C-1‴ versus C-3‴ determined to be *cis* by specific ROESY correlations [42] Absolute configuration at C-3 and C-3‴ established as (*S*) by oxidative degradation of crude extracts of *A. griffithii* (due to the exclusive formation of (3*S*)-aminobutyric acid) and, after isolation, of pure ancistrogriffithine A (**148**) [42, 80, 82] Absolute axial configuration of the two outer axes in the monomeric halves of ancistrogriffithine A (**148**) assigned as (*M*) due to specific long-range ROESY interactions supporting the relative configuration at centers versus axes in combination with the (1*R*,3*S*)-configurations in the two tetrahydroisoquinoline moieties as established by specific ROESY correlations and by oxidative degradation [17, 42, 82] LC-ECD coupling providing an experimental ECD spectrum of **148**, with strong bands between 200 and 300 nm derived from	Strong inhibitory activity in vitro against *Plasmodium falciparum* (K1 strain), comparable to that of chloroquine, but with cytotoxicity against L6 host cells [43] Moderate to low inhibitory activities in vitro against *Trypanosoma cruzi* and *T. brucei rhodesiense* [43]

(continued)

Table 17 (continued)

Name, formula (mol. wt.), and structure / Physical data [*M.p.*/°C, $[\alpha]_D^T$/°cm² g⁻¹ 10⁻¹ (solvent)]	Producing plants and their organs	Isolation, structure elucidation, and further investigations / Partial and/or total synthesis	Biological activities
		π,π*-transitions as expected for a stereochemically stable 7,8′-biaryl axis in the monomeric halves [82] cConstitution and absolute configurations at the stereogenic elements (centers and outer axes) of ancistrogriffithine A (**148**) previously deduced online, confirmed by HRESIMS, by ^{1}H and ^{13}C NMR (including HMQC, HMBC, and ROESY experiments) [43]	
3′,3″-dimerized naphthylisoquinoline alkaloids			
Shuangancistrotectorine A (262a) $C_{50}H_{56}N_2O_8$ (813.0) 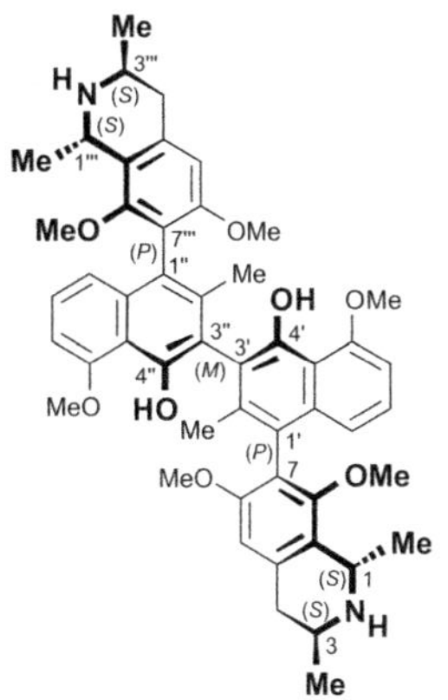 White amorphous powder: $[\alpha]_D^{20}$ +28.0 (*c* 0.26, MeOH) [31]	*A. tectorius* (PR China): twigs [31]	*Structural Elucidation*: Constitution deduced from HRESIMS, MALDI-TOF-MS, ^{1}H and ^{13}C NMR, and from HMBC and ROESY interactions, indicative of a C_2-symmetric dimeric naphthylisoquinoline alkaloid, with two homomorphous halves [17, 31] Coupling positions established by HMBC and ROESY interactions revealing the presence of a dimer consisting of two identical 7,1′-coupled naphthylisoquinoline monomers, linked via the sterically constrained C-3′ and C-3″ positions of the central binaphthalene core, supporting the presence of three consecutive stereogenic axes in shuangancistrotectorine A (**262a**) [17, 31] Relative configuration at the stereocenters at C-1 versus C-3 and at C-1‴ versus C-3‴ determined as *trans* by specific long-range ROESY correlations [17, 31] Absolute (*S*)-configuration both at C-1 and C-3, and at C-1‴ and C-3‴, assigned by oxidative degradation [31, 80] Absolute axial configuration of the two outer axes deduced as (*P*) from specific long-range ROESY correlations in combination with the (1*S*,3*S*)-configurations in the tetrahydroisoquinoline moieties as established by degradation [17, 31] Assignment of the absolute configuration at the central axis as (*M*) by comparison of the experimental LC-ECD spectrum (monitored in stopped-flow mode) with the ECD spectra obtained from quantum-chemical ECD calculations for the (*P*,*M*,*P*)- and (*P*,*P*,*P*)-atropo-diastereomers using the TD-B3LYP/6-31G* approach in combination with a conformational analysis by PM3 [17, 31]	Strong and—specific—inhibitory activities in vitro against the erythrocytic stages of *Plasmodium falciparum* (strains: K1 and NF54) [31] Pronounced inhibitory activity in vitro against *Trypanosoma brucei rhodesiense*, but only weak activity against *T. cruzi* [31] Virtually no inhibitory activity against *Leishmania donovani* [31]

(continued)

Table 17 (continued)

Name, formula (mol. wt.), and structure Physical data [*M.p.*/°C, $[\alpha]_D^T$/°cm^2 g^{-1} 10^{-1} (solvent)]	Producing plants and their organs	Isolation, structure elucidation, and further investigations Partial and/or total synthesis	Biological activities
		Configurational stability of shuangancistro-tectorine A (**262a**) at the central axis at ambient temperature and neutral pH—upon warming to 40°C, however, in the presence of acid, or upon storage in $CDCl_3$, atropisomerization to give a mixture of **262a** and shuangancistrotectorine B (**262b**) [31] Isolation from *A. tectorius* and purification of **262a** under the mildest-possible conditions (room temperature, avoiding light, drying by lyophylization), yielding shuangancistrotectorine A (**262a**) as stereochemically homogeneous material [31]	
Shuangancistrotectorine B (262b) $C_{50}H_{56}N_2O_8$ (813.0) 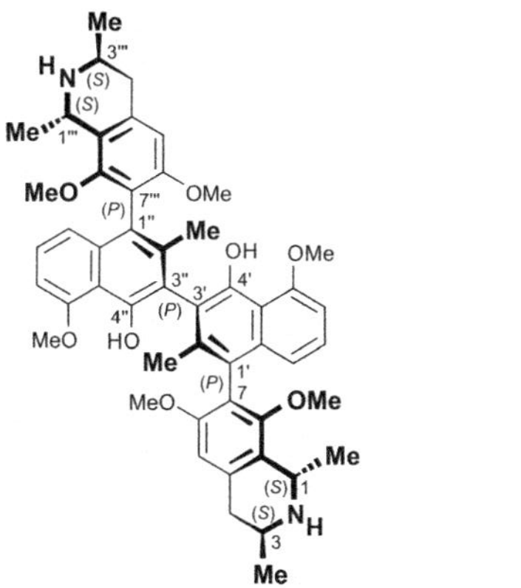 White amorphous powder: $[\alpha]_D^{20}$ −14.5 (*c* 0.30, MeOH) [31]	*A. tectorius* (PR China): twigs [31]	*Structural Elucidation*: Constitution deduced from HRESIMS and from ^{1}H and ^{13}C NMR, including HMBC, HMQC, and ROESY interactions, indicative of a C_2-symmetric naphthylisoquinoline dimer, consisting of two homomorphous halves and possessing the same constitution as shuangancistrotectorine A (**262a**) [17, 31] Relative configuration at C-1 versus C-3 and at C-1‴ versus C-3‴ determined to be *trans* by specific ROESY correlations [17, 31] Absolute configuration at C-1and C-3 and at C-1‴ and C-3‴ all established as (*S*) by oxidative degradation [31, 80] Absolute configuration of the two outer axes assigned as (*P*) from specific long-range ROESY correlations in combination with the (1*S*,3*S*)-configurations in the isoquinoline moieties as established by degradation—supporting **262b** to consist of two identical halves, all like for **262a** [17, 31] Absolute configuration at the central biaryl axis attributed as (*P*) due to a virtually opposite ECD spectrum of **262b** compared to that of **262a** [31] Configurational stability of **262b** at the central axis at ambient temperature and neutral pH—however, isolation from plant material performed under such mildest possible conditions, leading to shuangancistrotectorine A (**262a**) exclusively, thus, lack of evidence whether **262b** is an authentic (minor) natural product or an artifact formed by conversion of **262a** in the presence of acid,	Strong and—specific—inhibitory activities in vitro against the erythrocytic stages of *Plasmodium falciparum* (strains: K1 and NF54) [31] Pronounced inhibitory activity in vitro against *Trypanosoma brucei rhodesiense*, but only weak activity against *T. cruzi* [31] Virtually no inhibitory activity against *Leishmania donovani* [31]

(continued)

Table 17 (continued)

Name, formula (mol. wt.), and structure Physical data [*M.p.*/°C, $[\alpha]_D^T$/°cm^2 g^{-1} 10^{-1} (solvent)]	Producing plants and their organs	Isolation, structure elucidation, and further investigations Partial and/or total synthesis	Biological activities
		upon storage, or warming to >40°C, as experimentally evidenced by starting from pure **262a** to give a mixture of **262a** and **262b** after several days [31]	
Shuangancistrotectorine C (5) $C_{50}H_{52}N_2O_8$ (808.9) 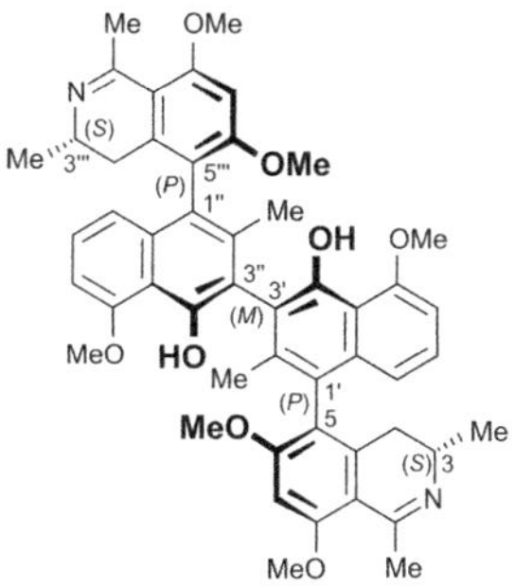 White amorphous powder: $[\alpha]_D^{20}$ −12.8 (*c* 0.05, MeOH) [31]	*A. tectorius* (PR China): twigs [31]	*Structural Elucidation*: Constitution deduced from HRESIMS, MALDI-TOF-MS as well as ^{1}H and ^{13}C NMR, including HMBC, HMQC, and ROESY interactions, supporting the presence of a naphthyldihydroisoquinoline dimer consisting of two identical 5,1′-coupled halves, linked through a 3′,3″-coupling in the binaphthalene core, with the central biaryl axis constituting an additional element of chirality, as in the case of the shuangancistrotectorines A (**262a**) and B (**262b**) [17, 31] Absolute configuration at C-3 and C-3‴ established as (*S*) by oxidative degradation [31, 80] Absolute configuration of the two outer axes in the monomeric halves assigned as (*P*) due to specific ROESY correlations in combination with the absolute (*S*)-configuration at C-3 and C-3‴ as established by degradation [17, 31] Assignment of the absolute axial configuration as (*P,M,P*) by comparison of the experimental ECD spectrum of **5** with the ECD curves calculated for the (*P,M,P*)- and the (*P,P,P*)-atropo-diastereomers using the TD-B3LYP/6-31G* approach in combination with a conformational analysis by PM3 [17, 31]	Strong—and specific—inhibitory activities in vitro against the erythrocytic stages of *Plasmodium falciparum* (strains: K1 and NF54) [31] Virtually no inhibitory activity against *Trypanosoma* and *Leishmania* parasites [31]

(continued)

Table 17 (continued)

Name, formula (mol. wt.), and structure Physical data [*M.p.*/°C, $[\alpha]_D^T$/°cm^2 g^{-1} 10^{-1} (solvent)]	Producing plants and their organs	Isolation, structure elucidation, and further investigations Partial and/or total synthesis	Biological activities
1′,1″-dimerized naphthylisoquinoline alkaloids			
Shuangancistrotectorine D (207a) $C_{50}H_{56}N_2O_8$ (813.0) 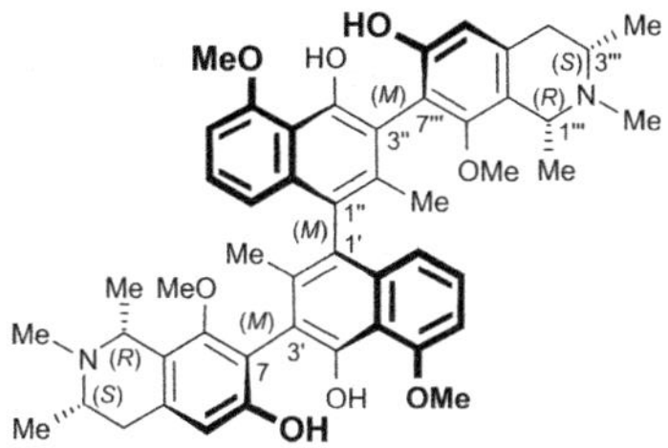 White amorphous powder: $[\alpha]_D^{20}$ +38.4 (*c* 0.09, MeOH) [31]	*A. tectorius* (PR China): twigs [31]	*Structural Elucidation*: Constitution deduced from HRESIMS, MALDI-TOF-MS as well as ^{1}H and ^{13}C NMR, including HMBC, HMQC, and ROESY interactions, indicative of the presence of a C_2-symmetric dimeric naphthylisoquinoline alkaloid, i.e., consisting of two homomorphous halves [17, 31] Coupling positions established by HMBC and ROESY interactions revealing the presence of a dimer consisting of two identical 7,3′-coupled N-methylated naphthyltetrahydroisoquinoline monomers, linked via the sterically constrained C-1′ and C-1″ positions of the central binaphthalene core, indicating the presence of three consecutive stereogenic axes in shuangancistrotectorine D (**207a**) [17, 31] Relative configuration at C-1 versus C-3 and at C-1‴ versus C-3‴ determined to be *cis* by specific ROESY correlations [17, 31] Absolute configurations established as (1*R*) and 1‴*R* and as (3*S*) and (3‴*S*) by oxidative degradation yielding *N*-methyl-D-alanine and *N*-methyl-(*S*)-3-amininobutyric acid [17, 31, 80] Absolute configuration of the two outer axes in the monomeric halves deduced as (*M*) due to specific ROESY correlations in combination with the absolute (*R*)-configuration at C-1 and C-1‴ in the isoquinoline moieties as established by degradation [17, 31] Confirmation of the stereochemical assignment as (*M,M,M*) by quantum-chemical ECD calculations of the (*M,M,M*)- and (*M,P,M*)-configured stereostructures (evidenced from ROESY interactions and oxidative degradation) by using the TD-B3LYP/6-31G* approach in combination with a PM3-based conformational analysis and comparison of the predicted ECD curves with the experimental ECD spectrum of shuangancistrotectorine D (**207a**) [17, 31]	Strong—and specific—inhibitory activity in vitro against *Plasmodium falciparum* [31] Weak inhibitory activities in vitro against *Trypanosoma brucei rhodesiense*, *T. cruzi*, and *Leishmania donovani* [31]

(continued)

Table 17 (continued)

Name, formula (mol. wt.), and structure Physical data [*M.p.*/°C, $[\alpha]_D^T$/°cm^2 g^{-1} 10^{-1} (solvent)]	Producing plants and their organs	Isolation, structure elucidation, and further investigations Partial and/or total synthesis	Biological activities
Shuangancistrotectorine E (207b) $C_{50}H_{56}N_2O_8$ (813.0) 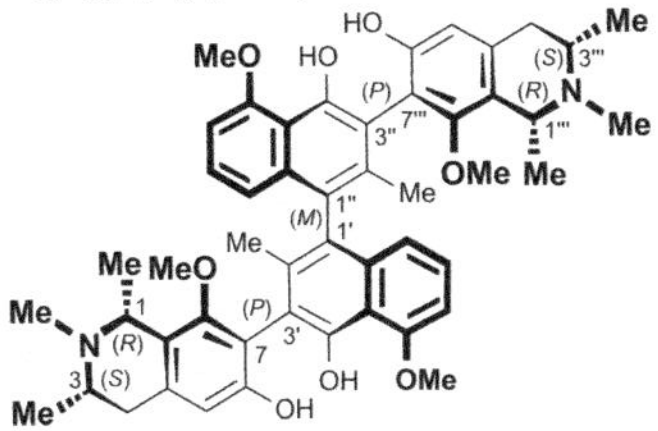 White amorphous powder: $[\alpha]_D^{20}$ −77.8 (*c* 0.09, MeOH) [31]	*A. tectorius* (PR China): twigs [31]	*Structural Elucidation*: Constitution deduced from HRESIMS and from ^{1}H and ^{13}C NMR, including HMBC, HMQC, and ROESY interactions, indicative of a C_2-symmetric naphthylisoquinoline dimer, consisting of two homomorphous halves and possessing the same constitution as shuangancistrotectorine D (**207a**) [17, 31] Relative configuration at C-1 versus C-3 and at C-1‴ versus C-3‴ determined to be *cis* by specific ROESY correlations—similar as for **207a** [17, 31] Absolute configurations established as (1*R*) and (1‴*R*) and as (3*S*) and (3‴*S*) by oxidative degradation—similar to **207a** [17, 31, 80] Absolute configuration at the central biaryl axis assigned as (*M*) due to the almost identical ECD curve of shuangancistrotectorine E (**207b**) compared to the ECD spectrum of shuangancistrotectorine D (**207a**) [17, 31] Confirmation of the stereochemical assignment as (*P*,*M*,*P*) by quantum-chemical ECD calculations of the (*P*,*M*,*P*)- and (*M*,*P*,*M*)-configured stereostructures (evidenced by ROESY interactions and oxidative degradation) by using the TD-B3LYP/6-31G* approach in combination with a PM3-based conformational analysis and comparison of the predicted ECD curves with the experimental ECD spectrum of shuangancistrotectorine E (**207b**) [17, 31]	Moderate inhibitory activity in vitro against *Plasmodium falciparum*, significantly weaker than that of its atropo-diastereomer **207a** and also distinctly weaker than that of shuangancistrotectorines A–C (**262a**, **262b**, and **5**) [31] Weak inhibitory activities in vitro against *Trypanosoma brucei rhodesiense*, *T. cruzi*, and *Leishmania donovani* [31]

Table 18 Naphthoquinones and tetralones related to naphthylisoquinoline alkaloids

Name, formula (mol. wt.), and structure Physical data [*M.p.*/°C, $[\alpha]_D^T$/°cm^2 g^{-1} 10^{-1} (solvent)]	Producing plants and their organs	Isolation, structure elucidation, and further investigations Partial and/or total synthesis	Biological activities
Ancistronaphthoside A (172) $C_{22}H_{28}O_{11}$ (468.2) 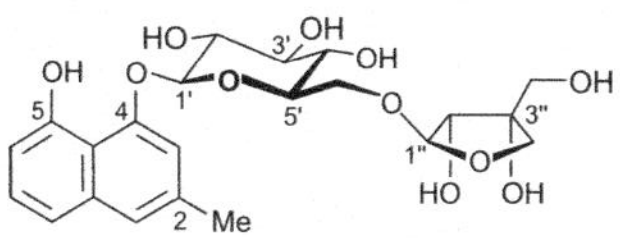 White amorphous powder (from MeOH): $[\alpha]_D^{20}$ −82.1 (*c* 0.20, MeOH) [48]	*A. cochinchinensis* (Vietnam): leaves [48]	Constitution deduced from HRESIMS, ^{1}H and ^{13}C NMR, and from HMBC, H,H-COSY, and ROESY interactions, indicative of a 2-methylnaphthalene equipped with a free hydroxy function and an oxygen function linked to a disaccharide unit [48] HMBC interactions revealing the phenolic hydroxy group to be located at C-5, and the oxygen function at C-4 to be linked to a *β*-D-apiofuranosyl-(1 → 6)-*O*-*β*-D-glucopyranosyl disaccharide moiety [48] Hydrolysis of ancistronaphthoside A (**172**) using methanolic HCl, yielding D-glucose and D-apiose, chromatographically identical with authentic reference samples [48]	No significant antiproliferative activity against several human cancer cell lines, among them LU-1 (lung carcinoma), HL-60 (promyelocytic leukemia), and SK-MEL-2 (melanoma) cells [48]
Plumbagin (37) $C_{11}H_8O_3$ (188.2) 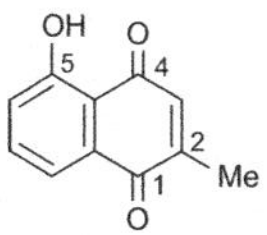 Compd. can also be addressed as 5-hydroxy-2-methyl-1,4-naphthoquinone Orange powder (from CH_2Cl_2): mp 78 [13] Orange crystals (from petroleum ether): mp 74 [497] Orange crystals (from *n*-hexane/$CHCl_3$): mp 73–74 [418] cf. Ref. [501]: Isolation from *Pera benensis* as orange needles (from $CHCl_3$): mp 78–79 cf. Ref. [253]: Orange needles: mp 77	*A. heyneanus* (India): roots, stem bark, and cell cultures [249, 250] *A. cochinchinensis* (Vietnam): leaves [46, 418, 419] *A. abbreviatus* (Ivory Coast): stem bark, twigs [13], and solid callus cultures [217] *A. barteri* (Ivory Coast): roots [13, 194, 502] *Triphyophyllum peltatum*: roots [13, 14, 503], cell cultures [217, 226, 504], and root cultures [216] *Dioncophyllum thollonii* (Gabon): roots	Plumbagin (**37**) as a polyketide-derived 1,4-naphthoquinone [254, 255] of widespread occurrence in various plant families, among them the Dioncophyllaceae [13, 14, 217, 226, 497, 503, 504] and the Ancistrocladaceae [13, 46, 194, 217, 249, 250, 418, 419, 502], but also closely related families such as the Plumbaginaceae [235, 237, 505, 506], the Drosophyllaceae [507], the Droseraceae [236, 508–510], and the Nepenthaceae [511, 512] *Further Investigations*: Formation of **37** from six acetate units via a *β*-pentaketo precursor according to the F-type folding mode—proof of the acetogenic origin of **37** by feeding experiments with $^{13}C_2$-labeled acetate administered to callus cultures of *A. heyneanus* [249] or *T. peltatum*	Strong antifeedant activity against *Spodoptera littoralis* [509] Pronounced inhibitory activity against the fungus *Penicillium citrinum*, strong inhibition of seed germination (test system: lettuce seeds), and strong ichthyotoxicity (fish toxicity) against male guppies of the species *Poecilia reticulata* [513] Highly potent inhibitory activity in vitro against *Trypanosoma brucei brucei* [514] Pronounced inhibitory activities against extracellular promastigotes of *Leishmania donovani*, *L. major*, and other strains, and against intracellular amastigotes of *L. donovani* [515, 516], as well as strong to good activities against lesion development in cutaneous leishmaniasis [501, 517]

(continued)

Table 18 (continued)

Name, formula (mol. wt.), and structure Physical data [*M.p.*/°C, $[\alpha]_D^T$/°cm^2 g^{-1} 10^{-1} (solvent)]	Producing plants and their organs	Isolation, structure elucidation, and further investigations Partial and/or total synthesis	Biological activities
	and stem bark [13, 14, 497] *Habropetalum dawei* (Sierra Leone): stem bark, and leaves [14]	[15, 131–134, 504], followed by 2D INADEQUATE NMR analysis, assisted by the cryoprobe methodology—after earlier hints at a biosynthesis of **37** from acetate units by feeding experiments on cell cultures of *Plumbago europaea* using ^{14}C-labeled acetate [505] Abundant amounts of **37**, apparently derived from the naphthalene precursor of naphthyliso-quinoline alkaloids, produced in Dioncophyllaceae and Ancistrocladaceae plants, and, in particular, in cell cultures of *Triphyophyllum peltatum*, *Ancistrocladus heyneanus*, and *A. abbreviatus*, when exposed to chemical, physical, or biotic stress [13, 14, 217, 226, 249, 250, 502, 504] Plumbagin (**37**) shown as the major antipathogenic factor involved in the incompatible relationship between the parasitic plant *Cuscuta reflexa* (Convolvulaceae) and *A. heyneanus*, leading to damage of the haustoria of the parasite and eventually to degeneration of the parasitic tissues [250]	Potent cytotoxic activities against multiple myeloma cells [218] Antiproliferation and antimetastasis effects of **37** against various cancer types both in vitro and in vivo; for detailed reviews on the cancer-related properties of **37**: see Refs. [227–232, 254, 255] High antibacterial activities against drug-sensitive and -resistant strains of *Mycobacterium tuberculosis* [432, 433] Pronounced inhibitory activities against sensitive and multidrug-resistant Gram-negative and Gram-positive bacteria, among them *Escherichia coli*, *Klebsiella pneumonia*, and *Staphylococcus aureus* [227, 229, 434, 435] Inhibition of biofilm-assisted infections caused by *Candida albicans* [227, 229, 434, 436] Neuroprotective [229, 441] and anti-inflammatory [229, 443] properties combined with the ability to improve memory performance [229, 443, 444]

(continued)

Table 18 (continued)

Name, formula (mol. wt.), and structure Physical data [$M.p./°C$, $[\alpha]_D^T/°cm^2\ g^{-1}\ 10^{-1}$ (solvent)]	Producing plants and their organs	Isolation, structure elucidation, and further investigations Partial and/or total synthesis	Biological activities
3,3′-Biplumbagin (185) $C_{22}H_{14}O_6$ (374.4) 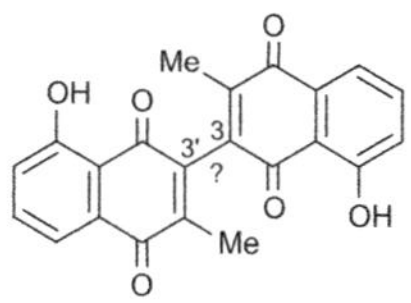 ? : absolute configuration and enantiomeric ratio unknown Orange-red crystals (from *n*-hexane/$CHCl_3$, 95:5): mp 203–206 [418] cf. Ref. [501]: Isolation from *Pera benensis* as orange-colored crystals (from $CHCl_3$): mp 213–214 cf. Ref. [253]: Orange-colored needles: mp 214–216 No α_D measured, no information about axial chirality [46, 418]	*A. cochinchinensis* (Vietnam): leaves [46, 418]	Occurrence of **185** along with its monomer plumbagin (**37**) in the leaves of *A. cochinchinensis* [46, 418], but also in other plants such as *Plumbago zeylanica* (Plumbaginaceae) [506], *Pera benensis* (Euphorbiaceae) [501], and *Aristea ecklonii* (Iridaceae) [425], and in many other plant species belonging to the Plumbaginaceae [254, 423, 424, 430] and Ebenaceae [254, 426–429, 431] families Treatment of cutaneous leishmaniasis by Chimane Indians in the Bolivian Amazonia using stem bark extracts of *Pera benensis*, the main constituents being plumbagin (**37**), as well as 3,3′-biplumbagin (**185**) and 8,8′-biplumbagin (**186**) [501, 517]	Moderate to strong inhibitory activities in vivo against lesion development in cutaneous leishmaniasis caused by *Leishmania braziliensis* [501, 517] Moderate to good inhibitory activities in vitro against *Leishmania* promastigotes and *Trypanosoma cruzi* epimastigotes [501, 517] Weak inhibitory activities in vitro against extracellular promastigotes of *Leishmania donovani*, *L. major*, and other strains, and against intracellular amastigote forms of *L. donovani* [515, 516] High antibacterial activities against drug-sensitive and -resistant strains of *Mycobacterium tuberculosis* [432, 433]
8,8′-Biplumbagin (186) (Maritinone) $C_{22}H_{14}O_6$ (374.4) 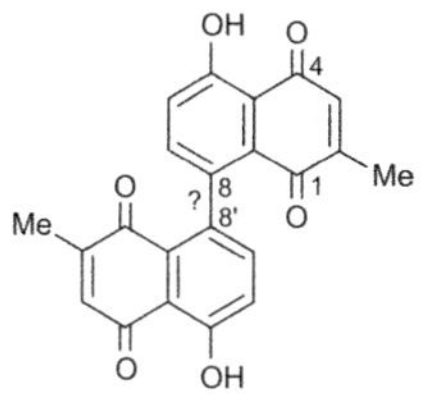 ? : absolute configuration and enantiomeric ratio unknown	*A. cochinchinensis* (Vietnam): leaves [46, 418]	Occurrence of **186** along with its monomer plumbagin (**37**) in the leaves of *A. cochinchinensis* [46, 418], but also in other plants such as *Plumbago zeylanica* (Plumbaginaceae) [506], *Pera benensis* (Euphorbiaceae) [501], and *Aristea ecklonii* (Iridaceae) [425], and in many other plant species belonging to the Plumbaginaceae [254, 423, 424, 430] and Ebenaceae [254, 426–429, 431] families Treatment of cutaneous leishmaniasis by Chimane Indians in the Bolivian Amazonia using stem bark extracts of *Pera benensis*, the main constituents being plumbagin (**37**), as well as 3,3′-biplumbagin (**185**) and 8,8′-biplumbagin (**186**) [501, 517]	Good-to-strong inhibitory activities in vivo against lesion development in cutaneous leishmaniasis caused by *Leishmania braziliensis* [501, 517] Moderate to good inhibitory activities in vitro against promastigote forms of *Leishmania* species and epimastigote forms of *Trypanosoma cruzi* [501, 517] Weak inhibitory activities in vitro against extracellular promastigotes of *Leishmania donovani*, *L. major*, and other strains, and against intracellular amastigote forms of *L. donovani* [515, 516]

(continued)

Table 18 (continued)

Name, formula (mol. wt.), and structure Physical data [*M.p.*/°C, $[\alpha]_D^T$/°cm^2 g^{-1} 10^{-1} (solvent)]	Producing plants and their organs	Isolation, structure elucidation, and further investigations Partial and/or total synthesis	Biological activities
Orange-red crystals (from *n*-hexane/$CHCl_3$, 95:5): mp 198–201 [418] cf. Ref. [501]: Isolation from *Pera benensis* as red needles (from $CHCl_3$): mp 190–191 cf. Ref. [253]: Red needles: mp 193–195 No α_D measured, no information about axial chirality [46, 418]			High antibacterial activities against drug-sensitive and -resistant strains of *Mycobacterium tuberculosis* [432, 433]
Droserone (38) $C_{11}H_8O_4$ (204.2) 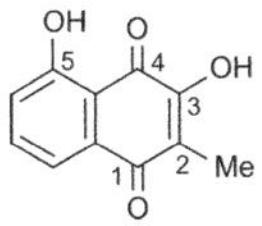 Compd. can also be addressed as 3,5-dihydroxy-2-methyl-1,4-naphthoquinone ——— Yellow solid (from MeOH): mp 179–181 [226] Yellow prisms (from MeOH): mp 178–180 [217] Orange-red needles (from diethyl ether): mp 172–173 [251, 486] Needle-shaped red–orange 'biogenic' crystals found in insect-dug holes in plant stems: mp 119 [251, 252] cf. Ref. [253]: yellow needles: mp 181	*A. heyneanus* (India): roots [486] *A. robertsoniorum* (Kenya): stem bark [251, 252] *A. abbreviatus* (Ivory Coast): solid callus cultures [217] *Triphyophyllum peltatum* (Ivory Coast): callus cultures [13, 14, 217, 218, 226] and root cultures [216] *Dioncophyllum thollonii* (Gabon): roots, stem bark, and twigs [14] *Habropetalum dawei* (Sierra Leone): twigs and stem bark [14]	Droserone (**38**) as a polyketide-derived 1,4-naphthoquinone [254, 255] of widespread occurrence in various plant families, among them the Dioncophyllaceae [13, 14, 217, 226, 504] and the Ancistrocladaceae [13, 251, 252, 486], but also closely related families such as the Plumbaginaceae [506], the Droseraceae [236, 510, 518], and the Nepenthaceae [512] *Further Investigations*: Pure orange-colored 'biogenic' crystals of **38** produced by the Kenyan liana *A. robertsoniorum*, well-suited for X-ray crystallography in their genuine form, found in small insect-dug holes in twigs and stems in the otherwise nearly white wood [251, 252] Formation of **38** from six acetate units via a *β*-pentaketo precursor according to the F-type folding mode [350]—proof of the acetogenic origin of **38** by feeding experiments with $^{13}C_2$-labeled acetate administered to callus cultures of *T. peltatum* [226], analyzed by the 2D INADEQUATE NMR method, assisted by cryoprobe methodology [15, 131–134, 504] Upon exposure to chemical, physical, or biotic stress, formation of abundant amounts	Moderate inhibition of seed germination (test system: lettuce seeds), and strong ichthyotoxicity against male guppies of the species *Poecilia reticulata* [513] No cytotoxic activity in vitro against the murine leukemia cell line P388 [510] No antimicrobial activity against *Candica albicans*, *Bacillus subtilis*, *Cladosporium resinae*, *Trichophyton mentagrophytes*, *Escherichia coli*, or *Pseudomonas aeruginosa* [510] Moderate antiviral activity against *Polio* virus type 1, but no inhibition of *Herpes simplex* type 1 [510] Considerable reduction of measles virus infection induced by droserone (**38**), but only upon addition of **38** during virus entry, not when added to the cells prior to virus uptake or after virus uptake, thus indicating interaction of **38** with the viral particles to reduce infectivity [519] Weak inhibitory activity in vitro against *Plasmodium falciparum* [520]

(continued)

Table 18 (continued)

Name, formula (mol. wt.), and structure Physical data [$M.p.$/°C, $[\alpha]_D^T$/°cm^2 g^{-1} 10^{-1} (solvent)]	Producing plants and their organs	Isolation, structure elucidation, and further investigations Partial and/or total synthesis	Biological activities
		of droserone (**38**), apparently from 1,8-dihydroxy-3-methylnaphthalene, the isocyclic precursor of naphthylisoquinoline alkaloids in Dioncophyllaceae and Ancistrocladaceae plants, and, in particular, in cell cultures of *T. peltatum*, *A. heyneanus*, and *A. abbreviatus* [13, 14, 217, 226, 504]	
Ancistroquinone A (81) $C_{12}H_{10}O_5$ (234.2) 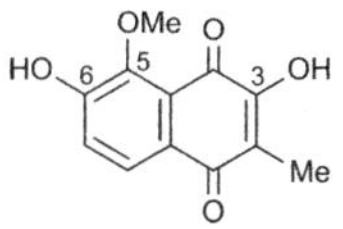 Compd. can also be addressed as 3,6-dihydroxy-5-methoxy-2-methyl-1,4-naphthoquinone Yellow needles (from acetone): mp 222–225 [315] cf.: Compound prepared by total synthesis [315]: Dark-yellow needles: mp 221–223	*A. heyneanus* (India): roots [315]	*Structure Elucidation*: Constitution deduced from MS, ^{1}H NMR, UV, and IR data, indicating the presence of a droserone derivative with an additional hydroxy function at C-6, and equipped with a methoxy group (instead of a hydroxy function) at C-5 [315] *Total Synthesis*: Total synthesis of **81** starting from 2,3-dimethoxyphenacetyl chloride: condensation with diethyl methylmalonate in the presence of magnesium ethoxide, and cyclization of the resulting ketoester in polyphosphoric acid, followed by selective O-demethylation using hydrobromic acid (48% in acetic acid) to give **81** [315]	Biological activities not determined for lack of isolated material
***cis*-Isoshinanolone** (*cis*-**67**) $C_{11}H_{12}O_3$ (192.2) 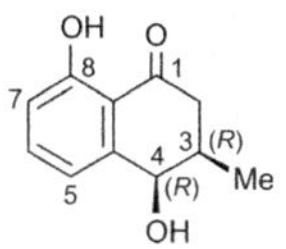	*A. heyneanus* (India): roots [13], cell cultures [249, 250] *A. cochinchinensis* (Vietnam): leaves [46, 418, 419] *A. abbreviatus* (Ivory Coast): roots [191] *A. barteri* (Ivory Coast): roots [13, 502] *A. likoko* (DR Congo): roots [167]	*cis*-Isoshinanolone (*cis*-**67**) as a widely distributed known natural tetralone isolated from various Dioncophyllaceae [13, 14, 81, 226, 247, 248, 504, 521, 522] and Ancistrocladaceae plant species [13, 167, 418, 458] *cis*-Isoshinanolone (*cis*-**67**) and its three further possible stereoisomeric forms (i.e., (3*R*,4*S*)-**67**, (3*S*,4*R*)-**67**, and (3*S*,4*S*)-**67**), as well known from other plant families like Plumbaginaceae [424, 523, 524],	Fish-stunning activity occurring at a concentration of 25 ppm of *cis*-**67**, stunning of the test fish *Barbus liberiensis* in a few minutes and killing within 45 min [248] Potent larvicidal activity against the yellow fever mosquito *Aedes aegypti* [523] Slight inhibitory activity in vitro against *Plasmodium falciparum* [520]

(continued)

Table 18 (continued)

Name, formula (mol. wt.), and structure Physical data [*M.p.*/°C, $[\alpha]_D^T$/°cm^2 g^{-1} 10^{-1} (solvent)]	Producing plants and their organs	Isolation, structure elucidation, and further investigations Partial and/or total synthesis	Biological activities
Compd. can also be addressed as (3*R*,4*R*)-4,8-dihydroxy-3-methyl-3,4-dihydro-2*H*-naphthalen-1-one Oil: $[\alpha]_D^{20}$ +20.0 (*c* 1.00, $CHCl_3$) [521] Oil: $[\alpha]_D^{25}$ +22.2 (*c* 1.00, $CHCl_3$) [247] Oil: bp 130°C at 0.8 mm Hg; $[\alpha]_D^{20}$ +33.0 (*c* 0.97, $CHCl_3$) [248] Oil: $[\alpha]_D^{20}$ +20.0 (*c* 1.00, $CHCl_3$) [13, 502] Oil: $[\alpha]_D^{25}$ +46.7 (*c* 0.60, $CHCl_3$) [168] Oil: $[\alpha]_D^{25}$ +25.4 (*c* 0.78, $CHCl_3$) [418] Semi-synthesis of the four possible stereoisomers of isoshinanolone by reduction of plumbagin (**37**) with $LiAlH_4$, giving rise to the *cis*- and *trans*-isomers (ratio ca. 1:4) as racemates [247, 428] *rac-cis*-Isoshinanolone: yellow semi-solid (from MeOH/water) [247]	*A. ileboensis* (DR Congo): roots [458] *Triphyophyllum peltatum* (Ivory Coast): cell cultures [226, 504] *Dioncophyllum thollonii* (Gabon): roots [14, 74, 247, 521, 522] *Habropetalum dawei* (Sierra Leone): stem bark [14, 81, 248]	Nepenthaceae [520, 525], Iridaceae [425], and Ebenaceae [428, 526–529] Identification in *Dioncophyllum thollonii* (Dioncophyllaceae) by HPLC-NMR online investigations on a root extract, applying time-slice experiments to distinguish between the two co-eluting diastereomers, *cis*- and *trans*-isoshinanolone (*cis*-**67** and *trans*-**67**) [74] Identification in *Habropetalum dawei* (Dioncophyllaceae): Constitution and complete stereostructure of *cis*-**67** established online, directly from crude extracts, by LC-MS/MS, LC-NMR, and LC-ECD, and by comparison of the data with those obtained from synthetic reference material [81] *Structural Elucidation and Stereoanalysis*: Attribution of the absolute configuration of the *cis*- and *trans*-diastereomers of isoshinanolone (and their enantiomers), from the known relative configuration at C-3 versus C-4 (through NMR), and from ruthenium-catalyzed oxidative degradation to give chiral 2-methylsuccinic acid as a stereochemically well known and easily analyzable degradation product [247] Ultimate proof of the absolute configuration of (+)-isoshinanolone as (3*R*,4*R*), by X-ray diffraction analysis of its 5,7-dibromo derivative [522]—in agreement with previous ECD investigations on a dibenzoate derivative [428] and with the results from the above-described oxidative degradation [247] Development of an HPLC-based stereoanalytical device for the rapid	

(continued)

Table 18 (continued)

Name, formula (mol. wt.), and structure Physical data [*M.p.*/°C, $[\alpha]_D^T$/°cm^2 g^{-1} 10^{-1} (solvent)]	Producing plants and their organs	Isolation, structure elucidation, and further investigations Partial and/or total synthesis	Biological activities
		discrimination between the four possible stereoisomeric forms of isoshinanolone (i.e., (3*R*,4*R*)-**67**, (3*R*,4*S*)-**67**, (3*S*,4*R*)-**67**, and (3*S*,4*S*)-**67**) by chromatography on a chiral stationary phase [247] Development of a chiroptical online stereoanalytical device for the discrimination of all four stereoisomeric forms of isoshinanolone by HPLC on a chiral phase coupled to ECD spectroscopy, providing the ECD spectra of the four stereoisomers directly from stereoisomeric mixtures (even applicable to the analysis of crude plant extracts) [522] *Further Investigations*: Formation of *cis*-**67** from six acetate units via a β-pentaketo precursor according to the F-type folding mode—proof of the acetogenic origin of *cis*-**67** by feeding experiments with $^{13}C_2$-labeled acetate administered to callus cultures of *A. heyneanus*, using 2D INADEQUATE NMR analysis, assisted by the cryoprobe methodology [249] Abundant amounts of *cis*-**67** (apparently arising from the naphthalene precursor of naphthylisoquinoline alkaloids, by reduction of plumbagin, **37**) produced by Dioncophyllaceae and Ancistrocladaceae plants, and, in particular, in cell cultures of *T. peltatum*, *A. heyneanus*, and *A. abbreviatus*, when exposed to chemical, physical, or biotic stress [13, 14, 504] *cis*-Isoshinanolone (*cis*-**67**) as an inducible pathogen repellent involved in the incompatible relationship between the	

(continued)

Table 18 (continued)

Name, formula (mol. wt.), and structure Physical data [*M.p.*/°C, $[\alpha]_D^T$/°cm² g⁻¹ 10⁻¹ (solvent)]	Producing plants and their organs	Isolation, structure elucidation, and further investigations Partial and/or total synthesis	Biological activities
		parasitic plant *Cuscuta reflexa* (Convolvulaceae) and *A. heyneanus*, leading to damage of the haustoria of the parasites and eventually to the degeneration of the parasitic tissues [250]	
(1*S*,3*R*)-1,8-Dihydroxy-3-methyltetralone (187) $C_{11}H_{12}O_3$ (192.2) 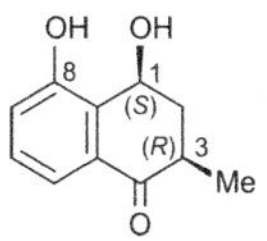 Oil: $[\alpha]_D^{25}$ +36.9 (*c* 0.71, $CHCl_3$) [418, 419]	*A. cochinchinensis* (Vietnam): leaves [418, 419]	*Structure Elucidation*: Constitution deduced by HRESIMS and by ^{1}H and ^{13}C NMR, including HMBC and H,H-COSY correlations, indicating the presence of a tetralone with a relative *cis*-configuration at C-1 versus C-3 as deduced from NOE interactions, and revealing a half-chair conformation, with the methyl group at C-3 occupying an *equatorial* position [418, 419]	Biological activities not determined for lack of isolated material
(1*R*,3*R*)-1,8-Dihydroxy-3-methyltetralone (188) $C_{11}H_{12}O_3$ (192.2) 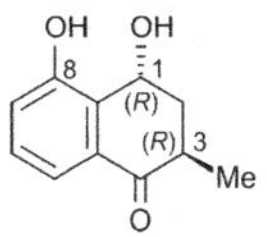 Oil: $[\alpha]_D^{25}$ +14.7 (*c* 0.34, $CHCl_3$) [418, 419]	*A. cochinchinensis* (Vietnam): leaves [418, 419]	*Structure Elucidation*: Constitution deduced by HRESIMS and by ^{1}H and ^{13}C NMR, including HMBC, H,H-COSY, and NOE correlations, indicating the presence of a tetralone with a relative *trans*-configuration at C-1 versus C-3 as deduced from NOE interactions, and revealing a half-chair conformation with both the methyl group at C-3 and the hydroxy group at C-1, occupying *equatorial* positions [418, 419]	Biological activities not determined for lack of isolated material

(continued)

Table 18 (continued)

Name, formula (mol. wt.), and structure Physical data [$M.p./°C$, $[\alpha]_D^T/°cm^2\ g^{-1}\ 10^{-1}$ (solvent)]	Producing plants and their organs	Isolation, structure elucidation, and further investigations Partial and/or total synthesis	Biological activities
(3*R*,4*S*)-3,4,8-Trihydroxy-3-methyltetralone (189) $C_{11}H_{12}O_4$ (208.4) 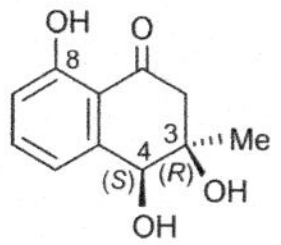 Amorphous solid: $[\alpha]_D^{25}$ +10.1 (*c* 0.33, $CHCl_3$) [418, 419]	*A. cochinchinensis* (Vietnam): leaves [418, 419]	*Structure Elucidation*: Constitution deduced by HRESIMS and by ^1H and ^{13}C NMR, including H,H-COSY, HMBC, and NOE correlations, indicating the presence of a trihydroxytetralone and revealing a half-chair conformation with both the methyl group at C-3 and the hydroxy group at C-4, occupying *equatorial* positions [418, 419]	Biological activities not determined for lack of isolated material
Ancistronaphthoside B (173) $C_{23}H_{32}O_{13}$ (516.2) 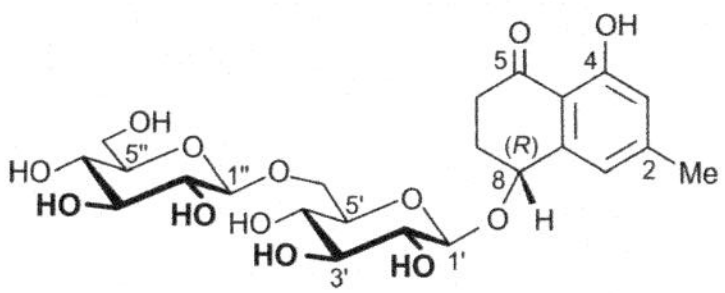 White amorphous powder (from MeOH): $[\alpha]_D^{25}$ −33.6 (*c* 0.20, MeOH) [48]	*A. cochinchinensis* (Vietnam): leaves [48]	*Structure Elucidation*: Constitution established by HRESIMS, ^1H and ^{13}C NMR, and by HMBC and HSQC interactions indicative of tetralone equipped with a carbonyl group located at C-5 and two further oxygen functions, one linked to two *β*-D-glucose residues [48] HMBC interactions revealing the free phenolic hydroxy group to be located at C-4 and the oxygen function at C-8 to be linked to a *β*-D-glucopyranosyl-(1 → 6)-*O*-*β*-D-glucopyranosyl disaccharide moiety [48] Absolute configuration at C-8 established as (*R*) by a positive Cotton effect observed in the ECD spectrum of **173**, in accordance with the ECD data of related known [247] tetralone-type compounds [48]	No significant antiproliferative activity against several human cancer cell lines, among them LU-1 (lung carcinoma), HL-60 (promyelocytic leukemia), and SK-MEL-2 (melanoma) cells [48]

Table 19 Naphthalene-devoid isoquinolines from Asian *Ancistrocladus* species

Name, formula (mol. wt.), and structure / Physical data [*M.p.*/°C, $[\alpha]_D^T$/°cm^2 g^{-1} 10^{-1} (solvent)]	Producing plants and their organs	Isolation, structure elucidation, and further investigations / Partial and/or total synthesis	Biological activities
(*S*)-6,8-Dimethoxy-1,3-dimethyl-3,4-dihydroisoquinoline (194) $C_{13}H_{17}NO_2$ (219.3) 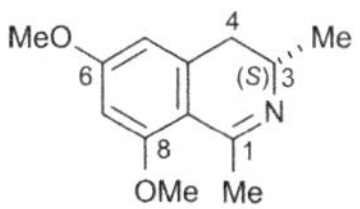 Amorphous solid, converted into the hydrobromide salt (from acetone): mp 197; $[\alpha]_D^{20}$ −138.0 (*c* 0.70, MeOH) [54] cf.: Stereoselective total synthesis of the title compd. [92]: Hydrobromide salt: white crystals (from EtOH): mp 202; $[\alpha]_D^{20}$ −141.0 (*c* 1.00, MeOH)	*A. tectorius* (Malaysia): stem bark [54]	Previously described as a synthetic compound [91], discovered for the first time in *A.* aff. *tectorius* from Malaysia as a genuine natural product [54] *Structure Elucidation*: Constitution deduced from ESIMS, and by ^{1}H and ^{13}C NMR, indicating the presence of a 6,8-dioxygenated 1,3-dimethyldihydroisoquinoline alkaloid [54] Absolute configuration at C-3 assumed to be (*S*) [54] (based on the observation that so far all naphthylisoquinoline alkaloids isolated from *A. tectorius* were (3*S*)-configured, exclusively) *Total Synthesis*: Stereoselective total synthesis by reductive amination of (*S*)-1-phenylethylamine and hydrogenolysis providing the respective enantiomerically pure primary amine, subsequently converted into the target molecule **194** by N-acetylation followed by Bischler–Napieralski cyclization [91]	Biological activities not determined for lack of isolated material

(continued)

Table 19 (continued)

Name, formula (mol. wt.), and structure Physical data [*M.p.*/°C, $[\alpha]_D^T$/°cm^2 g^{-1} 10^{-1} (solvent)]	Producing plants and their organs	Isolation, structure elucidation, and further investigations Partial and/or total synthesis	Biological activities
6,8-Dimethoxy-1,3-dimethylisoquinoline (195) $C_{13}H_{15}NO_2$ (217.3) 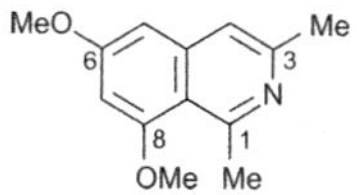 Crystalline solid (from diethyl ether): mp 71–73 [54] cf.: Compd. prepared by semi-synthesis: Crystalline solid (from diethyl ether): mp 85–90 [54] cf.: Compd. prepared by total synthesis: Colorless needles (from cyclohexane): mp 91 [453] Colorless needles (from diethyl ether): mp 65–67 [454]	*A. tectorius* (Malaysia): stem bark [54]	Previously described as a synthetic compound [453, 454], then discovered for the first time in *A.* aff. *tectorius* as a genuine natural product [54] *Structure Elucidation*: Constitution deduced by ESIMS, and by ^{1}H and ^{13}C NMR, indicating the presence of a non-hydrogenated 6,8-dioxygenated 1,3-dimethylisoquinoline alkaloid [54] *Total Synthesis*: Total synthesis (a) by Vilsmeier reaction of 3,5-dimethoxyphenylacetone with acetamide [453] or (b) starting from 3,5-dimethoxybenzaldehyde and nitroethane yielding the respective nitrostyrene, which was converted in several steps into 6,8-dimethoxy-1,3-dimethyldihydroisoquinoline, followed by dehydrogenation with Raney nickel giving rise to the target molecule **195** [454] Short and convenient novel total synthesis of **195** employing a C–H activation → alkenylation strategy, by Ce(III)-mediated methoximation of 2,4-dimethoxyacetophenone, followed by methoxime-directed Ru(III)-catalyzed allylation and an allyl-to-propenyl isomerization, and finally a microwave-promoted 6π-azaelectrocyclization, providing the target molecule **195** in just three steps [451, 452]	Moderate to weak cytotoxic activities against H1299 lung cancer cells and against the two breast carcinoma cell lines MDA-MB-468 and MCF7 [451]

(continued)

Table 19 (continued)

Name, formula (mol. wt.), and structure Physical data [*M.p.*/°C, $[\alpha]_D^T$/°cm^2 g^{-1} 10^{-1} (solvent)]	Producing plants and their organs	Isolation, structure elucidation, and further investigations Partial and/or total synthesis	Biological activities
6,8-Dimethoxy-3-hydroxymethyl-1-methylisoquinoline (160) 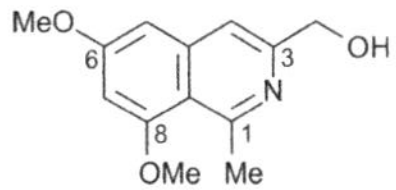 Compd. can also be addressed as 6-*O*-methylanciscochine ——— Crystalline solid (from MeOH/CH_2Cl_2): 164 [54] cf.: Compd. prepared by total synthesis: Off-white solid (from *n*-hexane/EtOAc): mp 162–163 [446]	*A. tectorius* (Malaysia): stem bark [54]	*Structure Elucidation*: Constitution established by HRESIMS, ^{1}H and ^{13}C NMR, and HMBC interactions indicative of a fully dehydrogenated 6,8-dioxygenated isoquinoline alkaloid equipped with a hydroxymethylene group at C-3 [54] Additional structural proof by conversion of **160** into the known [453, 454] 6,8-dimethoxy-1,3-dimethylisoquinoline **195**, by side-chain deoxygenation through successive mesylation (MsCl, NEt_3, CH_2Cl_2) and reduction with $LiAlH_4$ [54] ——— *Total Synthesis*: Total synthesis of **160** by Ce(III)-mediated oximation of commercially available 2,4-dimethoxyacetophenone, followed by oxime activation through *O*-pivaloylation, a one-pot rhodium(III)-catalyzed C–H activation/alkenylation with methyl acrylate, and C–N bond formation with concomitant in situ Ag(I)-catalyzed aromatization—reduction of the resulting isoquinoline 3-carboxylate with $NaBH_4/CaCl_2$ providing the target molecule **160** in four steps and in an overall yield of 43% [446]	No significant antiproliferative activity against several human cancer cell lines, among them HL-60 (promyelocytic leukemia), LU-1 (lung carcinoma), and SK-MEL-2 (melanoma) cells [48]

(continued)

Table 19 (continued)

Name, formula (mol. wt.), and structure Physical data [*M.p.*/°C, $[\alpha]_D^T$/°cm^2 g^{-1} 10^{-1} (solvent)]	Producing plants and their organs	Isolation, structure elucidation, and further investigations Partial and/or total synthesis	Biological activities
Anciscochine (161) $C_{12}H_{13}NO_3$ (219.1) 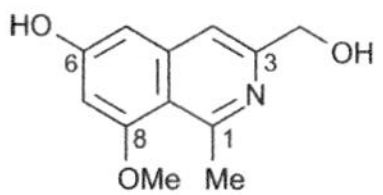 Compd. can also be addressed as 6-hydroxy-8methoxy-3-hydroxymethyl-1-methylisoquinoline Pale-yellow amorphous powder (from MeOH) [48]	*A. cochinchinensis* (Vietnam): leaves [48]	*Structure Elucidation*: Constitution assigned by HRESIMS as well as by ^{1}H and ^{13}C NMR, involving HMBC and NOESY interactions—by comparison of the data of anciscochine (**161**) with those of its parent compound **160**, revealing **161** to differ from **160** by the presence of a free hydroxy function at C-6 (instead of a methoxy group as in **160**) [48]	No significant antiproliferative activity against several human cancer cell lines, among them HL-60 (promyelocytic leukemia), LU-1 (lung carcinoma), and SK-MEL-2 (melanoma) cells [48]
Anciscochine-6-*O*-*β*-D-glucopyranoside (175) $C_{18}H_{23}NO_8$ (381.2) 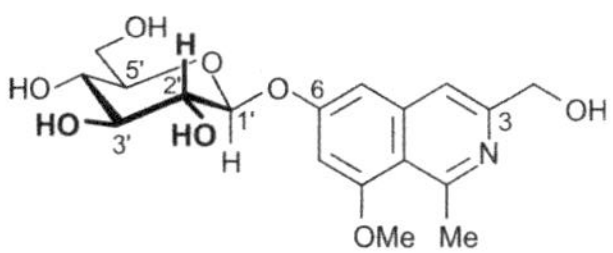 Pale-yellow amorphous powder (from MeOH); $[\alpha]_D^{25}$ −58.4 (*c* 0.20, MeOH) [48]	*A. cochinchinensis* (Vietnam): leaves [48]	*Structure Elucidation*: Constitution deduced by HRESIMS, ^{1}H and ^{13}C NMR, and from HMBC and H,H-COSY interactions, indicative of a non-hydrogenated isoquinoline derivative with a *β*-D-glucose residue at C-6 [48] Assignment of the configuration at the anomeric sugar carbon atom C-1′ to be *β* due to the large coupling constant from H-1′ to H-2′ ($^3J_{1'/2'}$ = 7.5 Hz), supporting their bis-axial arrangement in the six-membered ring [48] Hydrolysis of **175** using methanolic HCl yielding D-glucose and anciscochine (**161**) as the aglycone, which were both identified by comparison with authentic reference material [48]	No significant antiproliferative activity against several human cancer cell lines, among them HL-60 (promyelocytic leukemia), LU-1 (lung carcinoma), and SK-MEL-2 (melanoma) cells [48]

Acknowledgements The authors would like to thank those numerous skillful and highly committed students and coworkers of our group who contributed with great enthusiasm to the manifold developments during the past decades regarding the knowledge about naphthylisoquinoline alkaloids from Asian *Ancistrocladus* plants. We are in particular grateful (names given in an alphabetical order) to Dr. Stefan Busemann, Dr. Michael Dreyer, Dr. Torsten Geuder, Prof. Dr. Ralf God, Dr. Klaus-Peter Gulden, Prof. Dr. Tanja Gulder, Prof. Dr. Tobias A. M. Gulder, Dr. Christian Günther, Yasmin Hemberger, Dr. Barbara Hertlein-Amslinger, Dr. Jörg Holenz, Dr. Andreas Irmer, Dr. Johannes R. Jansen, Dr. Inga Kajahn, Dr. Dagmar Koppler, Dr. Katja Maksimenka, Dr. Kim Messer, Dr. Jörg Mühlbacher, Dr. Narasimhulu Manchala, Dr. Frank Pokorny, Dr. Matthias Reichert, Dr. Helmut Reuscher, Dr. Heiko Rischer, Dr. Markus Rückert, Dr. Raina Seupel, Dr. Anu Schaumlöffel, PD Dr. Jan Schlauer, Dr. Ralf Weirich, Barbara Wiesen, Dr. Michael Wohlfarth, Dr. Kristina Wolf, Dr. Minjuan Xu, Dr. Guoliang Zhang, and, in particular, Dr. Torsten Bruhn—the names of all involved coworkers can also be seen from the literature cited. Furthermore, we are indebted to Mrs. Lioba Kinzinger, Mr. Frank Meyer, Mrs. Manuela Michel, Mrs. Stefanie Schmitt, and Mrs. Michaela Schraut for their valuable technical assistance. We are grateful to Prof. Dr. Ulrich Zimmermann (Germany) for light and fluorescence microscopy investigations on *Ancistrocladus heyneanus*, Prof. Dr. Wolfgang Kiefer and Prof. Dr. Jürgen Popp (Germany) for FT-Raman investigations, and Prof. Dr. Axel Haase and Prof. Dr. Peter M. Jakob (Germany) for nuclear magnetic resonance microscopy studies. Moreover, we wish to thank Prof. Dr. Reto Brun, Dr. Marcel Kaiser (Switzerland), Prof. Dr. Thomas Dandekar, Prof. Dr. Thomas Efferth, Prof. Dr. August Stich, Prof. Dr. Heidrun Moll, Prof. Dr. Joachim Morschhäuser, PD Dr. Knut Ohlsen (Germany), Prof. Dr. Alicia Ponte-Sucre (Venezuela), Dr. Łukasz Adaszek (Poland), Prof. Dr. Michael R. Boyd (USA), and Prof. Dr. Suresh Awale (Japan) for intense, stimulating, and fruitful cooperations evaluating the bioactivities of genuine naphthylisoquinoline alkaloids and related, synthetic analogues.

We would further like to express our sincere compliments to the pioneer of naphthylisoquinoline alkaloids, late Prof. Dr. Tuticorin R. Govindachari, for his brilliant early work in the field, and to all of our Asian partners who have given us all sorts of support and cooperative reliability regarding botanical and ethnobotanical work, particularly Prof. Dr. A. S. Sankara Narayanan, Prof. Dr. M. R. Almeida, Prof. Dr. N. J. De Souza (India), Prof. Dr. Hamid A. Hadi (Malaysia), Prof. Dr. Luu Hoang Ngoc (Vietnam), and Prof. Dr. Jun Wu (PR China). Our particular thank goes to Prof. Dr. Günther Heubl (Germany) and Prof. Dr. Harald Meimberg (Austria) for their long-term cooperation focusing on the molecular phylogeny, chemotaxonomy, and botanical classification of the Asian *Ancistrocladus* species presented in this review. Moreover, we are grateful to Prof. Dr. Markus Riederer und Dr. Gerd Vogg (Germany) for their generous hospitality and competent support permitting us to cultivate and study more closely various *Ancistrocladus* plants from Asia in the Botanical Garden of the University of Würzburg, with an emphasis on the Indian liana *Ancistrocladus heyneanus* and the newly discovered taxon *Ancistrocladus benomensis* from Malaysia. We gratefully acknowledge the horticultural expertise and continuous support and commitment of all involved gardeners (in particular Andreas Kreiner and Friedrich Thiele) regarding the cultivation of these fascinating lianas from seeds to adult plants.

For generous financial support of this work, the authors would like to thank the Deutsche Forschungsgemeinschaft (projects Br 699/14-1 and 14-2; "Molecular Phylogeny and Chemotaxonomy of the Ancistrocladaceae Plant Family", Collaborative Research Center SFB 630 "Recognition, Preparation, and Functional Analysis of Agents against Infectious Diseases", Collaborative Research Center SFB 251 "Ecology, Physiology, and Biochemistry of Plants and Animals under Stress", Clinical Research Unit KFO 216 "Characterization of the Oncogenic Signaling Network in Multiple Myeloma: Development of Targeted Therapies", Priority Program SPP 1152 "Evolution of Metabolic Diversity", and Research Training Group "NMR in vivo and in vitro for Biological Basic Research"). This work was also funded by grants from the Bundesministerium für Bildung und Forschung (BMBF), project no. 0310722, the BASF AG, the Fonds der Chemischen Industrie (FCI), the Max Buchner Research Foundation, the German Academic Exchange Service (Deutscher Akademischer Austauschdienst, DAAD), and the Alexander von Humboldt Foundation.

References

1. Govindachari TR, Parthasarathy PC (1970) Ancistrocladine, a novel isoquinoline alkaloid from *Ancistrocladus heyneanus* Wall. Indian J Chem 8:567
2. Govindachari TR, Parthasarathy PC (1971) Ancistrocladine, a new type of isoquinoline alkaloid from *Ancistrocladus heyneanus*. Tetrahedron 27:1013
3. Govindachari TR, Parthasarathy PC, Desai HK (1971) Chemical investigation of *Ancistrocladus heyneanus* Wall: part III—further studies on ancistrocladine. Indian J Chem 9:931
4. Govindachari TR, Nagarajan K, Parthasarathy PC, Rajagopalan TG, Desai HK, Kartha G, Chen SML, Nakanishi K (1974) Absolute stereochemistry of ancistrocladine and ancistrocladinine. J Chem Soc Perkin Trans 1:1413
5. Govindachari TR, Parthasarathy PC (1977) Alkaloids of Ancistrocladaceae. Heterocycles 7:661
6. Bentley KW (1998) The isoquinoline alkaloids. In: Ravindranath B (ed) Chemistry and biochemistry of organic natural products. Harwood Academic Publishers, Amsterdam
7. Bentley KW (2006) β-Phenylethylamines and the isoquinoline alkaloids. Nat Prod Rep 23:444
8. Chrzanowska M, Grajewska A, Rozwadowska MD (2016) Asymmetric synthesis of isoquinoline alkaloids: 2004–2015. Chem Rev 116:12369
9. Iranshahy M, Quinn RJ, Iranshahi M (2014) Biologically active isoquinoline alkaloids with drug-like properties from the genus *Corydalis*. RSC Adv 4:15900
10. Weber C, Opatz T (2019) Bisbenzylisoquinoline alkaloids. In: Knölker HJ (ed) The alkaloids, vol 81. Academic Press, New York, p 1
11. Nguyen VK, Kou KGM (2021) The biology and total syntheses of bisbenzylisoquinoline alkaloids. Org Biomol Chem 19:7535
12. Shang XF, Yang CJ, Morris-Natschke SL, Li JC, Yin XD, Liu YQ, Peng JW, Goto M, Zhang JY, Lee KH (2020) Biologically active isoquinoline alkaloids covering 2014–2018. Med Res Rev 40:2212
13. Bringmann G, Pokorny F (1995) The naphthylisoquinoline alkaloids. In: Cordell GA (ed) The alkaloids: chemistry and pharmacology, vol 46. Academic Press, New York, p 127
14. Bringmann G, François G, Aké Assi L, Schlauer J (1998) The alkaloids of *Triphyophyllum peltatum* (Dioncophyllaceae). Chimia 52:18
15. Bringmann G, Günther C, Ochse M, Schupp O, Tasler S (2001) Biaryls in nature: a multi-facetted class of stereochemically, biosynthetically, and pharmacologically intriguing secondary metabolites. In: Herz W, Falk H, Kirby GW, Moore RE (eds) Progress in the chemistry of organic natural products, vol 82. Springer-Verlag, Wien, New York, p 1
16. Ibrahim SRM, Mohamed GA (2015) Naphthylisoquinoline alkaloids potential drug leads. Fitoterapia 106:194
17. Lombe BK, Feineis D, Bringmann G (2019) Dimeric naphthylisoquinoline alkaloids: polyketide-derived axially chiral bioactive quateraryls. Nat Prod Rep 36:1513
18. Tajuddeen N, Bringmann G (2021) N,C-Coupled naphthylisoquinoline alkaloids: a versatile new class of axially chiral natural products. Nat Prod Rep 38:2154
19. Gereau RE (1997) Typification of names of *Ancistrocladus* Wallich (Ancistrocladaceae). Novon 7:242
20. Cheek M (2000) A synoptic revision of *Ancistrocladus* (Ancistrocladaceae) in Africa, with a new species from Western Cameroon. Kew Bull 55:871
21. Taylor CM, Gereau RE, Walters GM (2005) Revision of *Ancistrocladus* Wall. (Ancistrocladaceae). Ann Missouri Bot Gard 92:360
22. Airy Shaw HK (1951) On the Dioncophyllaceae, a remarkable new family of flowering plants. Kew Bull 6:327

23. Govindachari TR, Parthasarathy PC, Rajagopalan TG, Desai HK, Ramachandran KS (1975) Hamatine, a new isoquinoline alkaloid from *Ancistrocladus hamatus* (Vahl) Gilg. Indian J Chem 13:641
24. Govindachari TR, Parthasarathy PC, Desai HK, Saindane MT (1977) On the absolute stereochemistry of hamatine. Indian J Chem 15B:871
25. Bringmann G, Teltschik F, Schäffer M, Haller R, Bär S, Robertson SA, Isahakia MA (1998) Ancistrobertsonine A and related naphthylisoquinoline alkaloids from *Ancistrocladus robertsoniorum*. Phytochemistry 47:31
26. Govindachari TR, Parthasarathy PC, Desai HK (1972) Chemical investigation of *Ancistrocladus heyneanus* Wall.: part VI—isolation & structure of ancistrocladisine, a novel alkaloid. Indian J Chem 10:1117
27. Govindachari TR, Parthasarathy PC, Rajagopalan TG, Desai HK, Ramachandran KS, Lee E (1975) Absolute configuration of ancistrocladisine and ancistrocladidine. J Chem Soc Perkin Trans 1:2134
28. Parthasarathy PC, Kartha G (1983) Rigorous structural proof for ancistrocladisine and ancistrocladidine, two minor alkaloids of *Ancistrocladus heyneanus* Wall. Indian J Chem 22B:590
29. Govindachari TR, Parthasarathy PC, Desai HK (1973) Chemical investigation of *Ancistrocladus heyneanus* Wall.—Ancistrocladidine, a new isoquinoline alkaloid. Indian J Chem 11:1190
30. Bringmann G, Dreyer M, Rischer H, Wolf K, Hadi HA, Brun R, Meimberg H, Heubl G (2004) Ancistrobenomine A, the first naphthylisoquinoline oxygenated at Me-3, and related 5,1′-coupled alkaloids, from the "new" plant species *Ancistrocladus benomensis*. J Nat Prod 67:2058
31. Xu M, Bruhn T, Hertlein B, Brun R, Stich A, Wu J, Bringmann G (2010) Shuangancistrotectorines A–E, dimeric naphthylisoquinoline alkaloids with three chiral biaryl axes, from the Chinese plant *Ancistrocladus tectorius*. Chem Eur J 16:4206
32. Bringmann G, Kajahn I, Reichert M, Pedersen SEH, Faber JH, Gulder T, Brun R, Christensen SB, Ponte-Sucre A, Moll H, Heubl G, Mudogo V (2006) Ancistrocladinium A and B, the first N,C-coupled naphthyldihydroisoquinoline alkaloids, from a Congolese *Ancistrocladus* species. J Org Chem 71:9348
33. Seupel R, Hemberger Y, Feineis D, Xu M, Seo EJ, Efferth T, Bringmann G (2018) Ancistrocyclinones A and B, unprecedented N,C-coupled naphthylisoquinoline alkaloids, from the Chinese liana *Ancistrocladus tectorius*. Org Biomol Chem 16:1581
34. Rischer H, Heubl G, Meimberg H, Dreyer M, Hadi HA, Bringmann G (2005) *Ancistrocladus benomensis* (Ancistrocladaceae): a new species from Peninsular Malaysia. Blumea 50:357
35. van Steenis CGGJ (1948) Ancistrocladaceae. In: van Steenis CGGJ (ed) Flora Malesiana, vol 4. Noordhoff-Kolff N.V., Djakarta, p 8
36. Meimberg H, Rischer H, Turini FG, Chamchumroon V, Dreyer M, Sommaro M, Bringmann G, Heubl G (2010) Evidence for species differentiation within the *Ancistrocladus tectorius* complex (Ancistrocladaceae) in Southeast Asia: a molecular approach. Plant Syst Evol 284:77
37. Govindachari TR, Parthasarathy PC, Desai HK (1971) Ancistrocladinine, a minor alkaloid from *Ancistrocladus heyneanus* Wall. Indian J Chem 9:1421
38. Bringmann G, Kinzinger L, Ortmann T, De Souza NJ (1994) Isoancistrocladine from *Ancistrocladus heyneanus*: the first naturally occurring *N*-unsubstituted *cis*-configurated naphthyltetrahydroisoquinoline alkaloid. Phytochemistry 35:259
39. Bringmann G, Koppler D, Wiesen B, François G, Sankara Narayanan AS, Almeida MR, Schneider H, Zimmermann U (1996) Ancistroheynine A, the first 7,8′-coupled naphthylisoquinoline alkaloid from *Ancistrocladus heyneanus*. Phytochemistry 43:1405
40. Yang LK, Glover RP, Yoganathan K, Sarnaik JP, Godbole AJ, Soejarto DD, Buss AD, Butler MS (2003) Ancisheynine, a novel naphthylisoquinoline alkaloid from *Ancistrocladus heyneanus*. Tetrahedron Lett 44:5827

41. Bringmann G, Dreyer M, Michel M, Tayman FSK, Brun R (2004) Ancistroheynine B and two further 7,3′-coupled naphthylisoquinoline alkaloids from *Ancistrocladus heyneanus* Wall. Phytochemistry 65:2903
42. Bringmann G, Wohlfarth M, Rischer H, Heubes M, Saeb W, Diem S, Herderich M, Schlauer J (2001) A photometric screening method for dimeric naphthylisoquinoline alkaloids and complete on-line structural elucidation of a dimer in crude plant extracts, by the LC-MS/LC-NMR/LC-CD triad. Anal Chem 73:2571
43. Bringmann G, Wohlfarth M, Rischer H, Schlauer J, Brun R (2002) Extract screening by HPLC coupled to MS-MS, NMR, and CD: a dimeric and three monomeric naphthylisoquinoline alkaloids from *Ancistrocladus griffithii*. Phytochemistry 61:195
44. Bringmann G, Dreyer M, Kopff H, Rischer H, Wohlfarth M, Hadi HA, Brun R, Meimberg H, Heubl G (2005) *ent*-Dioncophylleine A and related dehydrogenated naphthylisoquinoline alkaloids, the first Asian Dioncophyllaceae-type alkaloids, from the "new" plant species *Ancistrocladus benomensis*. J Nat Prod 68:686
45. Anh NH, Porzel A, Ripperger H, Bringmann G, Schäffer M, God R, Sung TV, Adam G (1997) Naphthylisoquinoline alkaloids from *Ancistrocladus cochinchinensis*. Phytochemistry 45:1287
46. Sung TV, Anh NH, Ripperger H, Porzel A, Adam G (2000) Chemical study on some endemic plants of Vietnam. ACGC Chem Res Commun 11:121
47. Bringmann G, Hertlein-Amslinger B, Kajahn I, Dreyer M, Brun R, Moll H, Stich A, Ndjoko Ioset K, Schmitz W, Ngoc LH (2011) Phenolic analogs of the N,C-coupled naphthylisoquinoline alkaloid ancistrocladinium A, from *Ancistrocladus cochinchinensis* (Ancistrocladaceae), with improved antiprotozoal activities. Phytochemistry 72:89
48. Lien LQ, Linh TM, Giang VH, Mai NC, Nhiem NX, Tai BH, Cuc NT, Anh LT, Ban NK, Minh CV, Kiem PV (2016) New naphthalene derivatives and isoquinoline alkaloids from *Ancistrocladus cochinchinensis* with their anti-proliferative activity on human cancer cells. Bioorg Med Chem Lett 26:3913
49. Foucher JP, Pousset JL, Cave A, Paris RR (1975) Chimiotaxonomie des Ancistrocladacées II.—sur le alcaloides de l'*Ancistrocladus tectorius* Loureiro, du Laos. Plantes Méd Phytothér 9:26
50. Chen Z, Wang B, Qin K, Zhang B, Su Q, Lin Q (1981) Isolation and identification of the alkaloids from *Ancistrocladus tectorius*. XaoXue XueBao (Acta Pharm Sin) 16:519
51. Ruangrungsi N, Wongpanich V, Tantivatana P, Cowe HJ, Cox PJ, Funayama S, Cordell GA (1985) Traditional medicinal plants of Thailand. V. Ancistrotectorine, a new naphthalene-isoquinoline alkaloid from *Ancistrocladus tectorius*. J Nat Prod 48:529
52. Meksuriyen D, Ruangrungsi N, Tantivatana P, Cordell GA (1990) NMR spectroscopic analysis of ancistrocladidine. Phytochemistry 29:2750
53. Bringmann G, Kinzinger L (1992) (+)-Ancistrocline, a naphthylisoquinoline alkaloid from *Ancistrocladus tectorius*. Phytochemistry 31:3297
54. Montagnac A, Hadi HA, Remy F, Païs M (1995) Isoquinoline alkaloids from *Ancistrocladus tectorius*. Phytochemistry 39:701
55. Manfredi KP, Britton M, Vissieche V, Pannell LL (1996) Three new naphthylisoquinoline alkaloids from *Ancistrocladus tectorius*. J Nat Prod 59:854
56. Tang CP, Yang YP, Zhong Y, Zhong QX, Wu HM, Ye Y (2000) Four new naphthylisoquinoline alkaloids from *Ancistrocladus tectorius*. J Nat Prod 63:1384
57. Su Z, Liu M, Li Z, Kang S, Hua H (2007) Chemical constituents of *Ancistrocladus tectorius*. Zhongguo Yaowu Huaxue Zazhi (Chin J Med Chem) 17:383
58. Liu M, Su Z, Zhang J, Li Z, Hua H (2008) Studies on constituents of the stems and branches of *Ancistrocladus tectorius*. Zhongguo Yaoxue Zazhi (Chin Pharm J) 43:1060
59. Tang CP, Xin ZQ, Li XQ, Ye Y (2010) Two new naphthylisoquinoline alkaloids from stems and leaves of *Ancistrocladus tectorius*. Nat Prod Res 24:989
60. Jiang C, Li ZL, Gong P, Kang SL, Liu MS, Pei YH, Jing YK, Hua HM (2013) Five novel naphthylisoquinoline alkaloids with growth inhibitory activities against human leukemia

cells HL-60, K562 and U937 from stems and leaves of *Ancistrocladus tectorius*. Fitoterapia 91:305
61. Bringmann G, Zhang G, Ölschläger T, Stich A, Wu J, Chatterjee M, Brun R (2013) Highly selective antiplasmodial naphthylisoquinoline alkaloids from *Ancistrocladus tectorius*. Phytochemistry 91:220
62. Bringmann G, Seupel R, Feineis D, Zhang G, Xu M, Wu J, Kaiser M, Brun R, Seo EJ, Efferth T (2016) Ancistectorine D, a naphthylisoquinoline alkaloid with antiprotozoal and antileukemic activities, and further 5,8′- and 7,1′-linked metabolites from the Chinese liana *Ancistrocladus tectorius*. Fitoterapia 115:1
63. Bringmann G, Xu M, Seupel R, Feineis D, Wu J (2016) Ancistrotectoquinones A and B, the first quinoid naphthylisoquinoline alkaloids, from the Chinese liana *Ancistrocladus tectorius*. Nat Prod Commun 11:971
64. Bringmann G, Seupel R, Feineis D, Xu M, Zhang G, Kaiser M, Brun R, Seo EJ, Efferth T (2017) Antileukemic ancistrobenomine B and related 5,1′-coupled naphthylisoquinoline alkaloids from the Chinese liana *Ancistrocladus tectorius*. Fitoterapia 121:76
65. Cecil A, Rikanović C, Ohlsen K, Liang C, Bernhardt J, Oelschlaeger TA, Gulder T, Bringmann G, Holzgrabe U, Unger M, Dandekar T (2011) Modeling antibiotic and cytotoxic effects of the dimeric isoquinoline IQ-143 on metabolism and its regulation in *Staphylococcus aureus*, *Staphylococcus epidermidis* and human cells. Genome Biol 12:R24
66. Ponte-Sucre A, Faber JH, Gulder T, Kajahn I, Pedersen SEH, Schultheis M, Bringmann G, Moll H (2007) Activities of naphthylisoquinoline alkaloids and synthetic analogs against *Leishmania major*. Antimicrob Agents Chemother 51:188
67. Ponte-Sucre A, Gulder T, Wegehaupt A, Albert C, Rikanović C, Schaeflein L, Frank A, Schultheis M, Unger M, Holzgrabe U, Bringmann G, Moll H (2009) Structure-activity relationship and studies on the molecular mechanism of leishmanicidal N,C-coupled arylisoquinolinium salts. J Med Chem 52:626
68. Ponte-Sucre A, Gulder T, Gulder TAM, Vollmers G, Bringmann G, Moll H (2010) Alterations to the structure of *Leishmania major* induced by *N*-arylisoquinolines correlate with compound accumulation and disposition. J Med Microbiol 59:69
69. Seupel R, Hertlein-Amslinger B, Gulder T, Stawski P, Kaiser M, Brun R, Bringmann G (2016) Directed synthesis of all four pure stereoisomers of the N,C-coupled naphthylisoquinoline alkaloid ancistrocladinium A. Org Lett 18:6508
70. Gulder T (2008) Neuartige Wirkstoffe gegen Infektionskrankheiten: N,C-gekuppelte Naphthylisochinolin-Alkaloide. PhD thesis, University of Würzburg
71. Amslinger B (2012) Isolierung, Totalsynthese, Stereostruktur und -dynamik neuartiger mono- und dimerer Naphthylisochinoline. PhD thesis, University of Würzburg
72. Seupel R (2018) Antiinfektive und antitumorale Naphthylisochinolin-Alkaloide: Isolierung und Strukturaufklärung, Totalsynthese und Untersuchungen zum Wirkmechanismus. PhD thesis, University of Würzburg
73. Bringmann G, Koppler D, Scheutzow D, Porzel A (1997) Determination of configuration at the biaryl axes of naphthylisoquinoline alkaloids by long-range NOE effects. Magn Reson Chem 35:297
74. Bringmann G, Rückert M, Messer K, Schupp O, Louis AM (1999) Use of on-line high-performance liquid chromatography—nuclear magenetic resonance spectrometry coupling in phytochemical screening studies: rapid identification of metabolites in *Dioncophyllum thollonii*. J Chromatogr A 837:267
75. Bringmann G, Günther C, Schlauer J, Rückert M (1998) HPLC-NMR online coupling including the ROESY technique: direct characterization of naphthylisoquinoline alkaloids in crude plant extracts. Anal Chem 70:2805
76. Bringmann G, Bruhn T, Maksimenka K, Hemberger Y (2009) The assignment of absolute stereostructures through quantum chemical circular dichroism calculations. Eur J Org Chem:2717

77. Bruhn T, Schaumlöffel A, Hemberger Y, Bringmann G (2013) SpecDis: quantifying the comparison of calculated and experimental electronic circular dichroism spectra. Chirality 25:243
78. Bringmann G, Götz D, Bruhn T (2012) The online stereochemical analysis of chiral compounds by HPLC-ECD coupling in combination with quantum chemical calculations. In: Berova N, Polavarapu PL, Nakanishi K, Woody RW (eds) Comprehensive chiroptical spectroscopy—applications in stereochemical analysis of synthetic compounds, natural products, and biomolecules, vol 2. Wiley, Hoboken, NJ, p 353
79. Bringmann G, Geuder T, Rübenacker M, Zagst R (1991) A facile degradation procedure for determination of absolute configuration in 1,3-dimethyltetra- and dihydroisoquinolines. Phytochemistry 30:2067
80. Bringmann G, God R, Schäffer M (1996) An improved degradation procedure for determination of the absolute configuration in chiral isoquinoline and β-carboline derivatives. Phytochemistry 43:1393
81. Bringmann G, Messer K, Wohlfarth M, Kraus J, Dumbuya K, Rückert M (1999) HPLC-CD on-line coupling in combination with HPLC-NMR and HPLC-MS/MS for the determination of the full absolute stereostructure of new metabolites in plant extracts. Anal Chem 71:2678
82. Bringmann G, Lang G (2003) Full absolute stereostructures of natural products directly from crude extracts: the HPLC-MS/MS-NMR-CD 'triad.' In: Müller WEG (ed) Marine molecular biotechnology. Springer Verlag, Berlin, Heidelberg, p 89
83. Bringmann G, Gulder TAM, Reichert M, Gulder T (2008) The online assignment of the absolute configuration of natural products: HPLC-CD in combination with quantum chemical CD calculations. Chirality 20:628
84. Bringmann G, Breuning M, Tasler S (1999) The lactone concept: an efficient pathway to axially chiral natural products and useful reagents. Synthesis 525
85. Bringmann G, Menche D (2001) Stereoselective total synthesis of axially chiral natural products via biaryl lactones. Acc Chem Res 34:615
86. Bringmann G, Tasler S (2001) Oxidative aryl coupling reactions: a biomimetic approach to configurationally unstable or axially chiral biaryl natural products and related bioactive compounds. Tetrahedron 57:331
87. Bringmann G, Tasler S, Pfeifer RM, Breuning M (2002) The directed synthesis of axially chiral ligands, reagents, catalysts, and natural products: through the 'lactone methodology.' J Organomet Chem 661:49
88. Bringmann G, Price Mortimer AJ, Keller PA, Gresser MJ, Breuning M (2005) Atroposelective synthesis of axially chiral biaryl compounds. Angew Chem 117:5518; Angew Chem Int Ed 44:5384
89. Bringmann G, Gulder T, Gulder TAM (2007) Asymmetric synthesis of biaryls by the 'lactone method.' In: Christmann M, Bräse S (eds) Asymmetric synthesis—the essentials. Wiley-VCH Verlag, Weinheim, p 246
90. Bringmann G, Gulder T, Gulder TAM, Breuning M (2011) Atroposelective total synthesis of axially chiral biaryl natural products. Chem Rev 111:563
91. Bringmann G, Jansen JR, Rink HP (1986) Regioselective and atropoisomeric-selective aryl coupling to give naphthylisoquinoline alkaloids: the first total synthesis of (–)-ancistrocladine. Angew Chem 98:917; Angew Chem Int Ed Engl 25:913
92. Bringmann G, Jansen JR (1989) Chiral economy with respect to rotational isomerism: rational synthesis of hamatine and (optionally) ancistrocladine from joint helical precursors. Heterocycles 28:137
93. Bringmann G, Reuscher H (1989) Aryl-coupling via "axially prostereogenic" lactones: first total synthesis of (+)-ancistrocladisine and (optionally) its atropisomer. Tetrahedron Lett 30:5249
94. Bringmann G, Reuscher H (1989) Atropodiastereoselective ring opening of bridged, "axial-prostereogenic" biaryls: directed synthesis of (+)-ancistrocladisine. Angew Chem 101:1725; Angew Chem Int Ed Engl 28:1672

95. Bringmann G, Saeb W, Rübenacker M (1999) Directed joint total synthesis of the three naphthylisoquinoline alkaloids dioncolactone A, dioncopeltine A, and 5′-*O*-demethyldioncophylline A. Tetrahedron 55:423
96. Bringmann G, Götz R, Harmsen S, Holenz J, Walter R (1996) Biomimetic total synthesis of michellamines A–C. Liebigs Ann 2045
97. Bringmann G, Holenz J, Weirich R, Rübenacker M, Funke C, Boyd MR, Gulakowski RJ, François G (1998) First synthesis of the antimalarial naphthylisoquinoline alkaloid dioncophylline C, and its unnatural anti-HIV dimer, jozimine C. Tetrahedron 54:497
98. Bringmann G, Ochse M, Götz R (2000) First atropo-divergent total synthesis of the antimalarial korupensamines A and B by the "lactone method." J Org Chem 65:2069
99. Bringmann G, Manchala N, Büttner T, Hertlein-Amslinger B, Seupel R (2016) First atroposelective total synthesis of enantiomerically pure ancistrocladidine and ancistrotectorine. Chem Eur J 22:9792
100. Schies C, Seupel R, Feineis D, Gehrold A, Schraut M, Kaiser M, Brun R, Bringmann G (2018) Biomimetic total synthesis of mbandakamine A and further antiplasmodial naphthylisoquinoline dimers. ChemistrySelect 3:940
101. Bringmann G, Gulder T, Hertlein B, Hemberger Y, Meyer F (2010) Total synthesis of the N,C-coupled naphthylisoquinoline alkaloids ancistrocladinium A and B and related analogues. J Am Chem Soc 132:1151
102. Rizzacasa MA (1998) Total synthesis of naphthylisoquinoline alkaloids. In: Atta-ur-Rahman (ed) Studies in natural products chemistry, vol 20. Elsevier Science B.V., Amsterdam, p 407
103. Rizzacasa MA, Sargent MV (1991) Synthetic approaches to the naphthylisoquinoline alkaloids, part 2: the total synthesis of (–)-*O*-methylancistrocladine and (+)-*O*-methylhamatine and their enantiomers. J Chem Soc Perkin Trans 1:845
104. Rizzacasa MA, Sargent MV (1991) Synthetic approaches to the alkaloids of the Ancistrocladaceae. Part 3. The total synthesis of (–)-ancistrocladinine: control of the diastereoisomer excess in the synthesis of axially chiral biaryls. J Chem Soc Perkin Trans 1:2773
105. Hobbs PD, Upender V, Liu J, Pollart DJ, Thomas DW, Dawson MI (1996) The first stereospecific synthesis of michellamine B. Chem Commun:923
106. Hobbs PD, Upender V, Dawson MI (1997) Stereospecific syntheses of michellamines A and C. Synlett:965
107. Bringmann G, Götz R, Keller PA, Walter R, Boyd MR, Lang F, Garcia A, Walsh JJ, Tellitu I, Bhaskar KV, Kelly TR (1998) A convergent total synthesis of the michellamines. J Org Chem 63:1090
108. Hoye TR, Chen M, Hoang B, Mi L, Priest OP (1999) Total synthesis of michellamines A–C, korupensamines A–D, and ancistrobrevine B. J Org Chem 64:7184
109. Watanabe T, Shakadou M, Uemura M (2000) Stereoselective synthesis of korupensamine A and *ent*-korupensamine B utilizing an identical planar chiral arene chromium complex. Synlett:1141
110. Watanabe T, Tanaka Y, Shoda R, Sakamoto R, Kamikawa K, Uemura M (2004) Stereoselective synthesis of atropisomeric korupensamines A and B utilizing planar chiral arene chromium complex. J Org Chem 69:4152
111. Lipshutz BH, Keith JM (1999) A stereospecific, intermolecular biaryl-coupling approach to korupensamine A—en route to the michellamines. Angew Chem 111:3743; Angew Chem Int Ed 38:3530
112. Huang S, Petersen TB, Lipshutz BH (2010) Total synthesis of (+)-korupensamine B via an atropselective intermolecular biaryl coupling. J Am Chem Soc 132:14021
113. Slack E, Seupel R, Aue D, Bringmann G, Lipshutz B (2019) Atroposelective total synthesis of the fourfold *ortho*-substituted naphthyltetrahydroisoquinoline biaryl *O*,*N*-dimethylhamatine. Chem Eur J 25:14237
114. Xu G, Fu W, Liu G, Senanayake CH, Tang W (2014) Efficient syntheses of korupensamines A, B and michellamine B by asymmetric Suzuki-Miyaura coupling reactions. J Am Chem Soc 136:570

115. Bungard CJ, Morris JC (2002) First total synthesis of the 7,3′-linked naphthylisoquinoline alkaloid ancistrocladidine. Org Lett 4:631
116. Bungard CJ, Morris JC (2006) Total synthesis of the 7,3′-linked naphthylisoquinoline alkaloid ancistrocladidine. J Org Chem 71:7354
117. Toop HD, Brusnahan JS, Morris JC (2017) Concise total synthesis of dioncophylline E through an *ortho*-arylation strategy. Angew Chem 129:8656; Angew Chem Int Ed 56:8546
118. Kim KH, Cheon CH (2016) Concise asymmetric total synthesis of *ent*-ancistrocladinium A. Adv Synth Catal 358:2883
119. Kim KH, Cheon CH (2017) Concise catalytic asymmetric total syntheses of ancistrocladinium A and its atropdiastereomer. Org Chem Front 4:1341
120. Jo YI, Lee CY, Cheon CH (2020) Asymmetric total syntheses of naphthylisoquinoline alkaloids via atroposelective coupling reaction using central chirality as atroposelectivity-controlling group. Org Lett 22:4653
121. Jo YI, Lee CY, Cheon CH (2020) Atroposelective total syntheses of naphthylisoquinoline alkaloids with (*P*)-configuration. J Org Chem 85:12770
122. Bringmann G, Zhang G, Büttner T, Bauckmann G, Kupfer T, Braunschweig H, Brun R, Mudogo V (2013) Jozimine A_2: the first dimeric Dioncophyllaceae-type naphthylisoquinoline alkaloid, with three chiral axes and high antiplasmodial activity. Chem Eur J 19:916
123. Li J, Seupel R, Bruhn T, Feineis D, Kaiser M, Brun R, Mudogo V, Awale S, Bringmann G (2017) Jozilebomines A and B, naphthylisoquinoline dimers from the Congolese liana *Ancistrocladus ileboensis*, with antiausterity activities against the PANC-1 human pancreatic cancer cell line. J Nat Prod 80:2807
124. Fayez S, Li J, Feineis D, Aké Assi L, Kaiser M, Brun R, Anany MA, Wajant H, Bringmann G (2019) A near-complete series of four atropisomeric jozimine-A_2 type naphthylisoquinoline dimers with antiplasmodial and cytotoxic activities and related alkaloids from *Ancistrocladus abbreviatus*. J Nat Prod 82:3033
125. Bringmann G (1996) Mono- and dimeric naphthylisoquinoline alkaloids—pharmaceutically and structurally exciting natural heterocycles with axial chirality. Bull Soc Chim Belg 105:601
126. Bringmann G, Saeb W, Koppler D, François G (1996) Jozimine A (‘dimeric’ dioncophylline A), a non-natural michellamine analog with high antimalarial activity. Tetrahedron 52:13409
127. Bringmann G, Götz R, François G (1996) Synthesis of pindikamine A, a michellamine-related dimer of a non-natural, ‘skew’ naphthylisoquinoline. Tetrahedron 52:13419
128. Bringmann G, Saeb W, Mies J, Messer K, Wohlfarth M, Brun R (2000) One-step oxidative dimerization of genuine, unprotected naphthylisoquinoline alkaloids to give michellamines and other bioactive quateraryls. Synthesis 1843
129. Bringmann G, Saeb W, Wohlfarth M, Messer K, Brun R (2000) Jozipeltine A, a novel, unnatural dimer of the highly hydroxylated naphthylisoquinoline alkaloid dioncopeltine A. Tetrahedron 56:5871
130. Hemberger Y, Zhang G, Brun R, Kaiser M, Bringmann G (2015) Highly antiplasmodial non-natural oxidative products of dioncophylline A: synthesis, absolute configuration, and conformational stability. Chem Eur J 21:14507
131. Bringmann G, Wohlfarth M, Rischer H, Grüne M, Schlauer J (2000) A new biosynthetic pathway to alkaloids in plants: acetogenic isoquinolines. Angew Chem 112:1523; Angew Chem Int Ed 39:1464
132. Bringmann G, Mutanyatta-Comar J, Greb M, Rüdenauer S, Noll TF, Irmer A (2007) Biosynthesis of naphthylisoquinoline alkaloids: synthesis and incorporation of an advanced $^{13}C_2$-labeled isoquinoline precursor. Tetrahedron 63:1755
133. Bringmann G, Irmer A, Rüdenauer S, Mutanyatta-Comar J, Seupel R, Feineis D (2016) 5′-*O*-Methyldioncophylline D, a 7,8′-coupled naphthylisoquinoline alkaloid from callus cultures of *Triphyophyllum peltatum*, and its biosynthesis from a late-stage tetrahydroisoquinoline precursor. Tetrahedron 72:2906

134. Bringmann G, Irmer A, Feineis D, Gulder TAM, Fiedler HP (2009) Convergence in the biosynthesis of acetogenic natural products from plants, fungi, and bacteria. Phytochemistry 70:1776
135. Staunton J (1979) Biosynthesis of isoquinoline alkaloids. Planta Med 36:1
136. Kutchan TM, Dittrich H, Bracher D, Zenk MH (1991) Enzymology and molecular biology of alkaloid biosynthesis. Tetrahedron 47:5945
137. O'Connor SE (2010) Alkaloids. In: Mander EL, Lui HW (eds) Comprehensive natural products II, vol 1. Elsevier, Oxford, p 977
138. O'Connor SE (2012) Alkaloids. In: Civjan N (ed) Natural products in chemical biology. Wiley, Hoboken, NJ, p 209
139. Stöckigt J, Chen Z, Ruppert M (2010) Enzymatic and chemo-enzymatic approaches towards natural and non-natural alkaloids: indoles, isoquinolines, and others. Top Curr Chem 297:67
140. Hagel JM, Facchini PJ (2013) Benzylisoquinoline alkaloid metabolism: a century of discovery and a brave new world. Plant Cell Physiol 54:647
141. Pal T, Pal A (1996) Oxidative phenol-coupling: a key step for the biomimetic synthesis of many important natural products. Curr Sci 71:106
142. Tajuddeen N, Van Heerden FR (2019) Antiplasmodial natural products: an update. Malar J 18:404
143. Simoben CV, Ntie-Kang F, Akone, SH, Sippl W (2018) Compounds from African medicinal plants with activities against selected parasitic diseases: schistosomiasis, trypanosomiasis and leishmaniasis. Nat Prod Bioprospect 8:151
144. Scotti MT, Scotti L, Ishiki H, Ribeiro FF, Duarte da Cruz RM, de Oliveira MP, Mendonça Jr FJB (2016) Natural products as a source for antileishmanial and antitrypanosomal agents. Comb Chem High Throughput Screen 19:537
145. Ntie-Kang F, Lifongo LL, Simoben CV, Babiaka SB, Sippl W, Meva'a Mbaze L (2014) The uniqueness and therapeutic value of natural products from West African medicinal plants. Part I: uniqueness and chemotaxonomy. RSC Adv 4:28728
146. Singh N, Mishra BB, Bajpai S, Singh RK, Tiwari VK (2014) Natural product based leads to fight against leishmaniasis. Bioorg Med Chem 22:18
147. Zofou D, Ntie-Kang F, Sippl W, Efange SMN (2013) Bioactive natural products derived from the Central African flora against neglected tropical diseases and HIV. Nat Prod Rep 30:1098
148. Onguéné PA, Ntie-Kang F, Lifongo LL, Ndom JC, Sippl W, Meva'a Mbaze L (2013) The potential of anti-malarial compounds derived from African medicinal plants. Part I: a pharmacological evaluation of alkaloids and terpenoids. Malar J 12:449
149. Izumi E, Ueda-Nakamura T, Dias Filho BP, Veiga JVF, Nakamura CV (2011) Natural products and Chagas' disease: a review of plant compounds studied for activity against *Trypanosoma cruzi*. Nat Prod Rep 28:809
150. Kumar V, Mahajan A, Chibale K (2009) Synthetic medicinal chemistry of selected antimalarial natural products. Bioorg Med Chem 17:2236
151. Kingston DGI, Cassera MB (2022) Antimalarial natural products. In: Kinghorn AD, Falk H, Gibbons S, Asakawa Y, Liu JK, Dirsch VM (eds) Progress in the chemistry of organic natural products, vol 117. Springer, Cham, p 40
152. Bringmann G, Ortmann T, Zagst R, Schöner B, Aké Assi L, Burschka C (1992) (±)-Dioncophyllacine A, a naphthylisoquinoline alkaloid with a 4-methoxy substituent from the leaves of *Triphyophyllum peltatum*. Phytochemistry 31:4015
153. Bringmann G, Günther C, Saeb W, Mies J, Brun R, Akè Assi L (2000) 8-*O*-Methyldioncophyllinol B and revised structures of other 7,6′-coupled naphthylisoquinoline alkaloids from *Triphyophyllum peltatum* (Dioncophyllaceae). Phytochemistry 54:337
154. Hallock YF, Cardellina II JH, Schäffer M, Stahl M, Bringmann G, François G, Boyd MR (1997) Yaoundamines A and B, new antimalarial naphthylisoquinoline alkaloids from *Ancistrocladus korupensis*. Tetrahedron 53:8121
155. Bringmann G, Dreyer M, Faber JH, Dalsgaard PW, Stærk D, Jarozewski JW, Ndangalasi H, Mbago F, Brun R, Reichert M, Maksimenka K, Christensen SB (2003) Ancistrotanzanine A,

the first 5,3′-coupled naphthylisoquinoline alkaloid, and two further, 5,8′-linked related compounds from the newly described species *Ancistrocladus tanzaniensis*. J Nat Prod 66:1159
156. Hallock YF, Manfredi KP, Blunt JW, Cardellina II JH, Schäffer M, Gulden KP, Bringmann G, Lee AY, Clardy J, François G, Boyd MR (1994) Korupensamines A–D, novel antimalarial alkaloids from *Ancistrocladus korupensis*. J Org Chem 59:6349
157. Bringmann G, Rübenacker M, Jansen JR, Scheutzow D, Aké Assi L (1990) On the structure of the Dioncophyllaceae alkaloids dioncophylline A ("triphyophylline") and "*O*-methyl--triphyophylline". Tetrahedron Lett 31:639
158. Bringmann G, Jansen JR, Reuscher H, Rübenacker M, Peters K, von Schnering HG (1990) First total synthesis of (–)-dioncophylline A ("triphyophylline") and of selected stereoisomers: complete (revised) stereostructure. Tetrahedron Lett 31:643
159. Bringmann G, Günther C, Busemann S, Schäffer M, Olowokudejo JD, Alo BI (1998) Ancistroguineines A and B as well as ancistrotectorine—naphthylisoquinoline alkaloids from *Ancistrocladus guineënsis*. Phytochemistry 47:37
160. Tshitenge DT, Bruhn T, Feineis D, Schmidt D, Mudogo V, Kaiser M, Brun R, Würthner F, Awale S, Bringmann G (2019) Ealamines A-H, a series of naphthylisoquinolines with the rare 7,8′-coupling site, from the Congolese liana *Ancistrocladus ealaensis*, targeting pancreatic cancer cells. J Nat Prod 82:3150
161. Kavatsurva SM, Lombe BK, Feineis D, Dibwe DF, Maharaj V, Awale S, Bringmann G (2018) Ancistroyafungines A–D, 5,8′- and 5,1′-coupled naphthylisoquinoline alkaloids from a Congolese *Ancistrocladus* species, with antiausterity activities against human PANC-1 pancreatic cancer cells. Fitoterapia 130:6
162. Heubl G, Turini F, Mudogo V, Kajahn I, Bringmann G (2010) *Ancistrocladus ileboensis* (DR Congo), a new liana with unique alkaloids. Plant Ecol Evol 143:63
163. Manfredi KP, Blunt JW, Cardellina II JH, McMahon JB, Pannell LL, Cragg GM, Boyd MR (1991) Novel alkaloids from the tropical plant *Ancistrocladus abbreviatus* inhibit cell killing by HIV-1 and HIV-2. J Med Chem 34:3402
164. Boyd MR, Hallock YF, Cardellina II JH, Manfredi KP, Blunt JW, McMahon JB, Buckheit Jr RW, Bringmann G, Schäffer M, Cragg GM, Thomas DW, Jato JG (1994) Anti-HIV michellamines from *Ancistrocladus korupensis*. J Med Chem 37:1740
165. Hallock YF, Manfredi KP, Dai JR, Cardellina II JH, Gulakowski RJ, McMahon JB, Schäffer M, Stahl M, Gulden KP, Bringmann G, François G, Boyd MR (1997) Michellamines D-F, new HIV inhibitory dimeric naphthylisoquinoline alkaloids, and korupensamine E, a new antimalarial monomer, from *Ancistrocladus korupensis*. J Nat Prod 60:677
166. Bringmann G, Günther C, Saeb W, Mies J, Wickramasinghe A, Mudogo V, Brun R (2000) Ancistrolikokines A–C: new 5,8′-coupled naphthylisoquinoline alkaloids from *Ancistrocladus likoko*. J Nat Prod 63:1333
167. Fayez S, Feineis D, Mudogo V, Awale S, Bringmann G (2017) Ancistrolikokines E-H and related 5,8′-coupled naphthylisoquinoline alkaloids from the Congolese liana *Ancistrocladus likoko* with antiausterity activities against PANC-1 human pancreatic cancer cells. RSC Adv 7:53740
168. Fayez S, Feineis D, Mudogo V, Seo EJ, Efferth T, Bringmann G (2018) Ancistrolikokine I and further 5,8′-coupled naphthylisoquinoline alkaloids from the Congolese liana *Ancistrocladus likoko* and their cytotoxic activities against drug-sensitive and multi-drug resistant human leukemia cells. Fitoterapia 129:114
169. Awale S, Dibwe DF, Balachandran C, Fayez S, Feineis D, Lombe BK, Bringmann G (2018) Ancistrolikokine E_3, a 5,8′-coupled naphthylisoquinoline alkaloid, eliminates the tolerance of cancer cells to nutrition starvation by inhibition of the Akt/mTOR/autophagy signaling pathway. J Nat Prod 81:2282
170. Lombe BK, Feineis D, Mudogo V, Brun R, Awale S, Bringmann G (2018) Michellamines A_6 and A_7, and further mono- and dimeric naphthylisoquinoline alkaloids from a Congolese *Ancistrocladus* liana and their antiausterity activities against pancreatic cancer cells. RSC Adv 8:5243

171. Bringmann G, Spuziak J, Faber JH, Gulder T, Kajahn I, Dreyer M, Heubl G, Brun R, Mudogo V (2008) Six naphthylisoquinoline alkaloids and a related benzopyranone from a Congolese *Ancistrocladus* species related to *Ancistrocladus congolensis*. Phytochemistry 69:1065
172. Bringmann G, Messer K, Brun R, Mudogo V (2002) Ancistrocongolines A–D, new naphthylisoquinoline alkaloids from *Ancistrocladus congolensis*. J Nat Prod 65:1096
173. Bringmann G, Steinert C, Feineis D, Mudogo V, Betzin J, Scheller C (2016) HIV-inhibitory michellamine-type dimeric naphthylisoquinoline alkaloids from the Central African liana *Ancistrocladus congolensis*. Phytochemistry 128:71
174. Bringmann G, Hamm A, Günther C, Michel M, Brun R, Mudogo V (2000) Ancistroealaine A and B, two new bioactive naphthylisoquinolines, and related naphthoic acids from *Ancistrocladus ealaensis*. J Nat Prod 63:1465
175. Bringmann G, Lombe BK, Steinert C, Ndjoko Ioset K, Brun R, Turini F, Heubl G, Mudogo V (2013) Mbandakamines A and B, unsymmetrically coupled dimeric naphthylisoquinoline alkaloids, from a Congolese *Ancistrocladus* species. Org Lett 15:2590
176. Tshitenge DT, Feineis D, Mudogo V, Kaiser M, Brun R, Seo EJ, Efferth T, Bringmann G (2018) Mbandakamine-type naphthylisoquinoline dimers and related alkaloids from the Central African liana *Ancistrocladus ealaensis* with antiparasitic and antileukemic activities. J Nat Prod 81:918
177. Mufusama JP, Feineis D, Mudogo V, Kaiser M, Brun R, Bringmann G (2019) Antiprotozoal dimeric naphthylisoquinolines, mbandakamines B_3 and B_4, and related 5,8′-coupled monomeric alkaloids, ikelacongolines A–D, from a Congolese *Ancistrocladus* liana. RSC Adv 9:12034
178. Lombe BK, Bruhn T, Feineis D, Mudogo V, Brun R, Bringmann G (2017) Cyclombandakamines A_1 and A_2, oxygen-bridged naphthylisoquinoline dimers from a Congolese *Ancistrocladus* liana. Org Lett 19:1342
179. Tshitenge DT, Bruhn T, Feineis D, Mudogo V, Kaiser M, Brun R, Bringmann G (2019) An unusually broad series of seven cyclombandakamines, bridged dimeric naphthylisoquinoline alkaloids from the Congolese liana *Ancistrocladus ealaensis*. Sci Rep 9:9812
180. Lombe BK, Bruhn T, Feineis D, Mudogo V, Brun R, Bringmann G (2017) Antiprotozoal spirombandakamines A_1 and A_2, fused naphthylisoquinoline dimers from a Congolese *Ancistrocladus* plant. Org Lett 19:6740
181. Lombe BK, Feineis D, Mudogo V, Kaiser M, Bringmann G (2021) Spirombandakamine A_3 and cyclombandakamines A_8 and A_9, polycyclic naphthylisoquinoline dimers, with antiprotozoal activity, from a Congolese *Ancistrocladus* plant. J Nat Prod 84:1335
182. Li J, Seupel R, Feineis D, Mudogo V, Kaiser M, Brun R, Brünnert D, Chatterjee M, Seo EJ, Efferth T, Bringmann G (2017) Dioncophyllines C_2, D_2, and F, and related naphthylisoquinoline alkaloids from the Congolese liana *Ancistrocladus ileboensis* with potent activities against *Plasmodium falciparum* and against multiple myeloma and leukemia cell lines. J Nat Prod 80:443
183. Hallock YF, Hughes CB, Cardellina II JH, Schäffer M, Gulden KP, Bringmann G, Boyd MR (1995) Dioncophylline A, the principal cytotoxin from *Ancistrocladus letestui*. Nat Prod Lett 6:315
184. Bringmann G, Lisch D, Reuscher H, Aké Assi L, Günther K (1991) Atrop-diastereomer separation by racemate resolution techniques: *N*-methyldioncophylline A and its 7-epimer from *Ancistrocladus abbreviatus*. Phytochemistry 30:1307
185. Bringmann G, Pokorny F, Stäblein M, Schäffer M, Aké Assi L (1993) Ancistrobrevine C from *Ancistrocladus abbreviatus*: the first mixed 'Ancistro-cladaceae/Dioncophyllaceae-type' naphthylisoquinoline alkaloid. Phytochemistry 33:1511
186. Fayez S, Feineis D, Aké Assi L, Seo EJ, Efferth T, Bringmann G (2019) Ancistrobreveines A–D and related dehydrogenated naphthylisoquinoline alkaloids with antiproliferative activities against leukemia cells, from the West African liana *Ancistrocladus abbreviatus*. RSC Adv 9:15738

187. Fayez S, Bruhn T, Feineis D, Aké Assi L, Awale S, Bringmann G (2020) Ancistrosecolines A–F, unprecedented *seco*-naphthylisoquinoline alkaloids from the roots of *Ancistrocladus abbreviatus*, with apoptosis-inducing potential against HeLa cancer cells. J Nat Prod 83:1139
188. Tshitenge DT, Feineis D, Mudogo V, Kaiser M, Brun R, Bringmann G (2017) Antiplasmodial ealapasamines A–C, 'mixed' naphthylisoquinoline dimers from the Central African liana *Ancistrocladus ealaensis*. Sci Rep 7:5767
189. Bringmann G, Dreyer M, Faber JH, Dalsgaard PW, Stærk D, Jaroszewski JW, Ndangalasi H, F, Brun R, Christensen SB (2004) Ancistrotanzanine C and related 5,1′- and 7,3′-coupled naphthylisoquinoline alkaloids from *Ancistrocladus tanzaniensis*. J Nat Prod 67:743
190. Bringmann G, Teltschik F, Michel M, Busemann S, Rückert M, Haller R, Bär S, Robertson SA, Kaminsky R (1999) Ancistrobertsonines B, C, and D as well as 1,2-didehydroancistrobertsonine D from *Ancistrocladus robertsoniorum*. Phytochemistry 52:321
191. Fayez S, Feineis D, Aké Assi L, Kaiser M, Brun R, Awale S, Bringmann G (2018) Ancistrobrevines E-J and related naphthylisoquinoline alkaloids from the West African liana *Ancistrocladus abbreviatus* with inhibitory activities against *Plasmodium falciparum* and PANC-1 human pancreatic cancer cells. Fitoterapia 131:245
192. Bringmann G, Zagst R, Reuscher H, Aké Assi L (1992) Ancistrobrevine B, the first naphthylisoquinoline alkaloid with 5,8'-coupling site, and related compounds from *Ancistrocladus abbreviatus*. Phytochemistry 31:4011
193. Fayez S, Cacciatore A, Sun S, Kim M, Aké Assi L, Feineis D, Awale S, Bringmann G (2021) Ancistrobrevidines A–C and related naphthylisoquinoline alkaloids with cytotoxic activities against HeLa and pancreatic cancer cells, from the liana *Ancistrocladus abbreviatus*. Bioorg Med Chem 30:115950
194. Bringmann G, Schneider C, Möhler U, Pfeifer RM, Götz R, Aké Assi L, Peters EM, Peters K (2003) Two atropisomeric *N*-methyldioncophyllines A and *N*-methylphylline, their naphthalene-free heterocyclic moiety, from *Ancistrocladus barteri*. Z Naturforsch 58b:577
195. Hallock YF, Cardellina II JH, Schäffer M, Bringmann G, François G, Boyd MR (1998) Korundamine A, a novel HIV-inhibitory and antimalarial "hybrid" naphthylisoquinoline alkaloid from *Ancistrocladus korupensis*. Bioorg Med Chem Lett 8:1729
196. Hutchinson J, Dalziel JM (1954) Dioncophyllaceae. In: Hutchinson J, Dalziel JM (eds, revised by Keay RWJ) Flora of West Tropical Africa, 2nd edn. Crown Agents Overseas Governments and Administration, London, p 191
197. Porembski S, Barthlott W (2002) Dioncophyllaceae. In: Kubitzki K, Bayer C (eds) The families and genera of vascular plants. V. Flowering plants—Dicotyledons, Malvales, Capparales and non-betalain Caryophyllales, vol 5. Springer, Heidelberg, p 178
198. François G, Bringmann G, Phillipson JD, Aké Assi L, Dochez C, Rübenacker M, Schneider C, Wéry M, Warhurst DC, Kirby GC (1994) Activity of extracts and naphthylisoquinoline alkaloids from *Triphyophyllum peltatum*, *Ancistrocladus abbreviatus* and *A. barteri* against *Plasmodium falciparum* in vitro. Phytochemistry 35:1461
199. François G, Bringmann G, Dochez C, Schneider C, Timperman G, Aké Assi L (1995) Activities of extracts and naphthylisoquinoline alkaloids from *Triphyophyllum peltatum*, *Ancistrocladus abbreviatus* and *Ancistrocladus barteri* against *Plasmodium berghei* (Anka strain) in vitro. J Ethnopharmacol 46:115
200. François G, Timperman G, Holenz J, Aké Assi L, Geuder T, Maes L, Dubois J, Hanocq M, Bringmann G (1996) Naphthylisoquinoline alkaloids exhibit strong growth-inhibiting activities against *Plasmodium falciparum* and *P. berghei* in vitro—structure-activity relationships of dioncophylline C. Ann Trop Med Parasitol 90:115
201. François G, Timperman G, Steenackers T, Aké Assi L, Holenz J, Bringmann G (1997) In vitro inhibition of liver forms of the rodent malaria parasite *Plasmodium berghei* by naphthylisoquinoline alkaloids—structure-activity relationships of dioncophyllines A and C and ancistrocladine. Parasitol Res 83:673

202. François G, Chimanuka B, Timperman G, Holenz J, Plaizier-Vercammen J, Aké Assi L, Bringmann G (1999) Differential sensitivity of erythrocytic stages of the rodent malaria parasite *Plasmodium chabaudi chabaudi* to dioncophylline B, a highly active naphthylisoquinoline alkaloid. Parasitol Res 85:935
203. Chimanuka B, François G, Timperman G, Vander Heyden Y, Holenz J, Plaizier-Vercammen J, Bringmann G (2001) A comparison of the stage-specific efficacy of chloroquine, artemether and dioncophylline B against the rodent malaria parasite *Plasmodium chabaudi chabaudi* in vivo. Parasitol Res 87:795
204. François G, Timperman G, Eling W, Aké Assi L, Holenz J, Bringmann G (1997) Naphthylisoquinoline alkaloids against malaria: evaluation of the curative potentials of dioncophylline C and dioncopeltine A against *Plasmodium berghei* in vivo. Antimicrob Agents Chemother 41:2533
205. Schwedhelm KF, Horstmann M, Faber JH, Reichert Y, Bringmann G, Faber C (2007) The novel antimalarial dioncophylline C forms a complex with heme in solution. ChemMedChem 2:541
206. Bringmann G, Bischof SK, Müller S, Gulder T, Winter C, Stich A, Moll H, Kaiser M, Brun R, Dreher J, Baumann K (2010) QSAR guided synthesis of simplified antiplasmodial analogs of naphthylisoquinoline alkaloids. Eur J Med Chem 45:5370
207. Moyo P, Shamburger W, van der Watt ME, Reader J, de Sousa ACC, Egan TJ, Maharaj VJ, Bringmann G, Birkholtz LM (2020) Naphthylisoquinoline alkaloids, validated as hit multistage antiplasmodial natural products. Int J Parasitol Drugs Drug Resist 13:51
208. Green S, Green TL, Heslop-Harrison Y (1979) Seasonal heterophylly and leaf gland features in *Triphyophyllum* (Dioncophyllaceae), a new carnivorous plant genus. Bot J Linn Soc 78:99
209. Marburger JE (1979) Glandular leaf structure of *Triphyophyllum peltatum* (Dioncophyllaceae): a "fly-paper" insect trapper. Am J Bot 66:404
210. Menninger EA (1965) An African vine with three kinds of leaves for three different jobs. J New York Bot Gard 15:30
211. Bringmann G, Schlauer J, Wolf K, Rischer H, Buschbom U, Kreiner A, Thiele F, Duschek M, Aké Assi L (1999) Cultivation of *Triphyophyllum peltatum* (Dioncophyllaceae), the part-time carnivorous plant. Carniv Pl Newslett 28:7
212. Bringmann G, Wenzel M, Bringmann HP, Schlauer J, Aké Assi L, Haas F (2001) Uptake of the amino acid alanine by digestive leaves: proof of carnivory in the tropical liana *Triphyophyllum peltatum* (Dioncophyllaceae). Carniv Pl Newslett 30:15
213. Bringmann G, Rischer H, Schlauer J, Wolf K, Kreiner A, Duschek M, Aké Assi L (2002) The tropical liana *Triphyophyllum peltatum* (Dioncophyllaceae): formation of carnivorous organs is only a prerequisite for shoot elongation. Carniv Pl Newslett 31:44
214. Rembold K, Irmer A, Poppinga S, Rischer H, Bringmann G (2010) Propagation of *Triphyophyllum peltatum* (Dioncophyllaceae) and observations on its carnivory. Carniv Pl Newslett 39:71
215. Bringmann G, Rischer H (2001) In vitro propagation of the alkaloid-producing rare African liana *Triphyophyllum peltatum* (Dioncophyllaceae). Plant Cell Rep 20:591
216. Bringmann G, Irmer A, Büttner T, Schaumlöffel A, Zhang G, Seupel R, Feineis D, Fester K (2016) Axially chiral dimeric naphthalene and naphthoquinone metabolites, from root cultures of the West African liana *Triphyophyllum peltatum*. J Nat Prod 79:2094
217. Bringmann G, Rüdenauer S, Irmer A, Bruhn T, Brun R, Heimberger T, Stühmer T, Bargou R, Chatterjee M (2008) Antitumoral and antileishmanial dioncoquinones and ancistroquinones from cell cultures of *Triphyophyllum peltatum* (Dioncophyllaceae) and *Ancistrocladus abbreviatus* (Ancistrocladaceae). Phytochemistry 69:2501
218. Bringmann G, Zhang G, Hager A, Moos M, Irmer A, Bargou R, Chatterjee M (2011) Antitumoral activities of dioncoquinones B and C and related naphthoquinones gained from total synthesis or isolation from plants. Eur J Med Chem 46:5778
219. Bringmann G, Rübenacker M, Geuder T, Aké Assi L (1991) Dioncophylline B, a naphthylisoquinoline alkaloid with a new coupling type from *Triphyophyllum peltatum*. Phytochemistry 30:3845

220. Bringmann G, Rübenacker M, Weirich R, Aké Assi L (1992) Dioncophylline C from the roots of *Triphyophyllum peltatum*, the first 5,1′-coupled Dioncophyllaceae alkaloid. Phytochemistry 41:4019
221. Bringmann G, Rübenacker M, Vogt P, Busse H, Aké Assi L, Peters K, von Schnering HG (1991) Dioncopeltine A and dioncolactone A: alkaloids from *Triphyophyllum peltatum*. Phytochemistry 30:1691
222. Bringmann G, Messer K, Schwöbel B, Brun R, Aké Assi L (2003) Habropetaline A, an antimalarial naphthylisoquinoline alkaloid from *Triphyophyllum peltatum*. Phytochemistry 62:345
223. Bringmann G, Saeb W, God R, Schäffer M, François G, Peters K, Peters EM, Proksch P, Hostettmann K, Akè Assi L (1998) 5’-*O*-Demethyldioncophylline A, a new antimalarial alkaloid from *Triphyophyllum peltatum*. Phytochemistry 49:1667
224. Bringmann G, Rübenacker M, Koch W, Koppler D, Ortmann T, Schäffer M, Aké Assi L (1994) 5′-*O*-Demethyl-8-*O*-methyl-7-*epi*-dioncophylline A and its ‘regularly’ configured atropisomer from *Triphyophyllum peltatum*. Phytochemistry 36:1057
225. Bringmann G, Messer K, Wolf K, Mühlbacher J, Grüne M, Louis AM (2002) Dioncophylline E from *Dioncophyllum thollonii*, the first 7,3′-coupled dioncophyllaceous naphthylisoquinoline alkaloid. Phytochemistry 60:389
226. Bringmann G, Rischer H, Wohlfarth M, Schlauer J, Aké Assi L (2000) Droserone from cell cultures of *Triphyophyllum peltatum* (Dioncophyllaceae) and its biosynthetic origin. Phytochemistry 53:339
227. Padhye S, Dandawate P, Yusufi M, Ahmad A, Sarkar FH (2012) Perspectives on medicinal properties of plumbagin and its analogs. Med Res Rev 32:1131
228. Yin Z, Zhang J, Chen L, Guo Q, Yang B, Zhang W, Kang W (2020) Anticancer effects and mechanisms of action of plumbagin: review of research advances. BioMed Res I:6940953
229. Badwaik HR, Kumari L, Nakhate K, Verma VS, Sakure K (2019) Phytoconstituent plumbagin: chemical, biotechnological and pharmaceutical aspects. In: Atta-ur-Rahman (ed) Studies in natural products chemistry, vol 63. Elsevier Science BV, Amsterdam, p 415
230. Lu JJ, Bao JL, Wu GS, Xu WS, Huang MQ, Chen XP, Wang YT (2013) Quinones derived from plant secondary metabolites as anti-cancer agents. Anti-Cancer Agents Med Chem 13:456
231. Roy A (2021) Plumbagin: a potential anti-cancer compound. Mini Rev Med Chem 21:731
232. Liu Y, Cai Y, He C, Chen M, Li H (2017) Anticancer properties and pharmaceutical applications of plumbagin: a review. Am J Chin Med 45:423
233. Kapoor N, Kandwal P, Sharma G, Gambhir L (2021) Redox ticklers and beyond: naphthoquinone repository in the spotlight against inflammation and associated maladies. Pharmacol Res 174:105968
234. Maruyama D, Iida S, Ogawa G, Fukuhara N, Seo S, Miyazaki K, Yoshimitsu M, Kuroda J, Tsukamoto N, Tsujimura H, Hangaishi A, Yamauchi T, Utsumi T, Mizuno I, Takamatsu Y, Nagata Y, Minauchi K, Ohtsuka E, Hanamura I, Yoshida S, Yamasaki S, Suehiro Y, Kamiyama Y, Tsukasaki K, Nagai H (2021) Randomised phase II study to optimise melphalan, prednisolone, and bortezomib in untreated multiple myeloma (JCOG1105). Br J Haematol 192:531
235. Yue J, Lin Z, Wang D, Feng Y, Sun H (1994) Plumbasides A–C, three naphthoquinones derivatives from *Ceratostigma minus*. Phytochemistry 35:1023
236. Kreher B, Neszmélyi A, Wagner H (1990) Naphthoquinones from *Dionaea muscipula*. Phytochemistry 29:605
237. Serrilli AM, Sanfilippo V, Ballero M, Sanna C, Poli F, Scartezzini P, Serafini M, Bianco A (2010) Polar and antioxidant fraction of *Plumbago europaea* L., a spontaneous plant of Sardinia. Nat Prod Res 24:633
238. Nwaka S, Ramirez B, Brun R, Maes L, Douglas F, Ridley R (2009) Advancing drug innovation for neglected diseases—criteria for lead progression. PLoS Negl Trop Dis 3:e440
239. Becke AD (1993) Density-functional thermochemistry. III. The role of exact exchange. J Chem Phys 98:5648

240. Lee C, Yang W, Parr RG (1988) Development of the Colle-Salvetti correlation-energy formula into a functional of the electron density. Phys Rev B 37:785
241. Hariharan PC, Pople JA (1973) The influence of polarization functions on molecular orbital hydrogenation energies. Theor Chim Acta 28:213
242. Bringmann G, Weirich R, Reuscher H, Jansen JR, Kinzinger L, Ortmann T (1993) The synthesis of all possible isomeric 6,8-dioxygenated 1,3-dimethyl-1,2,3,4-tetrahydroisoquinoline methyl ethers—useful chiral building blocks for naphthylisoquinoline alkaloids. Liebigs Ann Chem:877
243. Bringmann G (1986) The naphthylisoquinoline alkaloids. In: Brossi A (ed) The alkaloids: chemistry and pharmacology, vol 29. Academic Press, New York, p 141
244. Shamma M (1972) The isoquinoline alkaloids—chemistry and pharmacology. Academic Press, New York, p 501
245. Labanca F, Ovesna J, Milella L (2018) *Papaver somniferum* L. taxonomy, uses and new insight in poppy alkaloid pathways. Phytochem Rev 17:853
246. Wicks C, Hudlicky T, Rinner U (2021) Morphine alkaloids: history, biology, and synthesis. In: Knölker HJ (ed) The alkaloids, vol 86. Academic Press, New York, p 145
247. Bringmann G, Münchbach M, Messer K, Koppler D, Michel M, Schupp O, Wenzel M, Louis AM (1999) *Cis-* and *trans*-isoshinanolone from *Dioncophyllum thollonii*: absolute configuration of two 'known', wide-spread natural products. Phytochemistry 51:693
248. Hanson SW, Crawford M, Thanasingh DPJ (1981) (+)-Isoshinanolone and 2-methylbenzofuran-4-carbaldehyde from the fish-stunning plant *Habropetalum dawei*. Phytochemistry 20:1162
249. Bringmann G, Wohlfarth M, Rischer H, Rückert M, Schlauer J (1998) The polyketide folding mode in the biogenesis of isoshinanolone and plumbagin from *Ancistrocladus heyneanus* (Ancistrocladaceae). Tetrahedron Lett 39:8445
250. Bringmann G, Schlauer J, Rückert M, Wiesen B, Ehrenfeld K, Proksch P, Czygan FC (1999) Host-derived acetogenins involved in the incompatible parasitic relationship between *Cuscuta reflexa* (Convolvulaceae) and *Ancistrocladus heyneanus* (Ancistrocladaceae). Plant Biol 1:581
251. Bringmann G, Kehr C, Dauer U, Guldern KP, Haller RD, Bär S, Isahakia MA, Robertson SA, Peters K (1993) *Ancistrocladus robertsoniorum* "produces" pure crystalline droserone when wounded. Planta Med 93:A622
252. Peters K, Peters EM, von Schnering HG, Bringmann G, Kehr C, Haller RD, Bär S, Isahakia MA, Robertson SA (1995) Crystal structure of droserone in "biogenic crystals" found under the stem bark of *Ancistrocladus robertsoniorum*. Z Kristallogr 210:290
253. Thomson RH (1997) Naturally occurring quinones IV. Chapman and Hall, London
254. Babula P, Adam V, Havel L, Kizek R (2009) Noteworthy secondary metabolites naphthoquinones—their occurrence, pharmacological properties and analysis. Curr Pharm Anal 5:47
255. Hook I, Mills C, Sheridan H (2014) Bioactive naphthoquinones from higher plants. In: Atta-ur-Rahman (ed) Studies in natural products chemistry, vol 41. Elsevier Science BV, Amsterdam, p 119
256. Bringmann G, Gulder T, Reichert M, Meyer F (2006) Ancisheynine, the first N,C-coupled naphthylisoquinoline alkaloid: total synthesis and stereochemical analysis. Org Lett 8:1037
257. Bringmann G, Rückert M, Schlauer J, Herderich M (1998) Separation and identification of dimeric naphthylisoquinoline alkaloids by liquid chromatography coupled to electrospray ionization mass spectrometry. J Chromatogr A 810:231
258. Bringmann G, Saeb W, Peters K, Peters EM (1997) The absolute stereostructure of dioncophylline A by anomalous X-ray dispersion of a 5-bromo derivative. Phytochemistry 45:1283
259. Cheek M, Frimodt-Møller C, Hørlyck V (2000) A new submontane species of *Ancistrocladus* from Tanzania. Kew Bull 55:207

260. Gilg E (1895) Ancistrocladaceae. In: Engler A (ed) Die natürlichen Pflanzenfamilien nebst ihren Gattungen und wichtigeren Arten insbesondere den Nutzpflanzen, Teil 3, Abteilung 6. Verlag Wilhelm Engelmann, Leipzig, p 274
261. Gilg E (1925) Ancistrocladaceae. In: Engler A (ed) Die natürlichen Pflanzenfamilien nebst ihren Gattungen und wichtigeren Arten insbesondere den Nutzpflanzen, 2nd edn, vol 21. Duncker & Humblot, Berlin, p 589
262. Hutchinson J, Dalziel JM (1954) Ancistrocladaceae. In: Hutchinson J, Dalziel JM (eds, revised by Keay RWJ) Flora of West Tropical Africa, 2nd edn. Crown Agents Overseas Governments and Administration, London, p 233
263. Porembski S (2002) Ancistrocladaceae. In: Kubitzki K, Bayer C (eds) The families and genera of vascular plants. V. Flowering plants—Dicotyledons, Malvales, Capparales and Non-betalain Caryophyllales, vol 5. Springer, Heidelberg, p 25
264. Meimberg H, Dittrich P, Bringmann G, Schlauer J, Heubl G (2000) Molecular phylogeny of Caryophyllidae s.I. based on *matK* sequences with special emphasis on carnivorous taxa. Plant Biol 2:218
265. Cuénoud P, Savolainen V, Chatrou LW, Powell M, Grayer RJ, Chase MW (2002) Molecular phylogenetics of Caryophyllales based on nuclear 18S *r*DNA and plastid *rbc*L, *atp*B, and *mat*K DNA sequences. Am J Bot 89:132
266. Heubl G, Bringmann G, Meimberg H (2006) Molecular phylogeny and character evolution of carnivorous plant families in Caryophyllales—revisited. Plant Biol 8:821
267. Cameron KM, Chase MW, Swensen SM (1995) Molecular evidence for the relationships of *Triphyophyllum* (Dioncophyllaceae) and *Ancistrocladus* (Ancistrocladaceae). Am J Bot 82:117
268. de Loureiro (1790) Genus XXVII. Bembix. In: Flora Cochinchinensis, vol 1. Typis, Et Expensis Academics, Ulyssipone, p 282
269. Wallich N (1829) A numerical list of dried specimens of plants, in the East India Company's Museum collected under the superintendence of Dr. Wallich of the Company's botanic garden at Calcutta, nos. 1–2153. Lithographed manuscript, London
270. Planchon MJE (1849) Essai monographique d'une nouvelle famille de plantes propose sous le nom d'Ancistrocladacées. Ann Sci Nat Bot 3 Ser Bot 13:316
271. van Tieghem MP (1903) Sur les Ancistrocladacées. J Bot 17:151
272. Pham DP, Tran H (2009) Cytological activities and cytotoxicities of the alkaloidal components and naphthoquinones from three Vietnamese *Ancistrocladus* species. Tap Chi Duoc Hoc 49:23
273. Massart J (1896) Sur la morphologie du bourgeon. Ann Jard Bot Buitenzorg 13:121
274. Hallé N (1973) Crochets de lianes du Gabon: *Ancistrocladus* et *Anacolosa* (Ancistrocladacées et Olacacées). Adansonia 13:299
275. Graham J (1839) 206. *Ancistrocladus heyneanus*. In: Graham J (ed) A catalogue of the plants growing in Bombay and its vicinity. Government Press, Bombay, p 28
276. Van Rhede tot Draekestein H (1686) *Ancistrocladus heyneanus* Wallich ex J. Graham (as *Modira valli*). In: Hortus indicus malabaricus, vol 7. Van Someren & Dyck, Amstelodami, p 87, tab 46
277. Wight R (1853) *Ancistrocladus heyneanus* Wallich ex J. Graham. In: Icones plantarum Indiae orientalis (or figures of Indian plants), vol 6. American Mission Press, Madras, p 12, including plate no. 1987–88
278. De Candolle A (1864) Ancistrocladaceae. In: De Candolle A (ed) Prodromus systematis naturalis regni vegetabilis, vol 16. Victoris Masson et Filii, Paris, p 601
279. Thiselton Dyer WT (1874) XXV. Dipterocarpaceae. In: Hooker JD (ed) Flora of British India, vol 1. Lovell Reeve & Co. Ltd., Ashford, Kent, p 299
280. Brandis D (1906) Order XVIII. Ancistrocladaceae. In: Brandis D (ed) Indian trees—an account of trees, shrubs, woody climbers, bamboos and palms indigenous or commonly cultivated in the British Indian Empire. Bishen Singh Mahendra Pal Singh, Dehradun, p 73

281. Hutchinson J (1973) 149. Ancistrocladaceae. In: Hutchinson J (ed) The families of flowering plants—arranged according to a new system based on their probable phylogeny, vol 1, 3rd edn. Clarendon Press, Oxford, p 351
282. Santapau H (1967) The flora of Khandala on the Western Ghats of India. In: The manager of publication, civil lines, Calcutta (ed) Records of the botanical survey of India, vol 16, 3rd edn. Loyal Art Press Private Limited, Calcutta, p 15
283. Rao VS (1969) The floral anatomy of *Ancistrocladus*. Proc Indian Acad Sci Sect B 70:215
284. Ramamoorthy TP (1976) Ancistrocladaceae. In: Saldanha CJ, Nicolson DH (eds) Flora of Hassan District, Karnataka. Amerind Publishing, New Dehli, p 171
285. Das S (1993) Ancistrocladaceae. In: Sharma BD, Sanjappa M (eds) Flora of India, Portulacaceae-Ixonanthaceae, vol 5. Botanical Survey of India, Calcutta, p 252
286. Kundu SR (2005) A synopsis of Ancistrocladaceae in Indian subcontinent: its distribution and endemism. Geobios 32:221
287. Panday S, Roy DK, Odyuo N, Sinha BK (2012) Notes on a little known genus *Ancistrocladus* Wallich ex Wight & Arnott [Ancistrocladaceae] from Northeast India. Pleione 6:336
288. Datar MN, Lakshminarasimhan P (2013) Check list of wild angiosperms of Bhagwan Mahavir (Molem) National Park, Goa, India. Check List 9:186
289. Bringmann G, Pokorny F, Zinsmeister HD (1991) *Ancistrocladus*, eine botanisch und chemisch bemerkenswerte Gattung. Der Palmengarten 55(3):13
290. Bringmann G, Schneider C, Pokorny F, Lorenz HJ, Fleischmann H, Sankaranarayanan TR, Aké Assi L (1993) The cultivation of the genus *Ancistrocladus*. Planta Med 59:A623
291. Vahl M (1810) Beskrivelse over nye Planteslaegter—Wormia. Skr Naturhist Selsk 6:104
292. Walker-Arnott G (1836) Malphigiaceae. Nova Acta Acad Caes Leop-Carol German Nat Cur 18:325
293. Thwaites GHK (1849) Note on the genus *Ancistrocladus* Wallich. Trans Linn Soc Lond 21:225, Figs. 1–18
294. Trimen H (1893) Ancistrocladaceae. In: A handbook to the flora of Ceylon (containing descriptions of all the species of flowering plants indigenous to the island, and notes on their history, distribution, and uses), with an atlas of plates (illustrating some of the more interesting species), part 1. Dulau & Co., London, p 138, plate XVI
295. Boorsma WG (1899) Ancistrocladaceae. In: Onderzoek naar de Plantenstoffen van Nederlandsch-Indie. Mededeelingen uit 'S Lands Plantentuin, vol 31. G. Kolff & Co., Batavia, p 4
296. Abeywickrama BA (1959) A provisional check list of the flowering plants of Ceylon. Ceylon J Sci (Bio Sci) 2:119
297. Harriman NA (1987) Ancistrocladaceae. In: Dassanayake MD (ed) A revised handbook to the flora of Ceylon, vol 6. Amerind Publishing, New Dehli, p 1
298. Craib WG (1931) Ancistrocladaceae. In: Florae Siamensis enumeration—a list of plants known from Siam with records of their occurrence, vol 1. Siam Society, Bangkok, p 148
299. Parkinson CE (1972) Ancistrocladaceae. In: A forest flora of the Andaman Islands. Bishen Singh Mahendra Pal Singh, New Connaught Place, Dehradun, p 94
300. Dagar JC, Singh T (1999) Ancistrocladaceae. In: Plant resources of the Andaman & Nicobar Islands (enumeration and utilisation of vascular plants), vol 2. Bishen Singh Mahendra Pal Singh, New Connaught Place, Dehradun, p 308
301. Hajra PK, Rao PSN, Mudgal (1999) Ancistrocladaceae. In: Flora of Andaman and Nicobar Islands, vol 1. Botanical Survey of India, Calcutta, p 153
302. Pandey RP, Dilwakar PC (2008) An integrated check-list flora of Andaman and Nicobar Islands. J Econ Taxon Bot 32:403
303. Moore S, Tandy G (1927) Notes on two species of Loureiro's Flora Cochinchinensis. J Bot 65:279
304. Keng H (1967) Observations on *Ancistrocladus tectorius*. Gard Bull Singapore 22:113
305. Keng H (1970) Further observations on *Ancistrocladus tectorius* (Ancistrocladaceae). Gard Bull Singapore 25:235

306. Merrill ED (1928) A second supplementary list of Hainan plants. Lingnan Sci J 6:323
307. Mo SL, Zhong YC (1987) Plantae novae guangxiensis. Guihaia 7:287
308. Ku T (1999) Ancistrocladaceae. In: Flora Republicae Popularis Sinicae. Science Press, Beijing, p 267
309. Chayamarit K, Balslav H (2020) Ancistrocladaceae. In: Flora of Thailand, vol 14, part 4. The Forest Herbarium, National Park, Wildlife and Plant Conservation Department, Bangkok, p 511
310. Yunoh SMM (2007) *Ancistrocladus tectorius* (Lour.) Merill. In: Flora of Malaysia. MyCHM i-Newsletter Part 3. Forest Research Institute Malaysia (FRIM), p 46
311. Harada N, Nakanishi K (1972) The Exciton Chirality method and its application to configurational and conformational studies of natural products. Acc Chem Res 5:257
312. Harada N, Nakanishi K (1983) Circular dichroic spectroscopy—excition coupling in organic stereochemistry. University Science Books, Mill Valley, CA
313. Berova N, Nakanishi K (2000) Exciton Chirality method: principles and application. In: Nakanishi K, Berova N, Woody RW (eds) Circular dichroism: principles and application, 2nd edn. Wiley-VCH, New York, p 337
314. Harada N, Nakanishi K, Berova N (2012) Electronic CD Exciton Chirality method: principles and application. In: Berova N, Polavarapu PL, Nakanishi K, Woody RW (eds) Comprehensive chiroptical spectroscopy—applications in stereochemical analysis of synthetic compounds, natural products, and biomolecules, vol 2. Wiley, Hoboken, NJ, p 115
315. Govindachari TR, Parthasarathy PC, Modi JD (1971) Chemical investigation of *Ancistrocladus heyneanus* Wall.: part IV—structure & synthesis of ancistroquinone. Indian J Chem 9:1042
316. Desai HK, Gawad DH, Govindachari TR, Joshi BS, Parthasarathy PC, Ramachandran KS, Ravindranath KR, Sidhaye AR, Viswanathan N (1976) Chemical investigation of some Indian plants: part IX. Indian J Chem 14B:473
317. Rizzacasa MA, Sargent MV (1989) Synthetic approaches to the alkaloids of the Ancistrocladaceae: dehydroancistrocladisine. J Chem Soc Chem Commun:301
318. Rizzacasa MA, Sargent MV (1991) Synthetic approaches to the naphthylisoquinoline alkaloids. Part 1. Dehydroancistrocladisine. J Chem Soc Perkin Trans 1:841
319. Rizzacasa MA, Sargent MV (1990) Synthetic approaches to the alkaloids of the Ancistrocladaceae: (–)-*O*-methylancistrocladine and (+)-*O*-methylhamatine. J Chem Soc Chem Commun:894
320. Rizzacasa MA, Sargent MV (1991) Synthetic approaches to the alkaloids of the Ancistrocladaceae: control of the diastereoisomer excess in the synthesis of axially chiral biaryls: a synthesis of (–)-ancistrocladinine. J Chem Soc Chem Commun:278
321. Leighton BN, Rizzacasa MA (1995) Formal synthesis of (–)-*O*-methylancistrocladine. J Org Chem 60:5702
322. Chau P, Czuba IR, Rizzacasa MA, Bringmann G, Gulden KP, Schäffer M (1996) Convergent synthesis of naphthylisoquinoline alkaloids: total synthesis of (+)-*O*-methylancistrocline. J Org Chem 61:7101
323. Meyers AI, Knaus G, Kamata K, Ford ME (1976) Asymmetric synthesis of *R* and *S* α-alkylalkanoic acids from metalation and alkylation of chiral 2-oxazolines. J Am Chem Soc 98:567
324. Gant TG, Meyers AI (1994) The chemistry of 2-oxazolines (1985–present). Tetrahedron 50:2297
325. Wolfe JP (2005) Meyers oxazoline method. In: Li JJ, Corey EJ (eds) Name reactions in heterocyclic chemistry. Wiley, Hoboken, NJ, p 237
326. Li JJ (2009) Meyers oxazoline method. In: Li JJ (ed) Name reactions: a collection of detailed mechanisms and synthetic applications, 4th edn. Springer Verlag, Berlin, Heidelberg, p 351
327. Bringmann G, Pabst T, Henschel P, Kraus J, Peters K, Peters EM, Rycroft DS, Connolly JD (2000) Nondynamic and dynamic kinetic resolution of lactones with stereogenic centers and axes: stereoselective total synthesis of herbertenediol and mastigophorenes A and B. J Am Chem Soc 122:9127

328. Bringmann G, Menche D (2001) First, atropo-enantioselective total synthesis of the axially chiral phenylanthraquinone natural products knipholone and 6′-*O*-methylknipholone. Angew Chem 113:1733; Angew Chem Int Ed 40:1687
329. Bringmann G, Menche D, Kraus J, Mühlbacher J, Peters K, Peters EM, Brun R, Bezabih M, Abegaz BM (2002) Atropo-enantioselective total synthesis of knipholone and related antiplasmodial phenylanthraquinones. J Org Chem 67:5595
330. Bringmann G, Menche D, Mühlbacher J, Reichert M, Saito N, Pfeiffer SS, Lipshutz BH (2002) On the verge of axial chirality: atroposelective synthesis of the AB-biaryl fragment of vancomycin. Org Lett 4:2833
331. Bringmann G, Hinrichs J, Henschel P, Kraus J, Peters K, Peters EM (2002) Atropo-enantioselective synthesis of the natural bicoumarin (+)-isokotanin A via a configurationally stable biaryl lactone. Eur J Org Chem:1096
332. Kitamura M, Ohmori K, Kawase T, Suzuki K (1999) Total synthesis of pradimicinone, the common aglycon of the pradimicin-benanomicin antibiotics. Angew Chem 111:1308; Angew Chem Int Ed 38:1229
333. Ohmori K, Tamiya M, Kitamura M, Kato H, Oorui M, Suzuki K (2005) Regio- and stereocontrolled total synthesis of benanomicin B. Angew Chem 117:3939; Angew Chem Int Ed 44:3871
334. Tamiya M, Ohmori K, Kitamura M, Kato H, Arai T, Oorui M, Suzuki K (2007) General synthesis route to benanomicin-pradimicin antibiotics. Chem Eur J 13:9791
335. Bringmann G, Breuning M (1998) Enantioselective addition of diethylzinc to aldehydes using novel axially chiral 2-aminomethyl-1-(2′-hydroxy–phenyl)naphthalene catalysts. Tetrahedron Asymmetry 9:667
336. Bringmann G, Pfeifer RM, Rummey C, Hartner K, Breuning M (2003) Synthesis of enantiopure axially chiral C_3-symmetric tripodal ligands and their application as catalysts in the asymmetric addition of dialkylzinc to aldehydes. J Org Chem 68:6859
337. Bringmann G, Pfeifer RM, Schreiber P, Hartner K, Schraut M, Breuning M (2004) The 'lactone method': enantioselective preparation of novel P,N-biaryl ligands and their use in the synthesis of the biarylic alkaloids, ancistrotanzanine B and ancistroealaine A. Tetrahedron 60:4349
338. Mori K, Itakura T, Akiyama T (2016) Enantiodivergent atroposelective of chiral biaryls by asymmetric transfer hydrogenation: chiral phosphoric acid catalysed dynamic kinetic resolution. Angew Chem 128:11814; Angew Chem Int Ed 55:11642
339. Yu C, Huang H, Li X, Zhang Y, Wang W (2016) Dynamic kinetic resolution of biaryl lactones via a chiral bifunctional amine thiourea-catalyzed highly atropo-enantioselective transesterification. J Am Chem Soc 138:6956
340. Bringmann G, Hartung T, Göbel L, Schupp O, Ewers CLJ, Schöner B, Zagst R, Peters K, von Schnering HG, Burschka C (1992) Synthesis and structure of benzonaphthopyranones, useful bridged model precursors for stereoselective biaryl synthesis. Liebigs Ann Chem:225
341. Bringmann G, Hinrichs J, Kraus J, Wuzik A, Schulz T (2000) Nondynamic kinetic resolution of configurationally stable biaryl lactones by reduction with oxazaborolidine-activated borane: AM1 studies and experimental verification. J Org Chem 65:2517
342. Bringmann G, Breuning M, Pfeifer RM, Schenk WA, Kamikawa K, Uemura M (2002) The lactone concept—a novel approach to the metal-assisted atroposelective construction of axially chiral biaryl systems. J Organomet Chem 661:31
343. Bringmann G, Schneider C (1983) Improved methods for dehydration and hydroxy/halogen exchange using novel combinations of triphenylphosphine and halogenated ethanes. Synthesis:139
344. Johansson Seechurn CCC, Kitching MO, Colacot TJ, Snieckus V (2012) Palladium-catalyzed cross-coupling: a historical contextual perspective to the 2010 Nobel prize. Angew Chem 124:5150; Angew Chem Int Ed 51:5062
345. Meijere A, Bräse S, Oestreich M (2013) Metal-catalyzed cross-coupling reactions and more. Part 1–3. Wiley-VCH Verlag, Weinheim

346. Nicolaou KC, Bulger PG, Sarlah D (2005) Palladium-catalyzed cross-coupling reactions in total synthesis. Angew Chem 117:4516; Angew Chem Int Ed 44:4442
347. Barder TE, Walker SD, Martinelli JR, Buchwald SL (2005) Catalysts for Suzuki-Miyaura coupling processes: scope and studies of the effect of ligand structure. J Am Chem Soc 127:4685
348. Watanabe M, Hisamatsu S, Hotokezaka H, Furukawa S (1986) Reaction of lithiated senecioamide and related compounds with benzynes: efficient synthesis of naphthols and naphthoquinones. Chem Pharm Bull 34:2810
349. Urlaub E, Popp J, Kiefer W, Bringmann G, Koppler D, Schneider H, Zimmermann U, Schrader B (1998) FT-Raman investigation of alkaloids in the liana *Ancistrocladus heyneanus*. Biospectroscopy 4:113
350. Thomas R (2001) A biosynthetic classification of fungal and streptomycete fused-ring aromatic polyketides. ChemBioChem 2:612
351. Burza S, Croft SL, Boelaert M (2018) Leishmaniasis. Lancet 392:951
352. Hotez PJ (2018) The rise of leishmaniasis in the twenty-first century. Trans R Soc Trop Med Hyg 112:421
353. Kevric I, Cappel MA, Keeling JH (2015) New world and old world *Leishmania* infections—a practical review. Dermatol Clin 33:579
354. Den Boer M, Argaw JJ, Jannin J, Alvar J (2011) Leishmaniasis impact and treatment access. Clin Microbiol Infect 17:1471
355. Wheeler RJ, Gluenz E, Gull K (2011) The cell cycle of *Leishmania*: morphogenetic events and their implications for parasite biology. Mol Microbiol 79:647
356. Gantt KR, Schultz-Cherry S, Rodriguez N, Jeronimo SM, Nascimento ET, Goldman TL, Recker TJ, Miller MA, Wilson ME (2003) Activation of TGF-β by *Leishmania chagasi*: importance for parasite survival in macrophages. J Immunol 170:2613
357. Awasthi A, Marthur RK, Saha B (2004) Immune response to *Leishmania* infection. Indian J Med Res 119:238
358. Murphy MLL, Wille U, Villegas EN, Hunter CA, Farrell JP (2001) IL-10 mediates susceptibility to *Leishmania donovani* infection. Eur J Immunol 31:2848
359. Büscher P, Cecchi G, Jamonneau V, Priotto G (2017) Human African trypanosomiasis. Lancet 390:2397
360. Kennedy PGE, Rodgers J (2019) Clinical and neuropathogenetic aspects of human African trypanosomiasis. Front Immunol 10:39
361. Holanda-Freitas IT, do Carmo Cupertino M, dos Santos EC, Oliveira L, Geller M, Siqueira-Batista R (2020) Human African trypanosomiasis: current standing and challenges. J Trop Pathol 49:133
362. Aksoy S, Buscher P, Lehane M, Solano P, Van Den Abbeele J (2017) Human African trypanosomiasis control: achievements and challenges. PLoS Negl Trop Dis 11:e0005454
363. Sutherland CS, Yukich J, Goeree R, Tediosi F (2015) A literature review of economic evaluations for a neglected tropical disease: human African trypanosomiasis ("sleeping sickness"). PLoS Negl Trop Dis 9:e0003397
364. Alsford S, Eckert S, Baker N, Glover L, Sanchez-Flores A, Leung KF, Turner DJ, Field MC, Berriman M, Horn D (2012) High-throughput decoding of antitrypanosomal drug efficacy and resistance. Nature 482:232
365. Kumar A, Das S, Purkait B, Sardar AH, Ghosh AK, Dikhit MR, Abishek K, Das P (2014) Ascorbate peroxidase, a key molecule regulating amphotericin B resistance in clinical isolates of *Leishmania donovani*. Antimicrob Agents Chemother 58:6172
366. Shadab M, Jha B, Asad M, Deepthi M, Kamran M, Ali N (2017) Apoptosis-like cell death in *Leishmania donovani* treated with KalsomeTM10, a new liposomal amphotericin B. PLoS One 12:e0171306
367. Phillips MA, Burrows JN, Manyando C, van Huijsduijnen RH, van Voorhis WC, Wells NC (2017) Malaria. Nat Rev Dis Primers 3:17050
368. Cowman AF, Healer J, Marapana D, Marsh K (2016) Malaria: biology and disease. Cell 167:610

369. Moxon CA, Gibbins MP, McGuinness D, Milner Jr DA, Marti M (2020) New insights into malaria pathogenesis. Annu Rev Pathol Mech Dis 15:315
370. World Malaria Report 2020 (2020) 20 years of global progress and challenges. World Health Organization, Geneva
371. Kokwaro G (2009) Ongoing challenges in the management of malaria. Malar J 8:S2
372. Guyant P, Corbel V, Guérin PJ, Lautissier A, Nosten F, Boyer S, Coosemans M, Dondorp AM, Sinou V, Yeung S, White N (2015) Past and new challenges for malaria control and elimination: the role of operational research for innovation in designing interventions. Malar J 14:279
373. Blasco B, Leroy D, Fidock DA (2017) Antimalarial drug resistance: linking *Plasmodium falciparum* parasite biology to the clinic. Nat Med 23:917
374. Wicht KJ, Mok S, Fidock DA (2020) Molecular mechanisms of drug resistance in *Plasmodium falciparum* malaria. Annu Rev Microbiol 74:431
375. Duffey M, Blasco B, Burrows JN, Wells TNC, Fidock DA, Leroy D (2021) Assessing risks of *Plasmodium falciparum* resistance to select next-generation antimalarials. Trends Parasitol 37:709
376. Goldberg DE, Slater AFG (1992) The pathway to haemoglobin degradation in malaria parasites. Parasitol Today 8:280
377. Olliaro PL, Goldberg DE (1995) The *Plasmodium* digestive vacuole: metabolic headquarters and choice drug target. Parasitol Today 11:294
378. Sullivan DJ, Gluzman IY, Goldberg DE (1996) *Plasmodium* hemozoin formation mediated by histidine-rich proteins. Science 271:219
379. Dorn A, Vippagunta SR, Matile H, Jaquet C, Vennerstrom JL, Ridley RG (1998) An assessment of drug-haematin binding as a mechanism for inhibition of haematin polymerisation by quinolone antimalarials. Biochem Pharmacol 55:727
380. Pagola S, Stephens PW, Bohle DS, Kosar AD, Madsen SK (2000) The structure of malaria pigment β-haematin. Nature 404:307
381. de Villiers KA, Gildenhuys J, le Roex T (2012) Iron(III)protoporphyrin IX complexes of the antimalarial *Cinchona* alkaloids quinine and quinidine. ACS Chem Biol 7:666
382. Heller LE, Roepe PD (2018) Quantification of free ferriportoporphyrin IX heme and hemozoin for artimisinin sensitive versus delayed clearance phenotype *Plasmodium falciparum* malarial parasites. Biochemistry 57:6927
383. Blair JMA, Webber MA, Baylay AJ, Ogbolu DO, Piddock LJV (2015) Molecular mechanisms of antibiotic resistance. Nat Rev Microbiol 13:42
384. Peterson E, Kaur P (2018) Antibiotic resistance mechanisms in bacteria: relationships between resistance determinants of antiobiotic producers, environmental bacteria, and clinical pathogens. Front Microbiol 9:2928
385. Otto M (2009) *Staphylococcus epidermidis*—the 'accidental' pathogen. Nat Rev Microbiol 7:555
386. Brescó MS, Harris LG, Thompson K, Stanic B, Morgenstern M, O'Mahony LO, Richards RG, Moriarty TF (2017) Pathogenic mechanisms and host interactions in *Staphylococcus epidermidis* device-related infection. Front Microbiol 8:1401
387. Chambers HF, DeLeo FR (2009) Waves of resistance: *Staphylococcus aureus* in the antibiotic era. Nat Rev Microbiol 7:629
388. Turner NA, Sharma-Kuinkel BK, Maskarinec SA, Eichenberger EM, Shah PP, Carugati M, Holland TL, Fowler Jr VG (2019) Methicillin-resistant *Staphylococcus aureus*: an overview of basic and clinical research. Nat Rev Microbiol 17:203
389. De Oliveira GC, Vasconcelos CC, Lopes AJO, de Sousa Cartágenes MS, Filho AKDB, do Nascimento FRF, Ramos RM, Pires ERRB, de Andrade MS, Rocha FMG, de Andrade Monteiro C (2018) *Candida* infections and therapeutic strategies: mechanisms of action for traditional and alternative agents. Front Microbiol 9:1351
390. Pappas PC, Lionakis MS, Arendrup MC, Ostrosky-Zeichner L, Kullberg BJ (2018) Invasive candidiasis. Nat Rev Dis Primers 4:18026

391. Atriwal T, Azeem K, Husain FM, Hussain A, Khan MN, Alajmi MF, Abid M (2021) Mechanistic understanding of *Candida albicans* biofilm formation and approaches for its inhibition. Front Microbiol 12:638
392. Lohse MB, Gulati M, Johnson AD, Nobile CJ (2018) Development and regulation of single- and multi-species *Candida albicans* biofilms. Nat Rev Microbiol 16:19
393. Smallcombe SH, Patt SL, Keifer PA (1995) WET solvent suppression and its application to LC-NMR and high-resolution NMR spectroscopy. J Magn Reson Ser A 117:295
394. Nicholson JK, Foxall PJD, Spraul M, Farrant RD, Lindon JC (1995) 750 MHz ^{1}H and ^{13}C NMR spectroscopy of human blood plasma. Anal Chem 67:793
395. Berova N, Nakanishi K, Woody RW (2000) Circular dichroism: principles and applications, 2nd edn. Wiley-VCH, New York
396. Berova N, Di Bari L, Pescitelli G (2007) Application of electronic circular dichroism in configurational and conformational analysis of organic compounds. Chem Soc Rev 36:914
397. Di Bari L, Pescitelli G (2010) Electronic circular dichroism. In: Grunenberg J (ed) Computational spectroscopy—methods, experiments and applications. Wiley-VCH, Weinheim, p 241
398. Berova N, Woody RW, Polavarapu P, Nakanishi K (2012) Comprehensive chiroptical spectroscopy. Wiley-VCH, New York
399. Li XC, Ferreira D, Ding Y (2010) Determination of absolute configuration of natural products: theoretical calculation of electronic circular dichroism as a tool. Curr Org Chem 14:1678
400. Nugroho AE, Morita H (2014) Circular dichroism calculation for natural products. J Nat Med 68:1
401. Superchi S, Scafato P, Górecki GP (2018) Absolute configuration determination by quantum mechanical calculation of chiroptical spectra: basics and applications to fungal metabolites. Curr Med Chem 25:287
402. Liu L, Bruhn T, Guo L, Götz DCG, Brun R, Stich A, Che Y, Bringmann G (2011) Chloropupukeanolides C-E: cytotoxic pupukeanane chlorides with a spiroketal skeleton from *Pestalotiopsis fici*. Chem Eur J 17:2604
403. Li J, Li MY, Bruhn T, Götz DCG, Xiao Q, Satyanandamurty T, Wu J, Bringmann G (2012) Andraxylocarpins A–E: structurally intriguing limonoids from the true mangroves *Xylocarpus granatum* and *Xylocarpus moluccensis*. Chem Eur J 18:14342
404. Almeida C, Hemberger Y, Schmitt SM, Bouhired S, Natesan L, Kehraus S, Dimas K, Gütschow M, Bringmann G, König GM (2012) Marilines A–C: novel phthalimidines from the sponge-derived fungus *Stachylidium* sp. Chem Eur J 18:8827
405. Hemberger Y, Xu J, Wray V, Proksch P, Wu J, Bringmann G (2013) Pestalotiopens A and B: stereochemically challenging flexible sesquiterpene-cyclopaldic acid hybrids from *Pestalotiopsis* sp. Chem Eur J 19:15556
406. Tatsis EC, Schaumlöffel A, Warskulat AC, Massiot G, Schneider B, Bringmann G (2013) Nudicaulins, yellow flower pigments of *Papaver nudicaule*: revised constitution and assignment of absolute configuration. Org Lett 15:156
407. Blusch LK, Hemberger Y, Pröpper K, Dittrich B, Witterauf F, John M, Bringmann G, Brückner C, Meyer F (2013) Siamese-twin porphyrin: a pyrazole-based expanded porphyrin of persistent helical conformation. Chem Eur J 19:5868
408. Goel A, Kumar V, Hemberger Y, Singh FV, Nag P, Knauer M, Kant R, Raghunandan R, Maulik PR, Bringmann G (2016) Rotationally hindered biphenyls and terphenyls: synthesis, molecular dynamics, and configurational assignment. J Org Chem 81:10721
409. Gehrold AC, Bruhn T, Bringmann G (2016) Axial, helical, and planar chirality in directly linked basket-handle porphyrin arrays. J Org Chem 81:1075
410. Schies C, Alemayehu AB, Vazquez-Lima H, Thomas BK, Bruhn T, Bringmann G, Ghosh A (2017) Metallocorroles as inherently chiral chromophores: resolution and electronic circular dichroism spectroscopy of a tungsten biscorrole. Chem Commun 53:6121
411. Bringmann G, Günther C (1999) First total synthesis of dioncophylline B, a 7,6′-coupled naphthylisoquinoline alkaloid. Synlett 216

412. Bringmann G, Günther C, Peters EM, Peters K (2001) First total synthesis of 7,6′-coupled antifungal naphthylisoquinoline alkaloid dioncophylline B. Tetrahedron 57:1253
413. Bringmann G, Harmsen S, Holenz J, Geuder T, Götz R, Keller PA, Walter R, Hallock YF, Cardellina II JH, Boyd MR (1994) 'Biomimetic' oxidative dimerization of korupensamine A: completion of the first total synthesis of michellamines A, B, and C. Tetrahedron 50:9643
414. Fleischhauer J, Koslowski A, Kramer B, Zobel E, Bringmann G, Gulden KP, Ortmann T, Peter B (1993) Detection and calculation of the CD spectra from the biaryl alkaloids ancistrocladeine and dioncophylleine A. Z Naturforsch 48b:140
415. Gagnepain F (1909) Plantes nouvelles d'Indo-Chine. In: Lecomte H (ed) Notulae Systematicae, vol 4. J. Dumoulin, Paris, p 114
416. Pham-Hoang H (1991) Dipterocarpaceae. In: Câycỏ Việtnam—an illustrated flora of Vietnam, vol 1. Mekong Printing, Santa Ana, Canada, p 555
417. Savajol N, Toun V, Sam J (2011) *Ancistrocladus cochinchinensis*. In: Savajol N, Toun V, Sam J (eds) Traditional therapeutic knowledge of the Bunong people in Southeastern Cambodia. Nomad RSI Cambodia Edition, Phnom Penh, Cambodia, p 58
418. Anh NH, Sung TV, Ripperger H, Adam G (1996) Some results from chemical study on Vietnamese *Ancistrocladus cochinchinensis*. Tap Chi Hoa Hoc 34:89
419. Anh NH, Ripperger H, Porzel A, Sung TV, Adam G (1997) Tetralones from *Ancistrocladus cochinchinensis*. Phytochemistry 44:549
420. Bringmann G, Weirich R, Lisch D, Aké Assi L (1992) Ancistrobrevine D: an unusual alkaloid from *Ancistrocladus abbreviatus*. Planta Med 58:A703
421. Schäfer A, Horn H, Ahlrichs R (1992) Fully optimized contracted Gaussian basis sets for atoms Li to Kr. J Chem Phys 97:2571
422. Schäfer A, Huber C, Ahlrichs R (1994) Fully optimized contracted Gaussian basis sets of triple zeta valence quality for atoms Li to Kr. J Chem Phys 100:5829
423. Sidhu GS, Sankaram AVB (1971) A new biplumbagin and 3-chloroplumbagin from *Plumbago zeylanica*. Tetrahedron Lett 26:2385
424. Gunaherath GMKB, Gunatilaka AAL, Sultanbawa MUS, Balasubramaniam S (1983) 1,2(3)-Tetrahydro-3,3′-biplumbagin: a naphthalenone and other constituents from *Plumbago zeylanica*. Phytochemistry 22:1245
425. Kumar V, Meepagala KM, Balasubramaniam S (1985) Quinoid and other constituents of *Aristea ecklonii*. Phytochemistry 24:1118
426. Higa M, Noha N, Yokaryo H, Ogihara K, Yogi S (2002) Three new naphthoquinone derivatives from *Diospyros maritima* Blume. Chem Pharm Bull 50:590
427. Higa M, Takashima Y, Yokaryo H, Harie Y, Suzuka T, Ogihara K (2017) Naphthoquinone derivatives from *Diospyros maritima*. Chem Pharm Bull 65:739
428. Tezuka M, Takahashi C, Kuroyanagi M, Satake M, Yoshihira K, Natori S (1973) New naphthoquinones from *Diospyros*. Phytochemistry 12:175
429. Uc-Cachón AH, Molina-Salinas GM, Said-Fernández S, Méndez-Gonzáles M, Cáceres-Farfán M, Borges-Argáez R (2013) A new dimeric naphthoquinone from *Diospyros anisandra*. Nat Prod Res 27:1174
430. Gunaherath GMKB, Gunatilaka AAL, Thomson RH (1988) Studies on medicinal and related plants of Sri Lanka. Part 18. Structure of a new naphthoquinone from *Plumbago zeylanica*. J Chem Soc Perkin Trans 1:407
431. Gu JQ, Graf TN, Lee D, Chai HB, Mi Q, Kardono LBS, Setyowati FM, Ismail R, Riswan S, Farnsworth NR, Cordell GA, Pezzuto JM, Swanson SM, Kroll DJ, Falkinham JO III, Wall ME, Wani MC, Kinghorn AD, Oberlies NH (2004) Cytotoxic and antimicrobial constituents of the bark of *Diospyros maritima* collected in two geographical locations in Indonesia. J Nat Prod 67:1156
432. Porras G, Chassagne F, Lyles JT, Marquez L, Dettweiler M, Salam AM, Samarakoon T, Shabih S, Farrokhi R, Quave CL (2021) Ethnobotany and the role of plant natural products in antibiotic drug discovery. Chem Rev 121:3495
433. Uc-Cachón AH, Borges-Argáez R, Said-Fernández S, Vargas-Villarreal J, González-Salazar F, Méndez-Gonzáles M, Cáceres-Farfán M, Molina-Salinas GM (2014) Naphthoquinones

from *Diospyros anisandra* exhibit potent activity against pan-resistant first-line drugs *Mycobacterium tuberculosis* strains. Pulm Pharmacol Ther 27:114
434. Nair SV, Baranwal G, Chatterjee M, Sachu A, Vasudevan AK, Bose C, Banerji A, Biswas R (2016) Antimicrobial activity of plumbagin, a naturally occurring naphthoquinone from *Plumbago rosea*, against *Staphylococcus aureus* and *Candida albicans*. Int J Med Microbiol 306:237
435. Periasamy H, Iswarya S, Pavithra N, Senthilnathan S, Gnanamani A (2019) In vitro antibacterial activity of plumbagin isolated from *Plumbago zeylanica* L. against methicillin-resistant *Staphylococcus aureus*. Lett Appl Microbiol 69:41
436. Adusei EBA, Adosraku RK, Oppong-Kyekyeku J, Amengor CDK, Jibira Y (2019) Resistance modulation action, time-kill kinetics assay, and inhibition of biofilm formation effects of plumbagin from *Plumbago zeylanica* Linn. J Trop Med:1250645
437. Sharma N, Shukla AK, Das M, Dubey VK (2012) Evaluation of plumbagin and its derivative as potential modulators of redox thiol metabolism of *Leishmania* parasite. Parasitol Res 110:341
438. Awasthi BP, Kathuria M, Pant G, Kumari N, Mitra K (2016) Plumbagin, a plant-derived naphthoquinone metabolite induces mitochondria mediated apoptosis-like cell death in *Leishmania donovani*: an ultrastructural and physiological study. Apoptosis 21:941
439. Salomão K, De Santana NA, Molina MT, De Castro SL, Menna-Barreto RFS (2013) *Trypanosoma cruzi* mitochondrial swelling and membrane potential collapse as primary evidence of the mode of action of naphthoquinone analogues. BMC Microbiol 13:196
440. Sumsakul W, Plengsuriyakarn T, Chaijaroenkul W, Viyanant V, Karbwang J, Na-Bangchang K (2014) Antimalarial activity of plumbagin in vitro and in animal models. BMC Complement Altern Med 14:15
441. Jangra A, Chadha V, Kumar D, Kumar V, Arora MK (2021) Neuroprotective and acetylcholinesterase inhibitory activity of plumbagin in ICV-LPS-induced behavioural deficits in rats. Curr Res Behav Sci 2:100060
442. Checker R, Sharma D, Sandur SK, Khanam S, Poduval TB (2009) Anti-inflammatory effects of plumbagin are mediated by inhibition of NF-kappaB activation in lymphocytes. Int Immunopharmacol 9:949
443. Nakhate KT, Bharne AP, Verma VS, Aru DN, Kokare DM (2018) Plumbagin ameliorates memory dysfunction in streptozotocin induced Alzheimer's disease via activation of Nrf2/ARE pathway and inhibition of β-secretase. Biomed Pharmacother 101:379
444. Messeha SS, Zarnouh NO, Mendonca P, Kolta MG, Soliman KFA (2017) The attenuating effects of plumbagin on pro-inflammatory cytokine expression in LPS-activated BV-2 microglial cells. J Neuroimmunol 313:129
445. Hallock YF, Cardellina II JH, Kornek T, Gulden KP, Bringmann G, Boyd MR (1995) Gentrymine B, the first quaternary isoquinoline alkaloid from *Ancistrocladus korupensis*. Tetrahedron Lett 36:4753
446. Vargas DF, Romero BS, Larghi EL, Kaufman TS (2020) Rhodium(III)-catalyzed C–H activation-based first total synthesis of 6-*O*-methylanciscochine, an alkaloid isolated from *Ancistrocladus tectorius*. Synthesis 52:119
447. Lien LQ, Mai NC, Linh TM, Giang VH, Trung LQ, Van NT, Ban NK, Anh LDN, Kien PV, Min CV (2017) Phylogenetic analysis of *Ancistrocladus* species (Ancistrocladaceae) from Vietnam. Int J Agric Biol 19:1125
448. Turini FG, Steinert C, Heubl G, Bringmann G, Lombe BK, Mudogo V, Meimberg H (2014) Microsatellites facilitate species delimitation in Congolese *Ancistrocladus* (Ancistrocladaceae), a genus with pharmacologically potent naphthylisoquinoline alkaloids. Taxon 63:329
449. Hayata B (1913) Dipterocarpaceae. Ic Pl Formos 3:46
450. Craib WG (1925) Contributions to the flora of Siam—Additamentum XV. Bull Misc Inform Kew 7

451. Cortés I, Borini Etichetti CM, Girardini JE, Kaufman TS, Bracca ABJ (2019) Total synthesis and cytotoxic activity of 6,8-dimethoxy-1,3-dimethylisoquinoline isolated from *Ancistrocladus tectorius*: a 6-π-azaelectrocyclization approach. Synthesis 51:433
452. Fonzo S, Vargas DF, Kaufman TS (2019) A ruthenium-catalyzed C–H activation strategy as an efficient shortcut in the total synthesis of 6,8-dimethoxy-1,3-dimethylisoquinoline. Synthesis 51:3908
453. Hirota T, Koyama T, Nanba T, Yamato M, Matsumura T (1978) Polycyclic *N*-hetero compounds. XVI. Reactions of benzyl ketones with formamide or acetamide. Chem Pharm Bull 26:245
454. Rizzacasa MA, Sargent MV, Skelton BW, White AH (1990) The stereoisomers of 5-bromo-6,8-dimethoxy-1,2,3-trimethyl-1,2,3,4-tetrahydroisoquinoline: X-ray crystal structure of the *trans*-isomer. Aust J Chem 43:79
455. Mantell MA, Kampf JW, Sanford M (2018) Improved synthesis of $[Cp^{R}RhCl_2]_2$ complexes. Organometallics 37:3240
456. Zheng XI, Xing FW (2009) Ethnobotanical study on medicinal plants around Mt. Yinggeling, Hainan Island, China. J Ethnopharmacol 124:197
457. Zheng XI, Wei JH, Sun W, Li RT, Sb L, Dai HF (2013) Ethnobotanical study on medicinal plants around Limu Mountains of Hainan Island, China. J Ethnopharmacol 148:964
458. Unger M, Dreyer M, Specker S, Laug S, Pelzing M, Neusüß C, Holzgrabe U, Bringmann G (2004) Analytical characterization of crude extracts from an African *Ancistrocladus* species using high-performance liquid chromatography and capillary electrophoresis coupled to ion trap mass spectrometry. Phytochem Anal 15:21
459. Bringmann G, Fayez S, Shamburger W, Feineis D, Winiarczyk S, Janecki R, Adaszek Ł (2020) Naphthylisoquinoline alkaloids and their synthetic analogs as potent novel inhibitors against *Babesia canis* in vitro. Vet Parasitol 283:109177
460. Thorn CF, Oshiro C, Marsh S, Hernandez-Boussard T, McLeod H, Klein TE, Altman RB (2011) Doxorubicin pathways: pharmacodynamics and adverse effects. Pharmacogenet Genom 21:440
461. Sritharan S, Sivalingam N (2021) A comprehensive review on time-tested anticancer drug doxorubicin. Life Sci 278:119527
462. Bringmann G, Hamm A, Schraut M (2003) Atroposelective biaryl coupling with chiral catalysts: total synthesis of the antileishmanial naphthylisoquinoline alkaloids ancistrotanzanine B and ancistroealaine A. Org Lett 5:2805
463. Martin R, Buchwald SL (2008) Palladium-catalyzed Suzuki-Miyaura cross-coupling reactions employing dialkylbiaryl phosphine ligands. Acc Chem Res 41:1461
464. Surry DS, Buchwald SL (2008) Biaryl phosphane ligands in palladium-catalzyed amination. Angew Chem 120:6438; Angew Chem Int Ed 47:6338
465. Zhang G (2012) Phytochemical research on two *Ancistrocladus* species, semi-synthesis of dimeric naphthylisoquinoline alkaloids, and structure optimization of antitumoral naphthoquinones. PhD thesis, University of Würzburg
466. Nichols DE, Barfknecht CF, Rusterholz DB, Benington F, Morin RD (1973) Asymmetric synthesis of psychotomimetic phenylisopropylamines. J Med Chem 16:480
467. Hoye TR, Chen M (1996) Total synthesis of (*ent*)-korupensamine D. Tetrahedron Lett 37:3099
468. Hasegawa T, Yamamoto H (2003) Development of a new chiral auxiliary derived from (*S*)-phenylethylamine for a synthesis of enantiopure (*R*)-2-propyloctanoic acid. Synthesis 1181
469. Snieckus V (1990) Directed *ortho* metalation. Tertiary amide and *O*-carbamate directors in synthetic strategies for polysubstituted aromatics. Chem Rev 90:879
470. Snieckus V (1990) The directed *ortho* metalation reaction. Methodology, applications, synthetic links, and a non-aromatic ramification. Pure Appl Chem 62:2047
471. Pederson BF, Pederson B (1965) The stable conformation of *N*-methylacetanilide. Tetrahedron Lett 6:2995
472. Chupp JP, Olin JF (1967) Chemical and physical properties of some rotational isomers of α-haloacetanilides. A novel unreactive halogen system. J Org Chem 32:2297

473. Itai A, Toriumi Y, Tomioka N, Kagechika H, Azumaya I, Shudo K (1989) Stereochemistry of *N*-methylbenzanilide and benzanilide. Tetrahedron Lett 30:6177
474. Itai A, Toriumi Y, Saito S, Kagechika H, Shudo K (1992) Preference for *cis*-amide structure in *N*-acyl-*N*-methylanilines. J Am Chem Soc 114:10649
475. Saito S, Toriumi Y, Tomioka N, Itai A (1995) Theoretical studies on *cis*-amide preference in *N*-methylanilides. J Org Chem 60:4715
476. Kitagawa O, Yoshikawa M, Tanabe H, Morita T, Takahashi M, Dobashi Y, Taguchi T (2006) Highly enantioselective synthesis of atropisomeric anilide derivatives through catalytic asymmetric *N*-arylation: conformational analysis and application to asymmetric enolate chemistry. J Am Chem Soc 128:12923
477. Plasser F, Lischka H (2021) Multi-reference configuration interaction. In: Gonzáles L, Lindh R (eds) Quantum chemistry and dynamics of excited states: methods and applications. Wiley, Hoboken, NJ, p 277
478. Domingo MP, Pardo J, Cebolia V, Galvez EM (2010) Berberine: a fluorescent alkaloid with a variety of applications from medicine to chemistry. Mini-Rev Org Chem 7:335
479. Neag MA, Mocan A, Echeverria J, Pop RM, Bocsan CI, Crişan G, Buzoianu AD (2018) Berberine: botanical occurrence, traditional uses, extraction methods, and relevance in cardiovascular, metabolic, hepatic, and renal disorders. Front Pharmacol 9:557
480. Ai X, Yu P, Peng L, Luo L, Liu J, Li S, Lai X, Luan F, Meng X (2021) Berberine: a review of its pharmacokinetics properties and therapeutic potentials in diverse vascular diseases. Front Pharmacol 12:762654
481. Rauf A, Abu-Izneid T, Khalik AA, Imran M, Shah ZA, Bin Emran T, Mitra S, Khan Z, Alhumaydhi FA, Aljohani ASM, Khan I, Rahman MM, Jeandet P, Gondal TA (2021) Berberine as a potential anticancer agent: a comprehensive review. Molecules 26:7368
482. Becke AD (1993) A new mixing of Hartree-Fock and local density-functional theories. J Chem Phys 98:1372
483. Francl MM, Pietro WJ, Hehre WJ, Binkley JS, Gordon MS, DeFrees DJ, Pople JA (1982) Self-consistent molecular orbital methods. XXIII. A polarization-type basis set for second-row elements. J Chem Phys 77:3654
484. Bruhn T, Pescitelli G, Jurinovich S, Schaumlöffel A, Witterauf F, Ahrens J, Bröring M, Bringmann G (2014) Axially chiral BODIPY DYEmers: an apparent exception to the Exciton Chirality rule. Angew Chem Int Ed 53:14592
485. Gunatilaka AAL (1978) Alkaloids of some plants of Sri Lanka—chemistry and pharmacology. Natn Sci Coun Sri Lanka 6:39
486. Desai HK, Gawad DH, Govindachari TR, Joshi BS, Kamat VN, Modi JD, Parthasarathy PC, Radhakrishnan J, Shanbhag MN, Sidhaye AR, Viswanathan N (1973) Chemical investigations of Indian plants: part VII. Indian J Chem 11:840
487. Sharma SC, Shukla YN, Tandon JS (1975) Alkaloids and terpenoids of *Ancistrocladus heyneanus*, *Sagittaria sagitifolia*, *Lyonia formosa* and *Hedychium spicatum*. Phytochemistry 14:578
488. Bringmann G, Saeb W, Kraus J, Brun R, François G (2000) Jozimine B, a constitutionally unsymmetric, antiplasmodial 'dimer' of the naphthylisoquinoline alkaloid ancistrocladine. Tetrahedron 56:3523
489. Cai CH, Mei WL, Zuo WJ, Guo ZK, Wang H, Gu HG, Dai HF (2013) Antibacterial components from the branches of *Ancistrocladus tectorius* (Lour.) Merr. Redai Yaredi Zhiwu Xuebao (J Trop Subtrop Bot) 31:184
490. Bringmann G, Gulden KP, Busse H, Fleischhauer J, Kramer B, Zobel E (1993) Circular dichroism of naphthylisoquinoline alkaloids: calculation of CD spectra by semiempirical methods. Tetrahedron 49:3305
491. Giorgio E, Viglione RG, Rosini C (2004) Assignment of the absolute configuration of large molecules by ab initio calculation of the rotatory power within a small basis set scheme: the case of some biologically active natural products. Tetrahedron Asymmetry 15:1979

492. Fleischhauer J, Koslowski A, Repges C, Gulden KP, Bringmann G (1998) The absolute configuration of isoancistrocladine and 'isohamatine' by quantum chemical CD calculations. Z Naturforsch 53a:993
493. Bringmann G, Kinzinger L, Busse H, Zhao C (1992) Isolation, structure elucidation, and total synthesis of ancistrocline, an alkaloid of *Ancistrocladus tectorius*. Planta Med 58:A704
494. Kawami M, Yumoto R, Nagai J, Junyaprasert VB, Soonthornchareonnon N, Patanasethanont D, Sripandkulchai B, Takano M (2010) Effect of Thai plant extracts on P-glycoprotein function and viability in paclitaxel-resistant HepG2 cells. Drug Metab Pharmacokinet 25:155
495. Bringmann G, Rückert M, Saeb W, Mudogo V (1999) Characterization of metabolites in plant extracts of *Ancistrocladus likoko* by high-performance liquid chromatography coupled on-line with ^{1}H NMR spectroscopy. Magn Reson Chem 37:98
496. Bringmann G, Gramatzki S, Müller C, Proksch P (1993) Structure-activity relationships of naphthylisoquinoline alkaloids: feeding experiments with the herbivorous insect *Spodoptera littoralis*. Planta Med 59:A624
497. Lavault M, Bruneton J (1978) Sur trois alcaloïdes isolés de Dionchophyllacées. C R Séances Acad Sci Ser C 287:129
498. Aswathanarayan JB, Vittal RR (2013) In vitro evaluation of antioxidant and antibacterial activities of *Rotula aquatica* and *Ancistrocladus heyneanus*. J Pharm Res 6:31
499. Meininger M, Stowasser R, Jakob PM, Schneider H, Koppler D, Bringmann G, Zimmermann U, Haase A (1997) Nuclear magnetic microscopy of *Ancistrocladus heyneanus*. Protoplasma 198:210
500. Meininger M, Jakob PM, von Kienlin M, Koppler D, Bringmann G, Haase A (1997) Radial spectroscopic imaging. J Magn Reson 125:325
501. Fournet A, Angelo A, Muñoz V, Roblot F, Hocquemiller R, Cavé A (1992) Biological and chemical studies of *Pera benensis*, a Bolivian plant used in folk medicine as a treatment of cutaneous leishmaniasis. J Ethnopharmacol 37:159
502. Bringmann G, Schneider C, Aké Assi L (1991) Direct and modified bicyclic moieties of naphthylisoquinoline alkaloids from *Ancistrocladus barteri*: first biosynthetic indicators? Planta Med 57:A10
503. Bruneton J, Bouquet A, Fournet A, Cavé A (1976) La triphyophylline, nouvel alcaloïde isolé du *Triphyophyllum peltatum*. Phytochemistry 15:817
504. Bringmann G, Feineis D (2001) Stress-related polyketide metabolism of Dioncophyllaceae and Ancistrocladaceae. J Exp Bot 52:2015
505. Durand R, Zenk MH (1971) Biosynthesis of plumbagin (5-hydroxy-2-methyl-1,4-naphthoquinone) via the acetate pathway in higher plants. Tetrahedron Lett 32:3009
506. Chauhan R, Ruby K, Shori A, Dwivedi J (2012) *Plumbago zeylanica*, a potent herb for various ailments: a review. Int J Pharm Sci Rev Res 15:72
507. Budzianowski J, Budzianowska A, Kromer K (2002) Naphthalene glucoside and other phenolics from the shoot and callus cultures of *Drosophyllum lusitanicum*. Phytochemistry 61:421
508. Budzianowski J (1996) Naphthohydroquinone glucosides of *Drosera rotundifolia* and *D. intermedia* from in vitro cultures. Phytochemistry 42:1145
509. Tokunaga T, Dohmura A, Takada N, Ueda M (2004) Cytotoxic antifeedant from *Dionaea muscipula* Ellis: a defensive mechanism of carnivorous plants against predators. Bull Chem Soc Jpn 77:537
510. Timmers MA, Dias DA, Urban S (2013) HPLC-NMR chemical profiling of the Australian carnivorous plant, *Drosera erythrohiza* subspecies *magna*. Nat Prod J 3:35
511. Rischer H, Hamm A, Bringmann G (2002) *Nepenthes insignis* uses a C_2-portion of the carbon skeleton of L-alanine acquired via its carnivorous organs, to build up the allelochemical plumbagin. Phytochemistry 59:603
512. Cannon JR, Lojanapiwatna V, Raston CL, Sinchai W, White AH (1980) The quinones of *Nepenthes rafflesiana*. The crystal structure of 2,5-dihydroxy-3,8-dimethoxy-7-

methylnaphtho-1,4-quinone (nepenthone E) and a synthesis of 2,5-dihydroxy-3-methoxy-7-methylnaphtho-1,4-quinone (nepenthone C). Aust J Chem 33:1073
513. Higa M, Ogihara K, Yogi S (1998) Bioactive naphthoquinone derivatives from *Diospyros maritima* BLUME. Chem Pharm Bull 46:1189
514. Bringmann G, Holzgrabe U, Hoerr V, Stich A (2003) Antitrypanosomal naphthylisoquinoline alkaloids and related compounds. Pharmazie 58:343
515. Kayser O, Kiderlen AF, Laatsch H, Croft SL (2000) In vitro leishmanicidal activity of monomeric and dimeric naphthoquinones. Acta Trop 77:307
516. Ali A, Assimopoulou AN, Papageorgiou VP, Kolodziej H (2011) Structure/antileishmanial activity relationship study of naphthoquinones and dependency of the mode of action on the substitution patterns. Planta Med 77:2003
517. Fournet A, Barrios AA, Muñoz V, Hocquemiller R, Cavé A (1992) Effect of natural naphthoquinones in BALB7c mice infected with *Leishmania amazonensis* and *Leishmania venezuelensis*. Trop Med Parasitol 43:219
518. Budzianowski J (2000) Naphthoquinone glucosides of *Drosera gigantea* from in vitro cultures. Planta Med 66:667
519. Lieberherr C, Zhang G, Grafen A, Singethan K, Kendl S, Vogt V, Maier J, Bringmann G, Schneider-Schaulies J (2017) The plant-derived naphthoquinone droserone inhibits in vitro measles virus infection. Planta Med 83:232
520. Likhitwitayawuid K, Kaewamatawong R, Ruangrungsi N, Krungkrai J (1998) Antimalarial naphthoquinones from *Nepenthes thorelii*. Planta Med 64:237
521. Lavault M, Bruneton J (1980) Alcaloïdes du *Dionocophyllum thollonii*: isolement de deux nouveaux alcaloïdes, triphyopeltine et *O*-méthyl-5′-triphyopeltine. Planta Med 40 (Suppl 1):17
522. Bringmann G, Messer K, Saeb W, Peters EM, Peters K (2001) The absolute configuration of (+)-isoshinanolone and in situ LC-CD analysis of its stereoisomers from crude extracts. Phytochemistry 56:387
523. Sreelatha T, Hymavathi A, Madhusudhana Murthy J, Rani PU, Madhusudhana Rao J, Suresh Babu K (2010) Bioactivity-guided isolation of mosquitocidal constituents from the rhizomes of *Plumbago capensis* Thunb. Bioorg Med Chem Lett 20:2974
524. Bhattacharya J, De Carvalho VR (1986) *Epi*-Isoshinanolone from *Plumbago scandens*. Phytochemistry 255:764
525. Aung HH, Chia LS, Goh NK, Chia TF, Ahmed AA, Pare PW, Mabry TJ (2002) Phenolic constituents from the leaves of the carnivorous plant *Nepenthes gracilis*. Fitoterapia 73:445
526. Zhong SM, Waterman PG, Jeffreys JAD (1984) Naphthoquinones and triterpenes from African *Diospyros* species. Phytochemistry 23:1067
527. Bin Zakaria M, Jeffreys JAD, Waterman PG, Zhong SM (1984) Naphthoquinones and triterpenes from some Asian *Diospyros* species. Phytochemistry 23:1481
528. Richomme P, Papillon B, Cabalion P, Bruneton J (1991) Naphthoquinones de *Diospyros samoensis*. Pharm Acta Helv 66:88
529. Chang CI, Chen CR, Chiu HL, Kuo CL, Kuo YH (2009) Chemical constituents from the stems of *Diospyros maritima*. Molecules 14:5281

Doris Feineis graduated in Food Chemistry in 1987 at the University of Würzburg. She obtained her Ph.D. degree (1992) in analytical Natural Product Chemistry from the same university, under the supervision of Professor Gerhard Bringmann. She holds a permanent position as an administration and research associate at the Institute of Organic Chemistry in Würzburg. Her specialization is in the field of modern analytical methods. She has co-authored over 80 scientific publications, mainly dealing with the isolation, structure elucidation, biosynthesis, and pharmacological activities of anti-infective and antitumor naphthylisoquinoline alkaloids and related natural products.

Gerhard Bringmann studied chemistry and biology at the Universities of Gießen and Münster, and then worked with Prof. B. Franck for his Ph.D. degree (Münster, 1978). Postdoctoral studies conducted with Sir Derek H. R. Barton in Gif-sur-Yvette (France) were followed by his habilitation in 1984 at the University of Münster. In 1987, he became a full professor of Organic Chemistry at the University of Würzburg. Since his retirement in 2017, he has been further active as a "senior professor" there. His research interests lie in the field of Natural Products Chemistry with structural, biosynthetic, computational, and pharmacological facets, with special emphasis on stereochemical aspects of all sorts of stereogenic centers (C- or N-based), as well as axial (C,C or C,N, sp^2-sp^2 or sp^2-sp^3), planar, and helical chirality, both analytically and synthetically. A main recent focus has been the spectroscopic (LC-NMR, LC-MS, LC-ECD) and bioassay-guided search for new naphthylisoquinoline alkaloids and related compounds with antiplasmodial, antitrypanosomal, antileishmanial, or cytotoxic activities targeting leukemic or pancreatic cancer cells, from Asian and African Ancistrocladaceae and Dioncophyllaceae plants. He has published over 760 papers and patents. In 2008, he initiated the Excellence Scholarship Program BEBUC in DR Congo, presently supporting nearly 200 young academics from 25 Congolese institutions. For his scientific and humanitarian achievements, he received several prizes, among them the Adolf-Windaus Medal, the Paul J. Scheuer Award, the Médaille du Mérite Civique 1ère Classe of the Congolese People in Gold, the Gusi Peace Prize, the Medal Bene Merenti in Gold of the University of Würzburg, and the Order of Merit on Ribbon of the Federal Republic of Germany (Bundesverdienstkreuz am Bande). He has been awarded honorary doctorates from six Congolese Universities, and honorary professorships from Peking University and from Jinan University (both PR China), and from the University of Pretoria (South Africa). He is a Member of the European Academy of Sciences and Arts and of the African Academy of Sciences.

The manufacturer's authorised representative in the EU is Springer Nature Customer Service Centre GmbH, Europaplatz 3, 69115 Heidelberg, Germany. If you have any concerns regarding our products, please contact ProductSafety@springernature.com

Printed and bound by CPI Group (UK) Ltd, Croydon, CR0 4YY
30/09/2026
02225304-0012